Water Resources and Environmental Engineering

Water Resources and Environmental Engineering

Editor: Sarah Luck

R CALLISTO REFERENCE

www.callistoreference.com

Callisto Reference,
118-35 Queens Blvd., Suite 400,
Forest Hills, NY 11375, USA

Visit us on the World Wide Web at:
www.callistoreference.com

ISBN: 978-1-63239-841-3 (Hardback)

Cataloging-in-publication Data

Water resources and environmental engineering / edited by Sarah Luck.
  p. cm.
Includes bibliographical references and index.
ISBN 978-1-63239-841-3
1. Water-supply. 2. Environmental engineering. 3. Water--Pollution. 4. Drinking water--Health aspects.
5. Water resources development. I. Luck, Sarah.
TD345 .W38 2017
628.1--dc23

Table of Contents

Preface

Environment engineering is the branch of applied science and technology that addresses the issues of energy preservation, production asset and control of waste from human and animal activities. This book is a valuable compilation of topics, ranging from the basic to the most complex advancements in this field. The text strives to shed light on the varied applications and techniques used by the environmental engineers to combat water pollution like municipal water supply, industrial waste water treatment, etc. It will serve as an invaluable source of information for those seeking in-depth knowledge about the relationship between water resources and environmental engineering. As this field is emerging at a rapid pace, the contents of this book will help the readers understand the modern concepts and applications of the subject. It is a resource guide for experts as well as students.

In my initial years as a student, I used to run to the library at every possible instance to grab a book and learn something new. Books were my primary source of knowledge and I would not have come such a long way without all that I learnt from them. Thus, when I was approached to edit this book; I became understandably nostalgic. It was an absolute honor to be considered worthy of guiding the current generation as well as those to come. I put all my knowledge and hard work into making this book most beneficial for its readers.

I wish to thank my publisher for supporting me at every step. I would also like to thank all the authors who have contributed their researches in this book. I hope this book will be a valuable contribution to the progress of the field.

Editor

Environmental Influences on the Spatial Ecology of Juvenile Smalltooth Sawfish (*Pristis pectinata*): Results from Acoustic Monitoring

Colin A. Simpfendorfer[1][*][¤a], **Beau G. Yeiser**[1][¤b], **Tonya R. Wiley**[1][¤c], **Gregg R. Poulakis**[2], **Philip W. Stevens**[2], **Michelle R. Heupel**[1][¤d]

1 Center for Shark Research, Mote Marine Laboratory, Sarasota, Florida, United States of America, 2 Florida Fish and Wildlife Conservation Commission, Fish and Wildlife Research Institute, Charlotte Harbor Field Laboratory, Port Charlotte, Florida, United States of America

Abstract

To aid recovery efforts of smalltooth sawfish (*Pristis pectinata*) populations in U.S. waters a research project was developed to assess how changes in environmental conditions within estuarine areas affected the presence, movements, and activity space of this endangered species. Forty juvenile *P. pectinata* were fitted with acoustic tags and monitored within the lower 27 km of the Caloosahatchee River estuary, Florida, between 2005 and 2007. Sawfish were monitored within the study site from 1 to 473 days, and the number of consecutive days present ranged from 1 to 125. Residency index values for individuals varied considerably, with annual means highest in 2005 (0.95) and lowest in 2007 (0.73) when several *P. pectinata* moved upriver beyond detection range during drier conditions. Mean daily activity space was 1.42 km of river distance. The distance between 30-minute centers of activity was typically <0.1 km, suggesting limited movement over short time scales. Salinity electivity analysis demonstrated an affinity for salinities between 18 and at least 24 psu, suggesting movements are likely made in part, to remain within this range. Thus, freshwater flow from Lake Okeechobee (and its effect on salinity) affects the location of individuals within the estuary, although it remains unclear whether or not these movements are threatening recovery.

Editor: Brian Gratwicke, Smithsonian's National Zoological Park, United States of America

Funding: Financial support was provided by NOAA Fisheries Office of Protected Resources, South Florida Water Management District, National Fish and Wildlife Foundation, National Geographic Committee for Research and Exploration, The Disney Wildlife Conservation Fund, and the Florida Fish and Wildlife Conservation Commission. The funders had no role in study design, data collection and analysis, decision to publish, or preparation of the manuscript.

Competing Interests: The authors have declared that no competing interests exist.

* E-mail: colin.simpfendorfer@jcu.edu.au

¤a Current address: Fishing and Fisheries Research Centre, School of Earth and Environmental Sciences, James Cook University, Townsville, Australia
¤b Current address: Florida Fish and Wildlife Conservation Commission, Fish and Wildlife Research Institute, Tequesta Field Laboratory, Tequesta, Florida, United States of America
¤c Current address: Haven Worth Consulting, League City, Texas, United States of America
¤d Current address: School of Earth and Environmental Sciences, James Cook University, Townsville, Australia

Introduction

Environmental influences on the spatial ecology of elasmobranchs have been poorly investigated. The effect of seasonal temperature changes on the broad-scale distribution of species has been widely reported [1], although the mechanisms and specific tolerances have rarely been investigated [2,3]. Research has revealed that salinity plays an important role in the movement and distribution of nearshore and estuarine species [4]. Bull sharks (*Carcharhinus leucas*) are able to tolerate a wide range of salinities [5], but young juveniles have recently been shown to move so they remain at salinity levels between 7 psu and 20 psu [6,7,8]. Heithaus et al. [9] also demonstrated that for bull sharks in some estuarine habitats dissolved oxygen levels can influence movements and distribution more than salinity. Other nearshore species that have been shown to have movements affected by salinity include bonnetheads (*Sphyrna tiburo*) [10], sandbar sharks (*Carcharhinus plumbeus*) [11], and bat rays (*Myliobatis californica*) [12]. Given the importance of salinity, changes in freshwater flow regimes into estuaries as a result of climate change or water management practices will affect populations by potentially changing their distributions.

The sawfishes (Family Pristidae) were once common inhabitants of tropical and subtropical inshore, estuarine, and freshwater areas world-wide [13]. However, pressure from fishing and habitat loss have led to population declines [14] and all species are currently Critically Endangered on the IUCN Red List (see www.redlist.org); and some species are protected under national endangered species legislation. The smalltooth sawfish (*Pristis pectinata*) was listed as Endangered by the United States National Marine Fisheries Service (NMFS) and protected by the Endangered Species Act in 2003. Although once prevalent throughout Florida and commonly encountered from Texas to North Carolina, *P. pectinata* currently occurs mostly in south and southwest Florida [15,16,17]. It grows to over 500 cm STL (stretched total length) after being born in estuarine and nearshore areas at sizes between 69 and 81 cm STL [18]. One of the objectives of the recovery plan for *P. pectinata* is to protect or restore habitats for the juveniles that occur in estuarine and nearshore areas

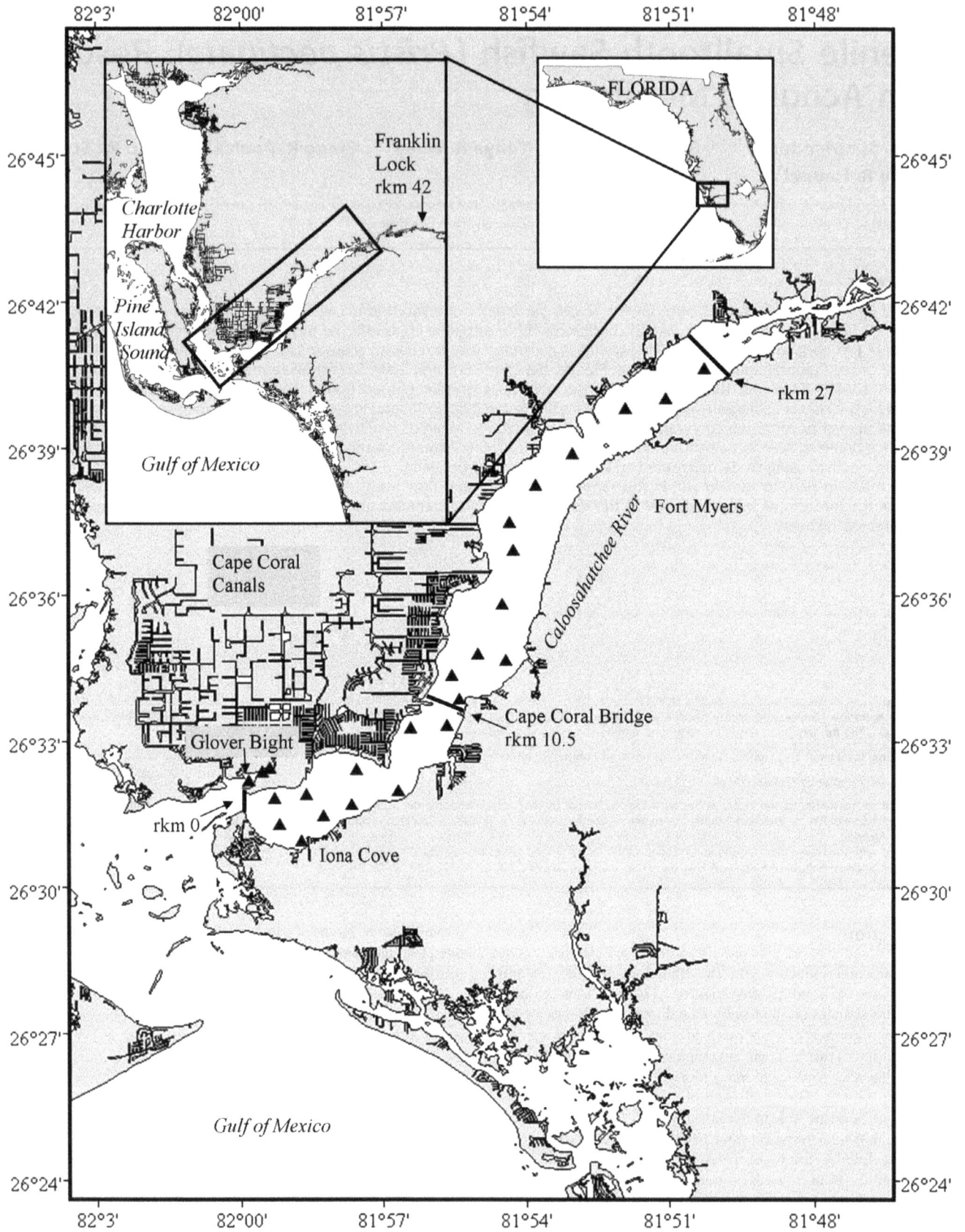

Figure 1. Study location in the Caloosahatchee River indicating locations of acoustic receiver stations (▲) within the river. Insets show study site location on the central Gulf of Mexico coast of Florida. rkm, river kilometer.

[19]. However, little is known about their long-term habitat use and movements, or how environmental factors affect these attributes, as large-scale spatial studies of *P. pectinata* ecology are lacking.

Studies of the movements of sawfish are limited. Thorson [5] reported tag and recapture results of largetooth sawfish (*Pristis perotteti*) showing that they moved between freshwater and saltwater in Nicaragua. Based on acoustic monitoring results, the freshwater sawfish (*P. microdon*), which spends the first few years of life in rivers [20], has been shown to ontogenetically partition depth within a river system in northern Australia [21]. Simpfendorfer et al. [22] used acoustic tracking and acoustic monitoring to investigate the short-term movements, site fidelity, and habitat use of juvenile *P. pectinata* in southern Florida. These studies have shown that for juvenile sawfish, very shallow depths are a critical factor, probably because of the protection that it can provide from predators, and that there are clear ontogenetic changes in habitat use. To date, there is no published information on how environmental factors can influence movement and distribution of *P. pectinata*. However, such information may be critical to recovery of this population as they occur adjacent to the Florida Everglades which is undergoing major restoration that is significantly altering freshwater flow patterns in southern Florida (see www.evergladesplan.org).

To investigate the role that changes in environmental conditions have on the movement and distribution of *P. pectinata*, an acoustic array was deployed within the estuarine portion of the Caloosahatchee River in southwest Florida to track their long-term movements. The specific aims of this study were to determine the level of residency, movement patterns and activity space within the

system, and investigate how these attributes were influenced by ontogeny and changes in environmental conditions.

Methods

Ethics statement

This research was conducted in accordance with National Marine Fisheries Service Endangered Species Permit numbers 1352 and 1475. Animal ethics approval was granted by Mote Marine Laboratory to Colin Simpfendorfer.

Study site

This study was conducted in the lower 27 km estuarine portion of the Caloosahatchee River in southwest Florida (Figure 1). The river connects Lake Okeechobee to the Gulf of Mexico and is the major source of freshwater to southern Charlotte Harbor. Water from Lake Okeechobee also flows to the east coast via the St Lucie River, and changes in the distribution between the two systems over time has affected the levels of freshwater flow. Freshwater flows in the Calooshatchee River during the current study were greatest during summer (Figure 2) and varied considerably between years depending on the magnitude of the wet season. The river has been substantially altered in the last 100 years [23], including an artificial link to Lake Okeechobee, extensive canal systems, three locks to permit boat passage, and dams to regulate water flow. The upper reaches of the study site had natural shoreline and native vegetation (primarily red mangroves *Rhizophora mangle*) while closer to the mouth the habitat was largely altered by urbanization including extensive canal develop-

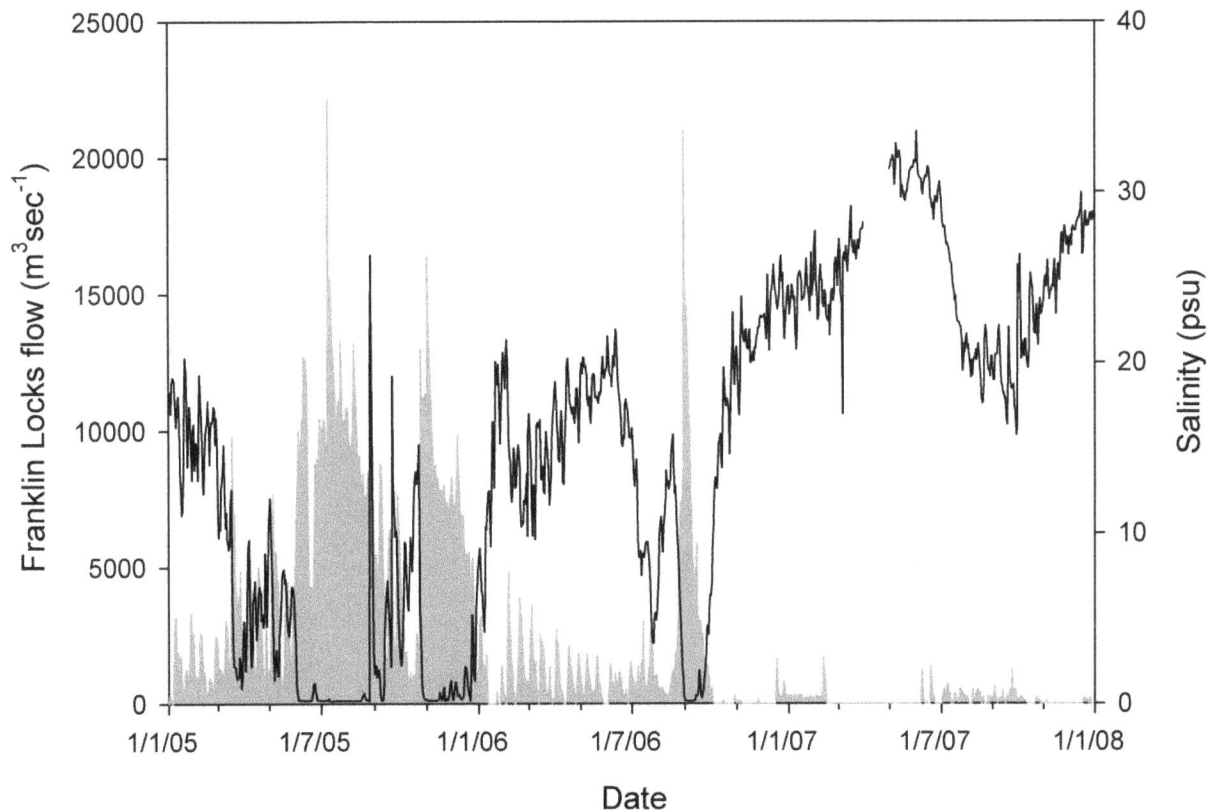

Figure 2. Environmental conditions within the study area. 2005 was a wet year with high freshwater flows through the Franklin Locks (grey shaded area) of long periods with salinity at Cape Coral (black line) below 10 psu. 2006 was a moderately wet year with a short period of high flow, and 2007 was a dry year with very small flows and high salinity.

ments and shoreline modification associated with the cities of Cape Coral and Fort Myers.

Field methods

A series of 25 VR2 acoustic receivers (Vemco Ltd.) were deployed within the study site to passively track the movements of *P. pectinata* fitted with acoustic tags (Figure 1). Methods for deploying receivers have previously been described [24,25]. Acoustic receivers were deployed in August 2003 and were continuously present within the study site until project completion in October 2007. These single frequency, omnidirectional data logging receivers recorded the time, date, and identity of *P. pectinata* fitted with acoustic transmitters that swam within the detection range. Receivers had a maximum detection range of approximately 800 m [26]. Detection range within the Caloosahatchee River allowed *P. pectinata* to often be detected on more than one receiver simultaneously and the receiver array allowed individuals to be continuously monitored while they were present within the study area. Approximately once per month data were downloaded from receivers and any required maintenance (e.g. removal of biofouling organisms and battery change) was performed.

Sampling for the capture of *P. pectinata* was conducted with longlines, gill nets, seine nets, and rod and reel (detailed field methods can be found in Simpfendorfer et al. [18]). All captured individuals were measured (stretch total length), sexed and fitted with an external identification tag. *Pristis pectinata* to be tracked were fitted with Vemco RCODE V9, V13 or V16 individually coded transmitters mounted on Rototags or Jumbo Rototags (Dalton, UK) and attached to the first or second dorsal fin. External attachment was required due to Endangered Species Act permitting limitations. Each transmitter was coded with a unique pulse series and operated on 69.0 kHz at random intervals between 60 and 180 seconds. Random signal transmission times prevented more than one signal continuously overlapping and blocking detection by a receiver. Expected battery life of transmitters was approximately 8 months (V9, V13) or 18 months (V16).

Data analysis

Occurrence of tagged *P. pectinata* in the array was determined on a daily basis and a presence history was plotted to provide a visually interpretable timeline of occurrence throughout the study period. The number of days individuals were detected on receivers

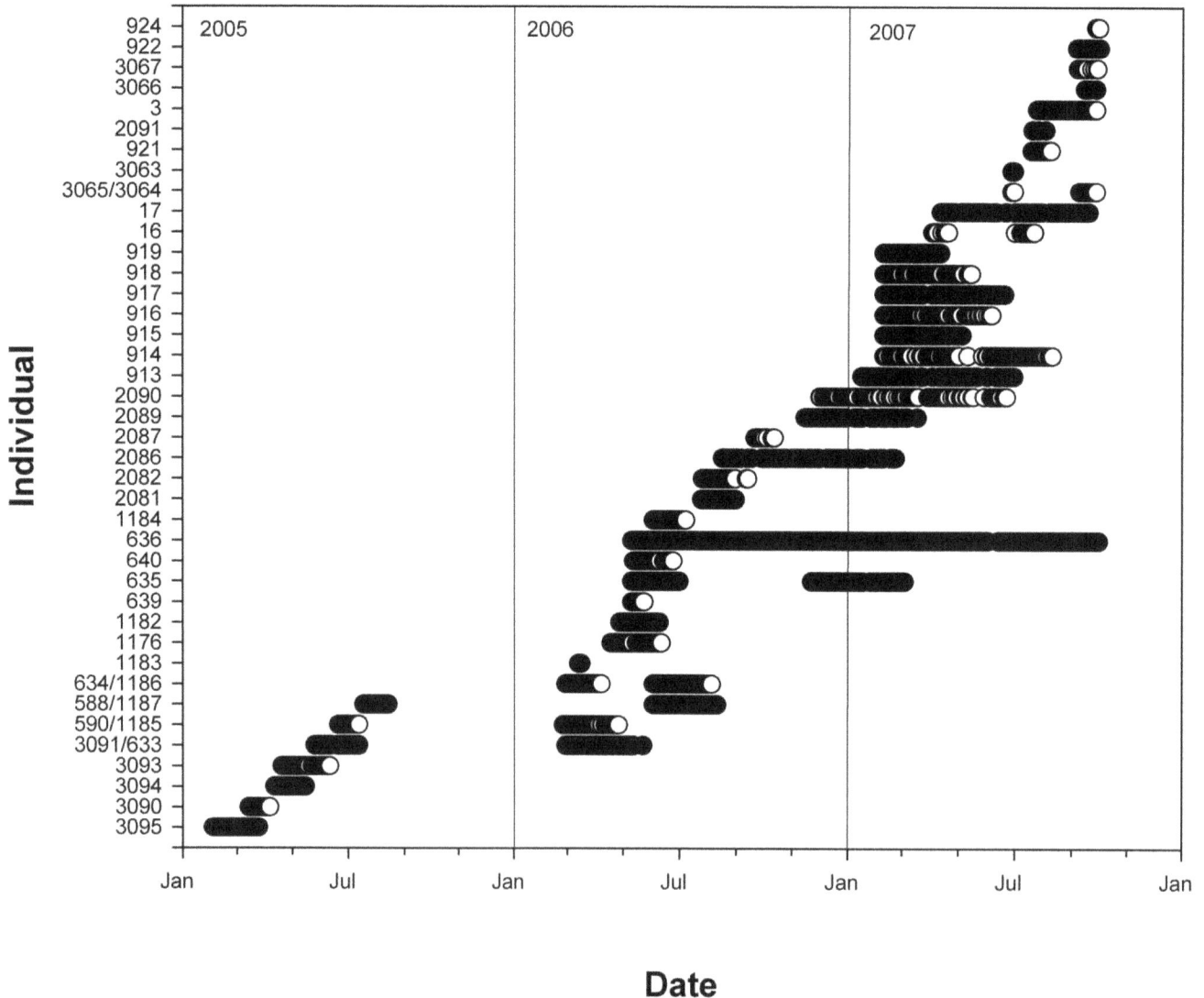

Figure 3. Presence history of the monitored *Pristis pectinata* in the Caloosahatchee River from 2005 to 2007.

in the study area, the total number of days from the first to last detection, and the maximum number of consecutive days present were calculated. A residency index was calculated as the ratio between the number of days an animal was detected to the number of days from the first to the last detection, with a value of one indicating it was detected every day and zero indicating it was never detected. Residency index values were compared between years with size and the total number of days monitored as covariates using analysis of covariance (ANCOVA). A post-hoc Tukeys unequal N Honest Significant Difference (HSD) test was used to determine years that were significantly different from each other.

The location of each tagged *P. pectinata* in the estuary was estimated every 30 minutes using the receiver distance algorithm described by Simpfendorfer *et al.* [24]. This algorithm used data from the receiver array to estimate the distance between the center of activity (COA) of each individual *P. pectinata* and the river mouth (river distance in river kilometers [rkm] = 0 km). The COA positions were used to generate daily minimum and maximum river distances, and mean daily river distances for individual *P. pectinata*. Daily, weekly, and monthly activity spaces were calculated as the difference between the maximum and minimum river distance for the relevant time period [26,27]. Analysis of covariance was used to test for differences in daily activity space between sexes and length (covariate). The pattern of movement of individuals between COA positions was assessed using a General Linear Model (GLM) with diel period (day or night), month, river distance (covariate) and the interaction between diel period and month.

To investigate how environmental factors influenced the distribution of *P. pectinata* within the estuary the daily mean river distance of each individual was regressed separately against temperature, salinity and freshwater inflow. Where necessary, mean river distance was log-transformed to meet conditions of normality. Daily mean values of salinity and temperature data were obtained from the South Florida Water Management District (SFWMD) Cape Coral Bridge station, approximately 10.5 km from the river mouth (in the middle of the study area). The salinity and temperature at the Cape Coral Bridge was used as an index of the salinity regime present in the river on each day. Daily freshwater inflow to the estuary was obtained from the SFWMD recording station at the Franklin Locks upstream of the study area (river distance = 42 km). Salinity was significantly negatively correlated with freshwater inflow ($R^2 = 0.591$, $p < 0.0001$), but also depended on rainfall [6]. The number of *P. pectinata* present each day in the main stem of the river was calculated and compared to the daily freshwater inflow through the Franklin Lock and salinity measured at the Cape Coral Bridge.

Electivity analysis was used to determine if *P. pectinata* exhibited affinity for, or avoidance of, specific salinity conditions within the river. To do this, the salinity in which individuals occurred each day was compared to those available in the river using Chesson's α [28]:

$$\alpha = (r_i/p_i) \Big/ \sum (r_i/p_i)$$

where r_i is the proportion of time an individual spent in salinity i and p_i is the proportion of salinity i available in the river. Since different years had different salinity regimes and not all salinities were available in each year, annual electivity values were standardised using the method described by Heupel and Simpfendorfer [6]. The salinity for a specific day and river distance ($s_{t,i}$) for any given location within the river (i) was

estimated using the equation from Heupel and Simpfendorfer [6]:

$$s_{t,i} = 48.593 - 4.034 \ln(flow_t) - 0.451i$$

where $flow_t$ is the freshwater inflow rate for day t into the estuary at the Franklin Lock. River distances for the electivity analysis were taken as the daily mean river distance of each individual.

To investigate the combined effects of environmental and other factors on the distribution of juvenile *P. pectinata* within the Caloosahatchee River estuary a Generalized Additive Model (GAM) was used to model the distribution of individuals on a daily basis within the river based on five factors: salinity, temperature, freshwater inflow, month and sawfish length. All factors were continuous except for month. Sawfish lengths of <100 cm corresponded to neonate individuals, those 100–140 cm were up to one year old, those 141–180 cm were one to two years old and those >180 cm were older than two years [18]. Twenty-five different models were constructed ranging from simple single factor models to multifactor models with interaction terms. The

Table 1. Comparison of General Additive Models constructed for predicting juvenile *Pristis pectinata* locations within the Caloosahatchee River (as defined by the distance from the river mouth (rkm)) in southwest Florida.

Index	Model	AIC	AIC weight
1	rkm~1	16944	<0.0001
2	rkm~s(sal)	13588	<0.0001
3	rkm~s(temp)	16566	<0.0001
4	rkm~s(len)	16406	<0.0001
5	rkm~s(mon)	16611	<0.0001
6	rkm~(ln(flow))	16281	<0.0001
7	rkm~s(sal)+s(temp)	13448	<0.0001
8	rkm~s(sal)+s(ln(flow))	13585	<0.0001
9	rkm~s(sal)+s(len)	13004	<0.0001
10	rkm~s(sal)+s(mon)	13327	<0.0001
11	rkm~s(sal)+s(ln(flow))	12995	<0.0001
12	rkm~s(sal)+s(ln(flow))+s(mon)	13326	<0.0001
13	rkm~s(mon)+s(ln(flow)+s(len)	15376	<0.0001
14	rkm~s(sal)+s(ln(flow))+s(len)+s(mon)	12800	<0.0001
15	rkm~s(sal)+s(ln(flow))+s(sal,ln(flow))	13524	<0.0001
16	rkm~s(sal)+s(temp)+s(sal,temp)	13414	<0.0001
17	rkm~s(sal)+s(len)+s(sal,len)	12387	<0.0001
18	rkm~s(sal)+s(mon)+s(sal,mon)	13267	<0.0001
19	rkm~s(sal)+s(temp)+s(sal,temp)	12648	<0.0001
20	rkm~s(sal)+s(len)+s(sal,len)+s(mon)	12180	<0.0001
21	rkm~s(sal)+s(len)+s(sal, len)+s(sal, temp)	11997	<0.0001
22	rkm~s(sal)+s(len)+s(sal,len)+s(len,mon)	11865	>0.9999
23	rkm~s(sal)+s(len)+s(sal,temp)+s(len,mon)	12308	<0.0001
24	rkm~s(sal)+s(len)+s(mon)+s(sal,temp)	12543	<0.0001
25	rkm~s(sal)+s(len)+s(mon)+s(sal,temp)+ s(len,mon)	12212	<0.0001

Models incorporate salinity (sal), water temperature (temp), length (len), month (mon) and freshwater flow (flow). Interaction terms are indicated by two factors within a term. Models with the lowest AIC value indicate the most plausible model, in this case number 22. AIC weight indicates the proportional support for the individual models.

model fit was determined using the Aikake Information Criteria (AIC) and the factors of the model with the lowest AIC value and highest AIC weight were considered to be those that best explained sawfish distribution within the river.

Results

Forty juvenile *P. pectinata* were tagged for monitoring in the Caloosahatchee River between 2005 and 2007 (Figure 3). Individuals ranged in size from 69 to 250 cm STL (mean = 149 cm STL) representing neonate, young-of-the-year, and juveniles (Table S1). Captures occurred in all months except October. Five individuals were recaptured during the study and fitted with a transmitter a second time. The total monitoring period from first to last detection for individuals ranged from 3 to 510 d, individuals were detected within the monitoring area from 2 to 473 d and maximum consecutive periods present ranged from 1 to 125 d.

Residency index values of individuals ranged from 0.23 to 1.0. Annual mean residency index values were significantly different between years (ANCOVA, $F_{2,35} = 4.65$, p = 0.016), with values highest in 2005 (0.95), moderate in 2006 (0.83) and lowest in 2007 (0.72). Residency index values were significantly related to the numbers of days present (ANCOVA, $F_{1,35} = 4.42$, p = 0.043), but not size (ANCOVA, $F_{1,35} = 1.95$, p = 0.171). Post-hoc tests showed that 2005 and 2007 were significantly different from each other, while 2006 was not significantly different from either 2005 or 2007. The lower value of residency index in 2007 was in part related to some tagged sawfish moving upstream out of the monitoring array. This

occurrence was evidenced by four individuals (915, 918, 2089, 2090) being detected on equipment maintained by the Florida Fish and Wildlife Conservation Commission (FWC) upstream from the study area for 2 to 55 d between March and June of 2007 (R. Taylor, unpublished data). Thus, these individuals were still in the river, but outside of the study area and not considered in the residency index.

Individual *P. pectinata* had relatively small daily activity spaces, but covered the entire study area over the long-term. Mean daily activity spaces ranged from 0.0 km (indicating that an individual was heard on only a single receiver for the whole day) to 3.88 km (Table 1); the mean across all individuals was 1.42 km. Individual daily activity space values were mostly less than 5 km, with estimates of less than 1 km common (2005: 44.9%; 2006: 54.3%; 2007: 56.3%). On rare occasions, values >10 km were observed. Mean daily activity space was not significantly different between males and females (ANCOVA, $F_{1,40} = 0.107$, p = 0.745) but was significantly related to individual size (ANCOVA $F_{1,40} = 14.0$, p<0.001), with larger individuals having larger activity space (Figure 4).

Mean weekly activity space of individual *P. pectinata* were larger than mean daily activity space (range 0–10.1 km), often by a factor of two or more. This suggests that individuals moved along the river over periods of several days rather than staying in the same location over longer periods. Monthly mean activity space of individuals was greater than mean weekly values, but by less than a factor of two (Table 1), suggesting that activity space was relatively stable over time frames >7 d. The largest individual monthly activity space estimates in 2006 and 2007 were 22.5 and 23.8 km, respectively, suggesting occasional use of nearly the entire study area within a month.

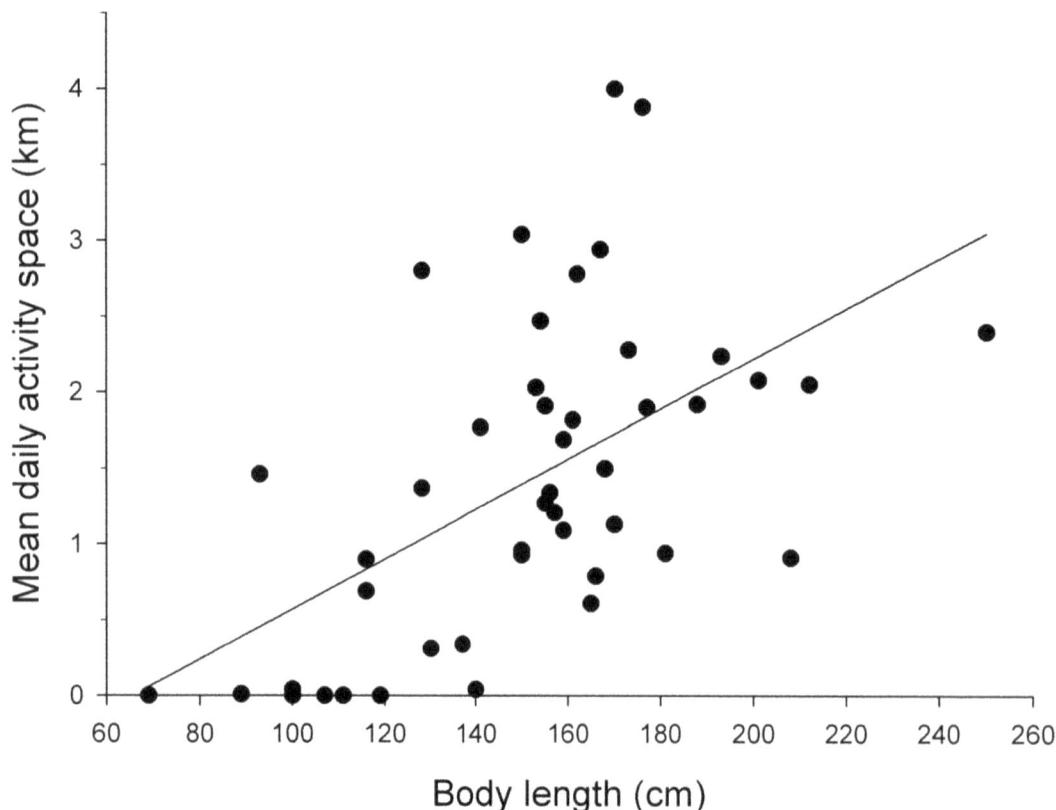

Figure 4. Relationship between body length and mean daily activity space of *Pristis pectinata* monitored in the Caloosahatchee River.

The majority of movements between 30 minute COA positions (91%) were less than 1.0 km, indicating that longer distance movements were rare. Movements between COA positions were significantly related to month (GLM, $F_{9,18800} = 2.06$, $p = 0.029$) and river distance (GLM, $F_{1,18800} = 86.4$, $p < 0.001$). Individuals that were located upstream were more likely to move downstream than those located downstream. There was no significant diel difference in the movement of individual *P. pectinata* between COA positions (GLM, $F_{1,18800} = 0.25$, $p = 0.62$), with overall movements of equal magnitude likely up or down the river either day or night. However, there was a significant interaction between diel period and month (GLM, $F_{9,18800} = 2.45$, $p < 0.001$). This interaction suggests that there were monthly differences in the diel movement patterns of *P. pectinata*. Large movements between COA positions (>5 km) occurred very rarely, mostly in 2007, but were not associated with major freshwater flow events.

During the study, water temperature ranged from 14.6 to 32.6 C, salinity ranged from 0.1 to 33.6 psu and freshwater inflow ranged from 0.0 to 627.4 m^3s^{-1}. *Pristis pectinata* were present throughout the entire range of these environmental conditions. Salinity was negatively correlated with flow ($r^2 = 0.646$, $p < 0.001$). There were positive correlations between *P. pectinata* log-transformed mean daily river location and salinity (Figure 5a; $r^2 = 0.126$, $p < 0.001$) and log-transformed mean daily river location and temperature (Figure 5c; $r^2 = 0.016$, $p < 0.05$). There was a negative relationship between log-transformed mean river location and flow (Figure 5b; $r^2 = 0.116$, $p < 0.001$). Distribution of *P. pectinata* within the river (Figure 6) indicated a significantly different proportion of detections by river location between years ($\chi^2 = 28766$, $df = 48$, $p < 0.001$). Differences between years were likely driven in part by differences in flow regime. During 2005, when flows were high, individuals were in the lower reaches of the river, while periods of little or no flow in 2007 corresponded to periods when individuals were far upriver. This suggests flow, in conjunction with physical factors such as depth, plays some role in individual location within the river, possibly through their influence on salinity. Electivity analysis demonstrated that *P. pectinata* had an affinity for salinity values between 18 and at least 24 psu (Figure 7). At salinities above 24 psu sample sizes were small and conclusions limited.

The best fitting GAM model included the factors salinity, length, and the interactions salinity*length, and month*length (Table 1). This model demonstrated that as salinity increased sawfish moved upriver, but that the salinities that different size classes moved at were different (Figure 8). The model predicted that neonate individuals (<100 cm) had limited movements in relation to salinity, while individuals between 100 cm and 140 cm (up to ~1 year old) moved further upriver and started moving at the lowest salinities. Individuals from 141 cm to 180 cm (1–2 years old) were predicted to move upriver at much higher salinities (i.e., they may be more tolerant of salinity changes), and not move as far as the 100 cm to 140 cm size class. Individuals >180 cm were predicted to have limited movements in relation to salinity, although sample sizes were relatively small. The interaction between month and length probably occurred because of the rapid growth of this species [18] that meant particular size classes were only available in particular months, and so accounting for this through the interaction increased the fit of the model.

Discussion

The results from acoustic monitoring within the Caloosahatchee River show that juvenile *P. pectinata* are often present within

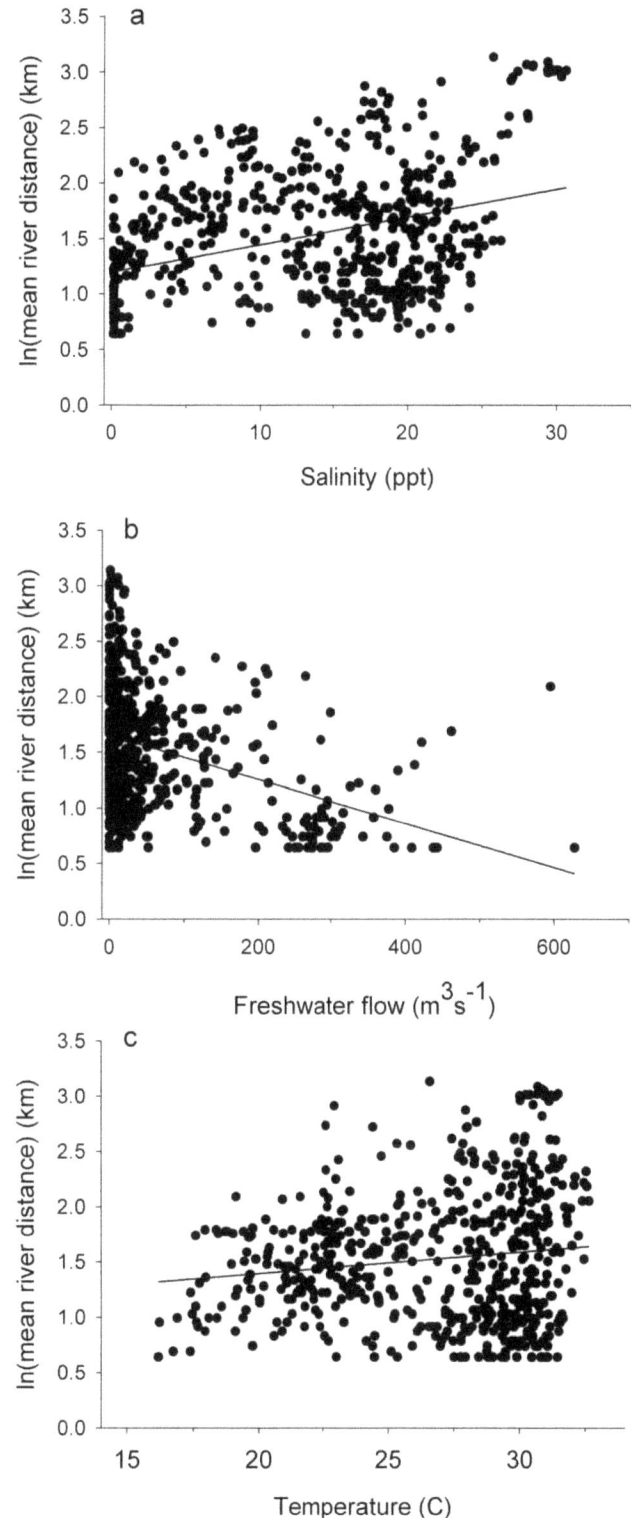

Figure 5. Relationship between distribution of tagged sawfish and environmental parameters. (a) salinity, (b) freshwater flow and (c) temperature.

estuarine areas for at least the first two years of life. This observation is consistent with results from short-term acoustic tracking [22] and encounter data [17] that indicate that individuals move away from shallow inshore habitats to deeper

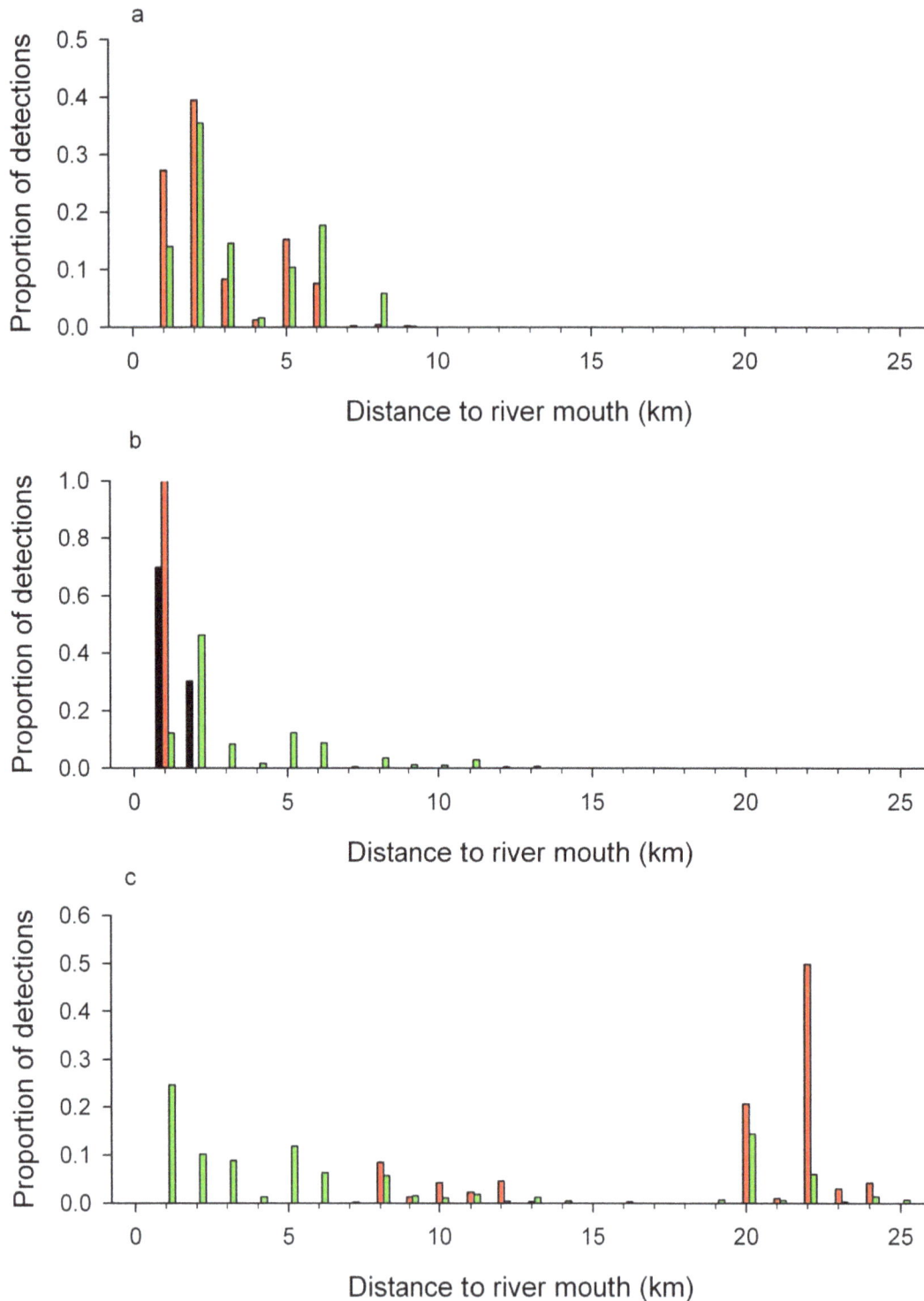

Figure 6. Distribution of neonate (black bars), young-of-the-year (red bars) and >1 year old (green bars) *Pristis pectinata* **within the Caloosahatchee River.** (a) 2005, (b) 2006 and (c) 2007 based on detections by acoustic receivers. Location was calculated as kilometers from the river mouth using a linear mean-position algorithm.

areas at sizes >250 cm (~2 yr old). Presence histories of individuals were mostly relatively short (~3 months), but this did not necessarily reflect animals moving out of the Caloosahatchee River estuarine system. Rather, the requirements to deploy acoustic tags externally to meet permitting restrictions meant that many were often prematurely shed due to rapid growth rates of

juvenile *P. pectinata* [18]. Juveniles are therefore likely to remain within the Caloosahatchee River estuary for most or all of the first few years of life (as suggested by high mean residency values), possibly only leaving when environmental conditions are less favorable. This observation is similar to that of Thorburn et al. [20] who reported that freshwater sawfish *Pristis microdon* remained

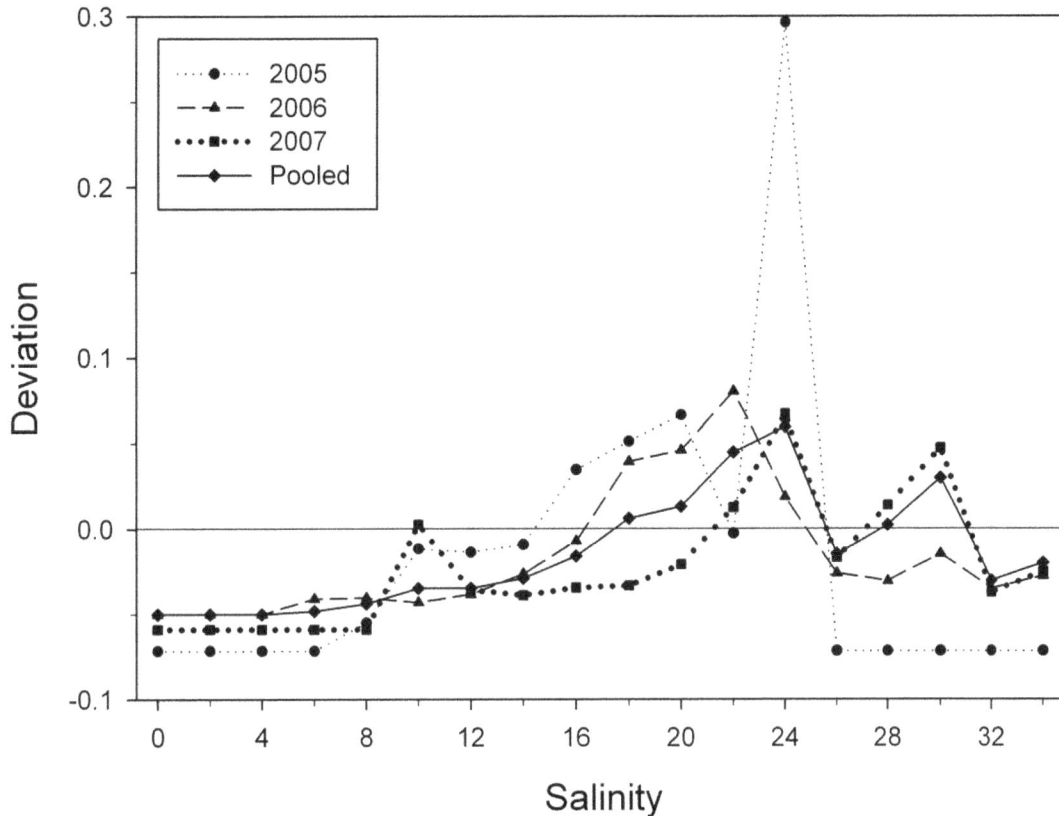

Figure 7. Salinity electivity of *Pristis pectinata* in the Caloosahatchee River estuary revealing a preference for salinities from 18 to at least 24 psu. Sample sizes above 24 psu were small, limiting the ability to make conclusions.

within the Fitzroy River in northern Australia for most of its juvenile life. However, *P. pectinata* does not appear to remain in the estuary until it matures, instead leaving after approximately two years, before the size at which they mature. The bull shark, which also uses the Caloosahatchee River estuary as a nursery area followed a similar pattern, leaving after about two years [6,7], well before they matured.

Within the estuary several factors influenced where *P. pectinata* occurred. The two most important factors were salinity regime and size. Changes in salinity could directly affect sawfish physiology or could be a proxy for indirect effects such as prey distribution that determine sawfish location in the river. During high freshwater flow conditions in 2005 and 2006, sawfish of all sizes remained in the lower portion of the river. Increases in salinity (measured at the mid-point of the study area) as a result of reductions in freshwater flow, likely caused individuals to move upriver in 2007 to meet their salinity requirements as drier conditions prevailed in the study area. Similar salinity effects have been observed in several other elasmobranch species that occur in estuarine and nearshore habitats [6,8,26,29]. However, for *P. pectinata* the response to changes in salinity regime differed among individuals of different ages. Individuals approaching 1 year of age (100–140cm) were most mobile, moving up river at the lowest salinity levels. The exact reasons for this sensitivity is unclear, but may relate to the energetic costs of osmoregulation [6], or the movement of preferred prey in response to salinity changes [8]. The youngest individuals (i.e., neonates) did not demonstrate this level of sensitivity despite a smaller body size which increases the cost of osmoregulation due to higher surface area to volume ratio. However, these animals are more vulnerable to predation due to

their small size, and fine-scale acoustic tracking has demonstrated that this size class has small activity spaces in very shallow habitats, probably to avoid predation [22]. Thus, the energetic cost of osmoregulation for this size class may be outweighed by the increase in survival that results from remaining in these protected habitats. Behavioral choices that lead to higher energetic costs are commonly observed in aquatic organisms [30] demonstrating the importance of predation risk in determining the movements and distribution of some species. Tracking data [22] also indicated that *P. pectinata* transition from restricted activity space to more broad use of space at ~100 cm, a result consistent with the predictions of the GAM for when individuals become most sensitive to changes in salinity regime. Tracking data indicate that this is a transition from remaining in very shallow areas, often mud banks in mangrove areas, to following mangrove shorelines. Individuals >1 yr of age may be less sensitive to salinity changes because of their larger body size, and so only respond to changes when freshwater flows have substantially decreased.

Though not the sole cause of sawfish movements, the influence of salinity was further supported by the results of electivity analysis, which demonstrated an affinity for areas with salinities between 18 and at least 24 psu. The movements of *P. pectinata* in response to changes in salinity may thus reflect individuals seeking to remain within this salinity range. As freshwater flows into the estuary decrease, the location of saltier water also moves further upriver as the marine influence increases through tidal movement. Research on bull sharks within the same system determined that they had an affinity for salinities between 7 and 20 psu [6,7]. This difference in salinity range may reduce predation on juvenile *P. pectinata* by facilitating a separation between these two species. While there is

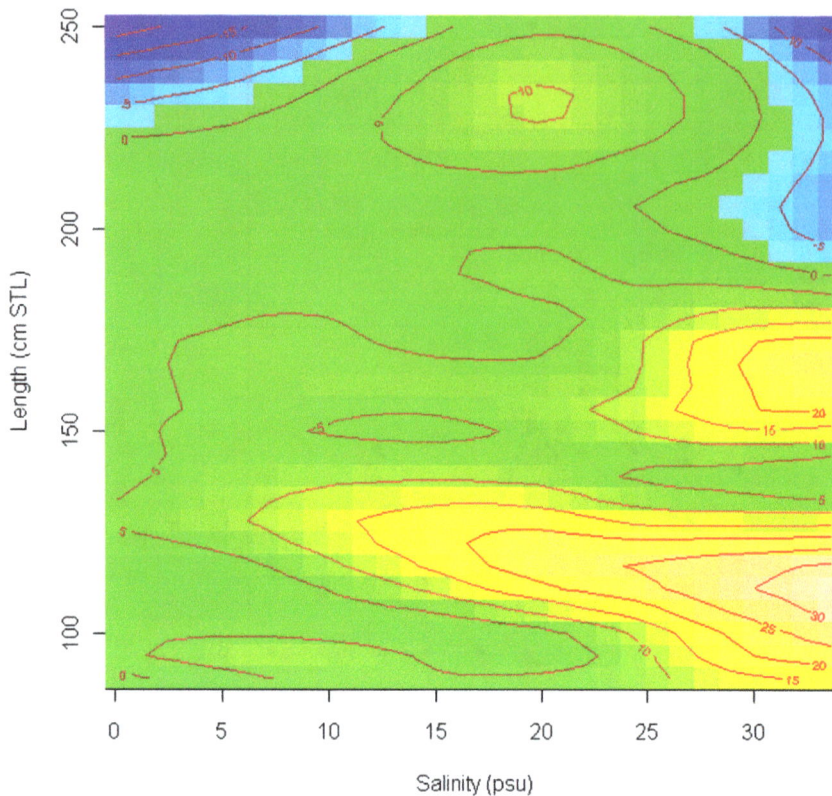

Figure 8. Predictions of river distance (contour lines, colors) based on salinity and sawfish length from the best fitting Generalized Additive Model for *Pristis pectinata* **in the Caloosahatchee River.** Blues indicate areas downstream of the study area, green indicates low reaches of the study area, yellow mid-reaches and white upper reaches.

no direct evidence for bull sharks consuming *P. pectinata*, predation has been observed on other species of *Pristis* [31]. An affinity for lower salinities than most marine predators also means that risk from that large potential source of predation is also lowered. Thus, juvenile *P. pectinata* appear to have multiple approaches to reducing predation risk.

Daily activity spaces of *P. pectinata* were relatively small compared to other species within the same system. Collins et al. [32] reported that cownose rays had an overall mean daily activity space of ~3.5 km of river, while Heupel et al. [27] reported daily activity spaces values of ~4 km for bull sharks. These values were >2 times larger than that for *P. pectinata*. This difference is attributed to disparate activity levels between these species. Both bull sharks and cownose rays are semi-pelagic species swimming almost continuously [8,26], while *P. pectinata* is a benthic species that spends considerable periods of time (~21% of time) resting on the bottom [22]. This conclusion is also supported by the observation in the current study that there was no movement between >90% of consecutive COA positions. Direct comparison to other studies on elasmobranchs, however, were inappropriate because most studies normally use area measurements [1] not linear measures as is often used in river studies. The observation of a size effect on activity space was consistent with results from active tracking, which demonstrated an increase in home range and rate of movement with size [22].

The movements between COA positions demonstrated that the short-term movement patterns observed using active tracking [22] occur through the juvenile life stages and were not an artifact of capture affecting behavior. The difference in movement direction between upriver and downriver locations was likely related to

behavior that enabled individuals to locate or remain in their preferred salinity. Behavioral mechanisms by which marine animals maintain the presence in locations with preferred attributes has been poorly studied in elasmobranchs. The use of behavior to achieve homeostasis has been suggested for both temperature [33,34] and salinity [6], but the mechanisms used to achieve it remain unknown [4]. The highly seasonal nature of rainfall (and hence freshwater inflow) in southern Florida was the most likely driver of the differences in movement patterns by month, with tendencies to move down river during wetter months during the summer and upriver during drier months, especially in autumn. Overall, the movements of *P. pectinata* were consistent with those of a species that had an affinity for a particular salinity range.

The results of this study demonstrate that water management practices will have effects on *P. pectinata*. The preference of juveniles for salinities between 18 and at least 24 psu means that as freshwater flows into systems decrease, they are likely to move higher into estuaries. The extent of these movements is related to the magnitude of salinity change. When flow patterns are changed, individuals may move to areas with their preferred salinity, but habitats within these areas may be less (or more) suitable than those previously occupied [35]. Within the Caloosahatchee River, increases in salinity that led to *P. pectinata* occurring upriver of the study area may be most problematic as the river becomes quite narrow with few shallow habitats that this species appears to use as a refuge from predation [22]. In addition, water management decisions related to the amount of flow from Lake Okeechobee via the Calooshatchee and St Lucie rivers may have important implications for conservation measures.

More broadly, flow regimes within any estuaries where *P. pectinata* occurs that result in animals being distributed in sub-optimal habitats may reduce survival and thus hinder the recovery of this population. Similarly, water management practices that result in repeated large changes in flow over short periods of time will result in large amounts of movement between different habitats which will increase energy expenditure, and may expose individuals to greater risks of predation. The most vulnerable portion of the population to effects from water management practices appear to be sawfish in their first year of life. Neonate animals remain in small patches of shallow habitat irrespective of salinity, suggesting that they may suffer greater osmotic stress if salinity within these areas falls outside their preferred range for long periods. Although neonate sawfish captured downriver during high flow periods exhibit fast growth [18], growth and survivorship of neonates located further upriver during drought conditions is unknown. Water management practices therefore need to be considered in relation to the recovery of the *P. pectinata* population. More information on the energetic costs associated with occupying different salinities and the effects of occupying sub-optimal habitats on survival of juvenile *P. pectinata* must be determined.

The results of this and similar studies [6,7,9,27,32] have demonstrated that water management practices can have significant effects on elasmobranchs that inhabit estuaries. Ensuring that water flows are managed to meet the physiological and ecological needs of these important species, especially those like *P. pectinata* that face conservation challenges, will ensure healthy estuarine ecosystems. Research that continues to increase the understanding of how environmental factors influence the movements and distribution of estuarine elasmobranchs will be required to enable water managers to effectively implement flow regimes that meet these needs.

Supporting Information

Table S1 Presence and activity space data for *Pristis pectinata* monitored in the Caloosahatchee River from 2005 to 2007. Transmitter numbers with identical numbered superscripts indicate individuals that were recaptured and fitted with an additional transmitter at a later date. Size, detection and activity space data reflect the two periods of monitoring for these individuals. STL, stretch total length; t_{det}, number of days detected; t_{max}, number of days from first to last detection; t_{con}, maximum number of consecutive days present; RI, residence index; AS_d, mean daily activity space; AS_w, mean weekly activity space; AS_m, mean monthly activity space.

Acknowledgments

This work was performed under US Endangered Species Act permit numbers 1352 (CAS) and 1475 (GRP) issued by the National Marine Fisheries Service. We gratefully acknowledge A. Timmers, N. Leonard, and the many college interns and volunteers for assisting with field work. P. Doering and K. Haunert from the South Florida Water Management District provided advice on water management practice and assistance with accessing environmental data.

Author Contributions

Conceived and designed the experiments: CAS TRW MRH. Performed the experiments: CAS BGY TRW GRP PWS MRH. Analyzed the data: CAS BGY TRW MRH. Contributed reagents/materials/analysis tools: CAS GRP PWS. Wrote the paper: BGY CAS TRW MRH GRP PWS.

References

1. Simpfendorfer CA, Heupel MR (2004) Assessing habitat use and movement. In: Carrier JC, Musick JA, Heithaus MR, eds. Biology of Sharks and Their Relatives. Boca Raton: CRC Press. pp 553–572.

2. Heupel MR (2007) Exiting Terra Ceia Bay: An examination of cues stimulating migration from a summer nursery area. In: C.T. M, Jr, PHL, Kohler NE, eds. Shark nursery grounds of the Gulf of Mexico and the East Coast waters of the United States: American Fisheries Society Symposium 50. pp 265–280.

3. Grubbs RD, Musick JA, Conrath CL, Romine JG (2007) Long-term movements, migration, and temporal delineation of a summer nursery for juvenile sandbar sharks in the Chesapeake Bay region. In: McCandless CT, Kohler NE, Pratt HL, Jr., eds. Shark nursery grounds of the Gulf of Mexico and the East Coast waters of the United States: American Fisheries Society Symposium 50. pp 87–108.

4. Dowd WW, Harris BN, Cech JJ, Kultz D (2010) Proteomic and physiological responses of leopard sharks (*Triakis semifasciata*) to salinity change. Journal of Experimental Biology 213: 210–224.

5. Thorson TB (1974) Occurrence of the sawfish, *Pristis perotteti*, in the Amazon River, with notes on *Pristis pectinatus*. Copeia 2: 560–564.

6. Heupel MR, Simpfendorfer CA (2008) Movement and distribution of young bull sharks *Carcharhinus leucas* in a variable estuarine environment. Aquatic Biology 1: 277–289.

7. Simpfendorfer CA, Freitas GG, Wiley TR, Heupel MR (2005) Distribution and habitat partitioning of immature bull sharks (*Carcharhinus leucas*) in a southwest Florida estuary. Estuaries 28: 78–85.

8. Ortega LA, Heupel MR, Van Beynen P, Motta PJ (2009) Movement patterns and water quality preferences of juvenile bull sharks (*Carcharhinus leucas*) in a Florida estuary. Environmental Biology of Fishes 84: 361–373.

9. Heithaus MR, Delius BK, Wirsing AJ, Dunphy-Daly MM (2009) Physical factors influencing the distribution of a top predator in a subtropical oligotrophic estuary. Limnology and Oceanography 54: 472–482.

10. Ubeda AJ, Simpfendorfer CA, Heupel MR (2009) Movements of bonnetheads, Sphyrna tiburo, as a response to salinity change in a Florida estuary. Environmental Biology of Fishes 84: 293–303.

11. Grubbs RD, Musick JA (2007) Spatial delineation of summer nursery areas for juvenile sandbar sharks in Chesapeake Bay, Virginia. In: McCandless CT, Kohler NE, Pratt HL, Jr., eds. Shark nursery grounds of the Gulf of Mexico and the East Coast waters of the United States: American Fisheries Society Symposium 50. pp 63–86.

12. Hopkins TE, Cech JJ (2003) The influence of environmental variables on the distribution and abundance of three elasmobranchs in Tomales Bay, California. Environmental Biology of Fishes 66: 279–291.

13. Compagno LJV, Cook SF (1995) The exploitation and conservation of freshwater elasmobranchs: status of taxa and prospects for the future. Journal of Aquariculture and Aquatic Sciences 7: 62–90.

14. Simpfendorfer CA (2000) Predicting population recovery rates for endangered western Atlantic sawfishes using demographic analysis. Environmental Biology of Fishes 58: 371–377.

15. Seitz JC, Poulakis GR (2002) Recent occurrence of sawfishes (Elasmobranchiomorphi: Pristidae) along the southwest coast of Florida (USA). Florida Scientist 65: 256–266.

16. Poulakis GR, Seitz JC (2004) Recent occurrence of the smalltooth sawfish, *Pristis pectinata* (Elasmobranchiomorphi: Pristidae), in Florida Bay and the Florida Keys, with comments on sawfish ecology. Florida Scientist 67: 27–35.

17. Wiley TR, Simpfendorfer CA (In press) Using public encounter data to direct recovery efforts for the endangered smalltooth sawfish (*Pristis pectinata*). Endangered Species Research.

18. Simpfendorfer CA, Poulakis GR, O'Donnell PM, Wiley TR (2008) Growth rates of juvenile smalltooth sawfish *Pristis pectinata* Latham in the western Atlantic. Journal of Fish Biology 72: 711–723.

19. NMFS (2009) Recovery Plan for smalltooth sawfish (*Pristis pectinata*). Silver SpringMaryland: National Marine Fisheries Service.

20. Thorburn DC, Morgan DL, Rowland AJ, Gill HS (2007) Freshwater sawfish *Pristis microdon* Latham, 1794 (Chondrichthyes: Pristidae) in the Kimberley region of Western Australia. Zootaxa 1471: 27–41.

21. Whitty JM, Morgan DL, Peverell SC, Thorburn DC, Beatty SJ (2009) Ontogenetic depth partitioning by juvenile freshwater sawfish (*Pristis microdon*: Pristidae) in a riverine environment. Marine and Freshwater Research 60: 306–316.

22. Simpfendorfer CA, Wiley TR, Yeiser BG (2010) Improving conservation planning for an endangered sawfish using data from acoustic telemetry. Biological Conservation 143: 1460–1469.

23. Doering PH, Chamberlain RH (1988) Water quality in the Caloosahatchee Estuary, San Carlos Bay and Pine Island Sound. Charlotte Harbor National Estuary Program Technical Report 98-02. pp 229–240.

24. Simpfendorfer CA, Heupel MR, Collins AB (2008) Variation in the performance of acoustic receivers and its implication for positioning algorithms in a riverine setting. Canadian Journal of Fisheries and Aquatic Sciences 65: 482–492.

25. Heupel MR, Hueter RE (2001) Use of an automated acoustic telemetry system to passively track juvenile blacktip shark movements. In: Sibert JR, Nielsen JL, eds. Electronic Tagging and Tracking in Marine Fisheries. The Netherlands: Kluwer Academic.

26. Collins AB, Heupel MR, Motta PJ (2007) Residence and movement patterns of cownose rays *Rhinoptera bonasus* within a south-west Florida estuary. Journal of Fish Biology 71: 1159–1178.

27. Heupel MR, Yeiser BG, Collins AB, Ortega L, Simpfendorfer CA (2010) Long-term presence and movement patterns of juvenile bull sharks, *Carcharhinus leucas*, in an estuarine river system. Marine and Freshwater Research 61: 1–10.

28. Chesson J (1978) Measuring preference in selective predation. Ecology 59: 211–215.

29. Yeiser BG, Heupel MR, Simpfendorfer CA (2008) Occurrence, home range and movement patterns of juvenile bull (*Carcharhinus leucas*) and lemon (*Negaprion brevirostris*) sharks within a Florida estuary. Marine and Freshwater Research 59: 489–501.

30. Heithaus MR (2004) Predator-prey interactions. In: Carrier JC, Musick JA, Heithaus MR, eds. Biology of Sharks and Their Relatives. Boca Raton: CRC Press. pp 487–521.

31. Thorburn DC (2006) Biology, ecology and trophic interactions of elasmobranchs and other fishes in riverine waters of Northern Australia. Perth: Murdoch University. 135 p.

32. Collins AB, Heupel MR, Simpfendorfer CA (2008) Spatial distribution and long-term movement patterns of cownose rays *Rhinoptera bonasus* within an estuarine river. Estuaries and Coasts 31: 1174–1184.

33. Matern SA, Cech JJ, Hopkins TE (2000) Diel movements of bat rays, *Myliobatis californica*, in Tomales Bay, California: Evidence for behavioral thermoregulation ? Environmental Biology of Fishes 58: 173–182.

34. Hight BV, Lowe CG (2007) Elevated body temperatures of adult female leopard sharks, *Triakis semifasciata*, while aggregating in shallow nearshore embayments: Evidence for behavioral thermoregulation? Journal of Experimental Marine Biology and Ecology 352: 114–128.

35. Sklar FH, Browder JA (1998) Coastal environmental impacts brought about by alterations to freshwater flow in the Gulf of Mexico. Environmental Management 22: 547–562.

2

Diarrhea Outbreak during U.S. Military Training in El Salvador

Matthew R. Kasper[1]*, **Andres G. Lescano**[1,4], **Carmen Lucas**[1], **Duncan Gilles**[2], **Brian J. Biese**[3], **Gary Stolovitz**[3], **Erik J. Reaves**[1]

1 U.S. Naval Medical Research Unit 6, Lima, Peru, 2 Madigan Healthcare System, Tacoma, Washington, United States of America, 3 452nd Combat Support Hospital, U.S. Army Reserve, Milwaukee, Wisconsin, United States of America, 4 Universidad Peruana Cayetano Heredia, Lima, Peru

Abstract

Infectious diarrhea remains a major risk to deployed military units worldwide in addition to their impact on travelers and populations living in the developing world. This report describes an outbreak of diarrheal illness in the U.S. military's 130th Maneuver Enhancement Brigade deployed in San Vicente, El Salvador during a training and humanitarian assistance mission. An outbreak investigation team from U.S. Naval Medical Research Unit – Six conducted an epidemiologic survey and environmental assessment, patient interviews, and collected stool samples for analysis in an at risk population of 287 personnel from May 31st to June 3rd, 2011. Personnel (n = 241) completed an epidemiological survey (87% response rate) and 67 (27%) reported diarrhea and/or vomiting during the past two weeks. The median duration of illness was reported to be 3 days (IQR 2–4 days) and abdominal pain was reported among 30 (49%) individuals. Presentation to the medical aid station was sought by (62%) individuals and 9 (15%) had to stop or significantly reduce work for at least one day. Microscopy and PCR analysis of 14 stool samples collected from previously symptomatic patients, *Shigella* (7), *Cryptosporidium* (5), and *Cyclospora* (4) were the most prevalent pathogens detected. Consumption of food from on-base local vendors (RR = 4.01, 95% CI = 1.53–10.5, p-value <0.001) and arriving on base within the past two weeks (RR = 2.79, 95% confidence [CI] = 1.35–5.76, p-value = 0.001) were associated with increased risk of developing diarrheal disease. The risk of infectious diarrhea is great among reserve military personnel during two week training exercises. The consumption of local food, prepared without proper monitoring, is a risk factor for deployed personnel developing diarrheal illness. Additional information is needed to better understand disease risks to personnel conducting humanitarian assistance activities in the Latin America Region.

Editor: Martyn Kirk, The Australian National University, Australia

Funding: This work was funded by the U.S. Department of Defense Global Emerging Infectious Systems (DoD-GEIS), a division of the Armed Forces Health Surveillance Center. The participation of Dr. Lescano in this activity was partially funded by the National Institutes of Health/Fogarty International Center training grant 2D43 TW0007393 "Peru Infectious Diseases Epidemiology Research Training Consortium". The funders had no role in study design, data collection and analysis, decision to publish, or preparation of the manuscript.

Competing Interests: The authors have declared that no competing interests exist.

* E-mail: Matthew.kasper@med.navy.mil

Introduction

Infectious diarrhea remains a global health problem and a risk to travelers and military personnel deploying to developing regions. Existing epidemiologic data indicates that enterotoxigenic *E. coli* (ETEC), *Campylobacter jejuni*, and *Shigella* spp. (particularly *S. flexneri* and *S. sonnei*) are the most common causes of diarrheal disease among adults and children who live in the developing world as well as among U.S. military personnel deployed to these areas [1,2,3].

Diarrheal illness is one of the most common infectious risks among short-term travelers to the developing world, with some studies indicating over 50% of travelers being affected during a two week visit to an endemic country [4,5]. In a series of 784 American tourists traveling in the developing world for a median 19 days, 46% reported at least one episode of diarrhea (Hill, 2000), while Scottish tourists in Central and South America reported comparable rates of diarrhea (39.5%) [6]. On the other hand, a cohort of 36 Peace Corps volunteers in Guatemala developed 4.7 episodes of diarrhea over a mean 1.8 years of follow-up; 6.1 episodes/person-year occurred in the first 6 months, declining to 3.6 episodes/person-year after 12 months [7].

Among military populations, there were diarrheal disease studies conducted in the Middle East during Operation Bright Star. In 1989, up to 44% of personnel reported diarrheal disease with ETEC (49%) as the predominant pathogen identified [8]. During surveillance activities in 2001, 9.3% of troops reported a diarrheal episode and in 2005, diarrheal disease was prevalent with 35 cases of diarrhea/100 person-months and contributed to 17 non-combat related illnesses/100 person-months [9]. In personnel deploying to Iraq or Afghanistan in 2003 to 2004, 78.6% of troops in Iraq and 54.4% of those in Afghanistan experienced diarrhea, with 80% seeking care from their unit medic; eating local food from non-U.S. sources was associated with an increased risk of illness [10]. Outpatient medical surveillance of U.S. forces during missions conducted in Latin America showed an overall attack rate of 26%, with off-base travel and ice consumption being associated with higher reported disease rates [11]. These studies have demonstrated the risk that diarrheal illness presents to military operations and the risks associated with local food sources.

As part of building partnerships with Latin American nations, the U.S. military historically has embarked on a number of humanitarian operations in the region. Closer interaction with local populations may be necessary in humanitarian operations, increasing potential risks for disease transmission. *Beyond the Horizons* is a U.S. Southern Command sponsored 16-week joint, humanitarian and civic assistance exercise conducted by various U.S. military components scheduled from February to June each year in Latin America. In 2011, activities included engineering, dental, and medical projects to aid citizens in the local vicinity of San Vicente, El Salvador, in areas affected by natural disaster in late 2009. This report describes an outbreak of diarrhea among U.S. military personnel deployed to *Beyond the Horizons* in May 2011.

Methods

Study Site and Subjects

Approximately 300 individuals from the 130th Maneuver Enhancement Brigade (MEB) were detached to the Base Poligono in San Vicente, El Salvador as part of a military training and humanitarian assistance mission. The mission took place between February and June 2011, and approximately 75% of the personnel rotated every two weeks. On May 27th, 2011, the health unit of the base detected an increase in the number of diarrhea cases presenting for care at the base medical aid station. The U.S. Naval Medical Research Unit Six (NAMRU-6) provided assistance to determine the etiology and mechanism of the outbreak. An outbreak investigation team composed of five personnel was arranged and arrived to El Salvador on May 31st. Working in coordination with the health unit on base, the outbreak team reviewed local records and medical charts, conducted an epidemiologic survey and environmental assessment, patient interviews, and collected stool sample for further laboratory analysis.

Epidemiologic Surveys

After reviewing records and medical charts, the NAMRU-6 team developed a 61-question survey regarding demographics, health status, clinical symptoms, and food consumption habits during the last two week deployment period (May 21st to June 4th, 2011). The survey was administered to all brigade personnel during a scheduled daily informational meeting. The surveys were completed by each individual voluntarily. The case definition for diarrhea was broadened to one or more loose stool episodes (compared to the usual definition of 3 or more loose stools in 24 hrs or at least 2 stools accompanied by fever, blood, etc.) to increase its sensitivity and to capture all possible gastrointestinal illness cases for epidemiologic analysis and targeted stool collection.

An additional survey was administered to three units with the highest prevalence of diarrheal illness to further assess daily food consumption habits during the prior two weeks of the deployment. This survey was developed after discussion with personnel on base about dining options and specific food choices available.

Environmental Assessment

The environmental assessment conducted on Base Poligono included tent city (living quarters), latrines, hygiene facilities, the potable water distribution system, dining facility (DFAC), on-base vendor locations, and medical aid station. Logistic difficulties prevented the assessment of off-base work sites and food providers. No environmental sampling was conducted. Vendor food sources were not available after May 27th, 2011 because on-base vendor facilities and off-base work site food selling practices were stopped following the diarrheal outbreak as a prevention measure. Testing of the base water distribution system and point-of-use locations were performed by the Base personnel in accordance with U.S. Army standards [12] and before the arrival of the NAMRU-6 team.

Specimen Collection and Laboratory Testing

Stool samples were collected from volunteers in the at-risk population between June 1st and June 3rd, 2011. Samples were coded and linked to questionnaires. Fecal smears were made for immediate on-site field microscopy analysis. In addition, sample was preserved in sodium acetate, acetic acid and formalin (SAF) from all volunteers providing a sample, and in potassium dichromate from those volunteers providing a sample with a soft or loose consistency before transportation and further microbiological analysis by microscopy and PCR at NAMRU-6 in Lima, Peru.

Polymerase Chain Reaction Amplification

Stool samples were tested by PCR for the presence of enteropathogen nucleic acid. PCR testing was conducted for the following virulence markers or bacteria as previously described: (1) ipaH (*Shigella sp.*/Enteroinvasive *E. Coli*) [13] (2) *Campylobacter sp.* [14]. PCR testing was conducted as previously described for the following parasites: (1) *Cyclospora cayetanensis* [15], (2) *Cryptosporidium parvum* [16], (3) *Entamoeba histolytica/dispar* [17], (4) *Giardia lamblia* [18]. All Amplified PCR products were identified by gel electrophoresis, 1.5% or 2% agarose stained with ethidium bromide and viewed under ultra-violet light using the Bio-Rad Gel Doc XR Universal Hood II.

Real Time RT-PCR testing was conducted to detect norovirus using primers and probes for the polymerase gene of both genotypes I and II (GI and GII), as previously described [19].

Statistical Analysis

All data was entered into MS Access (Microsoft Inc., Redmond, WA, USA). Data was imported into SAS v9.2 (SAS, Cary, NC), which was used for all statistical analyses. Simple and multiple generalized linear models regression for binomial family and logarithmic link were applied to calculate the relative risk (RR) for factors associated with the cumulative incidence of diarrhea. The Akaike Information Criterion (AIC) was used to determine model fitness. First, we analyzed the data from the first questionnaire applied to all subjects and then we analyzed the results of the second questionnaire applied to only a subset with higher incidence. For the multivariate model using data collected from all personnel, we included the questions that had a statistically significant association with diarrheal illness. All confidence intervals are calculated with 95% confidence and statistical significance was determined at 0.05.

Ethical Considerations

This activity was deemed an outbreak investigation and determined to not be human subjects research by the Institutional Review Board of NAMRU-6. No informed consent was requested because the activity was conducted fulfilling a required public health mandate. Survey responses and stool samples were voluntary and analyzed de-identified.

Results

Diarrheal Illness

Based on passive surveillance data from the medical aid station, the epidemic curve shows an increase in the number of diarrhea cases starting at the end of April and going through the beginning of June, 2011 (Figure 1). Over this time period, 11, 25, and 39 personnel reported to the medical aid station during the two week rotation periods of April 26th – May 8th, May 9th – May 22nd, and May 23rd – June 4th, respectively. The peak of diarrheal illnesses occurred on May 27th, with 18 personnel reporting to the medical aid station. Cases were treated with Ciprofloxacin 500 mg twice daily for three days and metronidazole 500 mg thrice daily for seven days after the diagnosis of *E. histolytica* was made by a local laboratory.

The epidemiological survey had a completion rate of 87% (241/287 personnel) and 67 individuals met the diarrheal outbreak case definition of one or more loose stools in the two-week deployment period, a diarrheal attack rate of 27.8%. Descriptive analyses of the cases revealed that 83.8% (202/241) were male and the median age was 27 years (IQR 22–37) (Table 1). Among patients reporting diarrhea, symptoms included abdominal cramping (49.2%), headache (32.8%), nausea (27.9%), dehydration (23.0%), vomiting (8.2%), fever (8.2%), and one patient reported the presence of blood in their stool. The median duration of illness was 3 days (IQR 2–4 days). Medical care was sought by 62.3% of cases. Among those not seeking care (37.7%), 17.4% self-medicated. Nine personnel with diarrhea (14.8%) reported stopping or significantly reducing work for at least one day. Pre-deployment preventive medicine information was reported to have been received by 35.3% of personnel, with the majority recalling the topics of malaria, personal hygiene and diarrheal illness.

The bivariate analysis of the completed surveys indicates that the individuals arriving on base within the past two weeks were at higher risk for developing diarrheal disease (estimated risk ratio [RR] = 2.79, 95% confidence [CI] = 1.35–5.76, p-value = .001) (Table 2). In addition, the consumption of meals from on-base local vendors was statistically associated with illness (RR = 4.01, 95% CI = 1.53–10.5, p-value <.001) (Table 2). Consumption of food off base that was cooked (RR = 0.35, 95% CI = 0.22–0.54, p-value = .005) or served hot (RR = 0.41, 95% CI = 0.25–0.67, p-value = .007) was considered to be protective from diarrheal disease. Other factors such as adding hot sauces to off-base foods, eating at street vendors off base and drinking non-bottled beverages off-base were not associated with risk of diarrhea. Multiple regression analysis using the significant variables from bivariate analysis showed that only arriving on base within the past two weeks (RR = 2.85, 95% confidence CI = 1.31–6.18, p-value = .008) and eating at local vendors on base (RR = 3.91, 95% confidence CI = 1.59–10.18, p-value = .005) were significantly associated with diarrheal risk (Table 2).

Two units (n = 80) accounted for more than half of the cases of diarrhea (45%, 36/80). The detailed survey among this group indicated that separate consumption of either tacos or papusas on May 23rd or May 25th from vendors on base were statistically associated with a higher incidence of diarrheal illness (Table 3). Individuals who ate neither tacos on May 23rd nor papusas on the 25th, reported lower than average diarrheal rates (10%, 4/39). Among personnel who only ate either tacos on May 23rd or papusas on the 25th, 80% (12/15) and 76% (16/21) reported diarrheal illness, respectively. Personnel who ate both tacos on May 23rd and papusas on May 25th, 80% (4/5) reported diarrheal illness. Overall, 63.8% (23/36) of personnel from these two units who developed diarrheal illness ate either tacos on May 23rd or papusas on May 25th. A similar but less striking pattern was observed analyzing in detail eating tacos on the 25th or papusas on the 23rd, suggesting that reporting of these foods may be mainly correlated with the actual vehicles of infection but not necessarily a causative factor.

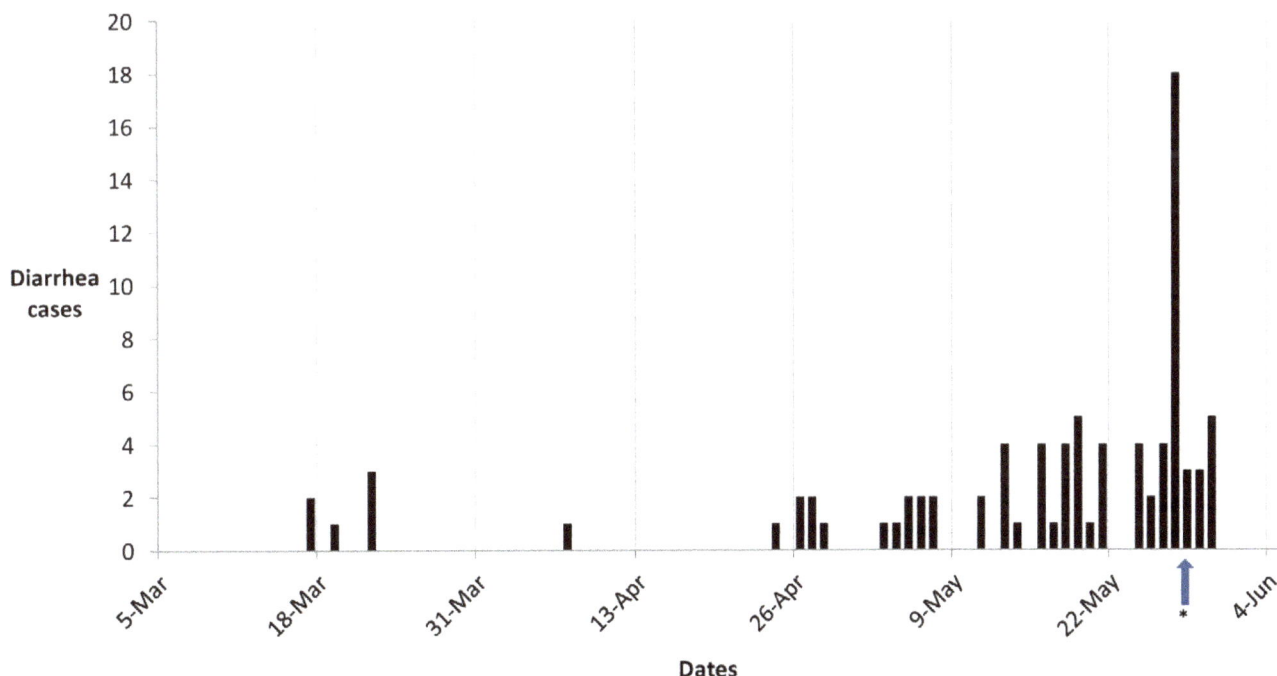

Figure 1. Epidemic curve of personnel reporting to the Medical Aid Station with diarrhea, San Vicente, El Salvador, 2011. The start of each two week deployment periods is indicated by the date. *Indicates the time at which vendors were no longer allowed on base to serve food.

Table 1. Demographic and Clinical Characteristics of Study Population.

Clinical and Demographic information	Overall (n = 241)	%
Age - years (median - IQR)	27	IQR 22–37
Gender - male	202	83.8
Have you been in a developing country in the last year?	41	17.1
What country?		
- Latin America/Caribbean	23	55.2
- Middle East	17	40.8
- Other	1	2.4
Diarrhea or vomiting in the last two weeks?	67	27.8
Duration of diarrhea or vomiting? (days)	3	IQR 2–4
Fever	5	8.2
Blood in Stool	1	1.0
Abdominal Pain	30	49.2
Dehydration	14	23.0
Nausea	17	27.9
Headache	20	32.8
Present at sick call due to diarrhea or vomiting?	38	62.3
Stop or signifcantly reduce work at least one day?	9	14.8
how many days miss work?		
- one day	5	55.5
- two days	4	44.4
Receive IV fluids?	4	6.6
Receive antibiotics?	35	57.4
Antibiotic self-medication?	4	6.6
Preventive Medicine Information before Deployment	85	35.3
Method of Delivery		
- paper-based	25	29.1
- Lecture	39	45.4
- online training	6	6.9
- other	7	8.1
Topics covered in training		
- Malaria	64	74.4
- dengue	33	38.3
- diarrhea	51	59.3
- injuries	40	46.5
- healthcare access	33	38.4
- insurance	11	12.8
- personal hygiene	62	72.1
- other	7	8.1

Laboratory Results

Fifty-one subjects provided stool samples and twelve with a soft or loose consistency were analyzed on-site by field microscopy for parasites. *Cyclospora* spp. was identified in one symptomatic patient receiving ciprofloxacin and metronidazole. In consultation with the senior medical officer, this patient's treatment regimen was changed to include an antiparasitic agent with effectiveness against *Cyclospora*.

PCR testing was performed on 14 stool samples from patients who had presented with a soft or loose stool and had reported diarrhea over the past two weeks. PCR analysis identified pathogens in 78% (11/14) of samples tested with 43% (6/14) testing positive for multiple pathogens. The majority of samples (7/14) were positive for a *Shigella* spp. or enteroinvasive *E. coli* virulence marker (ipaH) (Table 4).

The original sample from which *E. histolytica* had been identified by a local laboratory was not available for further analysis. Testing of a subsequent stool sample of the same case collected after treatment with metronidazole had begun was negative for *E. histolytica* by microscopy and PCR analysis.

Table 2. Association of risk factors with acquisition of illness among all survey takers (n = 241) and multiple regression analysis of risk factors for acquisition of diarrheal illness.

			Univariate Analysis		Multivariate Analysis		
Question	Yes/No	Diarrhea	Risk Ratio	p-value	Risk Ratio	95% CL	p-value
Arriving on base within past two weeks	Yes	60 (33.7%)	2.79	0.001	2.85	1.31–6.18	0.008
	No	7 (12.1%)					
Eat off base in last two weeks?	Yes	50 (31.8%)	1.51	0.077			
	No	17 (20.9%)					
Were off base foods cooked?	Yes	43 (29.2%)	0.35	0.005*			
	No	5 (83.3%)					
Were off base foods served hot?	Yes	42 (29.1%)	0.41	0.007*			
	No	7 (70.0%)					
Did you add hot sauces to off-base foods?	Yes	11 (23.4%)	0.67	0.157			
	No	37 (34.9%)					
Eat at street vendors off base in last two weeks?	Yes	28 (28.2%)	1.01	0.938			
	No	37 (27.8%)					
Eat at local vendors (non-DFAC) on-base?	Yes	59 (32.7%)	4.01	<0.001	3.91	1.50–10.18	0.0051
	No	4 (8.1%)					
Drink non-bottled beverages off-base in last two weeks?	Yes	15 (37.5%)	1.41	0.157			
	No	51 (26.4%)					

Food and Beverage Assessment

Food and beverage sources for U.S. personnel were available from the on-base Dining Facility (DFAC), Meals-Ready-To-Eat (MREs), and local vendors both on and off-base. The DFAC prepared three meals daily and was established on base in accordance with Army regulations [20]. Meals were standard prepackaged food approved by the U.S. Army, heated on site, opened, and served directly to service members. Disposable plates and utensils were utilized and waste from all food products was disposed daily.

Table 3. Food Consumption among three units with highest incidence of diarrheal disease (n = 80).

		Univariate Analysis	
Question	Yes/No Diarrhea	Risk Ratio	p-value
Ate Tacos for dinner on the 23rd?	Yes	12 (33.3%)	2.13	0.0029
	No	3 (6.9%)		
Ate Tacos for dinner on the 25th?	Yes	14 (37.8%)	2.09	0.0023
	No	4 (9.3%)		
Ate Papusas for dinner on the 23rd?	Yes	14 (38.9%)	1.87	0.01
	No	6 (13.9%)		
Ate Papusas for dinner on the 25th?	Yes	16 (43.2%)	2.14	0.0014
	No	5 (11.6%)		

The only option for breakfast was either MRE or DFAC. Food options for lunch and dinner included DFAC, MRE, or local vendors. There were two food vendors on base, named "upper" and "lower" vendors, that prepared meals daily from fresh food purchased locally and brought to base daily by local personnel. Local vendors were not contracted nor assessed for food safety practices prior to providing services. Off-base vendors were local street sellers who took lunch orders from U.S. personnel at work-sites and prepared meals in personal home or local restaurant kitchens.

Observation and assessment of vendor food preparation practices could not be completed because vendors were closed as a mitigation strategy for this diarrheal outbreak. Key informant interviews reported on-base vendor food was brought to base raw or uncooked in buckets, left unrefrigerated outside, and prepared under outdoor open, covered structures on grills and wood tables. The same cooking utensils were used to prepare raw food and serve cooked meat products and vegetables. Meats were reported to not be cooked completely prior to serving and many raw vegetables were served without knowledge of cleaning from potable water sources. Off-base work site vendors brought cooked meals to U.S. personnel presumably prepared in local homes or restaurants.

Potable Water Assessment

On-base water was provided by the primary local city source. The base Reverse Osmosis Water Purification Unit (ROWPU) was not functional. Water was stored in a 10,000 gallon water storage bladder (blivits) and made potable by chlorination in accordance with U.S. Army standards [12]. Potable water was distributed by soft hose and made available for consumption at two water tanks. This potable water was served at multiple cooler distribution points in the DFAC. Local bottled water was also available for purchase on base. After identification of this diarrheal

Table 4. PCR detection of enteropathogens (n = 14) among 14 samples collected from personnel reporting diarrhea in El Salvador.

Sample	Cyclospora cayetanensis	Cryptosporidium	Entamoeba spp.	Giardia lamblia	norovirus	IpaH	Campylobacter
1	**Positive**	Negative	Negative	Negative	Negative	**Positive**	**Positive** - *C. jejuni*
2	**Positive**	Negative	Negative	**Positive**	Negative	**Positive**	Negative
3	Negative	Negative	Negative	Negative	Negative	**Positive**	Negative
4	**Positive**	**Positive**	Negative	Negative	Negative	Negative	Negative
5	**Positive**	Negative	Negative	**Positive**	Negative	**Positive**	Negative
6	Negative	Negative	**Positive**	Negative	Negative	Negative	Negative
7	**Positive**	**Positive**	Negative	Negative	Negative	Negative	Negative
8	Negative	**Positive**	Negative	Negative	Negative	**Positive**	Negative
9	Negative	**Positive**	Negative	Negative	Negative	Negative	Negative
10	Negative	Negative	Negative	Negative	Negative	Negative	Negative
11	Negative	Negative	Negative	Negative	Negative	Negative	Negative
12	Negative	Negative	Negative	Negative	**Positive**	Negative	Negative
13	Negative	Negative	Negative	Negative	Negative	**Positive**	Negative
14	Negative	Negative	Negative	Negative	Negative	**Positive**	Negative
Total	5 (35.7%)	4 (28.6%)	1 (7.1%)	2 (14.3%)	1 (7.1%)	7 (50.0%)	1 (7.1%)

outbreak, both water tanks were drained, shock chlorinated, and re-filled, and additional multiple water quality checks were conducted within the base water distribution system.

Sanitation Assessment

Base latrines were contracted Portalet-type facilities that were drained and cleaned daily. Each latrine had a hand-washing sink with water foot pump and soap. There were two hand-washing stations at the entrance to the DFAC. After identification of this diarrhea outbreak, each individual latrine hand washing sink and DFAC hand-washing stations were routinely chlorinated. In addition, an individual was assigned to the DFAC entrance to ensure hand washing by all personnel during meal hours. Personal hygiene facilities were cleaned daily.

Discussion

This study describes an outbreak of diarrheal illness among deployed U.S. military personnel during Operation *Beyond the Horizon* in El Salvador. We identified an increasing trend of diarrheal illness occurring throughout the course of the operation and an association with the consumption of food sold from local vendors on-base. Diarrheal illness is a risk to deployed U.S. military units and travelers throughout the world. In the humanitarian and civic assistance mission setting, operating off base, working in local communities, and sharing meals are important cultural components that present infectious disease risks to deployed personnel [8,10].

The burden of diarrheal illness represented by the epidemiologic curve highlights a recurring increase in diarrheal cases during each two-week deployment period, especially during the last three deployment periods. This data is representative of other studies showing diarrheal illness being a common burden of illness in short-term deployed U.S. military units [21,22,23,24]. The shorter duration of this deployment and inability to follow-up with personnel that had already completed their two week rotation may, in part, explain the lower prevalence of diarrheal illness compared to studies in Iraq or Afghanistan. However, there have been few studies on the epidemiology of diarrheal illness and the

effectiveness of preventive strategies in deployed military units in Central and South America [11,25,26].

With the majority of personnel for these operations rotating every two weeks, it is critical to ensure that public health measures are maintained throughout the duration of the operation. While situations such as local food sources may not be covered under governing regulations, preventive medicine principles should be applied to the greatest extent possible according to the situation (e.g. training and monitoring of local vendors). In addition, part of maintaining appropriate public health measures includes preventive medicine education, which the majority of personnel report they did not receive. Ensuring the continuity of public health principles can help prevent diarrheal disease.

Epidemiologic and statistical analysis of food consumption histories obtained from surveys suggests that the consumption of food from local vendors, specifically tacos and papusas, on base was most significantly associated with illness. These findings are consistent with environmental evidence indicating that improper cooking procedures at local vendors could have resulted in the growth of microbiological organisms. However, vendors were not inspected nor trained on proper food handling practices prior to operating on base, and these hypotheses could not be confirmed in our investigation.

An important feature of stopping this diarrheal outbreak was the prompt recognition of increasing cases by U.S. military medical personnel and the implementation of mitigation strategies that decreased incidence before the arrival of the NAMRU-6 team. The banning of local vendor food is an important prevention strategy for diarrheal disease outbreaks; however, the availability of alternative food options during deployment is important for troop morale and building partnerships. The offering of local food by host nation people is a sign of hospitality. Under such circumstances, the refusal of food offerings could be perceived as rude and counter to the U.S. diplomatic and training goals. However, diarrheal illness among U.S. military personnel caused on average the loss of 1.5 duty days and 27% of personnel were affected during a 2-week deployment period. Such impact can affect mission readiness and the ability to complete projects on time. As part of the efforts to promote readiness and shared cultural experiences, proper food handling and hygiene instruction should be conducted for vendors that are coming on base to serve foods and at off-base locations providing meals to military personnel in the field.

The laboratory results identified numerous bacterial and parasitic pathogens present in the stool samples from individuals that had reported diarrheal illness. The illness observed in this outbreak was characterized primarily by diarrhea and abdominal pain, with very few cases of vomiting or fever. These clinical features closely resembled those of a bacterial etiology, and the predominant pathogen identified was for ipaH, present in all four *Shigella* spp. and enteroinvasive *E. coli*. An extensive series of laboratory tests for bacteria, viruses and parasites was performed on samples from this outbreak investigation. The polymicrobial findings of the stools that were tested suggest that heavy contamination of vendor prepared food from fecal material may have played a role in this outbreak. The use of molecular methods may represent an increased sensitivity in the detection of pathogens compared to conventional culture and microscopy based methods [27]. We were unable to perform bacterial culture on the remaining sample and microscopic examination could confirm only one of the *Cyclospora* positive samples. However, given the inability to collect control samples, the identified etiologies should be cautiously interpreted in the context of this outbreak investigation. In addition, we were not able to assay for the most common cause of deployment-associated diarrhea, diarrheagenic *E. coli* (e.g. enterotoxigenic *E. coli*, enteroaggregative *E. coli*). The limitations in pathogen identification in this type of setting highlight the need for improved field diagnostics.

Previous studies have also demonstrated the importance of bacterial etiologies in diarrheal disease among deployed military units [1,8,10,21,28]. Due to the potential of diarrheal illness to disrupt military missions, the U.S. military has placed a high priority on the development of effective vaccines and other prophylactic measures against the most common enteropathogens like ETEC, *Campylobacter* and *Shigella* species [29,30,31,32]. Short-term travelers and deployed military units may serve as important populations for testing preventive strategies.

This outbreak investigation in El Salvador highlights the need for more information on infectious disease risks to deployed military personnel in Latin America to help guide prevention measures and empiric treatment when routine laboratory or diagnostic support is not available. Information gathered from infectious disease surveillance activities during these operations will help prepare military personnel and travelers visiting the region.

Acknowledgments

The authors are grateful to the medical staff of the 130th Maneuver Enhancement Brigade and U.S. Southern Command Surgeons office and support of Operation *Beyond the Horizons*, El Salvador. The authors would like to thank Julio Ventocilla, Maruja Bernal, and Giannina Luna for laboratory support. The authors would also like to thank Mark Riddle for review of this manuscript.

The views expressed in this article are those of the author and do not necessarily reflect the official policy or position of the Department of the Navy, Department of the Army, Department of Defense, nor the U.S. Government.

Several authors of this manuscript are military service members or employees of the U.S. government. This work was prepared as part of their official duties. Title 17 U.S.C. §105 provides that 'Copyright protection under this title is not available for any work of the United States Government.' Title 17 U.S.C. §101 defines a U.S. Government work as a work prepared by a military service member or employee of the U.S. Government as part of that person's official duties.

Author Contributions

Conceived and designed the experiments: GS BB AGL MRK EJR DG. Performed the experiments: GS BB CL AGL MRK EJR DG. Analyzed the data: MRK AGL. Contributed reagents/materials/analysis tools: MRK AGL EJR. Wrote the paper: MRK AGL EJR.

References

1. Hyams KC, Bourgeois AL, Merrell BR, Rozmajzl P, Escamilla J, et al. (1991) Diarrheal disease during Operation Desert Shield. N Engl J Med 325: 1423–1428.

2. Ochoa TJ, Ecker L, Barletta F, Mispireta ML, Gil AI, et al. (2009) Age-related susceptibility to infection with diarrheagenic Escherichia coli among infants from Periurban areas in Lima, Peru. Clin Infect Dis 49: 1694–1702.

3. World Health Organization (WHO) (2005) Guidelines for the control of shigellosis, including epidemics due to *Shigella Dysenteriae*.

4. von Sonnenburg F, Tornieporth N, Waiyaki P, Lowe B, Peruski LF Jr, et al. (2000) Risk and aetiology of diarrhoea at various tourist destinations. Lancet 356: 133–134.

5. Castelli F, Pezzoli C, Tomasoni L (2001) Epidemiology of travelers' diarrhea. J Travel Med 8: S26–30.

6. Redman CA, Maclennan A, Wilson E, Walker E (2006) Diarrhea and respiratory symptoms among travelers to Asia, Africa, and South and Central America from Scotland. J Travel Med 13: 203–211.

7. Hoge CW, Shlim DR, Echeverria P, Rajah R, Herrmann JE, et al. (1996) Epidemiology of diarrhea among expatriate residents living in a highly endemic environment. JAMA 275: 533–538.

8. Haberberger RL Jr, Mikhail IA, Burans JP, Hyams KC, Glenn JC, et al. (1991) Travelers' diarrhea among United States military personnel during joint American-Egyptian armed forces exercises in Cairo, Egypt. Mil Med 156: 27–30.

9. Riddle MS, Halvorson HA, Shiau D, Althoff J, Monteville MR, et al. (2007) Acute gastrointestinal infection, respiratory illness, and noncombat injury among US military personnel during Operation Bright Star 2005, in Northern Egypt. J Travel Med 14: 392–401.

10. Putnam SD, Sanders JW, Frenck RW, Monteville M, Riddle MS, et al. (2006) Self-reported description of diarrhea among military populations in operations Iraqi Freedom and Enduring Freedom. J Travel Med 13: 92–99.

11. Sanchez JL, Gelnett J, Petruccelli BP, Defraites RF, Taylor DN (1998) Diarrheal disease incidence and morbidity among United States military personnel during short-term missions overseas. Am J Trop Med Hyg 58: 299–304.

12. Army Dot (2007) Preventive Medicine. Army Regulation 40–5.

13. Vu DT, Sethabutr O, Von Seidlein L, Tran VT, Do GC, et al. (2004) Detection of Shigella by a PCR assay targeting the ipaH gene suggests increased prevalence of shigellosis in Nha Trang, Vietnam. J Clin Microbiol 42: 2031–2035.

14. Klena JD, Parker CT, Knibb K, Ibbitt JC, Devane PM, et al. (2004) Differentiation of Campylobacter coli, Campylobacter jejuni, Campylobacter lari, and Campylobacter upsaliensis by a multiplex PCR developed from the nucleotide sequence of the lipid A gene lpxA. J Clin Microbiol 42: 5549–5557.

15. Orlandi PA, Carter L, Brinker AM, da Silva AJ, Chu DM, et al. (2003) Targeting single-nucleotide polymorphisms in the 18S rRNA gene to differentiate Cyclospora species from Eimeria species by multiplex PCR. Appl Environ Microbiol 69: 4806–4813.

16. Sturbaum GD, Reed C, Hoover PJ, Jost BH, Marshall MM, et al. (2001) Species-specific, nested PCR-restriction fragment length polymorphism detection of single Cryptosporidium parvum oocysts. Appl Environ Microbiol 67: 2665–2668.

17. Fotedar R, Stark D, Beebe N, Marriott D, Ellis J, et al. (2007) PCR detection of Entamoeba histolytica, Entamoeba dispar, and Entamoeba moshkovskii in stool samples from Sydney, Australia. J Clin Microbiol 45: 1035–1037.

18. Minvielle MC, Molina NB, Polverino D, Basualdo JA (2008) First genotyping of Giardia lamblia from human and animal feces in Argentina, South America. Mem Inst Oswaldo Cruz 103: 98–103.

19. Trujillo AA, McCaustland KA, Zheng DP, Hadley LA, Vaughn G, et al. (2006) Use of TaqMan real-time reverse transcription-PCR for rapid detection, quantification, and typing of norovirus. J Clin Microbiol 44: 1405–1412.

20. Army Dot (2005) The Army Food Program. Army Regulation 30–22.

21. Riddle MS, Rockabrand DM, Schlett C, Monteville MR, Frenck RW, et al. (2011) A prospective study of acute diarrhea in a cohort of United States military personnel on deployment to the Multinational Force and Observers, Sinai, Egypt. Am J Trop Med Hyg 84: 59–64.

22. Sanders JW, Putnam SD, Riddle MS, Tribble DR, Jobanputra NK, et al. (2004) The epidemiology of self-reported diarrhea in operations Iraqi freedom and enduring freedom. Diagn Microbiol Infect Dis 50: 89–93.

23. Sanders JW, Putnam SD, Gould P, Kolisnyk J, Merced N, et al. (2005) Diarrheal illness among deployed U.S. military personnel during Operation Bright Star 2001–Egypt. Diagn Microbiol Infect Dis 52: 85–90.

24. Riddle MS, Tribble DR, Jobanputra NK, Jones JJ, Putnam SD, et al. (2005) Knowledge, attitudes, and practices regarding epidemiology and management of travelers' diarrhea: a survey of front-line providers in Iraq and Afghanistan. Mil Med 170: 492–495.

25. Thornton SA, Wignall SF, Kilpatrick ME, Bourgeois AL, Gardiner C, et al. (1992) Norfloxacin compared to trimethoprim/sulfamethoxazole for the treatment of travelers' diarrhea among U.S. military personnel deployed to South America and West Africa. Mil Med 157: 55–58.

26. Bourgeois AL, Gardiner CH, Thornton SA, Batchelor RA, Burr DH, et al. (1993) Etiology of acute diarrhea among United States military personnel deployed to South America and west Africa. Am J Trop Med Hyg 48: 243–248.

27. de Boer RF, Ott A, Kesztyus B, Kooistra-Smid AM (2010) Improved detection of five major gastrointestinal pathogens by use of a molecular screening approach. J Clin Microbiol 48: 4140–4146.

28. Sanders JW, Isenbarger DW, Walz SE, Pang LW, Scott DA, et al. (2002) An observational clinic-based study of diarrheal illness in deployed United States military personnel in Thailand: presentation and outcome of Campylobacter infection. Am J Trop Med Hyg 67: 533–538.

29. McKenzie R, Porter CK, Cantrell JA, Denearing B, O'Dowd A, et al. (2011) Volunteer challenge with enterotoxigenic Escherichia coli that express intestinal colonization factor fimbriae CS17 and CS19. J Infect Dis 204: 60–64.

30. Porter CK, Riddle MS, Tribble DR, Louis Bougeois A, McKenzie R, et al. (2011) A systematic review of experimental infections with enterotoxigenic Escherichia coli (ETEC). Vaccine 29: 5869–5885.

31. Riddle MS, Tribble DR, Cachafiero SP, Putnam SD, Hooper TI (2008) Development of a travelers' diarrhea vaccine for the military: how much is an ounce of prevention really worth? Vaccine 26: 2490–2502.

32. Riddle MS, Kaminski RW, Williams C, Porter C, Baqar S, et al. (2011) Safety and immunogenicity of an intranasal Shigella flexneri 2a Invaplex 50 vaccine. Vaccine 29: 7009–7019.

3

Potential Impacts of Climate Warming on Water Supply Reliability in the Tuolumne and Merced River Basins, California

Michael Kiparsky[1]*, **Brian Joyce**[2], **David Purkey**[2], **Charles Young**[2]

1 Wheeler Institute for Water Law & Policy, University of California, Berkeley, California, United States of America, **2** Stockholm Environment Institute, Davis, California, United States of America

Abstract

We present an integrated hydrology/water operations simulation model of the Tuolumne and Merced River Basins, California, using the Water Evaluation and Planning (WEAP) platform. The model represents hydrology as well as water operations, which together influence water supplied for agricultural, urban, and environmental uses. The model is developed for impacts assessment using scenarios for climate change and other drivers of water system behavior. In this paper, we describe the model structure, its representation of historical streamflow, agricultural and urban water demands, and water operations. We describe projected impacts of climate change on hydrology and water supply to the major irrigation districts in the area, using uniform 2°C, 4°C, and 6°C increases applied to climate inputs from the calibration period. Consistent with other studies, we find that the timing of hydrology shifts earlier in the water year in response to temperature warming (5–21 days). The integrated agricultural model responds with increased water demands 2°C (1.4–2.0%), 4°C (2.8–3.9%), and 6°C (4.2–5.8%). In this sensitivity analysis, the combination of altered hydrology and increased demands results in decreased reliability of surface water supplied for agricultural purposes, with modeled quantity-based reliability metrics decreasing from a range of 0.84–0.90 under historical conditions to 0.75–0.79 under 6°C warming scenario.

Editor: Juan A. Añel, University of Oxford, United Kingdom

Funding: This study received funding from an NSF Graduate Research Fellowship, an NSF Doctoral Dissertation Research Improvement Grant, a CALFED Bay-Delta Program Fellowship, and the California Energy Commission PIER program. The funders had no role in study design, data collection and analysis, decision to publish, or preparation of the manuscript.

Competing Interests: The authors have declared that no competing interests exist.

* E-mail: kiparsky@berkeley.edu

Introduction

There is a near consensus among scientists that the Earth's climate is changing, and that under even the best-case scenarios of emissions and climate sensitivity, climate impacts are virtually certain [1,2]. Climate change is a global environmental problem, but humans will be most concerned with the local and regional effects. One of the most robust findings in climate impacts research is that climate change will alter hydrology and water resources around the globe. In California, two decades of studies of projected climatic impacts on water systems have progressed from hydrologic systems, to agricultural systems, to water storage and conveyance systems [3]. Impacts on hydrology will cascade directly into human and ecological systems at all scales. In California, as in other snow dominated watersheds, climate change will result in reduced snowpack storage, reduced streamflow, and changing seasonal flow patterns that will challenge the resilience of coupled water, energy, agricultural, and ecological systems [4,5].

To the extent that scenarios of future water supply reliability model water deliveries, such modeling has often been derived from historical climate [6] that may neither accurately represent past [7] nor future [3,8] climatic conditions. This disconnect has motivated interest in formal integration of climate change into water planning, and the uncertainties inherent in both water systems

modeling and climate modeling suggest that incorporating climate impacts on hydrology and water resources could help define potential anticipatory responses.

Projected hydrologic impacts of climate change have been the subject of many studies, and assessment of the potential impacts on water rights holders and environmental flows are starting to appear in the literature. Vulnerability assessments can use 'outcome oriented' approaches [9] to describe potential impacts and adaptation options quantitatively by integrating results from multiple models. This can be done using a cascade of modeling information, from large-scale, coarse-grained models, to finer resolution models that cover less spatial or conceptual area but represent specific processes of interest in more detail.

This paper describes the development of a modeling tool and its application to a sensitivity analysis for temperature warming in the Merced and Tuolumne River Basins in California's Central Valley (Figure 1). To evaluate impacts on hydrology and water supply reliability, we used the Water Evaluation and Planning (WEAP) [10,11] as a framework to model hydrology and water operations in the three case study basins.

This paper presents the model structure and calibration procedure for the WEAP model of these three basins, along with results from a sensitivity analysis for temperature increases of 2°C, 4°C and 6°C. The model represents seasonal and inter-annual historical hydrologic variability, as well as variation in reservoir

Figure 1. Map of project area for case study. The Stanislaus, Tuolumne, and Merced Rivers flow from the Sierra Nevada Mountains to the San Joaquin River, and thence north towards the Sacramento-San Joaquin Delta, the "hub" of California's water supply system. Figure source: [73].

levels and surface water deliveries to the major irrigation districts (IDs) in the region. Our findings show hydrologic impacts of temperature change in the form of shifts in timing of streamflow and increased evapotranspiration demands for irrigation water. When interacting with the modeled representation of the managed water system, the impacts manifest as decreased water supply reliability and lower reservoir storage volumes. Thus, while system attenuation of the climate change signal may occur through the capacity of large reservoir and conveyance systems to buffer altered hydrology, with modeled operation regimes unchanged, temperature-driven warming and its influence on modeled hydrology translates into decreased surface water supply reliability in these basins. These results suggest avenues for further work with more detailed analysis of climate and other future stressors for water supply in these basins.

Study Area

As elsewhere, water is important in California economically, politically, and socially. California has been well-studied with respect to future climatic impacts on water [3]. Hydrology in California, as in many mountainous regions, is dominated by the dynamics of snow accumulation and melting. Gleick (1987) demonstrated the sensitivity of this hydrology to climate warming, projecting earlier and higher hydrograph peaks under climate warming scenarios. This general conclusion has proven robust after two decades and dozens of peer-reviewed studies [3], and exemplifies a hydrologic response of the type that threatens snowmelt-dominated hydrologic systems providing water supply to one-sixth of the world's population [4].

The state has ongoing conflict over water, even without the perturbation of climate change. However, California has traditionally been a national leader in the development of innovative environmental policies [12], and climate change is no exception [13,14]. Recently, legal motivation to consider climate change in planning decisions has increased in the state, primarily on mitigation, but increasingly on adaptation [15], resulting in path-breaking studies and efforts to apply cutting-edge science. Efforts to integrate research into actual anticipatory changes to water management, however, may require more in-depth studies on scales relevant to local decision-makers.

The Tuolumne and Merced River Basins (TM) in California's Central Valley (Figure 1) lie on the western slope of the Sierra Nevada Mountains, where hydrology will be sensitive to climate change [16,17,18,19]. The terminal reservoirs on each system (New Don Pedro Reservoir and Lake McClure, respectively) each have total storage capacity approximately equal to one year of average annual flow. Agricultural diversions dominate use in both basins, with domestic, municipal, and industrial use a minor fraction. Downstream flow in both cases goes through the San Joaquin River to the Sacramento-San Joaquin Delta. Water allocation in these basins is run through a variety of institutions, much of which revolves around legal and regulatory constraints (e.g., water rights and water quality regulations). Notably, regulatory water quality requirements at Vernalis are key drivers of water releases. Decisions made by water organizations such as federal and state agencies, and local IDs are key to system operation.

With some bounding assumptions, and the inclusion of elements of the Stanislaus River system not described in detail here, the basins can be modeled as a distinct hydrologic unit.

Methods

WEAP model structure

The WEAP model consists of interlinked modules for both physical hydrology and operations to calculate demands and allocate water at each time step, and has been described in detail elsewhere [10,20]. A description of a related application to the hydrology of the Sierra Nevada Mountains has been previously described [21], from which the hydrology module described below is derived. We give a brief overview of the model structure and algorithms in this section, and refer the interested reader to these publications for more details.

WEAP's physical hydrology water balance representation consists of several components designed to represent variability in the key hydrologic components relevant to a study at this temporal and spatial resolution. A one-dimensional soil water accounting scheme routes moisture through two soil layers, with empirical functions describing evapotranspiration, surface runoff, sub-surface runoff, and deep percolation.

WEAP's snowmelt model computes effective liquid water input in each time step as the sum of rain plus snowmelt. To get the latter term, snow water equivalent and snow melt are computed using a temperature index snow accumulation model. Assigned melting and freezing thresholds are used to determine a melting coefficient that specifies snowmelt based on available melting energy and the latent heat of fusion. Available melting energy is a function of net solar radiation and a lumped term comprising other available forms of energy that is adjusted during calibration.

Within each catchment, a water balance is computed based on each unique combination of soil and land cover using a continuous mass balance equation. Evapotranspiration from each fractional area is computed using the Penman-Montieth reference crop potential evapotranspiration equation, using crop/plant coefficients assigned to each land cover type. Surface runoff is calculated using a term scaled by a runoff resistance factor that represents surface characteristics such as roughness, Leaf and Stem Area Index, average slope, porosity, etc. In the two-layer soil moisture scheme, interflow and deep percolation are adjusted using a conductivity parameter, which represents an estimate of upper storage conductivity, and a tuning parameter that partitions flow between horizontal and vertical. Alluvial aquifers are represented in the valley portion of the model, and in these catchments the deep water storage layer is removed and the deep percolation term replaced by percolation from the upper layer directly to the aquifer.

Climate inputs affecting evapotranspiration include temperature, relative humidity, wind speed, and insolation (a function of latitude, Julian day, and cloud cover). Catchments containing irrigated agriculture use the upper soil water store as a trigger for irrigation demands. Threshold values assigned to each crop model irrigation efficiency by determining the level of soil moisture reduction that triggers an irrigation demand.

The WEAP model uses a preference- and priority-driven logic to determine allocation of water to agricultural, urban, and in-stream demands. A node-and-link structure connects sources and supplies. Within each time step (monthly in this implementation), a linear program satisfies demands first to nodes with highest priority, then sequentially allocates water to lower priority users until either demands are satisfied or specified constraints preclude further allocation of water. Each demand node may be supplied by multiple water sources. Water supply preferences can be assigned to simulate user behavior when multiple sources are available, such as in a case where surface water is preferred to groundwater.

Reservoirs are simulated based on their physical characteristics as well as operation parameters that reflect decisions based on balancing flood control, water supply, and carryover storage. A *conservation zone* reflects required space for flood control. A *buffer zone* specifies reservoir levels below which releases are limited in each time step to a specified percentage of the existing water in the reservoir. This approach reduces the complex conditional logic by which actual operational decisions are made to an analogue for conservatism of reservoir operators.

The dynamically interconnected model integrates a rainfall-runoff model with infrastructure and operations logic. There are conceptual and operational differences between the modeled representations of the upper watersheds (areas above the large dam on each river) and the valley floor (agriculturally dominated areas below these dams) that reflect the differences between the two areas. In the upper watersheds, land use is predominantly native vegetation, while in the valley floor agriculture dominates and urban centers are larger. In the upper watersheds, terrain is complex, with individual watersheds spanning large elevation ranges, while the lower watersheds are relatively homogenous.

These differences result in different emphasis in the modeling of the upper and lower watersheds within the model. The upper watersheds are modeled with the primarily goal of representing inflows to the major reservoirs in the system, and the sensitivity of those inflows to future changes in climate. The lower watersheds are modeled primarily to represent agricultural and urban demands, the storage and conveyance facilities that deliver water to satisfy those demands, flows of water thought the managed part of the system, and the sensitivity of both demands and deliveries to changes in climate and other variables in future projections. We describe below the component parts of these overlapping and integrated analyses.

Data and model implementation

This section describes data used for initial parameterization and calibration of the model, and the following section describes calibration procedures. The model relies heavily on previously published work, with some modifications. The hydrology component of the present model has been adapted for the current study to build on previous work in several ways: it has a monthly time step, larger elevation bands, and a different sub-watershed structure than the version we presented previously [21,22]. Similarly, the agricultural component relies on methods developed for nearby areas of California [20,23], and we refer the user to these previous reports for more information.

Watershed characteristics. We developed a GIS model of the physical and institutional aspects of the study basins to enable representation of spatially explicit watershed characteristics. The study area was first divided geographically at nested scales based on topography using ArcGIS [24]. *Watersheds* were defined hydrologically by the major dam on each of the three rivers that drain the Sierra mountains. Each watershed was further divided into *sub-watersheds* based on *pour points* at which streamflow is simulated (e.g. locations of gages with historical data). Each sub-watershed is further divided into 500 m elevation bands (*catchments*), coarser than Young et al. [21] for computational efficiency. The total number of catchments in the upper watersheds was condensed from 248 to 80, resulting in a range of effective catchment areas from about 2–600 km^2. Parameters were adjusted to retain seasonal and annual hydrologic variability as described below. Sub-watersheds and catchments were defined in the area below the upper watersheds and bounded by the San Joaquin River, here by institutional boundaries rather than topography. Sub-watersheds define hydrologically distinct units

that enable calibration through comparison of known hydrologic response of smaller model sections to modeled response, as well as describing areas meaningful to managers.

Land surface and subsurface characteristics were derived from United States Geological Survey (USGS) digital elevation models [25] at 10 m (upper watersheds) and 30 m (valley floor), and the ArcHydro toolkit [26] was used to delineate the network of streams and rivers. Each sub-watershed in the valley floor is represented as single elevation catchment.

Within each catchment, we defined classifications of land use, land cover, and soil type, and determined the fractional area of each combination. Vegetation and landcover estimates for the upper, mostly non-agricultural watersheds were based on the National Land Cover Dataset (NLCD) [27] (Table S1 in File S1). Land cover, including cropping patterns, for the agriculturally intensive valley floor were based on the California Land and Water Use survey [28], mapped onto 15 categories (Table S2 in File S1). Intersecting cropping patterns with irrigation district locations [29,30,31] resulted in estimated percentages of each land cover and crop type. Historical or future changes in cropping patterns over time were not simulated here. Soils were classified based on the SSURGO and STATSGO datasets [32,33], as described in Young et al. [21].

Dams [34], canals and other conveyances [35] streamflow gage data, and locations [36] were also incorporated, with physical characteristics taken from published sources (Table S3 in File S1). Reservoir evaporation rate is based on available historical average monthly values [37]. Volume-elevation curves used to approximate surface area were derived from data available from the California Data Exchange Center [38] and the United States Geological Survey [39], and Merced ID storage and rating tables. There is no substantial water infrastructure above the New Exchequer Dam on the Merced River. On the Tuolumne River infrastructure exists for water supply and hydropower purposes above the terminal reservoirs. We simplified the schematic and operational regime in each of these basins, and used generalized representations of water storage and release for hydropower production and water diversions outside each basin based on available historical data. Water infrastructure in the Tuolumne River Basin is operated for local irrigation districts and the San Francisco Public Utilities Commission (SFPUC) to satisfy demands for irrigation and urban uses in the San Francisco Bay Area, governed by the Raker Act [40] and we approximate the daily flow-based logic of this agreement with adjusted monthly parameters for joint reservoir storage and an external monthly demand function for San Francisco based on historical and 'typical' monthly flows in the San Joaquin Pipeline. Don Pedro Reservoir is represented as two objects to depict the 'virtual' storage by SFPUC in that reservoir.

The Sierra crest forms the upper boundary of the three main watersheds in the model. All three rivers flow into the San Joaquin River (SJR), which in turn flows north to the Sacramento-San Joaquin Delta. Because of upstream diversions, these three basins form a somewhat isolated hydrologic unit: the "section of the SJR between Gravelly Ford and Mendota Pool, a reach of approximately 27 km, is generally dry except when releases are made from Friant Dam for flood control" [41]. Thus, we assume Gravelly Ford constitutes an upper boundary of the model, while acknowledging that future policies may alter this assumption. The lower boundaries of each basin are at the confluences of each river with the San Joaquin. Future work could connect this model with a representation of the Sacramento-San Joaquin Delta and west side tributaries to the San Joaquin River. Groundwater basins are based on California Department of Water Resources (DWR)

Table 1. Groundwater use by district (10^6 m^3).

Model node	Modeled groundwater use			Estimated historical minimum pumping	
	Average	Min	Max	Low	High
Modesto ID	107	47	190	48	80
Turlock ID	296	183	444	195	234
Merced ID	249	159	376	10*	226*

*Note that range of groundwater use minimums include the sum of district and non-district pumping, except in the case of Merced ID, which includes district pumping only. Sources: [52,59].

Bulletin 118 [42]. Future work could incorporate more detailed groundwater modeling.

Demand centers. The modeling effort described here focuses on the major irrigation districts [29,30,31] and urban areas. We classified adjacent areas of land that receive minimal or no surface water supply as separate from the major irrigation districts, although there are technically irrigation districts within the adjacent land area. Contracts for water supply are represented as lower priority transmission links to invoke WEAP's priority-driven allocation logic as described in Section 0.

Don Pedro Reservoir on the Tuolumne River supplies water to Turlock ID and Modesto ID. Surface water from the Merced River supplies water to the Merced Irrigation District and other entities within its Sphere of Influence. Contract water transfers to these 'non-district' areas are represented through a low-priority transmission link. The Merced National Wildlife Refuge also receives surface water through the Merced Main Canal, at a priority higher than Merced ID supplies.

Historical climate. A 1/8° gridded observed meteorological historical climate dataset [43] was used for historical climate inputs. For each catchment, we selected the gridpoint closest to the catchment centroid, and used the corresponding time series for temperature, precipitation, and average daily wind speed adjusted for the model's monthly time step. To account for possible bias in upper elevation watersheds, we adjusted the 1/8° inputs to actual elevation of each catchment centroid using a lapse rate of 6.5°C per 1000 m elevation change, altering the temperature input for each catchment by the difference between the midpoint elevation of each catchment and the elevation of the corresponding climate input grid point, multiplied by lapse rate. Overall, these adjustments resulted in a slight average decrease in modeled temperature inputs.

As the goal of this study focuses on streamflow into the terminal reservoirs, and because making selective fine-scale adjustments to precipitation estimates would introduce additional uncertainty and bias onto modeled and/or downscaled projections, we left precipitation inputs unmodified. Note that this approach is in effect the same as most other studies of this type that focus calibration efforts on terminal streamflow without examining or reporting sub-watershed bias. In this, as in other mountainous areas, daily precipitation totals can vary greatly between measurement instruments located within a basin [44]. This can be reflected in hydrologic analysis where isolated, but significant, precipitation events are not captured by an existing precipitation measurement network. For example, detailed studies found stream responses that could not be explained by precipitation measurements alone [44]. Similar effects from bias in input climate data have been observed previously [21,45,46].

Seasonal wind speed patterns were based roughly on monthly averages for two years of data at the Merced CIMIS station that overlap with the calibration period. For the valley floor, average relative humidity was interpolated between a peak of 90% in January and a low of 45% in July. For the upper watersheds, relative humidity was estimated from DAYMET, a model that generates estimates of historical weather parameters in complex terrain [47], interpolating between an average high humidity of 60% in January and an average low humidity of 23% in September.

Urban demands and supplies. We represented urban demands by aggregating urban population projections based on agricultural district boundaries and representing each as a lumped demand node. To do so, we clipped spatially explicit population projection grids [48] to catchment node areas in the valley floor (Table S4 in File S1).

Urban demands are modeled by multiplying estimated population in a given urban node by an estimated per capita water use level. Per capita water use estimates are taken as the 'baseline' 1995–2005 values for the San Joaquin Valley from the State Water Resources Control Board's 20x2020 efforts [49], as 939 liters per capita per day. Consumptive use in urban areas was assumed to be 30%. Urban supplies are mostly met by groundwater, or through arrangements with Irrigation Districts for surface supplies.

Table 2. Parameters for groundwater/surface water allocation of supply to Districts.

Watershed	Model Node	System losses, %	SW constraint, % of total demand	Canal evaporation, %	Max surface water diversion, m^3/s (cfs)
TUO	Modesto Main	38	85	0	
TUO	Turlock Main	30	75	0	
MER	Merced ID N	33	75	2	2.8 (100)
MER	Merced ID Main	27	70 before 1991, 60 after	2	56.6 (2000)

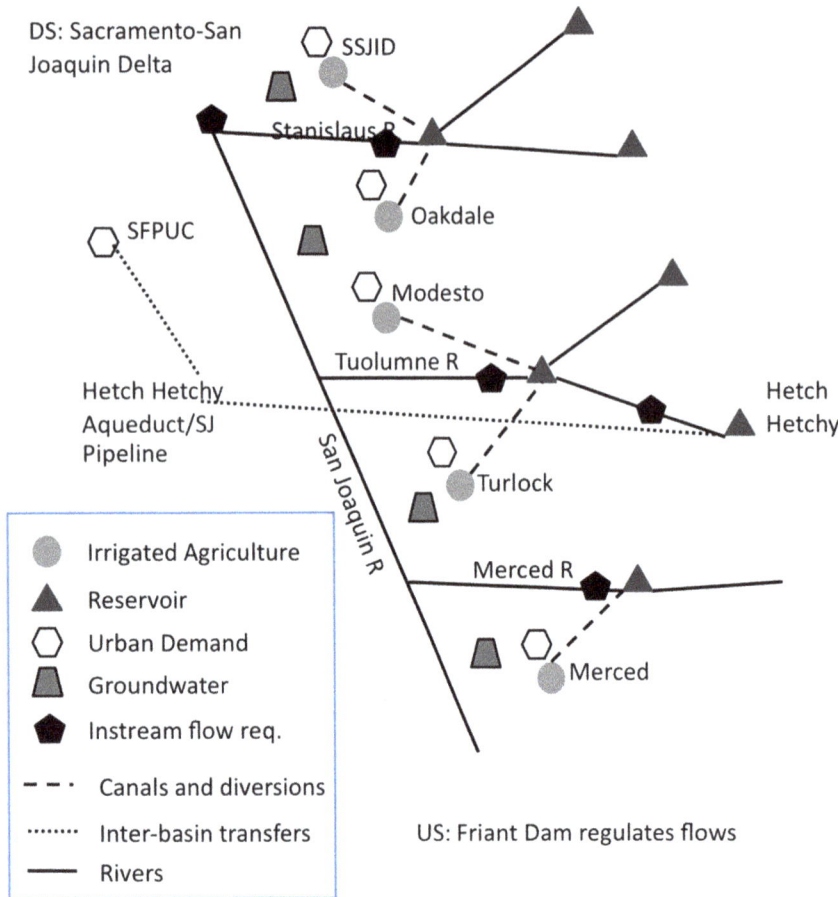

Figure 2. Model schematic. Simplified conceptual schematic of the more detailed model used to represent the river basins.

Other water uses. In-stream flows and hydropower releases are modeled based on internally consistent heuristics. Environmental flow requirements are conditional on modeled representations of year type and/or snowpack-based forecasts (Tables S5–S11 in File S1). Delta flow requirements downstream of the model boundary are based on a proxy for flows released to meet water quality requirements [50,51]. We detail logic for instream flows for each river basin in Material S1 in File S1. We simulated summer hydropower releases in each stream based on approximate historical summer flows and calibrating to observed flows and reservoir levels.

Groundwater use often occurs within irrigation districts in the region even when surface water supplies are seemingly plentiful. To simulate this, we first assigned a supply preference to each agricultural catchment for surface water, then constrained the total amount of demands that can be supplied by surface water to a

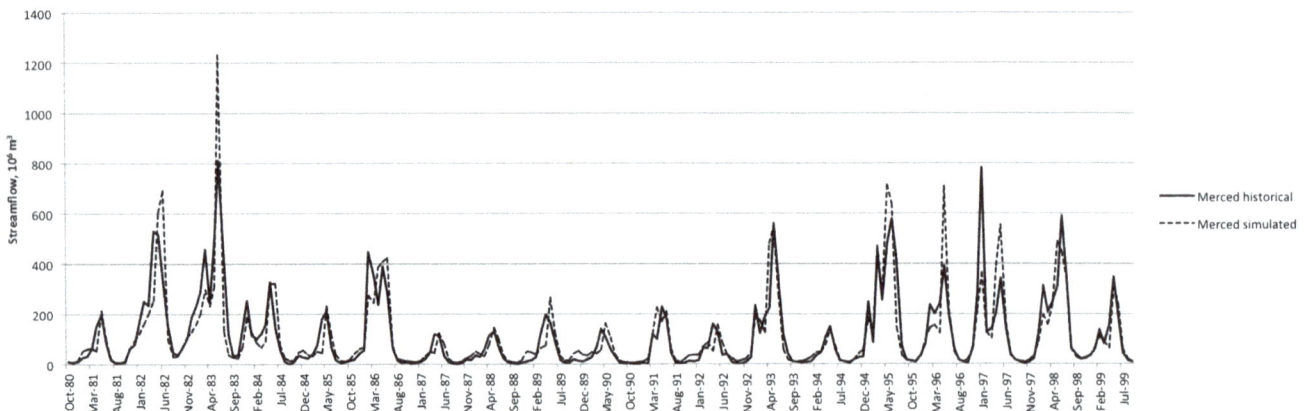

Figure 3. Unimpaired Tuolumne River streamflow. Simulated (dotted) and historical (solid) calculated Full Natural Flows (CDEC gage TLG).

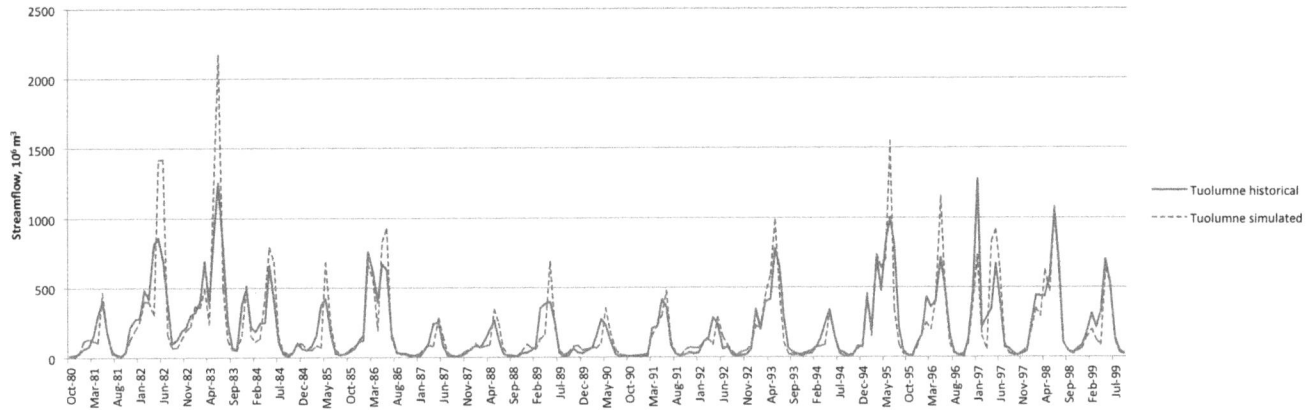

Figure 4. Merced River streamflow (CDEC gage MRC).

percentage that reflects estimates [52] of groundwater use in the region. This in effect forces a minimum amount of groundwater pumping even when sufficient surface water is available, and allows increased groundwater pumping to meet demands when surface water deliveries are constrained by hydrology or operations. In areas with no surface water supply, or limited surface water supply, groundwater use accounts for all demands, and no limits on groundwater use are currently modeled.

Groundwater usage is based on unconstrained access to groundwater resources, given other priorities and preferences specified in the model. The model represents groundwater simply as a stock. The model draws from this stock to satisfy demands in accordance with defined preference and priorities, and recharges it based on hydrologic conditions. Groundwater is represented based on sub-basins of the San Joaquin Valley Groundwater Basin, as defined by the California Department of Water Resources [53].

Our representation of groundwater use is within the range suggested by other studies. Table 1 shows average groundwater use within the primary irrigation districts, compared to the range suggested by previous studies. Note that we used data from a water balance conducted by DWR and USBR [52] to parameterize the current model. Both our approach and the DWR water balance partition canal flows between deliveries and system losses (Table 2), calibrating to other known historical values such as gauged surface water flows. Actual groundwater use and system losses are unknown, and should be treated as estimates.

Other model elements.   In the present analysis, we represent as static some system elements that will clearly change in future, including land use/land cover, urbanization, cropping patterns, and institutional elements. Thus, the modeling presented here can be considered a sensitivity analysis for the climatic variable of temperature [54], and future work will report on efforts to incorporate variability in these elements.

Amount and timing of water rights are represented coarsely. If the total diversions to the main canals in the Merced Basin over the course of a water year exceed total annual water rights, no more surface water diversions are allowed for the rest of the water year. Timing of all diversions is limited to the current irrigation months of April to September, per water rights and historical patterns of diversion, except where water is delivered through the Merced Main Canal to the Merced National Wildlife Refuge during some winter months.

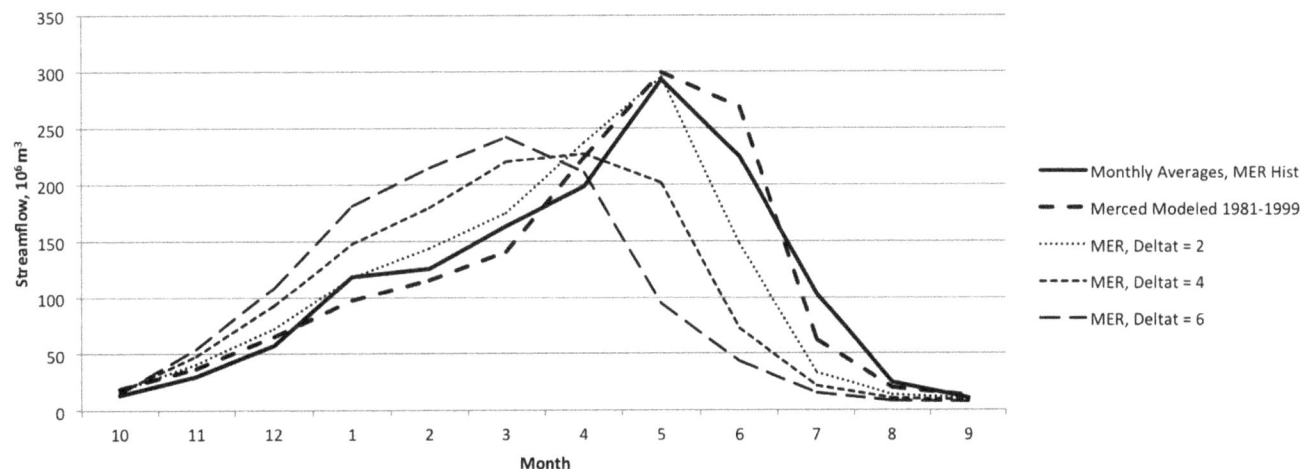

Figure 5. Simulated and historical average monthly unimpaired streamflow. Results (10^6 m^3) for the Tuolumne River, as calibrated and in response to temperature increases of 2°C, 4°C, and 6°C.

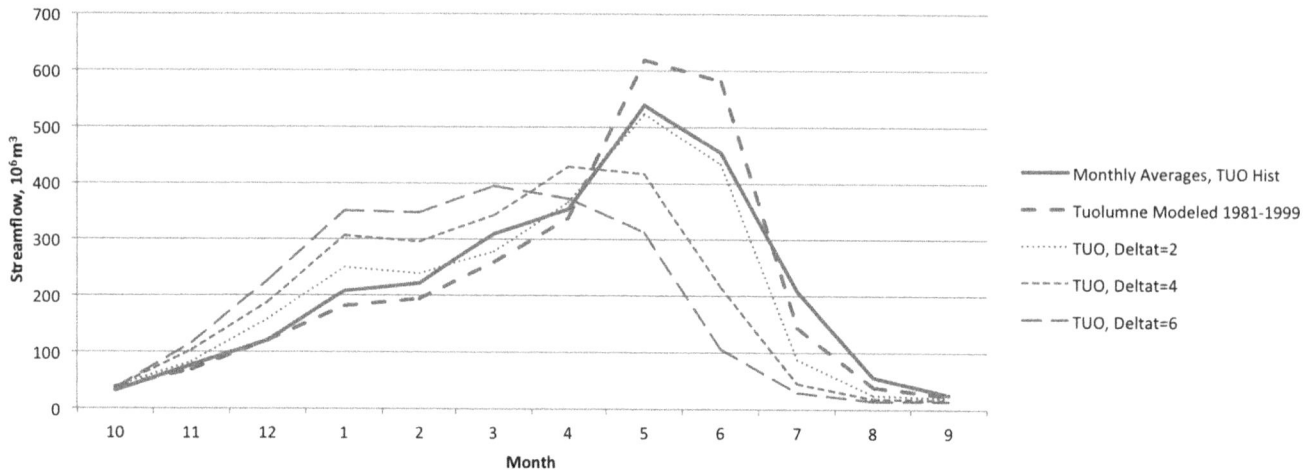

Figure 6. Simulated and historical average monthly unimpaired streamflow. Results (10^6 m^3) for the Merced River, as calibrated and in response to temperature increases of 2°C, 4°C, and 6°C.

Representation of historical hydrology and water operations

The WEAP model represents the above-described area with four main streams, 10 reservoir objects, 10 diversions, 4 groundwater basins, 18 distinct agricultural catchments, 88 catchments in the upper watersheds, 15 demands sites, and 19 streamflow requirements. Of these, we focus in this article on those most relevant for the purpose of investigating surface water supplies to agricultural areas, as shown in the highly simplified schematic in Figure 2.

Calibration of this model was guided by the primary study goal of assessment of water supply reliability. The primary calibration point for hydrology was unimpaired streamflow to each major reservoir, as this is the primary hydrologic influence on water reliability. We used historical data for streamflow and reservoir levels from US Geological Survey, California Data Exchange Center, and district sources, as cited in Table S3 in File S1. We also referenced other data sources including sub-catchment streamflow gages and snow surveys for general consistency, but they were not the focus of detailed calibration.

Calibration of unimpaired streamflow, crop demands, and system operation were performed sequentially based on historical data from water years (WY, October–September) 1981–1999. Calibration was performed iteratively, comparing model output with historical data for each watershed. Calibration was aided by Latin Hypercube sampling across parameter space, using Computer Aided Reasoning System software [55]. Further details on calibration and model representation follow.

Unimpaired surface water hydrology. Figure 3 and Figure 4 depict the model's representation of historical hydrology,

comparing WEAP outputs at the pour points representing the large dams at the base of each upper watershed with DWR reconstructed full natural flows. Figure 5 and Figure 6 show average monthly results over the same time period. The model captures historical annual and seasonal variation of flow patterns reasonably well (Table 3). Bias (−2.5% and −0.1%) and goodness of fit (Root Mean Squared Error (RMSE) 67% and 75%) statistics are in the range reported for previous modeling efforts in the region using the Variable Infiltration Capacity model (VIC) [43] and WEAP [20]. Nash-Sutcliffe efficiency index [56] suggests reasonable predictive power for the model.

As in previous efforts to model mountainous, snow-driven hydrology with WEAP [21,57], key parameters adjusted during calibration of streamflow in the upper watersheds include those influencing soil water flux, snowmelt and freezing, along with others as listed in Table 4. These ranges are similar to previously published efforts with WEAP. We used a version of WEAP [11] with an updated value for the latent heat of fusion in the snowmelt

Table 3. Goodness-of-fit statistics for unimpaired hydrology representation for the period from WY 1981–1999.

	TUO	MER
Nash-Sutcliffe	0.67	0.66
Bias (%)	0.1	−0.1
RMSE (%)	66	75

Table 4. Ranges of hydrologic parameters used in the WEAP model.

Parameter	Range
Deep Soil Water Capacity (mm)	732–1339
Root Zone Soil Water Capacity (mm)	288–527
Root Zone Soil Water Capacity (mm)	250
Deep Water Capacity (mm)	200
Runoff Resistance Factors (land-cover dependent)	4–20
Root Zone Conductivity, Deep Soils (mm/month)	43
Root Zone Conductivity, Shallow Soils (mm/month)	331
Deep Conductivity (mm/month)	129–193
Melting threshold (°C)	1
Freezing threshold (°C)	0
Radiation factor	3.5–5.5
Albedo, new snow	0.7
Albedo, old snow	0.03

Table 5. Goodness of fit statistics for diversions in major canals.

Watershed	Node	Result	RMSE (%)	Nash-Sutcliffe Efficiency Index	Bias (%)
TUO	Modesto Cn	Diversions	46	0.65	4.3
TUO	Turlock Cn	Diversions	46	0.68	2.3
MER	Merced ID N Cn	Diversions	52	0.69	−18.8
MER	Merced ID Main Cn	Diversions	58	0.66	3.0

algorithm. This enables the use of more physically realistic parameters for melting point than previously published WEAP models.

Agricultural demands. Agricultural demands are modeled as a function of climate and crop type using the Penman-Monteith equation and empirically derived crop coefficients [20]. Demand for irrigation water is modeled as a function of reference evapotranspiration (ETo) and non-dimensional crop coefficients (Kc) for a given crop. Values for Kc were estimated for each crop, tuned from initial estimates for the region found in Bulletin 113-3 [58]. Since observed data in Bulletin 113-3 were produced for the purpose of estimating irrigation requirements rather than modeling year-round hydrology, these data ignore winter-time evaporation, with zero values during non-irrigated months. Missing wintertime Kc values were estimated as 0.5. The model simulates irrigation patterns using soil moisture as a trigger for irrigation demands [10]. Flood-irrigated rice uses a separate function to simulate ponding and flushing.

Demands were compared to a metric of Total Applied Water Demand (TAWD), based on estimates for representative irrigation districts in the region [59], where

$$TAWD =$$
$$Consumptive\ Use\ of\ Applied\ Water/Irrigation\ Efficiency, \quad (1)$$

with initial figures taken from Bulletin 113-3 and data from Merced Irrigation District [37]. The above calculations collectively define the amount of water demand at the crop. A key unknown in this region is values for conveyance losses to seepage and evaporation. As in other studies in the region [48], we represent this using a loss factor calibrated to close the water balance, and reflect the total demand for water at the diversion point in each time step. Thus, surface water diversions are modeled as a function of demands for water application, alternative sources of water supply such as groundwater, conveyance capacity, conveyance losses, institutional constraints such as water rights, and reservoir operations [20].

Managed water system. Surface water deliveries at the ID or sub-District level are measured at points of diversion from the rivers into the canals that convey water to irrigated and urban demand centers. Modeled deliveries are a function of demands, priorities, preferences, reservoir operations, and available water in a given time step, as described above.

Reservoir operations mediate deliveries to satisfy agricultural water demands for surface water, and in particular to mediate inter-annual variability by reserving water in wet years for use in dry years. The large, terminal reservoirs on these streams also serve flood control functions. Modeling the balance between the two functions was accomplished by adjusting the parameters described in Section 0 above. The top of the buffer zone is defined as a coefficient b times the total available storage (conservation level), or as the inactive storage, whichever is greater, with b varied during calibration.

An institutional shift took place during the calibration period that influenced operation parameters. All of the reservoirs in this model produce hydropower, and in most cases much of the resulting electricity is sold outside the service areas with revenues benefiting the organizations that also supply water for agricultural purposes. During and after the dry period from 1987–1992, "water first" policies were enacted with the goal of ensuring that system operation hews to its nominal top priority of delivering reliable water supply for constituents. In order to represent this shift to a water first policy, reservoir operations parameters on the Merced and Tuolumne were adjusted before and after 1991. Better fits to historical data were obtained with marginally stricter operating parameters (e.g., higher levels for the buffer zone and/or lower values for the buffer parameter), in keeping with expectations that operational policies would result in greater tendency to store water.

The model represents the average annual diversions with reasonable goodness of fit (RMSE ranging from 46–58%, Table 5). While the model exhibits some error in reproducing individual events, it does capture the overall patterns for diversions over the calibration time period, including the shift to lower diversions during the string of critical water years during the drought from

Table 6. Annual historical and modeled diversions to irrigation districts, WY 1981–2000 (10^6 m^3).

Basin	Node	Historical average diversions		Modeled average diversions		Model bias (%)	
		1981–1999	1987–1992	1981–1999	1987–1992	1981–1999	1987–1992
TUO	Modesto Canal	379	318	380	312	0	−2
TUO	Turlock Canal	720	581	712	611	−1	5
MER	Merced ID North Canal	26	25	26	20	−2	−23
MER	Merced ID Main Canal	603	482	607	466	1	−3

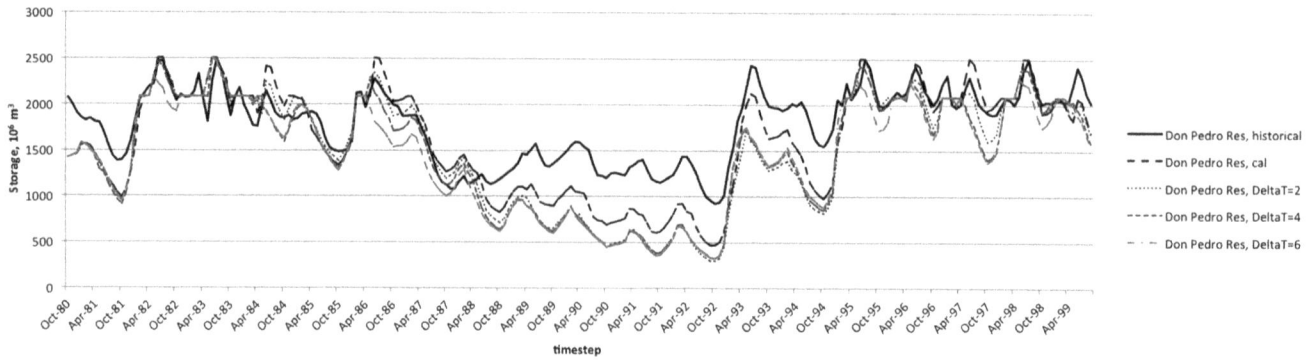

Figure 7. Modeled reservoir levels at Don Pedro terminal reservoir. Results as calibrated to historical observations and modeled with perturbed temperatures of 2°C, 4°C, and 6°C.

1987–1992 (Table 6). The model captures seasonal and interannual variability in water levels at the terminal reservoirs, with some under-estimation of levels Don Pedro Reservoir (Figure 7, Table 7).

Temperature warming scenarios. Climate change assessments can choose from a range of methods for representing modeled sensitivity to future climate change along a continuum of resource intensity [54]. For the present study, we choose to apply a sensitivity analysis to temperature warming. A temperature sensitivity analysis is useful as a first order representation of model behavior, and has been applied in many similar studies in this region and elsewhere. It also serves as a prelude to more detailed analysis in future work. In California, however, there is a reasonable *a priori* expectation that such studies will be meaningful in the face of anthropogenic climate change. First, while climate projections are consistent in their projection of temperature warming for the region, their representation of future precipitation regimes is less so [60], and thus focusing on temperature alone makes sense. Second, since the first studies of mountain hydrology and climate warming in the region, a consistent projection of sensitivity of streamflow to warming has been made in many studies [3], suggesting the usefulness of studying temperature impacts on warming.

We generate a sensitivity analysis using warming scenarios of 2°C, 4°C, and 6°C to bracket expected changes in temperature. As described in Cayan et al. [61] for the Sacramento region, in all projected climate scenarios California retains its Mediterranean temperature and precipitation patterns, with cool wet winters and warm dry summers. Temperature increases between 1°C to 3°C by mid-century, and 2°C to 5°C by end of century, with greater increases in the summer months than in winter. High variability in precipitation evident in the historical and paleoclimate records for the region is also visible in projected climate futures, and the majority of models also show drying trends relative to historical precipitation under each emissions scenario, but we leave exploration of changes in variability to future work.

Results and Discussion

This section describes the response of modeled hydrology and water operations to the temperature warming scenarios of 2°C, 4°C, and 6°C.

Hydrology

Under climate warming scenarios, the hydrology of the upper watersheds responds with a shift in timing and magnitude of seasonal flows. As in previous modeling efforts [3,21], response to simulated warming includes earlier timing of snowmelt and a resulting shift of peak flows earlier in the water year (Figure 5 and Figure 6). Table 8 illustrates this sensitivity through the shift in timing of the center of mass of streamflow. The present model is

Table 7. Goodness of fit statistics for reservoir inflows, storage, and releases.

Watershed	Node	Result	RMSE (%)	Nash-Sutcliffe Efficiency Index	Bias (%)
TUO	CE	Inflows	101	0.27	−22.5
TUO	CE	Storage	41	−0.27	−0.7
TUO	CE	Releases	103	−0.54	−21.2
TUO	HH	Inflows	84	0.62	−15.0
TUO	HH	Storage	32	0.16	12.2
TUO	HH	Releases	245	−0.17	59.6
TUO	DPR	Inflows	67	0.63	2.5
TUO	DPR	Storage	18	0.37	−8.9
TUO	DPR	Releases	76	0.29	2.0
MER	NE	Inflows	75	0.66	−0.1
MER	NE	Storage	15	0.91	1.0
MER	NE	Releases	76	0.32	0.6

Table 8. Simulated shift in hydrograph center of mass (COM).

| Watershed | Historical COM (1981–1999) | Simulated COM (1981–99) | Simulated shift in COM, months (days) | | |
			2°C	4°C	6°C
TUO	June 1	June 5	0.17 (5)	0.42 (13)	0.51 (16)
MER	May 28	June 1	0.30 (9)	0.51 (15)	0.68 (21)

Shifts earlier in the water year, with uniform temperature increase applied to historical climate inputs over the reference period from 1981–1999.

somewhat less sensitive to temperature by this measure than other efforts [21], and thus may have a relatively muted climate response when compared to other such studies. In addition to the effects on timing, increased temperature decreases magnitude of streamflow (Table 9) through its effect on increased evapotranspiration above the terminal reservoirs (data not shown).

Water supply reliability

For the agricultural districts modeled here, surface supply reliability is reduced under uniform temperature warming of 2°C, 4°C, and 6°C. This is a function of changes in agricultural water demands and hydrology mediated by the modeled behavior of storage, conveyance and irrigation systems.

Reliability metrics can assign a binary metric for each iteration, where a given time point is determined either to a failure or success state based on a threshold condition, and reliability is a probabilistic measure of rate of success [62,63]. We use the quantity-based reliability measures suggested by Dracup et al. [64], which measure degree of failure based on the amount of shortfall below the threshold:

$$R_{ij} = 1 - \frac{(Demand_{ij} - Delivery_{ij})}{Demand_{ij}} \; if \; Demand_{ij} \geq Delivery_{ij};$$

$$if \; not \; R_{ij} = 1. \tag{2}$$

This metric is collapsed into an a overall reliability measure

$$R_i = 1 - \frac{\sum_j (Demand_{ij} - Delivery_{ij})^+}{\sum_j Demand_{ij}} \tag{3}$$

where i represents a given demand point or group of demand points, and j represents timesteps. The $^+$ indicates that negative values are replaced with zero values so as not to bias the analysis in the case of overdelivery at a given time step.

As described above, the model includes a historically based preference for some groundwater supply for each agricultural demand node. We normalized the reliability calculations to exclude this groundwater supply such that groundwater use forced by historical preference does not reduce reliability statistics for

surface water. This may skew estimates of overall reliability, as we do not separate preference for groundwater by users within an irrigation district from use of groundwater in response to shortages.

Table 10 shows the trajectory of modeled supply reliability at each of the major irrigation districts in the basins under the temperature warming scenarios. Reliability decreases marginally with increasing temperature in each case, even in absence of changes in precipitation patterns. These changes are driven in part by the changes in streamflow described above. In addition, modeled demands increase for 2°C (1.4–2.0%), 4°C (2.8–3.9%), and 6°C (4.2–5.8%). The values for this sensitivity analysis result from representation of agricultural demands via the Penman-Montieth equation, which is sensitive to temperature. These results do not take into account the potential for other physical (e.g., plant physiological response to increases in CO_2 concentrations) or behavioral changes (e.g., changes in cropping patterns or irrigation technology), and could be refined in future efforts.

The influence of temperature-driven changes in hydrology and agricultural water demands are apparent in modeled reservoir levels (Figure 7 and Figure 8). In each of the terminal reservoirs, increasing temperature inputs results in lower reservoir levels on average over the calibration period (Table 11). The effect is different in each basin, however. In New Exchequer, the storage decreases are minimal during drought years, while the reverse is true in Don Pedro. This may be a result of the operations behavior during this time period. For example, New Exchequer's smaller buffer zone allows it to be drawn down low initially during the drought period, leaving less potential for meeting demands later in the irrigation season, while Don Pedro has somewhat more gradual initial drawdown and more room for change. Note that the underestimation of Don Pedro Reservoir levels in calibrated results suggests that reliability impacts may be overestimated in the modeled scenarios, although this is likely not a major issue in light of the size of the reservoirs relative to demands. While shifts in streamflow timing and magnitude result from the modeled temperature changes, the large storage volume in these reservoirs allows them to meet demands with a fairly small loss of reliability. Thus, reliability is somewhat, but not dramatically, sensitive to the modeled temperature changes. Since temperature change is only one of the many relevant climate variables projected to affect water resources [9], future work in this system will examine system

Table 9. Mean annual unimpaired streamflow and its sensitivity to modeled temperature warming (10^6 m^3).

Watershed	Historical (1981–1999)	Simulated (1981–99)	Simulated (2°C)	Simulated (4°C)	Simulated (6°C)
TUO	2,608	2,609	2,506 (−3.9%)	2,418 (−7.3%)	2,323 (−11%)
MER	1,363	1,361	1,303 (−4.3%)	1,245 (−8.5%)	1,195 (−12.1%)

Table 10. Modeled surface supply reliability decreases with increasing temperature, based on the reliability metric described in the text.

	ΔT			
	0°C	**2°C**	**4°C**	**6°C**
Modesto ID	0.84	0.82	0.79	0.75
Turlock ID	0.86	0.85	0.82	0.79
Merced ID	0.90	0.86	0.81	0.75

Table 11. Mean reservoir levels and response to temperature increases.

ΔT	New Exchequer Res	Don Pedro Res
Historical	666	1784
Modeled	671	1626
2°C	625 (−6.8%)	1520 (−6.6%)
4°C	590 (−12.1%)	1483 (−8.8%)
6°C	550 (−17.9%)	1448 (−10.9%)

Table shows average reservoir storage volumes (10^6 m³) at the terminal reservoir for each stream. Percent decrease from calibrated reservoir volume is shown in parentheses.

sensitivity to models forced by General Circulation Models that incorporate changes in variability of temperature and precipitation in more nuanced ways.

Water resources infrastructure and policies are designed in part to reduce the risks inherent to climatically driven, variable water resources systems. In short, they are built for adaptation to climate variability. A crucial question for water resources managers is how water resources systems built with the purpose of reducing the impacts of climate variability such as droughts and floods will help these water systems reduce the risks from changing climatic conditions.

While the conceptual question is a general one, the answers will necessarily be place-based and situation-specific. Extensive work has been done investigating and characterizing the potential changes that may result in California's snowmelt-driven hydrology under future climatic non-stationary. The present study joins others in developing a case study with integrated modeling techniques capable of analyzing the impacts of such streamflow changes on the combination of physically-driven streamflow with built infrastructure and its operation.

The results presented above suggest that increased temperatures within the range of what is expected over the course of the 21st century could affect the chief nominal goal of the modeled water resources system, namely the provision of reliable water supply for use in agriculture. Whether these representations are inherent characteristics of institutional policies or artifacts of the model structure and parameterization, hydrologic change introduced by increasing temperatures results in decreases to modeled reliability of water supply.

In sum, modeled impacts of warming scenarios from 2–6°C affect streamflow magnitude (decreases from 4–12%, Table 9), timing (shifts from 5–21 days earlier, Table 8), while increasing

agricultural demands by 1.4–5.8%. The net effect of these changes is that modeled surface water supply reliability decreases in each district, but less than might be expected were the reliability response a simple summation of supply and demand changes (Table 10). The substantial reservoirs providing storage intended to buffer the effects of climate variability serve to reduce, but not eliminate, the hydrologic impacts of climate change in the same way as they offset short-term hydrologic droughts.

While the built system in this model contributes to attenuation of the signal of climate change, it does not do so completely. Impacts are still apparent in the form of reduced surface water supplies in dry years. While the reliability of the system remains high under temperature warming, such a shift may be enough to cause concern among water managers.

Model appropriateness and limitations

The strengths and limitations of a related WEAP application have been discussed by Yates et al. [20], including static land use, the origin of the climate inputs, and some simplifying assumptions in representing agriculture and reservoir operations. The limitations discussed by Yates et al. [20] apply generally to the present work as well, which is based on similar methods. We expand on some of these observations here, while noting that given a conceptual continuum from oversimplified and lacking nuance to a 1:1 scale map that is highly representative but cumbersome and possibly intractable, the present exercise attempts to strike a suitable balance for the intended purpose of long-range sensitivity analysis to major system-scale stressors such as climate change.

To simulate changing deliveries under varying water availability, reservoir operations rules limit surface supplies when reservoir

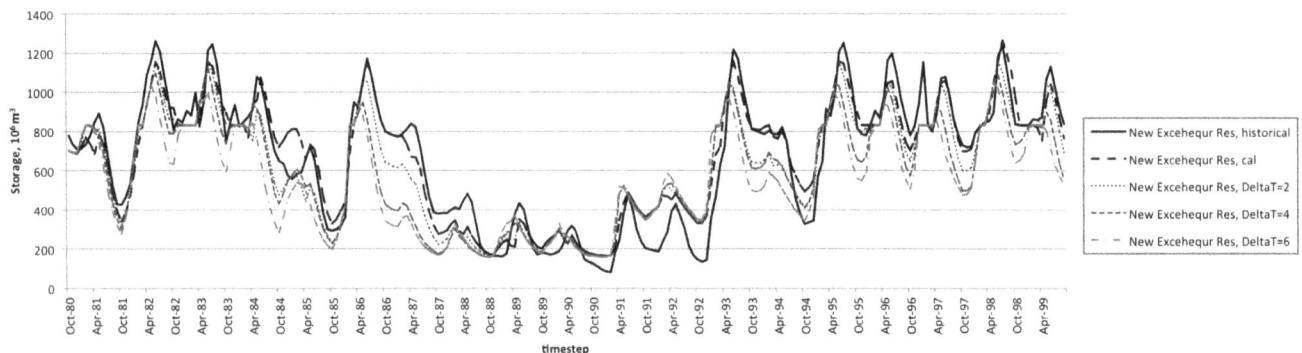

Figure 8. Modeled reservoir levels at New Exchequer terminal reservoir. Results as calibrated to historical observations and modeled with perturbed temperatures of 2°C, 4°C, and 6°C.

storage falls below a specified level. When surface water use is curtailed, for example when low reservoir levels trigger delivery limitations, simulating concern for carryover storage for the next year, the remaining demands are supplied through groundwater pumping. Currently, there is no constraint on groundwater use in the model. Incorporating better data on district-wide pumping capacity, or assumptions of future pumping capacity, could better describe groundwater pumping limits.

Reproducing fine-scale weather patterns is notoriously difficult, especially in mountainous terrain. With data at 1/8° resolution, nuance can be expected to be lost, as described above in the climate inputs section. While the model performs well overall in reproducing historical streamflow, in some areas bias might result from input data.

Spatial and temporal scales are relatively coarse. We chose to implement this model on a monthly time step because of the management relevance for long term planning applications, because some data are only available at monthly time steps, and because computational limitations would make long-range projections intractable using smaller timesteps. The monthly time step limits the resolution for individual events. Thus, analysis of flood risk, sub-monthly instream-flow requirements such as pulse flows, and other applications requiring finer-scale hydrology must be left to other platforms. Hydrology is modeled in WEAP using a quasi-physical lumped parameter approach, whereby land classes within each catchment object are combined and assigned common hydrologic responses. A more detailed approach might, for example, assign land classes and hydrologic response parameters to each segment of a grid covering the model domain, and route water between points on the grid [65]. However, others have argued that more detail inherently comes with the disadvantage of increasing uncertainty intrinsic to larger numbers of estimated parameters [66,67].

Water operations in WEAP enable the model to represent satisfaction of competing current and future demands for water. Operations are defined through a combination of logical constraints (e.g. minimum instream flow requirements) and more general characteristics (e.g. reservoir operations parameters). The latter represents what in reality is a complex set of decisions that even in historical representation includes factors outside the model domain such as economics, long-range weather forecasts, political decisions, changing legal constraints, and so forth. One can think of the buffer concept as a way to represent the general degree of 'conservatism' in operations decisions. It has the advantage of flexibility, and the disadvantage that details of operational decisions and changes in reservoir operations logic can only be represented in a broad-brush sense. For long range planning scenarios, this approach is a valid one given the tremendous uncertainty that exists in the details of future policy decisions. In addition, a strength of this approach is the flexibility with which general changes in policy such as adaptive changes in flood rules or reservoir tolerances could be modeled as part of future efforts, suggesting that the tool described here provides the basis for meeting a recognized need for future climate impact and adaptation studies [3].

Future work could implement changes in agricultural water use efficiency [68], response of cropping patterns to drought [23,69], and change in crops to favor products with higher economic value [70]. Also, the modeling framework is designed for future implementation of more detailed scenario analyses driven by downscaled GCM data [50], although such analysis is beyond the scope of the current paper. We also leave model structure uncertainty and hydrology parameter uncertainty [71] for later work.

Conclusion

In this paper, we have presented a tool for studying climate impacts and adaptation in California water resources. The model represents the hydrology and water operations of the Tuolumne and Merced River Basins in California's San Joaquin Valley. Although climate change is a global environmental catastrophe, the impacts of greatest interest to humans will be its local manifestations, and particularly those manifestations that result in direct consequences to the systems that directly support natural and man-made processes vital to life on earth. Translation of climate impacts most relevant to human decision-making will involve case studies at local levels. Thus, a contribution of this paper lies in its value as a case study for such translation not only in terms of scale, but of topic, from proxies for climate warming to reservoir operations and agricultural water supply reliability.

The WEAP model described in this paper represents annual and seasonal variability in hydrology and water operations, and enables the development of analysis of future water conditions using projections of climate change, land use change, and population growth. The results presented here illustrate system attenuation of the climate change signal: as impacts move to higher order impacts, flexibility in the system buffers the response.

Whether reliability changes of the magnitude estimated here are viewed as significant will depend on the perspective of water managers and water users. For example, water managers often tend to be averse to downside risk [72], and such risk aversion in effect amplifies the felt impacts of supply shortfalls. Formal approaches to answering these questions could include addressing in more detail formal elicitation of the risk preferences and value functions of water users and water managers, and detailing the ultimate impacts felt by users such as crop failure or additional costs incurred for pumping. However, reliability estimates such as those described here may inform anticipatory adaptation actions such as investment in increased water use efficiency measures, the use of crop insurance, and other measures [9].

Nonetheless, impacts on reliability are visible with temperature warming, suggesting value in future work that will also move along the continuum of climate analyses described by Wilby et al. [54]. Such efforts will incorporate downscaled GCM models, and enable us to reflect changes in variability of temperature and precipitation for more detailed climate impacts analysis.

Supporting Information

File S1 Supporting methods and tables, including logic for operations, forecasting, and flow constraints such as in-stream flow requirements and hydropower.

Acknowledgments

David Groves, Sebastian Vicuna and Luke Harmon generously provided assistance with scripting, and Jack Sieber gave useful advice on WEAP. Bruce McGurk and Peter Drekmeier provided helpful context. Neil Stephenson at SuperSpeed provided RamDisk software that greatly increased the speed of model runs. We thank two anonymous reviewers for suggestions that improved the paper. Naturally, responsibility for any errors naturally lies with the authors.

Author Contributions

Conceived and designed the experiments: MK DP. Performed the experiments: MK. Analyzed the data: MK BJ CY. Contributed reagents/materials/analysis tools: MK BJ CY DP. Wrote the paper: MK.

References

1. Oreskes N (2004) The Scientific Consensus on Climate Change. Science 306: 1686.

2. IPCC (2007) Climate Change 2007: The Physical Science Basis. Contribution of Working Group I to the Fourth Assessment Report of the Intergovernmental Panel on Climate Change; Solomon S, Qin D, Manning M, Chen Z, Marquis M et al., editors. Cambridge, United Kingdom and New YorkNY, USA: Cambridge University Press. 996 p.

3. Vicuna S, Dracup J (2007) The evolution of climate change impact studies on hydrology and water resources in California. Climatic Change 82: 327–350.

4. Barnett TP, Adam JC, Lettenmaier DP (2005) Potential impacts of a warming climate on water availability in snow-dominated regions. Nature 438: 303–309.

5. Barnett TP, Pierce DW, Hidalgo HG, Bonfils C, Santer BD, et al. (2008) Human-Induced Changes in the Hydrology of the Western United States. Science 319: 1080–1083.

6. Draper AJ, Munévar A, Arora SK, Reyes E, Parker NL, et al. (2004) CalSim: Generalized Model for Reservoir System Analysis. J Water Res Pl-ASCE 130: 480–489.

7. Stine S (1994) Extreme and Persistent Drought in California and Patagonia During Medieval Time. Nature 369: 546–549.

8. Kiparsky M, Gleick PH (2003) Climate Change and California Water Resources: A Survey and Summary of the Literature. California Energy Commission. CEC-500-2004-073-ED2 CEC-500-2004-073-ED2. 35 p.

9. Kiparsky M, Milman A, Vicuña S (2012) Climate and Water: Knowledge of Impacts to Action on Adaptation. Annu Rev Energ Env 37: 163–194.

10. Yates D, Sieber J, Purkey D, Huber-Lee A (2005) WEAP21 - A demand-, priority-, and preference-driven water planning model Part 1: Model characteristics. Water Int 30: 487–500.

11. Stockholm Environment Institute (2011) Water Evaluation and Planning version 3.2 (WEAP). Stockholm: Stockholm Environment Institute.

12. Getches DH (2003) Constraints of Law and Policy on the Management of Western Water. In: William M. . Lewis Jr., editor. Water and Climate in the Western United States. Boulder: University Press of Colorado. pp. 183–234.

13. Kiparsky M, Gleick PH (2005) Climate Change and California Water Resources. In: Gleick PH, editor. The World's Water 2004–2005. Washington, D.C.: Island Press.

14. Hanemann WM (2007) How California Came to Pass AB 32, the Global Warming Solutions Act of 2006. Department of Agricultural & Resource Economics, U.C. Berkeley. 30 p.

15. Schwarzenegger A (2005) Executive Order S-3-05. State of California.

16. Gleick PH (1987) Regional Hydrologic Consequences of Increases in Atmospheric Carbon Dioxide and Other Trace Gases. Climatic Change 10: 137–161.

17. Stewart IT, Cayan DR, Dettinger MD (2004) Changes in snowmelt runoff timing in western North America under a 'business as usual' climate change scenario. Climatic Change 62: 217–232.

18. Dettinger MD, Cayan DR, Meyer M, Jeton AE (2004) Simulated hydrologic responses to climate variations and change in the Merced, Carson, and American River basins, Sierra Nevada, California, 1900-2099. Climatic Change 62: 283–317.

19. Knowles N, Cayan DR (2002) Potential effects of global warming on the Sacramento/San Joaquin watershed and the San Francisco estuary. Geophys Res Lett 29: 1891.

20. Yates D, Purkey D, Sieber J, Huber-Lee A, Galbraith H, et al. (2009) Climate Driven Water Resources Model of the Sacramento Basin, California. J Water Res Pl-ASCE 135: 303–313.

21. Young CA, Escobar-Arias MI, Fernandes M, Joyce B, Kiparsky M, et al. (2009) Modeling the Hydrology of Climate Change in California's Sierra Nevada for Subwatershed Scale Adaptation. J Am Water Resour As 45: 1409–1423.

22. Null S, Viers J, Mount J (2010) Hydrologic Response and Watershed Sensitivity to Climate Warming in California's Sierra Nevada. PLoS ONE 5(4): e9932 doi:101371/journalpone0009932

23. Joyce B, Vicuna S, Dale L, Dracup J, Hanemann M, et al. (2006) Climate Change Impacts on Water for Agriculture in California: A Case Study in the Sacramento Valley - FINAL REPORT. Sacramento, CA: CEC PIER.

24. ESRI (2007) ArcGIS 9.2. Redlands, CA: ESRI.

25. United States Geological Survey (2007) National Elevation Dataset. Reston, VA: United States Geological Survey.

26. ESRI (2007) ArcHydro Toolkit. Redlands, CA: ESRI.

27. Multi-Resolution Land Characteristics Consortium (2001) National Land Cover Database. Sioux Falls, SD: U.S. Geological Survey.

28. California Department of Water Resources (2006) California Land and Water Use Database. Sacramento, CA: Department of Water Resources, California Resources Agency.

29. United States Bureau of Reclamation (2006) Federal Water Districts - Mid-Pacific Region version 4.0. Sacramento, CA: U.S. Bureau of Reclamation

30. United States Bureau of Reclamation (2003) State Water Districts for California version 1.1. Sacramento, CA: U.S. Bureau of Reclamation in coordination with the California Department of Water Resources

31. United States Bureau of Reclamation (2003) Private Water Districts for California 1:24,000-scale version 1.6. Sacramento, CA: U.S. Bureau of

Reclamation in coordination with the California Department of Water Resources

32. United States Department of Agriculture (2006) Soil Survey Geographic (SSURGO) Database. Washington, DC: United States Department of Agriculture.

33. Natural Resources Conservation Service (2006) U.S. General Soil Map (STATSGO) for California. Washington, DC: United States Department of Agriculture.

34. National Atlas of the United States (2006) Major Dams of the United States. National Atlas of the United States.

35. United States Bureau of Reclamation (2003) Canals for California USBR at 1:24,000-scale, 10th edition, 2003-12-01. Sacramento, CA: U.S. Bureau of Reclamation, Mid-Pacific Region, MPGIS Service Center.

36. United States Geological Survey (2013) National Water Information System. United States Geological Survey.

37. MBK Engineers (2001) Merced Water Supply Plan Update Final Status Report Exhibit 1: MRSIM. Sacramento, CA: City of Merced, Merced ID, UC Merced.

38. California Department of Water Resources (2011) California Data Exchange Center (CDEC).

39. USGS (2007) National Water Information System. United States Geological Survey.

40. United States Senate (1913) An Act Granting To The City And County Of San Francisco Certain Rights Of Way In, Over, And Through Certain Public Lands, The Yosemite National Park, And Stanislaus National Forest, And Certain Lands In The Yosemite National Park, The Stanislaus National Forest, And The Public Lands In The State Of California, And For Other Purposes [Raker Act]. In: Hetch Hetchy Reservoir Site Hearing Before The Committee On Public Lands United States Senate Sixty-Third Congress First Session On H. R 7207, editor.

41. URS (2008) Delta-Mendota Canal Recirculation Initial Alternatives Information Report Sacramento, CA: US Bureau of Reclamation and CA DWR. 06CS204097A 06CS204097A.

42. DWR (2003) California's Groundwater, Bulletin 118, Update 2003. Sacramento, CA: Department of Water Resources, California Resources Agency.

43. Maurer EP, Wood AW, Adam JC, Lettenmaier DP, Nijssen B (2002) A Long-Term Hydrologically Based Dataset of Land Surface Fluxes and States for the Conterminous United States. J Climate 15: 3237–3251.

44. Lundquist J, Huggett B, Roop H, Low N (2009) Use of spatially distributed stream stage recorders to augment rain gages by identifying locations of thunderstorm precipitation and distinguishing rain from snow. Water Resour Res 45: doi: 10.1029/2008WR006995

45. Knowles N (2000) Hydroclimate of the San Francisco Bay-Delta Estuary and Watershed: University of California, San Diego, La Jolla, CA. 292 p.

46. Koczot KM, Jeton AE, McGurk BJ, Dettinger MD (2005) Precipitation-runoff processes in the Feather River Basin, northeastern California, with prospects for streamflow predictability, water years 1971–97. U.S. Geological Survey Scientific Investigations Report 2004–5202. 82 p.

47. Thornton PE, Running SW, White MA (1997) Generating surfaces of daily meteorological variables over large regions of complex terrain. Journal of Hydrology 190: 214–251.

48. Sanstad AH, Johnson H, Goldstein N, Franco G (2009) Long-run socioeconomic and demographic scenarios for California, Draft Report. Sacramento, CA: California Climate Change Center

49. State Water Resources Control Board (2009) 20x2020 Water Conservation Plan Draft Final Report. Sacramento, CA: California Environmental Protection Agency.

50. State Water Resources Control Board (1999) Final Environmental Impact Report for Implementation of the 1995 Bay/Delta Water Quality Control Plan: Volume 2, Technical Appendices. Sacramento, CA: State Water Resources Control Board, California Environmental Protection Agency.

51. State Water Resources Control Board (1999) Final Environmental Impact Report for Implementation of the 1995 Bay/Delta Water Quality Control Plan: Volume 1. Sacramento, CA: State Water Resources Control Board, California Environmental Protection Agency.

52. USBR (2005) CALSIM II San Joaquin River Model (DRAFT). Sacramento, CA: U.S. Bureau of Reclamation, Mid Pacific Region.

53. DWR (2004) Bulletin 118 Update, California's Groundwater, San Joaquin Valley Groundwater Basin, Merced Subbasin. Sacramento, CA: Department of Water Resources, California Resources Agency.

54. Wilby RL, Troni J, Biot Y, Tedd L, Hewitson BC, et al. (2009) A review of climate risk information for adaptation and development planning. Int J Climatol 29: 1193–1215.

55. Evolving Logic (2010) Computer Aided Reasoning Software version 09.09.2007. Topanga, CA: Evolving Logic.

56. Nash JE, Sutcliffe JV (1970) River flow forecasting through conceptual models part I — A discussion of principles. J Hydrol 10: 282–290.

57. Vicuña S, Garreaud RD, McPhee J (2010) Climate change impacts on the hydrology of a snowmelt driven basin in semiarid Chile. Climatic Change.

58. DWR (1975) Bulletin 113-3: Vegetative Water Use in California. Sacramento, CA: California Department of Water Resources.

59. CH2M Hill (2001) Merced Water Supply Plan Update Final Status Report - Merced Water Supply Plan Land Use and Water Demand Model. Merced, CA: City of Merced, Merced ID, UC Merced.

60. Dettinger MD (2005) From Climate-Change Spaghetti to Climate-Change Distributions for 21st Century California. San Francisco Estuary and Watershed Science 3: Article 4, Available: http://repositories.cdlib.org/jmie/sfews/vol3/iss1/art4.

61. Cayan D, Tyree M, Dettinger M, Hidalgo H, Das T, et al. (2009) Climate Change Scenarios And Sea Level Rise Estimates For The California 2009 Climate Change Scenarios Assessment. Sacramento, CA: California Energy Commission, California Climate Change Center.

62. Loucks DP, Van Beek E (2005) Water Resources Systems Planning and Management: An Introduction to Methods, Models and Applications. Paris: UNESCO. 680 pp. p.

63. Hashimoto T, Stedinger JR, Loucks DP (1982) Reliability, resiliency, and vulnerability criteria for water resource system performance evaluation. Water Resour Res 18: 14–20.

64. Dracup JA, Vicuna S, Leonardson R, Dale L, Hanneman M (2005) Climate Change and Water Supply Reliability. Sacramento, CA: California Energy Commission, PIER Energy- Related Environmental Research. CEC-500-2005-053.

65. Flint LE, Flint AL (2007) Ground-Water Recharge in the Arid and Semiarid Southwestern United States - Chapter B Regional Analysis of Ground-Water Recharge USGS Professional Paper 1703-B.

66. Beven K (1993) Prophecy, Reality and Uncertainty in Distributed Hydrological Modeling. Adv in Water Resour 16: 41–51.

67. Beven K (2001) How far can we go in distributed hydrological modelling? Hydrol Earth Syst Sc 5: 1–12.

68. Cooley H, Christian-Smith J, Gleick PH (2008) More with Less: Agricultural Water Conservation and Efficiency in California: A Special Focus on the Delta. OaklandCA: Pacific Institute. 69 p.

69. Purkey D, Joyce B, Vicuna S, Hanemann M, Dale L, et al. (2008) Robust analysis of future climate change impacts on water for agriculture and other sectors: a case study in the Sacramento Valley. Climatic Change 87: 109–122.

70. Groves DG, Matyac S, Hawkins T (2005) Quantified Scenarios of 2030 California Water Demand. Sacramento, CA: Department of Water Resources, California Resources Agency.

71. Ajami NK, Duan Q, Sorooshian S (2007) An integrated hydrologic Bayesian multimodel combination framework: Confronting input, parameter, and model structural uncertainty in hydrologic prediction. Water Resour Res 43: W01403, doi:01410.01029/02005WR004745

72. Kiparsky M, Hanemann WM (2011) How conservative are water managers? Measuring risk aversion and loss aversion for water supply under climate uncertainty. In: Dunning CM, Olson JR, Sehlke G Eds., editors. Proceedings of the AWRA 2011 Spring Specialty Conference on Managing Climate Change Impacts on Water Resources: Adaptation Issues, Options, and Strategies Middleberg, VA. American Water Resources Association.

73. San Joaquin River Group Authority (2001) Additional Water for the San Joaquin River Agreement, 2001–2010 Supplemental EIS/EIR.

4

Water Level Flux in Household Containers in Vietnam - A Key Determinant of *Aedes aegypti* Population Dynamics

Jason A. L. Jeffery[1], **Archie C. A. Clements**[1,2], **Yen Thi Nguyen**[3], **Le Hoang Nguyen**[3], **Son Hai Tran**[3], **Nghia Trung Le**[4], **Nam Sinh Vu**[5], **Peter A. Ryan**[1], **Brian H. Kay**[1]*

1 Queensland Institute of Medical Research, PO Royal Brisbane Hospital, Brisbane, Queensland, Australia, 2 Infectious Disease Epidemiology Unit, School of Population Health, University of Queensland, Herston, Queensland, Australia, 3 National Institute of Hygiene and Epidemiology, Hanoi, Vietnam, 4 Institute Pasteur, Nha Trang, Vietnam, 5 General Department of Preventive Medicine and Environmental Health, Ministry of Health, Hanoi, Vietnam

Abstract

We examined changes in the abundance of immature *Aedes aegypti* at the household and water storage container level during the dry-season (June-July, 2008) in Tri Nguyen village, central Vietnam. We conducted quantitative immature mosquito surveys of 171 containers in the same 41 households, with replacement of samples, every two days during a 29-day period. We developed multi-level mixed effects regression models to investigate container and household variability in pupal abundance. The percentage of houses that were positive for I/II instars, III/IV instars and pupae during any one survey ranged from 19.5–43.9%, 48.8–75.6% and 17.1–53.7%, respectively. The mean numbers of *Ae. aegypti* pupae per house ranged between 1.9–12.6 over the study period. Estimates of absolute pupal abundance were highly variable over the 29-day period despite relatively stable weather conditions. Most variability in pupal abundance occurred at the container rather than the household level. A key determinant of *Ae. aegypti* production was the frequent filling of the containers with water, which caused asynchronous hatching of *Ae. aegypti* eggs and development of cohorts of immatures. We calculated the probability of the water volume of a large container (>500L) increasing or decreasing by ≥20% to be 0.05 and 0.07 per day, respectively, and for small containers (<500L) to be 0.11 and 0.13 per day, respectively. These human water-management behaviors are important determinants of *Ae. aegypti* production during the dry season. This has implications for choosing a suitable *Wolbachia* strain for release as it appears that prolonged egg desiccation does not occur in this village.

Editor: Luciano A. Moreira, Centro de Pesquisas René Rachou, Brazil

Funding: This study was funded by a grant from the Bill and Melinda Gates Foundation "Grand Challenges in Global Health" #55, administered by FNIH. The funders had no role in the study design, data collection and analysis, decision to publish, or preparation of this manuscript.

Competing Interests: The authors have declared that no competing interests exist.

* E-mail: brian.kay@qimr.edu.au

Introduction

Dengue affects 50 million people annually with approximately 20,000 deaths [1]. Four antigenically related but distinct viruses are transmitted principally by the mosquito *Aedes aegypti* (L.). Because there is no vaccine available, vector control remains the cornerstone of epidemic prevention and control [2,3].

Control of *Ae. aegypti* using virulent strains of *Wolbachia* has gained impetus due to life-shortening [4] and/or viral interference [5] phenotypes observed after successful micro-injection of *Ae. aegypti* with *w*Mel and *w*MelPop strains in the laboratory. In the summer of 2011, *w*Mel-infected *Ae. aegypti* adults were successfully released into two field localities around Cairns, Australia [6], demonstrating the feasibility of *Wolbachia*-based dengue control strategies under field conditions.

To trial a *Wolbachia*-based control strategy in Vietnam, the village of Tri Nguyen has been selected by the Ministry of Health as a potential release site for *w*MelPop-CLA (a mosquito cell-line adapted isolate of *w*MelPop) transinfected *Ae. aegypti*. This village is located on Hon Mieu Island, approximately 1 km off the coast of central Vietnam. This *w*MelPop-CLA strain causes several fitness effects on *Ae. aegypti* [3] including reduced fecundity due to life-shortening and reduced ability of eggs to withstand desiccation [7],

two phenotypes that could hinder the establishment of the *w*MelPop-CLA transinfected *Ae. aegypti*, particularly over the dry season. It has been suggested that in tropical areas such as Thailand and Vietnam, where abundant breeding sites are regularly filled by rainfall and/or by human manipulation, the *w*MelPop-CLA strain may spread and persist in *Ae. aegypti* [7]. Thus one of the goals of this paper was to examine the effect of human water manipulation behaviours on *Ae. aegypti* populations in this village during the dry season.

From the results of nine entomologic surveys conducted over 14 months, we determined that village-wide spatial patterns in *Ae. aegypti* presence and abundance in houses were considerably heterogeneous [8]. Importantly, key premises were present with high numbers of mosquitoes, although in contrast to North Queensland [9] and Trinidad [10], these were not temporally stable. In Vietnam, the pattern observed over 14 months suggested that at the household level, *Ae. aegypti* production displayed a cohort or pulse effect, rather than production being continuous and overlapping between mosquito generations. Surprisingly, there was no clear association between season and the prevalence or abundance of *Ae. aegypti* immatures (larvae or pupae) or adults, even though central Vietnam experiences distinct wet (September – December) and dry seasons (February – August) [8]. This lack of a clear association between season and *Ae. aegypti* abundance was noted in the analysis of long-term *Ae. aegypti* data from Puerto Rico and Thailand. Mosquito populations in those two countries were sensitive to different environmental factors (rainfall in Puerto Rico, temperature in Thailand), probably a reflection of local habitat differences and adaptation to unique seasonal environments (distinct wet- and dry-seasons in Thailand, but not in Puerto Rico) [11].

Our study aimed to examine the abundance of *Ae. aegypti* in a village in central Vietnam during the dry season in relation to householder water storage management. Specifically, we wanted to know how frequently householders were filling or emptying their containers, and whether this was sufficient to maintain *Ae. aegypti* populations during a period when little rainfall was expected. This information will be useful in a control program based on the use of *w*MelPop-CLA transinfected *Ae. aegypti*, which has a phenotypic disadvantage whereby the eggs have reduced desiccation resistance [7]. We sampled a cohort of 171 containers at 41 houses, measuring *Ae. aegypti* production every 2 d for almost a month and recording water volume changes in containers. We then used multi-level models to determine whether household- or container-level factors contributed more to *Ae. aegypti* abundance.

Materials and Methods

Ethics

All necessary permits were obtained for the described field studies. Informed verbal consent was obtained from the head of each household according to the Institute Pasteur Nha Trang ethics policy. Although samples had to be returned to the containers after each survey, all mosquito immatures were discarded after the survey on day 29.

Study Site

The area chosen for this study was the village of Tri Nguyen, on Hon Mieu island (12°18'N, 109°14'E), Khanh Hoa province, central Vietnam. A description of the features of the village, cultural practices, occupations and a map, can be found in a report of our long-term entomological survey of this island [8]. In 9 surveys from November 2006–December 2007, large water storage containers (moulded tanks, cylindrical tanks, box tanks and large jars) contained 97–100% of the standing crop of third and fourth instars, and 93–100% of pupae. Small containers such as discards, vases and ant traps contained <5% of production. House type, education level, occupation, income and water management behavior were recorded by the survey staff at this time.

Entomologic Surveys

The current study was undertaken in June-July 2008 to define the short-term temporal variability in immature *Ae. aegypti* production at the household and container level. Forty-one houses (6.7%) were randomly selected from a geo-referenced database containing information on the 611 houses in Tri Nguyen village [8]. These houses were then surveyed every 2 d, for a total of 29 d, by two teams of 2–3 people. The 2-day sampling period was chosen as it matched the minimum duration of the pupal stage [8,12] so that cohorts were not missed. On the day of the first survey, containers located in and around each house were marked with a unique identification number so these could be tracked throughout the study, and any new ones recognized. Every 2 d, the volume, source and use of water, location of the container, and lid status (full, partial or no cover) was recorded for each container. All wet containers were then sampled (either with the 5 sweep net

method or pipette) [13]. For both *Aedes* and *Culex* spp. immatures, presence/absence of I/II instars and the approximate number of III/IV instars (0, 1–10, 11–100, 101–1000, 1000+) were recorded. The number of pupae were counted. Presence or absence of potential mosquito predators (principally *Mesocyclops* spp., *Micronecta* spp. and fish) was also recorded. The sample was then returned to the container. On each sampling occasion, the height of the water level in each container was estimated using a graduated rule. The height was then used to calculate the volume of water in each container and this was expressed as a percentage of the total capacity. Percentage change between successive surveys was used to measure water flux.

Because of the large size and configuration of the containers, plus their usage patterns [13], it was not possible to estimate oviposition and egg hatching using paper strips [14,15]. Consequently, a modelling approach was adopted using pupal counts. We assumed that the main factor influencing pupal abundance is water level changes triggering egg hatching, although we acknowledge that unmeasured variables such as overcrowding, nutrient levels and egg-laying behaviours can influence immature abundance. We also acknowledge that water-level changes may have occurred without our knowledge in the period between visits every 2 d and that these may have also influenced immature abundance.

For the water volume analyses, 171 containers within the 41 properties were classified as either large (>500 L, n = 119) or small (<500 L, n = 52). A variable was created to indicate water flux, categorized according to whether the volume of water had increased by >20%, decreased by >20% or neither increased or decreased by >20%, relative to the previous survey. Daily rainfall (mm) and daily minimum, mean and maximum temperatures (°C) were obtained from Nha Trang city weather station (Figure 1). June/July is the dry and hot season in central Vietnam and there were only three rain events >5 mm during the survey period.

Multi-level Model of Pupal Abundance

Pupal abundance was highly aggregated because a large proportion (range 76.8–94.8%) of the containers during any one survey was negative for pupae. We investigated different types of models of the pupal counts, including a Poisson model, a negative binomial model and a zero-inflated Poisson (ZIP) model and found that the ZIP model provided the best fit to the data (using the Akaike Information Criterion and the Vuong statistic). Consequently, we used a ZIP mixed effects model to examine the relationship between water volume changes and pupal abundance. This allowed us to incorporate both the Poisson structure of the distribution of pupal counts (which includes some zeroes) and a zero-inflated component that modelled the excess negative containers.

The model took the form:

$$Y_{ijk} \sim Pois(\mu_{ijk});$$

$$\mu_{ijk} = Z_{ijk} * \lambda_{ijk};$$

$$Z_{ijk} \sim bern(p_{ijk});$$

$$Logit(p_{ijk}) = \beta_0 + \beta_1 C_{1,ijk} + \beta_2 C_{2,ijk} + \beta_3 t_k + u_i + v_j;$$
$$Log(\lambda_{ijk}) = \delta_0 + \delta_1 C_{1,ijk} + \delta_2 C_{2,ijk} + \delta_3 t_k + w_i + z_j;$$

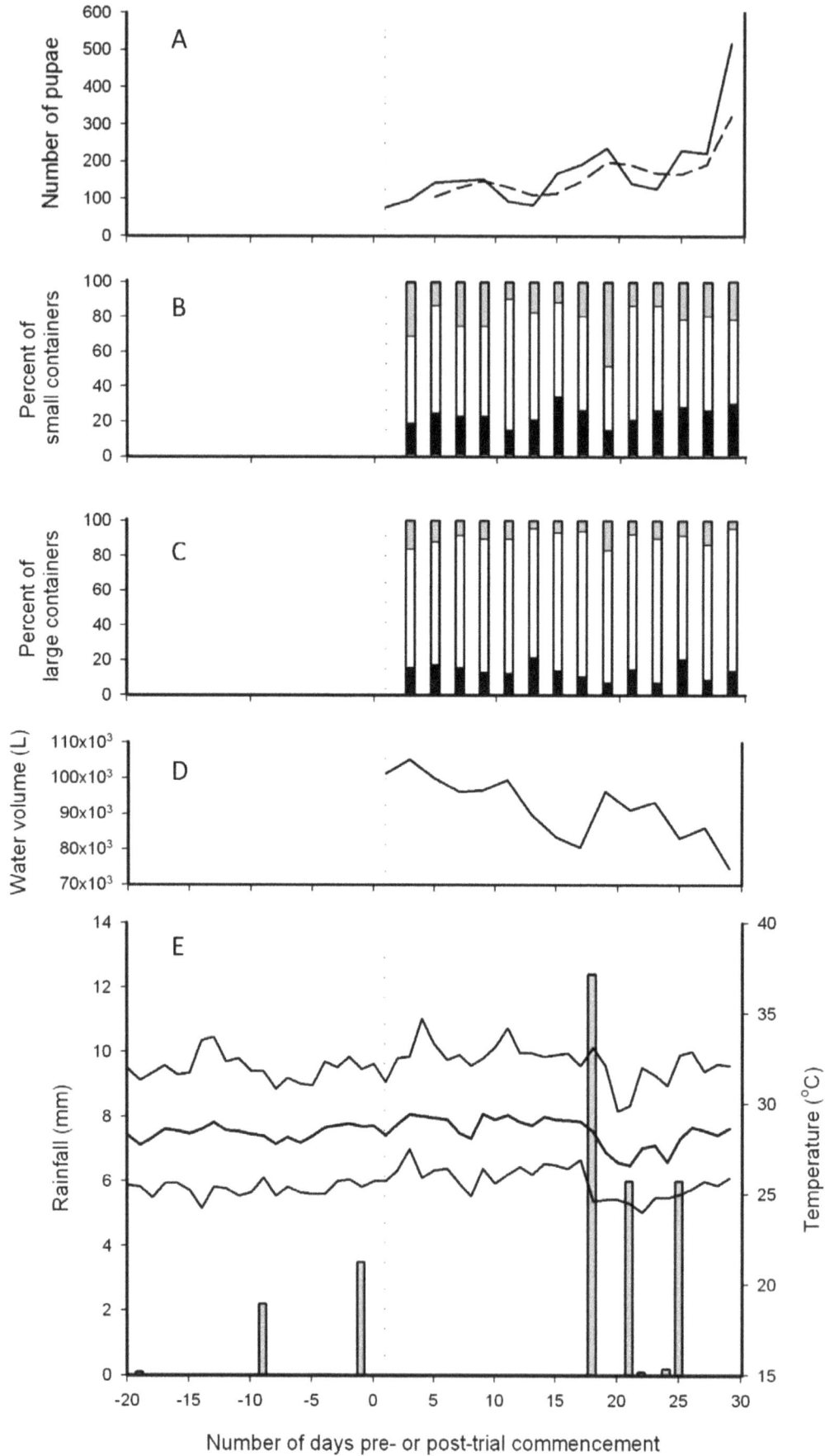

Figure 1. Absolute counts of pupae every 2 days, in relation to water flux, water storage and weather conditions.

where Y was the observed number of pupae in container i, household j, survey k, μ was the modelled mean number of pupae, $\mathcal{Z}$ represented the excess zeroes in the observed pupal distribution, and λ the count of pupae. For the zero-inflated part of the model, $\beta_{0...2}$ were the intercept and coefficients for the fixed effects (two dummy variables representing categories of water level change: C_1 an increase in water volume $\geq 20\%$ relative to the last survey and C_2, a decrease in water volume $\geq 20\%$ relative to the last survey, with the reference category being a change in water level of $<20\%$, and a term for temporal trend), u_i was a container-level random effect and v_j was a household-level random effect. Similarly, for the count part of the model, $\delta_{0...2}$ were the intercept and coefficients for the fixed effects (as above), w_i was a container-level random effect and z_j was a household-level random effect. Because change in container water volume was assessed relative to the previous survey, we only used data from surveys 2–15 in the models. We specified non-informative priors for the intercepts and coefficients (normal priors with a mean of 0 and a precision, the inverse of variance, of 1/10,000). All of the random effects were assumed to have a normal distribution centred on zero, and with an unknown precision modelled with non-informative gamma priors (having shape and scale parameters = 0.01).

We used a Markov chain Monte Carlo simulation to fit the models. A burn-in of 5,000 iterations was allowed, followed by 100,000 iterations where values for the intercept, coefficients, and the means and variance of the random effects were monitored and stored. To reduce autocorrelation in the chains, only every 10th iteration was stored, giving a total of 10,000 iterations for the posterior distribution of each monitored variable. Convergence was checked by visual examination of density and history plots. To ensure that sufficient iterations were performed to adequately describe the posterior distributions, Monte Carlo error (MCE)/SD was calculated for each variable. If this value was <0.05 for each parameter, we considered the number of iterations to be sufficient [16]. Random effects were considered significant at the 5% level when the 95% credible intervals excluded zero. All analyses were performed using WinBUGS version 1.4 (Imperial College, London, and Medical Research Council, UK).

Results

Descriptive Analyses

The mean number of people per house was 5.6 (range 1–13). Water was stored in containers ranging in size from <100–10,000 L. Containers were replenished by purchasing water or by channelling rain water from the roof into selected containers. There were 195 containers identified in the 41 households (mean of 4.8 per household) of which 171 were examined at every time-point (i.e. 15 times). The remaining 24 containers were excluded from further analyses because they were not surveyed 15 times. This was due to householders switching their containers to non-water storage uses at some stage during the survey period. Any residual water was usually poured into another container. There were no new containers introduced during the study period.

Of the 171 containers, 17 (9.9%) were moulded tanks (2000 L capacity), 91 (53.2%) were cylindrical tanks (1000–2000 L capacity), 11 (6.4%) were box tanks (100–10,000 L capacity), 42 (24.6%) were standard jars and drums (>100 L capacity), and 7 (5.0%) were small jars (including buckets) (<100 L capacity). Almost all of the containers (98%) were located outdoors with those indoors mainly used during routine daily activity. As with

previous surveys [8], numbers of predators were low, ranging from 0–4%.

Temporal Patterns

The total estimated number of pupae collected from all houses ranged from 77 on day 1 (mean of 1.9 per house) to 517 on day 29 (mean of 12.6 per house) (Figure 1). The percentage of containers that were wet during each survey ranged from 72.8–82.1%. The trend in volume of water stored over the 29 days was generally decreasing (a reduction from 102,000 to 75,000 L) but regular water replenishment was evident in both large and small containers (Figure 1). Increases of $>20\%$ in the volume of water in containers, compared with water volume observed during the previous survey period, could be seen during all sampling periods. This was due to householders either purchasing water from vendors or by consolidating stored water into fewer containers. Two of the three rain events on days 21 (5.6 mm) and 25 (6.1 mm) had a negligible effect on the filling of containers (only 13% of small and 8% of large containers had increases in water volumes of $\geq 20\%$ on day 21, and 21% of small and 8% of large containers increased by $\geq 20\%$ on day 25). However, the rainfall event on day 18 (12.5 mm) resulted in the highest rates of container filling, with increases in water volumes $\geq 20\%$ in 48% of small containers and 17% of large containers. The latter event resulted in a 2.3 to 6.6-fold rise in pupal abundance 11 d later, from between 1.9 and 5.6 pupae per house on days 1–18, to 12.6 pupae per house on day 29.

In terms of container-level water volume changes observed during each survey, there were 304 events where water volume increased $\geq 20\%$, 410 events where water volume decreased $\geq 20\%$, and 1851 events in which water volumes did not vary by $\geq 20\%$, compared to the previous survey. Water volume changes occurred every day, with a range of 9.6–48.1% of small containers being filled and 15.4–34.6% being reduced by $\geq 20\%$, respectively, and 4.2–16.8% and 7.6–21.8% of large containers being filled or reduced by $\geq 20\%$, respectively. The percentage of small and large containers whose volume did not change by $\geq 20\%$ compared with the previous survey, ranged from 36.5–75.0% and 68.0–83.2%, respectively (Figure 1). The probability of the water volume of a large container (>500 L) increasing $\geq 20\%$ was 0.05 and decreasing $\geq 20\%$ was 0.07 per day. For small containers (<500 L) it was 0.11 (increasing) and 0.13 (decreasing) per day.

The percentage of houses and containers that was positive for pupae during any one survey ranged from 17.1–53.7 and 5.2–21.4%, respectively. However, by the end of the 29 d, 87.8% of houses and 46.2% of containers had been recorded as positive (Table 1), indicating that even in a short time frame, most houses and almost half of all containers had or were producing pupae. Similar patterns were observed for I/II and III/IV instars. Interestingly, seven (17.1%) houses were positive for III/IV instars at every time point but no houses were consistently positive for I/II instars or pupae. Only one container was consistently positive for III/IV instars at every time point and no containers were consistently positive for I/II instars or pupae. There were also a proportion of houses (5%) and containers (40%) that remained consistently negative for immature *Ae. aegypti* throughout the entire study (Table 1).

Statistical Analysis

The Bayesian hierarchical model showed that there was more variation in pupal abundance at the container level compared with

Table 1. Percentage of houses and containers positive or negative for immature Ae. Aegypti.

	Houses (%)			Containers (%)		
	Range +ve during the surveys	Cumulative +ve	Always –ve	Range +ve during the surveys	Cumulative +ve	Always –ve
I/II instars	19.5–43.9	90.2	9.8	6.3–20.0	56.1	43.9
III/IV instars	48.8–75.6	92.7	7.3	20.3–37.0	63.7	36.3
Pupae	17.1–53.7	87.8	12.2	5.2–21.4	46.2	53.8
All stages			5.0			40.0

the house level for both the zero-inflated and count components of the model (Table 2). Containers where the water volume increased relative to the previous survey had a significantly higher count of pupae (if pupae were present) and the counts of pupae showed a significantly increasing trend over the study period.

A significant container-level random effect indicated that there were unmeasured variables acting at the container level that influenced pupal presence and abundance (note, values of the random effects are not shown). For the zero-inflated component of the model, no containers had random effects significantly lower than the overall mean and 21 containers had random effects significantly greater than the overall mean (i.e. they were more likely than average to have a zero count). For the count component of the model, 5 containers had random effects significantly lower than the overall mean and 20 containers had random effects significantly greater than the overall mean (i.e. they had a significantly higher than average. count). None of the household-level random effects were significantly different from the mean (they all had 95% Bayesian credible interval limits that included zero).

Discussion

Because piped water was unavailable, villagers in Tri Nguyen relied on water management, water purchase and occasionally rainfall to fill a variety of containers ranging from 100–10,000 L. Not surprisingly, our data provide evidence that frequent filling of containers is positively associated with the abundance of Ae. aegypti pupae. Despite the relative lack of rainfall and a reducing but fluctuating water volume, there were enough filling events (304 over 29 days) to support hatching of the desiccation-resistant eggs of Ae. aegypti inside these containers and thus ensure survival and, sometimes, population growth, although the most significant increase in pupal abundance followed 12.5 mm of rainfall on day 18. Overall, this suggests that if wMelPop-CLA transinfected Ae.

aegypti are released in Tri Nguyen village, there will be sufficient water filling events from everyday householder behaviours to minimise the effect of the reduced desiccation resistant phenotype. Thus, this might support the release of wMelPop-CLA mosquitoes during the wet season, and their survival through the dry season.

Over the 14-month study period reported in our previous work, the percentage of houses positive for III/IV instars and pupae ranged from 54–81 and 13–48%, respectively [8]. This is similar to the equivalent ranges we observed over the one-month period of the current study (49–76 and 17–54%, respectively). In terms of container positivity for III/IV instars and pupae, the 14-month range (26–49 and 6–22%, respectively) was similar to the one-month range found in the current study (20–37 and 5–21%, respectively).

Stoddard [17] indicated that human behavior is an understudied aspect of vector control and disease management. Our work suggests that water storage behavior, particularly in relation to human-driven water volume changes at the container-level and over small temporal scales, is an important driver of Ae. aegypti population dynamics. Although other studies have been undertaken on water storage practices and behaviors in Vietnam, these have focussed primarily on changes in human perceptions of water supply and changes in water storage behavior subsequent to the provision of new water supply infrastructure [18]. Our study was concerned with existing infrastructure and cultural practices of water storage. Upon questioning of householders after changes in water level ≥20%, it was apparent that we had observed a continuous practice of householder transfer of water, and not surprisingly, asynchronous hatching of Ae. aegypti eggs and subsequent development of cohorts of immatures. The relatively low percentage of immature positivity on any one day for houses and, to a lesser extent, containers, compared to the high cumulative house or container positivity at the end of the 29 day

Table 2. Results from the analysis of Ae. aegypti pupal abundance using a zero-inflated Poisson model in a Bayesian framework.

Variable	Zero-inflated component	Count component
Intercept	–3.58 (–4.18– –3.02)	1.45 (1.17–1.72)
Coefficient: increasing volume	–0.48 (–1.04–0.06)	0.95 (0.79–1.11)*
Coefficient: decreasing volume	0.16 (–0.30–0.60)	0.06 (–0.07–0.19)
Coefficient: temporal trend	0.04 (–9.9×10^{-4}–0.08)	0.06 (0.04–0.07)*
Variance container RE	2.43 (1.50–4.14)	0.78 (0.53–1.24)
Variance household RE	0.06 (0.01–1.19)	0.03 (0.01–0.32)

*Significant with ≥95% probability; RE = random effect; estimates show the mean and 95% Bayesian credible interval.

period, is indicative of the asynchrony of *Ae. aegypti* cohorts across the village.

The frequency of container filling events (n = 304) was 26% fewer than water draw-down events (n = 410). Although we selected 20% as a definite and observable water volume change, we acknowledge that water level increases smaller than this would also be capable of causing egg hatching, indicating one limitation of the study. We chose 20% as a cut-off because we were evaluating broad patterns of water management, and whether or not these were sufficient to maintain *Ae. aegypti* populations during a period when little rainfall was expected. Since we observed that there were enough water volume changes to support *Ae. aegypti* populations, any smaller changes that we may have overlooked would most likely have an additive effect and so our measurements of water flux are most likely underestimates. Hatching could even be initiated by disruption of the water surface during the retrieval process by householders, but this also could not be measured with any precision. Despite this, it appears unlikely that *Ae. aegypti* relies on prolonged desiccation resistance at Tri Nguyen through the dry season.

Although further investigation of this effect is required, it would seem that such behavior should be incorporated into *Ae. aegypti* population models such as CIMSiM [19,20] and Skeeter Buster [21] to ensure they are realistic, accurate and location specific. This concurs with findings in Colombia [22] and Puerto Rico [23]. Given that more variability in pupal abundance occurred at the container level, any pre-release vector control needs to focus on all containers in the target area, and not just on key containers or high-mosquito burden households, because we saw little evidence

for their existence. As in Iquitos [24], high productivity, whether in containers or households, was transient. Our data also suggest that in Tri Nguyen village, container level variability was more important than household level data. As in Iquitos [24], we have previously demonstrated that the correlation between pupal and adult abundance is the strongest [25], so we believe our estimates are robust. Although we acknowledge that we only studied a small number of houses, our models also suggest that there were significantly more unmeasured variables at the container level influencing pupal abundance, compared to the household level. We expect that this would be applicable to other immature stages. Thus as with other studies [22,23], human water management practices would seem to be a previously underrated factor in driving container productivity of *Ae. aegypti*.

Acknowledgments

We thank the people of Tri Nguyen village, Vietnam, for their hospitality and for allowing us repeated access to their premises. We thank Bui Tan Phu, Ton Nu Van Anh, and Bui Thi Cam Nhung (Institute Pasteur, Nha Trang) for their assistance with the entomologic surveys. We also thank Simon Kutcher (Australian Foundation for the Peoples of Asia and the Pacific) for providing logistic and administrative support.

Author Contributions

Conceived and designed the experiments: JALJ BHK PAR NSV. Performed the experiments: JALJ YTN LHN SHT NTL. Analyzed the data: JALJ ACAC PAR. Contributed reagents/materials/analysis tools: YTN NTL NSV. Wrote the paper: JALJ ACAC PAR BHK.

References

1. Beatty ME, Beutels P, Meltzer MI, Shepard DS, Hombach J, et al. (2011) Health economics of dengue: a systematic literature review and expert panel's assessment. Am J Trop Med Hyg 84: 473–488.
2. Gubler D (1997) Dengue and dengue hemorrhagic fever: its history and resurgence as a global public health problem. In: Gubler D, Kuno G, editors. Dengue and Dengue Hemorrhagic Fever. New York: CAB International.
3. Iturbe-Ormaetxe I, Walker T, O'Neill SL (2011) Wolbachia and the biological control of mosquito-borne disease. EMBO Rep 12: 508–518.
4. McMeniman CJ, Lane RV, Cass BN, Fong AW, Sidhu M, et al. (2009) Stable introduction of a life-shortening Wolbachia infection into the mosquito Aedes aegypti. Science 323: 141–144.
5. Moreira LA, Iturbe-Ormaetxe I, Jeffery JA, Lu G, Pyke AT, et al. (2009) A Wolbachia symbiont in Aedes aegypti limits infection with dengue, Chikungunya, and Plasmodium. Cell 139: 1268–1278.
6. Hoffmann AA, Montgomery BL, Popovici J, Iturbe-Ormaetxe I, Johnson PH, et al. (2011) Successful establishment of Wolbachia in Aedes populations to suppress dengue transmission. Nature 476: 454–457.
7. McMeniman CJ, O'Neill SL (2010) A virulent Wolbachia infection decreases the viability of the dengue vector Aedes aegypti during periods of embryonic quiescence. PLoS Negl Trop Dis 4: e748.
8. Jeffery JAL, Yen NT, Nam VS, Nghia LT, Hoffmann AA, et al. (2009) Characterizing the Aedes aegypti population in a Vietnamese village in preparation for a Wolbachia-based mosquito control strategy to eliminate dengue. PLoS Negl Trop Dis 3: e552.
9. Tun-Lin W, Kay BH, Barnes A (1995) Understanding productivity, a key to Aedes aegypti surveillance. Am J Trop Med Hyg 53: 595–601.
10. Chadee DD (2004) Key premises, a guide to Aedes aegypti (Diptera: Culicidae) surveillance and control. Bull Entomol Res 94: 201–207.
11. Chaves LF, Morrison AC, Kitron UD, Scott TW (2012) Nonlinear impacts of climatic variability on the density-dependent regulation of an insect vector of disease. Glob Change Biol 18: 457–468.
12. Southwood TR, Murdie G, Yasuno M, Tonn RJ, Reader PM (1972) Studies on the life budget of Aedes aegypti in Wat Samphaya, Bangkok, Thailand. Bull World Health Organ 46: 211–226.
13. Knox TB, Yen NT, Nam VS, Gatton ML, Kay BH, et al. (2007) Critical evaluation of quantitative sampling methods for Aedes aegypti (Diptera: Culicidae) immatures in water storage containers in Vietnam. J Med Entomol 44: 192–204.
14. Wong J, Astete H, Morrison AC, Scott TW (2011) Sampling considerations for designing Aedes aegypti (Diptera:Culicidae) oviposition studies in Iquitos, Peru:

substrate preference, diurnal periodicity, and gonotrophic cycle length. J Med Entomol 48: 45–52.
15. Wong J, Stoddard ST, Astete H, Morrison AC, Scott TW (2011) Oviposition site selection by the dengue vector Aedes aegypti and its implications for dengue control. PLoS Negl Trop Dis 5: e1015.
16. Clements AC, Moyeed R, Brooker S (2006) Bayesian geostatistical prediction of the intensity of infection with Schistosoma mansoni in East Africa. Parasitology 133: 711–719.
17. Stoddard ST, Morrison AC, Vazquez-Prokopec GM, Paz Soldan V, Kochel TJ, et al. (2009) The role of human movement in the transmission of vector-borne pathogens. PLoS Negl Trop Dis 3: e481.
18. Tran HP, Adams J, Jeffery JAL, Nguyen YT, Vu NS, et al. (2010) Householder perspectives and preferences on water storage and use, with reference to dengue, in the Mekong Delta, southern Vietnam. Int Health 2: 136–142.
19. Focks DA, Daniels E, Haile DG, Keesling JE (1995) A simulation model of the epidemiology of urban dengue fever: literature analysis, model development, preliminary validation, and samples of simulation results. Am J Trop Med Hyg 53: 489–506.
20. Focks DA, Haile DG, Daniels E, Mount GA (1993) Dynamic life table model for Aedes aegypti (Diptera: Culicidae): analysis of the literature and model development. J Med Entomol 30: 1003–1017.
21. Magori K, Legros M, Puente ME, Focks DA, Scott TW, et al. (2009) Skeeter Buster: a stochastic, spatially explicit modeling tool for studying Aedes aegypti population replacement and population suppression strategies. PLoS Negl Trop Dis 3: e508.
22. Padmanabha H, Soto E, Mosquera M, Lord CC, Lounibos LP (2010) Ecological links between water storage behaviors and Aedes aegypti production: implications for dengue vector control in variable climates. Ecohealth 7: 78–90.
23. Barrera R, Amador M, MacKay AJ (2011) Population dynamics of Aedes aegypti and dengue as influenced by weather and human behavior in San Juan, Puerto Rico. PLoS Negl Trop Dis 5: e1378.
24. Getis A, Morrison AC, Gray K, Scott TW (2003) Characteristics of the spatial pattern of the dengue vector, Aedes aegypti, in Iquitos, Peru. Am J Trop Med Hyg 69: 494–505.
25. Knox TB, Nguyen YT, Vu NS, Kay BH, Ryan PA (2010) Quantitative relationships between immature and emergent adult Aedes aegypti (Diptera: Culicidae) populations in water storage container habitats. J Med Entomol 47: 748–758.

5

Biological Instability in a Chlorinated Drinking Water Distribution Network

Alina Nescerecka[1,2], Janis Rubulis[1], Marius Vital[2], Talis Juhna[1], Frederik Hammes[2]*

1 Department of Water Engineering and Technology, Riga Technical University, Riga, Latvia, **2** Department of Environmental Microbiology, Eawag, Swiss Federal Institute for Aquatic Science and Technology, Dübendorf, Switzerland

Abstract

The purpose of a drinking water distribution system is to deliver drinking water to the consumer, preferably with the same quality as when it left the treatment plant. In this context, the maintenance of good microbiological quality is often referred to as biological stability, and the addition of sufficient chlorine residuals is regarded as one way to achieve this. The full-scale drinking water distribution system of Riga (Latvia) was investigated with respect to biological stability in chlorinated drinking water. Flow cytometric (FCM) intact cell concentrations, intracellular adenosine tri-phosphate (ATP), heterotrophic plate counts and residual chlorine measurements were performed to evaluate the drinking water quality and stability at 49 sampling points throughout the distribution network. Cell viability methods were compared and the importance of extracellular ATP measurements was examined as well. FCM intact cell concentrations varied from 5×10^3 cells mL^{-1} to 4.66×10^5 cells mL^{-1} in the network. While this parameter did not exceed 2.1×10^4 cells mL^{-1} in the effluent from any water treatment plant, 50% of all the network samples contained more than 1.06×10^5 cells mL^{-1}. This indisputably demonstrates biological instability in this particular drinking water distribution system, which was ascribed to a loss of disinfectant residuals and concomitant bacterial growth. The study highlights the potential of using cultivation-independent methods for the assessment of chlorinated water samples. In addition, it underlines the complexity of full-scale drinking water distribution systems, and the resulting challenges to establish the causes of biological instability.

Editor: Jose Luis Balcazar, Catalan Institute for Water Research (ICRA), Spain

Funding: The authors acknowledge the financial support of the EU project "TECHNEAU" (Nr. 018320) and the NMS-CH project "BioWater: Assessment of biological stability in drinking water distribution networks with chlorine residuals" (Sciex-N-7 12.265). The funders had no role in study design, data collection and analysis, decision to publish, or preparation of the manuscript.

Competing Interests: The authors have declared that no competing interests exist.

* E-mail: frederik.hammes@eawag.ch

Introduction

The goal of public drinking water supply systems is to produce water of acceptable aesthetic and hygienic quality and to maintain that quality throughout distribution until the point of consumption. From a microbiological perspective, the quality of treated water can deteriorate as a result of excessive bacterial growth, which can lead to problems such as a sensory deterioration of water quality (e.g. taste, odor, turbidity, discoloration) as well as pathogen proliferation [1–10]. To avoid this, biological stability during distribution can be achieved by maintaining sufficient residual disinfectants in the water, and/or through nutrient limitations [3,7,11,12]. However, drinking water systems should not be viewed as sterile; complex indigenous bacterial communities have been shown to inhabit both chlorinated and non-chlorinated drinking water distribution systems [5,13–17].

The concept of biological stability and its impact on a system's microbiology has been discussed extensively in the framework of non-chlorinated drinking water distribution systems [3,7,17–20]. However, many treatment plants worldwide employ a final disinfection step to ensure that no viable bacteria enter the distribution system. The latter is often achieved by oxidative disinfection, usually by chlorination [21]. Disinfection has a number of implications for a biological system. During chlorination, one can expect that a considerable fraction of bacteria in the

water are killed or damaged, while some residual chlorine may remain in the water (Figure 1). This could be visible through numerous microbial monitoring methods. For example, the number of cultivable bacteria, measured with heterotrophic plate counts, would reduce dramatically [22,23]. Secondly, bacteria cells are likely to display measurable membrane damage irrespective of their cultivability [24], though the rate and extent of damage may differ between different communities. This would be detectable with several staining techniques coupled with epifluorescence microscopy or flow cytometry (FCM). Also, adenosine tri-phosphate (ATP), often used as a cultivation-independent viability method [19,22,25] will be severely affected. Based on data from Hammes and co-workers [4] one may reasonably expect increased levels of extracellular ATP (so-called free ATP) and decreased concentrations of intracellular ATP (bacterial ATP) following oxidative disinfection. Irrespective of the detection method, the overall consequence of disinfection is a considerable decrease in the viable biomass, potentially opening a niche for microorganisms to occupy downstream of the treatment process. Following initial disinfection, residual chlorine might provoke undesirable changes during drinking water distribution. Disinfectants target not only bacteria, but it also react with natural organic matter, pipe surfaces and particles in the network, thus potentially forming/releasing assimilable organic carbon (AOC) [26–30]. AOC can easily be consumed by bacteria, and is therefore seen as a main contributor to biological instability. Moreover, chlorine decay within the network negatively affects its ability to inhibit microbial growth at the far ends of the network [12]. If all factors were considered, the presence of nutrients, a reduction in the number of competing bacteria, and the lack of residual disinfectant would potentially lead to biological instability in the distribution network, manifesting in a subsequent bacterial growth (Figure 1). Besides the importance of nutrients, the extent of bacterial growth will be influenced by a number of factors. For example, increased water temperature can accelerate chlorine decay and favor bacteria growth [19,31], while changes in hydraulic conditions can alter nutrient supply for microorganisms in biofilms and/or bacteria detachment from the pipe surfaces [32,33]. Finally, the quality of materials in contact with drinking water, as well as the presence of sediments and loose deposits, can both affect the general microbial quality of the water [6,34,35].

In the present study we examined some of the above-discussed concepts in a full-scale, chlorinated distribution system in the city of Riga (Latvia) with a number of microbiological methods. The purpose was a detailed investigation of the entire city's distribution network, asking the basic question whether evidence of spatial and/or temporal biological instability exists, and if so, to which degree. Additional goals were to evaluate the use of fluorescent staining coupled with FCM, as well as ATP analysis, for the assessment of chlorinated drinking water in a distribution network with disinfectant residuals.

Materials and Methods

Ethics statement

Permission for sampling at all locations in the present study was obtained from the local water utility (Rīgas Ūdens).

Description of study site

Sampling was performed in the full-scale distribution network of Riga (Latvia) with a total length of about 1400 km. The city is supplied with drinking water from six water treatment plants (WTP) produced from both surface and groundwater (150 000 m^3 d^{-1}). Only the three major WTP, which are continuously operated, were included in the sampling campaign. Average WTP effluent water quality parameters for each treatment plant are shown in Table 1. The distribution network mainly consists of cast iron (80%) and unlined iron (15%) pipes as old as 50 years.

The diameters of pipes ranged from 100 to 1200 mm. Three reservoirs are operated in the network to compensate for fluctuations in the daily water demand, while four high-pressure zones are maintained in some distal areas of the network. The high-pressure zones were excluded from the present study. A total of 49 sampling sites were selected across the city to cover the network broadly and to include both proximal and distal zones relative to the treatment plants. The sampling sites were selected according to the approximate water retention times obtained from a validated hydraulic model made in EPANET 2.0 [36,37] based on a total length of 538 km (39% of the total length of the network). Apart from the effluents of the three treatment plants, the sampling sites were in all cases fire hydrants in order to attain some degree of reproducibility between sampling and to avoid localized effects (e.g. household growth). The exact locations of sampled fire hydrants can be obtained from the authors after agreement from the local water utility.

Sampling protocol

A specific sampling protocol was designed and followed in order to avoid artifacts due to water stagnation in unused fire hydrants. Each hydrant was pre-flushed at a high velocity (never exceeding 1.6 m s^{-1}) for no more than 60 s, then immediately adjusted to a low velocity of 0.015–0.25 m s^{-1} and connected to an online system for monitoring pH, temperature, redox potential, electroconductivity and turbidity. The low sampling velocity was specifically used to ensure a minimal possible impact of cell wall erosion and detachment from biofilms on the samples and measurements. Readings of all parameters were taken at 5–10 minute intervals, and water was only sampled for microbiological analysis once all of the parameters stabilized. The impact of this hydrant flushing is demonstrated in an example in Figure 2 and discussed in detail in the results section. Samples were kept in cold storage ($\approx$5°C) and analyzed within four hours of sampling.

Chemical analysis

Determination of free chlorine was performed according to standard method EN ISO 7393-1, based on the direct reaction with N,N-diethyl-1,4-phenylenediamine (DPD) and subsequent formation of a red compound at pH 6.2–6.5. Afterwards titration by means of a standard solution of ammonium iron (III) sulfate until disappearance of the red color was performed. Determination of total chlorine was performed according to EN ISO 7393-1,

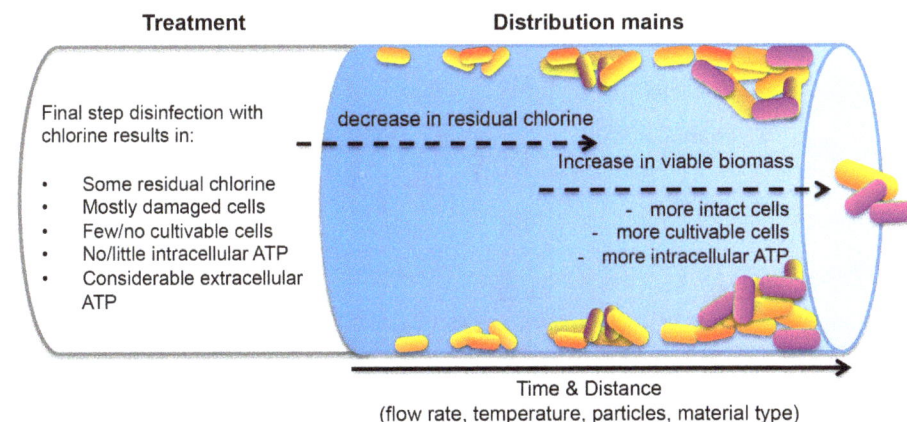

Figure 1. A worst-case-scenario in an unstable, chlorinated distribution network. Prediction of changes in the microbiological state of the water due to the depletion of residual chlorine and the concomitant growth of bacteria, potentially resulting in hygienic and sensory deterioration of the water quality.

Table 1. Average water quality parameters for the final effluents of the the three main treatment plants of Riga (Latvia).

	WTP 1	WTP 2	WTP 3
Source water	surface water	artificially recharged groundwater	groundwater
Final treatment step[a]	Cl_2 (0.5–3 mg L^{-1})	Cl_2 (ca. 1.5 mg L^{-1})	N.A.
Residual chlorine (mg L^{-1})[b]	0.44±0.11	0.51±0.01	0.42±0.26
Total organic carbon (TOC) (mg L^{-1})[a]	6±1	9±3	3
Assimilable organic carbon (AOC) (µg L^{-1})[a]	213±37	209±59	N.A.
Total cell concentration (cells mL^{-1})[b]	$5.31±0.97×10^5$	$5.45±0.47×10^5$	$1.69±0.18×10^5$
Intact cell concentration (cells mL^{-1})[b]	$1.83±1.18×10^4$	$1.4±0.86×10^4$	$1.03±0.68×10^4$
Total ATP (nM)[b]	0.015±0.005	0.029±0.004	0.011±0.002
Intracellular ATP (nM)[b]	0.007±0.003	0.000±0.004	0.001±0.002
HPC 22°C (CFU mL^{-1})[b]	23±24	4±2	4
HPC 36°C (CFU mL^{-1})[b]	16±16	4±2	1
Conductivity (µS cm^{-1}), 25°C[a]	468±101	625±4	272±25
pH[a]	6.63±0.18	7.41±0.04	7.5±0.05

[a]Data supplied by the water utility or measured in previous sampling campaigns.
[b]Data from present study.

based on the reaction with DPD in the presence of an excess of potassium iodide, and then titration as described above.

Fluorescent staining and flow cytometry (FCM) of water samples

Staining and FCM analysis was done as described previously [4,38]. In short, for a working solution, SYBR® Green I (SG) stock (Invitrogen AG, Basel, Switzerland) was diluted 100x in anhydrous dimethylsulfoxide (DMSO) and propidium iodide (PI; 30 mM) was mixed with the SYBR® Green I working solution to a final PI concentration of 0.6 mM. This working solution was stored at −20°C until use. From every water sample, 1 mL was stained with SGPI at 10 µL mL^{-1}. Before analysis, samples were incubated in the dark for 15 minutes. Prior to FCM analysis, the water samples were diluted with 0.22 µm filtered bottled water (Evian, France) to

10% v/v of the initial concentration. FCM was performed using a Partec CyFlow SL instrument (Partec GmbH, Münster, Germany), equipped with a blue 25 mW solid state laser emitting light at a fixed wavelength of 488 nm. Green fluorescence was collected at 520±10 nm, red fluorescence above 630 nm, and high angle sideward scatter (SSC) at 488 nm. The trigger was set on the green fluorescence channel and data were acquired on two-parameter density plots while no compensation was used for any of the measurements. The CyFlow SL instrument is equipped with volumetric counting hardware and has an experimentally determined quantification limit of 1000 cells mL^{-1} [4].

Adenosine tri-phosphate (ATP) analysis

Total ATP was determined using the BacTiter-Glo reagent (Promega Corporation, Madison, WI, USA) and a luminometer

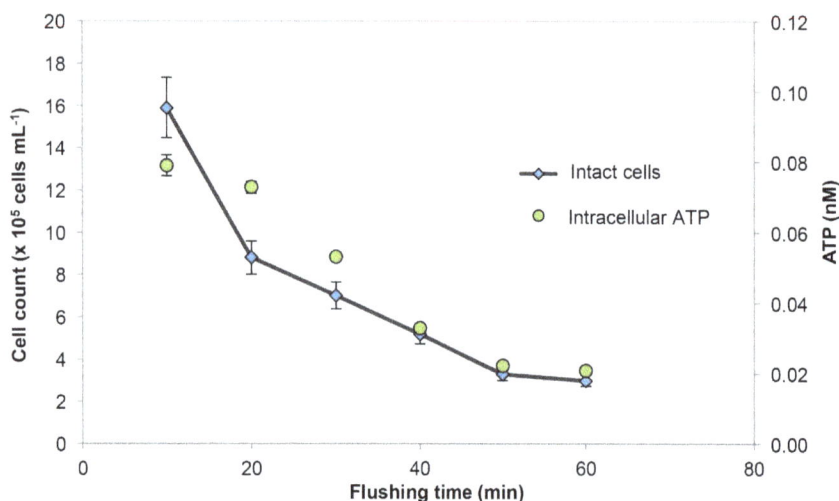

Figure 2. The impact of low velocity flushing on the water quality in a newly-opened fire hydrant. Intracellular adenosine tri-phosphate (ATP) data points were derived from duplicate measurements of extracellular and total ATP concentrations. FCM intact cells (after SYBR Green I and propidium iodide staining) were single measurements, with a relative standard deviation of 9% calculated from all data in the present study.

(Glomax, Turner Biosystems, Sunnyvale, CA, USA) as described elsewhere [25]. A water sample (500 µl) and the ATP reagent (50 µl) were warmed to 38°C simultaneously in separate sterile Eppendorf tubes. The sample and the reagent were then combined and then the luminescence was measured after 20 s reaction time at 38°C. The data were collected as relative light units (RLU) and converted to ATP (nM) by means of a calibration curve made with a known ATP standard (Promega). For extracellular ATP analysis, each sample was filtered through a 0.1 µm sterile syringe filter (Millex-GP, Millipore, Billerica, MA, USA), followed by analysis as described above. The intracellular ATP was calculated by subtracting the extracellular ATP from the total ATP for each individual sample. ATP was measured in duplicate for all samples.

Heterotrophic plate counts

To obtain heterotrophic plate counts (HPC), samples were serially diluted in sterile distilled water and then inoculated onto nutrient yeast agar plates using the spread plate technique. All plates were incubated in dark at 22°C or 36°C for 3 and 7 days, respectively. Results were expressed as colony forming units (CFU) per ml of water sample.

Statistical analysis

Statistical data evaluation was performed with the MS Excel Data Analysis tool (Descriptive statistics, Regression). The reproducibility for indirect/calculated data (e.g., intracellular ATP) was calculated by a propagation-of-uncertainty method. FCM data was not always measured in duplicate, due to practical constraints. In these cases, a 9% error (average coefficient of variation (CV) (n = 39)) was applied for representing FCM data. The residual chlorine concentration distribution box plot was created using on-line calculator on http://www.physics.csbsju.edu/stats/.

Results and Discussion

The importance of correct sampling

Sample collection during this study elucidated some of the problems specific for this network and highlighted the broader importance of correct sampling procedures. Fire hydrants were selected as sampling points to enable direct access to the distribution network and avoid potential household effects [39]. We opted for a low velocity water flow in combination with online monitoring to achieve comparable samples. In some cases, the water initially emerging from the fire hydrants were visibly turbid and/or discolored (data not shown). Turbid water is clearly unwanted and serves as a first visual confirmation of some form of system failure. In this regard, a recent study in the Netherlands has established an important link between suspended solids and microbial growth and biological instability [6]. Hence in some instances continuous low velocity flushing of up to 60 minutes was required before stable values for chemical and physical parameters as well as microbiological parameters were obtained (Figure 2; Table S1). The data in Figure 2 demonstrate clearly the need for a carefully planned sampling protocol when assessing full-scale systems. It should be noted that Figure 2 represents an example of some of the worst sampling points in the system. Data from other hydrants often showed less fluctuation during flushing (Figure S1). One potential problem during the sampling procedure is the re-suspension of sediments/particles and sloughing of biofilms from the pipes, causing artifacts in the measurements. In this respect, we specifically employed a low velocity (0.015–0.25 m s^{-1}) pre-sampling flushing procedure. The latter differs from extreme flushing applied for network cleaning, which is operated with high velocities of 1.5–1.8 m s^{-1} [40,41]. According to Antoun and co-workers [40] low-veocity flushing (below 0.3 m s^{-1}) does not cause any scouring actions. However, it should be considered that part of the samples, especially during the first minutes of the flushing, can cointain biofilm bacteria detached in a result of pre-flushing [35].

The concept of detecting instability: a single point in the distribution network

In the introduction we proposed the straightforward hypothesis that biological parameters would show an increase between the point of treatment and a point during distribution in case of biological instability (Figure 1). Before the relation between different parameters and the impact on the entire network are discussed in detail below, a single sampling point is compared to its source water as an example to illustrate the concept (Figure 3A). The point was selected on the basis of (1) hydraulic data linking it with a specific WTP, (2) its medial distance from WTP (neither too close and nor too far from the WTP) and, (3) the fact that all microbiological parameters (FCM, ATP and HPC) as well as residual chlorine measurements were performed on this sample. For the purpose of clarity, the data was normalized to the values of the treated water and expressed as the relative change (the raw data and standard deviations for the data in Figure 3A are shown in Figure S2). Evidently the data from Figure 3A supports the basic hypothesis. The microbial parameters such as intact cell concentration, ATP and colony forming units all show a considerable increase in their values. Simultaneously, only 12% (0.06 mg L^{-1}) of the initial residual chlorine concentration (0.5 mg L^{-1}) was left in the water sample. The data suggests that the residual chlorine in the network was not sufficient to inhibit microbial growth, concurring with earlier report from Prévost and colleagues [42] showing increased HPC, total direct and direct viable bacteria counts in a distribution network coinciding with chlorine depletion. Other studies also showed the presence of viable bacteria in water with chlorine concentration lower than 0.1 mg L^{-1} [23] and that residual chlorine levels below 0.07 mg L^{-1} allows bacterial growth [12]. Data of residual chlorine concentrations in the drinking water network is summarized in Figure S3. Evidently a considerable fraction of samples (18%) had residual chlorine concentrations below 0.1 mg L^{-1}.

Staining of bacteria with fluorescent dyes was previously suggested as a way to distinguish between viable and damaged bacteria in real water samples [43,44], and the application of this approach has been successfully demonstrated in laboratory scale chlorination studies [24,45]. One focus point of the present study was to determine whether FCM combined with viability staining can be used for a fast and meaningful assessment of viable bacteria in chlorinated drinking water systems. The same samples from Figure 3A, stained with SYBR Green and propidium iodide (SGPI), are shown as density plots obtained with FCM (Figure 3B). The theory behind the staining method and the interpretation of such data are discussed in detail elsewhere [20,24,38,43,46]. In the treatment plant sample, where the water was recently exposed to chlorine, 98% of all cells were measured as membrane compromised, seen by absence of events inside the gated area of the plot (Figure 3B). In the distribution network (DN) sample, a high concentration of intact cells appeared (Figure 3B). Since these intact cells were clearly not present in the influent, the plausible conclusion is that the bacterial growth occurred during distribution.

A

B

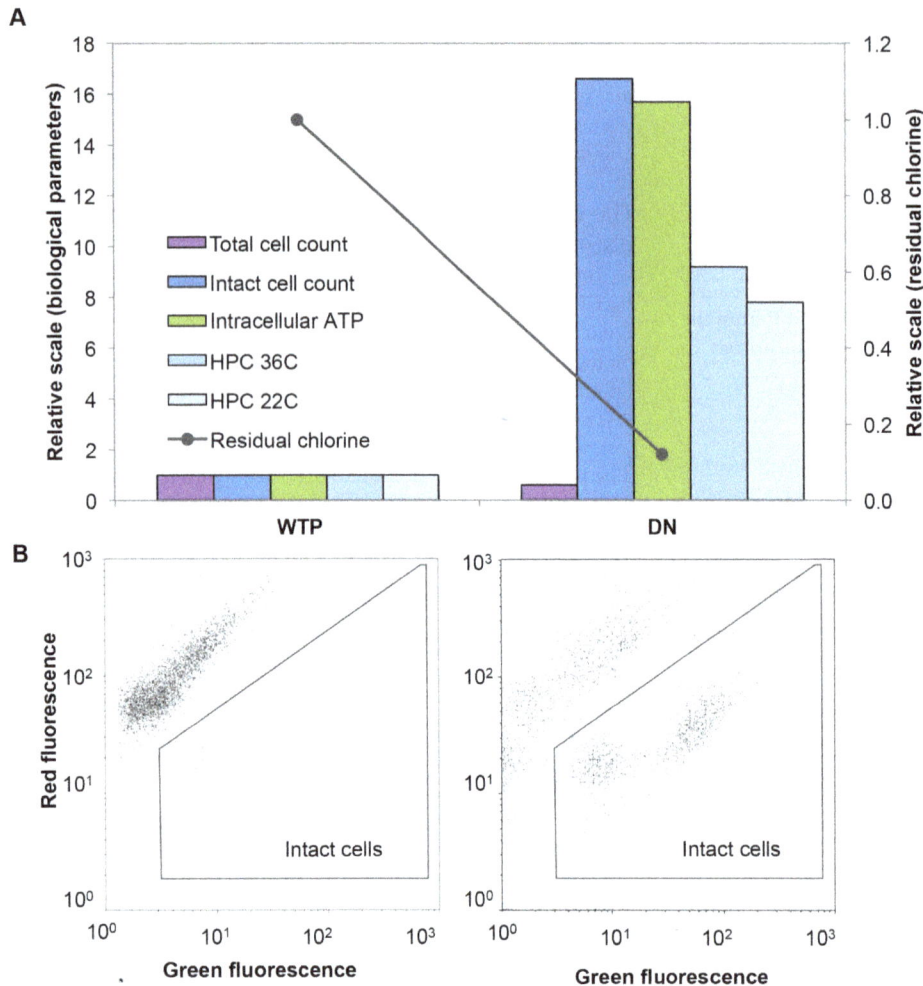

Figure 3. Changes in bacterial parameters between water treatment plant (WTP) and distribution network (DN) sampling points. (A) For comparison, all values at the WTP were set to 1, and values in the DN were expressed relative to their values at the WTP. The original raw data for these samples are shown in Figure S2. Data points are average values for duplicate FCM and ATP measurements and triplicate HPC measurements; (B) Flow cytometric density plots of samples stained with SYBR Green I and propidium iodide, showing the intact cell concentration at the plant and in the specific network point.

Detailed assessment of dynamic changes in a single point

High frequency monitoring of a single sampling point revealed temporal instability in the distribution network. We monitored the effluent of one treatment plant and one point in the network with 1-hour intervals during a day (ca. 21 h). The sampling was arranged in such a way that the network sampling started 15 hours after the treatment plant sampling, which corresponded with the estimated water residence time (WRT) for this location. Figure 4 displays the changes of intracellular ATP and intact cell concentrations in the network and the water treatment plant. Values for both parameters were low in the water samples from the treatment plant ($n = 19$): intracellular ATP varied from 0.0025 nM to 0.0096 nM (mean $= 0.0061 \pm 0.002$ nM) and the intact cell concentration amongst 19 samples varied from 7.5×10^3 to 6.3×10^4 cells mL^{-1} (mean $= 1.6 \times 10^4 \pm 1.2 \times 10^4$ cells mL^{-1} in average). In turn, the values from the distribution network point ($n = 23$) were significantly higher: intact cell concentrations ranged from 1.37×10^5 to 4.66×10^5 cells mL^{-1} (mean $= 2.5 \times 10^5 \pm 9.9 \times 10^4$ cells mL^{-1}), and the ATP concentrations from 0.021 to 0.063 nM (mean $= 0.038 \pm 0.012$ nM). More-

over, a distinct pattern was apparent in the distribution network data, with values peaking at about 05:00–07:00 and again at 12:00–13:00. During both these events, the intracellular ATP data followed a similar pattern as the intact cell concentration data, with a good overall correlation ($R^2 = 0.81$; $p < 0.005$). Although it is not evident exactly why the bacterial concentrations peaked at these specific time periods, a plausible explanation is a change in the flow velocity due to diurnal changes in water consumption by both industrial and domestic consumers. It was previously shown in laboratory scale experiments that increased flow velocity could lead to increased bacterial detachment from biofilms and a re-suspension of loose deposits, thus leading to an increase in suspended cell concentrations [32,33,47]. In addition, it is possible that lower water consumption overnight resulted in considerably reduced flow rates, and consequently a faster decay of chlorine and increased bacterial growth [42,48].

Detailed data sets of diurnal changes in the microbial quality of water mains, such as Figure 4, are particularly scarce in literature. Importantly, this clearly demonstrated temporal instability in the network for which the exact cause remains uncertain. Moreover, it

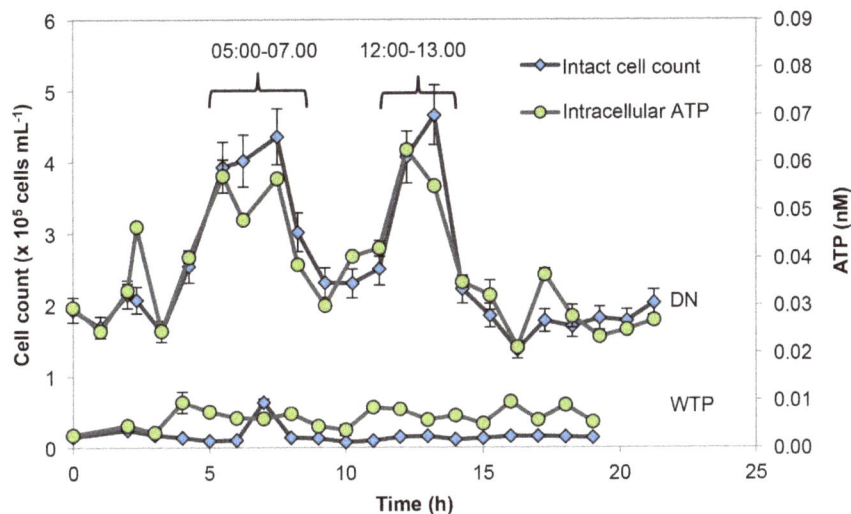

Figure 4. Diurnal changes in bacterial parameters of WTP and DN points. Intensive sampling of one WTP (n = 19) and one point in the DN (n = 23) during 21 hours reveals steady cell concentrations at the treatment plant but clear variations in the distribution network. Intracellular adenosine tri-phosphate (ATP) data points were derived from duplicate measurements of extracellular and total ATP concentrations. FCM intact cells (after SYBR Green I and propidium iodide staining) were single measurements, with a relative standard deviation of 9% calculated from all data in the present study.

shows that the absolute cell concentrations at any sampling point may be influenced by the time of sampling.

Instability data for the entire network

Full-scale distribution networks are complicated systems, not restricted to a single source or a straight distribution line [17]. The Riga distribution network is supplied with drinking water from several separate treatment plants (Table 1). One plant treats surface water from the Daugava River (WTP 1) and the others supply natural groundwater (WTP 3) and artificially recharged groundwater (WTP 2). Chlorination is applied as the final disinfection step at all plants, resulting in low concentrations of intact cells, intracellular ATP and cultivable bacteria in the effluents (Table 1). A large fraction of the active chorine is rapidly consumed due to relatively high levels of organic matter. Despite the fact that the purpose of chlorination and residual chlorine is to limit microbial growth during distribution, a considerable increase in the concentration of intact cells was detected throughout the distribution network. Figure 5A shows the range of intact cell concentrations arranged in ascending order. Treated water contained between 1.84×10^5–5.63×10^5 total cells mL^{-1} and between 9.7×10^3–2.13×10^4 intact cells mL^{-1} (hence 2–5% intact cells) depending on WTP. The data confirms effective final disinfection in all treatment plants. The total cell concentration values of the drinking water samples from the distribution network (n = 49) varied from 1.62×10^5 cells mL^{-1} to 1.07×10^6 cells mL^{-1} and the range of the intact cell concentration was from 5.28×10^3 cells mL^{-1} to 4.66×10^5 cells mL^{-1} (3–59% intact cells). Notably, 50% of all samples contained more than 1.06×10^5 intact cells mL^{-1} corresponding to an increase of at least one order of magnitude in those samples compared to effluent water, which clearly shows that bacterial growth in the distribution network was not an isolated occurrence. The observed increase in intact cell concentration is likely related to the presence of assimilable organic carbon (AOC) in the distributed water. While AOC was not measured in the present study, previous data for two of the treatment plants were high (in the range of 200 µg L^{-1}; Table 1), and nutrient availability in the water is generally regarded as a key

factor that promotes microbial growth [29,49]. It cannot be excluded that some variability in the data resulted from bacteria detached from biofilms or re-suspended from sediments during the fire hydrant sampling procedure. However, the potential adverse impact of this was minimized by the low velocity sampling protocol (see above), while the systematic increase in cell concentrations in the network clearly suggests the occurrence of biological instability rather than sampling artifacts. In contrast to these findings, several studies analyzing drinking water distribution systems without any additional residual disinfectants showed no (or only minute) changes in bacterial parameters during distribution [17,19,20]. These distributions systems rely on nutrient limitation to achieve biological stability, and while intact cell concentrations are often relatively high (ca. 1×10^5 cells mL^{-1}) [17,20], changes during distribution tend to be negligible.

To examine the spatial distribution of the growth/instability in the network, the data was divided into four broad categories based on the extent of growth (Figure 5A). These were visualized on the sampling map (Figure 5B). The sampling points with the lowest intact cell concentration (less than 5×10^4 cells mL^{-1}) are marked with green bullets. Yellow and orange colored bullets indicate higher concentrations, while the points with the highest values (over 2×10^5 cells mL^{-1}) are shown as red bullets. As could be expected, the map shows that the points with the lowest cell concentrations are mostly concentrated in areas close to the water treatment plants. Low intact cell concentrations in those areas could be ascribed to (1) disinfection during treatment and (2) growth inhibition from sufficient residual chlorine. Also the flow rate in the outgoing pipes closest to the treatment plants is high, which likely prevents water stagnation, sedimentation and cell adhesion on the pipe surface, and, consequently, biofilm formation and further bacterial growth. A different situation is observed in the distant areas from the water treatment plants and particularly in the so-called mixing zones, where the water from three different water treatment plants potentially mix. The map displays different color points spread in these zones without any visible order. The prevalence of the samples with higher cell concentrations there compared to the areas close to WTPs also corroborates the

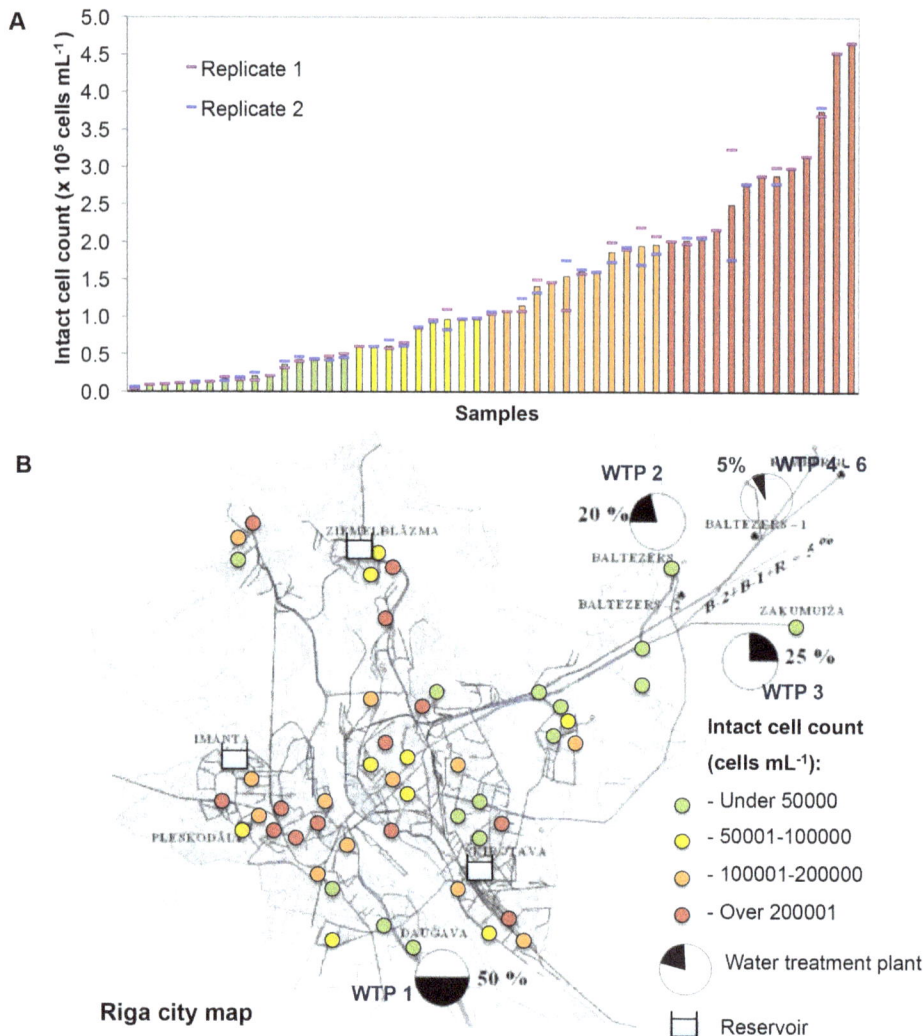

Figure 5. Intact cell concentrations of all samples measured from the distribution network (n = 49). (A) Intact cell concentrations arranged in ascending order and categorized into four main classes (colored bars) according to increasing concentrations. Data points are average values of duplicate measurements. Blue and purple stripes above and below data bars show the measured values. (B) Actual distribution of the classes of intact cells (colored circles) throughout the drinking water distribution network. WTP 1, WTP 2, WTP 3 represent location and productivity of the main water treatment plans supplying the city: WTP 1 operates using surface water, WTP 2 – artificially recharged ground water, WTP 3 – natural groundwater. WTP 4 – 6 indicates on other three pump stations with less significance for the city water supply.

argument that increasing distance and water residence time could lead to chlorine decay with concomitant oxidation of dissolved organic matter; both these events would favor bacterial growth. Moreover, mixing zones are potential hot-spots for bacterial growth, as one water might well contain the nutrients that are growth limiting in the other.

The uneven spatial distribution of the samples with different intact cell concentrations is noteworthy, highlighted for example by the three points in upper-left corner of the map. Based on the long distance from the WTPs, high intact cell concentrations were expected, but the samples taken from the hydrants located in this small area rather show variability (respectively 1.82×10^4, 1.87×10^5, 2.51×10^5 intact cells mL^{-1}). Such different intact cell concentrations could be due to several reasons: the time the samples were taken, which is linked to water consumption and the potential impact of which is shown in Figure 4, the condition of the pipes in this specific area (unknown), the way water flows from the treatment plant, and/or the relative proximity of these sample

points to one of the reservoirs (Figure 5B), etc. Other authors showed a decrease in AOC [13] and ATP [19] in the some distal points of the distribution networks. Decrease of AOC concentration was explained by its consumption by bacteria within the network. These authors argued that an insufficient amount of nutrients led to starvation and a decrease in bacterial parameters at the end of the pipelines. However, it is an unlikely reason in the present study, because this phenomenon seems more occasional than systematic.

The combined data demonstrates clearly biological instability throughout the distribution network. However, despite the relative simplicity of the concept (Figure 1; Figure 3A), a complex interplay of chemical, physical and biological parameters and hydraulic conditions should be taken into account for characterization of each particular case of instability.

Comparison of different microbiological parameters

FCM and ATP data showed clear correlations, but these data did not correlate well with conventional HPC data. A total amount of 49 different samples was measured in duplicate with ATP (total and extracellular) and FCM (total and intact cell concentration) analyses, while 38 of those samples were further analyzed with HPC. The significant linear correlation ($R^2 = 0.77$; n = 49) between intracellular ATP and FCM intact cell concentration is shown in Figure 6A. This corroborates previous studies that showed good results comparing total ATP with total cell concentration [19,50] and intracellular ATP with intact cell count as well [20,25]. The strong correlation is encouraging, since FCM and ATP analysis are independent viability parameters – integrity of the cell membrane (FCM) and cellular energy (ATP). A correlation between these parameters during disinfection is not necessarily a given fact. The membrane integrity based PI staining method implies that PI positive cells are damaged and thus considered as inactive, yet extreme examples where living cells became permeable for propidium iodide have been described [46]. In turn, Nocker and co-workers [51] showed that after UV-C exposure cells became inactivated, while their membranes remained essentially intact. Discrepancies between intracellular ATP and intact cell concentration can also result from cell morphology, bacterial species and physiological state, that was discussed in detail previously [25]. The results provided by FCM provide information on single cell level, whereas during ATP analyses the values are evaluated per volume. Hence, intracellular ATP-per-cell was calculated for characterization of biomass activity. In the present study intracellular ATP-per-cell ranges from zero (no cell-bound ATP observed) to 5.92×10^{-10} nM cell^{-1} ($= 3 \times 10^{-17}$ g cell^{-1}) with the average value of 1.68×10^{-10} nM cell^{-1} ($= 8.52 \times 10^{-18}$ g cell^{-1}) (stdev $= 9.58 \times 10^{-11}$ nM cell^{-1}, n = 49). The result is in the same range as ATP-per-cell values obtained from various water sources, which were analyzed with the same methods [20,25]. This suggests that bacterial activity (ATP values) in the intact cells was not affected by any remaining chlorine residuals, and that membrane damage (SGPI values) was in this case reflective of viability in the sample. The good correlation between these two independent parameters is an optimistic prospect for applying these methods for chlorinated water analyses in future studies.

The conventional HPC results were compared with the FCM intact cell concentration values. A weak correlation ($R^2 = 0.18$, n = 38) was observed between HPC (at 22°C) and intact cell concentrations (Figure 6B), similar to reports in previous studies [38,50]. It could be explained by the often described phenomenon, that less than 1% of drinking water bacteria are cultivable on conventional agar plates [25,50,52]. In addition, Mezule and co-workers [53] demonstrated evidence of the presence of so called viable-but-not-cultivable (VNBC) bacterial state, in both drinking water and biofilms for the network investigated here, thus indicating further limitations in the HPC method. Since intracellular ATP showed a good correlation with intact cell concentration, but intact cell count correlated weakly with HPC, it was expected that intracellular ATP and HPC would not correlate well (Figure 6C; $R^2 = 0.11$, n = 38). Various studies were performed to compare ATP and HPC parameters from water samples, but good correlations were never observed e.g., $R^2 = 0.20$ [19], $R^2 = 0.36$ [22] and $R^2 = 0.31$ [50]. Our results combined with those from previous studies cast further doubts on the value of using the HPC method for general microbiological drinking water quality control. In our opinion, the clear correlation between two methodologically independent viability parameters (intracellular ATP and FCM intact cell counts), and the absence of any correlations with

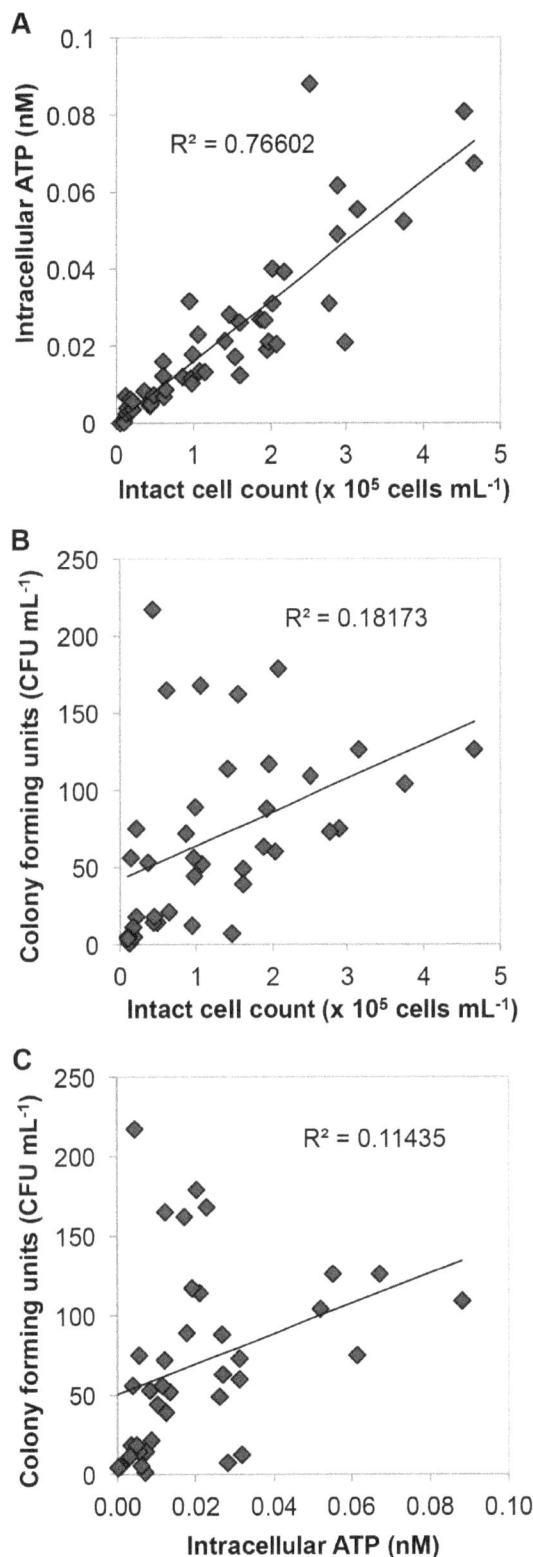

Figure 6. Comparison of the various microbiological parameters. Clear correlations were observed between intact cells and intracellular ATP (n = 49) (A), but no obvious correlations between these two parameters and heterotrophic plate counts at 22°C (n = 38) (B) (C).

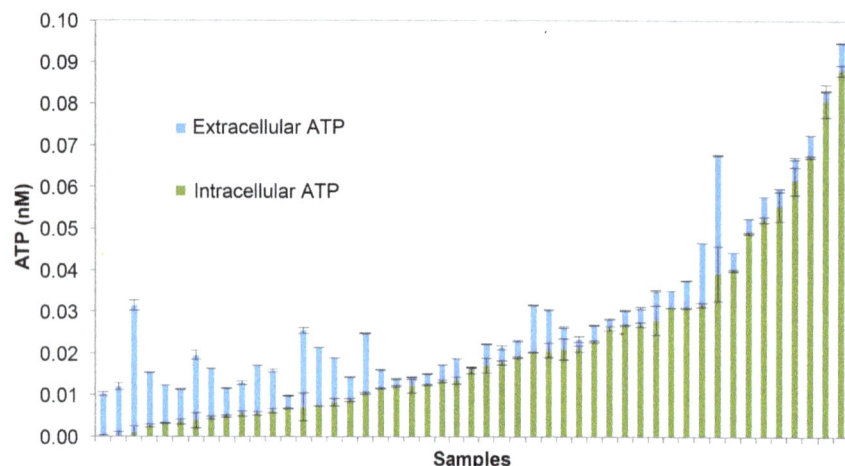

Figure 7. Distribution of intracellular and extracellular ATP in the water samples. In general, higher concentrations and relative percentages of extracellular ATP were measured in samples that exhibited lower intracellular ATP concentrations (n = 49).

two different HPC methods, renders the former methods more meaningful for assessing and understanding biological instability, particularly in chlorinated environments.

Importance of measuring extracellular ATP

Arguments for and against the concept and importance of measuring extracellular ATP have been made [4,19,20,25,54,55]. To understand this better, we arranged our data according to increasing intracellular ATP concentrations, after which the measured extracellular ATP values were added to each corresponding sample (Figure 7). It is evident that extracellular ATP constitutes a considerable fraction of the total ATP amount in some samples – varying from 3% up to 100% – with an average contribution of 36% (n = 49). Moreover, 33% of the samples contain more that 50% of extracellular ATP. This data supports other studies, where analyses showed high extracellular ATP in drinking water samples from the distribution networks [20,25]. Interestingly, the highest extracellular ATP ratio is mostly observed in the samples with relatively low intracellular ATP, in this case samples with close proximity to the treatment plant. In the case of chlorinated water, this could potentially be explained by the oxidative effect of chlorine on bacterial cells. Previous studies have shown extensive damage to bacterial membranes during chlorination [24,45], after which a release of extracellular ATP from the damaged bacteria can occur. This membrane damage was also clearly detected in the present study (e.g., Figure 3B). Although, there is lack of detailed data considering the release of extracellular ATP in water samples affected by chlorination, strong evidence of ATP release during oxidation was presented in previous studies [4,20]. Both these works showed a significant decrease in cell concentrations and intracellular ATP after ozonation, whereas extracellular ATP comprised 83–100% of the total ATP. Moreover, Figure 7 shows that samples with increased intracellular ATP concentrations, which we linked to bacterial growth during distribution, often had considerably less extracellular ATP in relation to total ATP. This could be due to the fact that extracellular ATP can be biodegraded by bacteria or extracellular enzymes in the network [54,56–58]. However, it cannot be excluded that a decrease in extracellular ATP during distribution occurs due to oxidation by residual chlorine present in the network.

Conclusions

- An investigation of a full-scale chlorinated drinking water distribution network with various microbiological methods clearly demonstrated both spatial and temporal biological instability in the network.

- Fluorescent staining with SGPI in combination with ATP measurements provided reliable and descriptive information about bacterial density and viability in chlorinated drinking water samples.

- A good correlation was observed between intracellular ATP and intact cell counts ($R^2 = 0.77$), whereas HPC showed poor correlations with both parameters ($R^2 = 0.18$ with intact cell concentration and $R^2 = 0.11$ with intracellular ATP).

- Extracellular ATP constituted on average 36% of total ATP in the present study, which confirms the necessity of extracellular ATP subtraction from total ATP measurements during chlorinated drinking water analyses.

- Overall the results raise questions with respect to the offset between increased biological safety gained from disinfection opposed to increased risk from instability (uncontrolled bacterial growth). While an improvement of the chlorination procedure could be a solution, the data suggests looking beyond only disinfection for achieving biological stability of drinking water.

Supporting Information

Figure S1   Additional examples of hydrant flushing. Changes in intact cell concentration and intracellular ATP during flushing in 6 newly-opened fire hydrants. Intact cell concentration values are shown as solid lines with blue markers, whereas intracellular ATP results displayed as single green bullets.

Figure S2   Actual data for Figure 3A. Changes in various bacterial parameters between one water treatment plant and a randomly selected point in the distribution network (actual values for Figure 3A).

Figure S3   Residual chlorine concentration in the distri-

bution network. 50% of residual chlorine concentration in the network was between 0.12 (first quartile) and 0.23 (third quartile) mg mL^{-1}, with a mean value of 0.17 mg mL^{-1} (n = 27). The whiskers indicate on minimum and maximum values, whereas bullets show outliers of the population.

Table S1 Physical and chemical parameters of water measured on-line during low velocity flushing of newly-opened fire hydrant. Some measurements were omitted during the first 20 minutes of flushing due to the high fluctuation in measuring tools readings.

References

1. Bartram J, Cotruvo J, Exner M, Fricker C, Glasmacher A (2003) Heterotrophic plate count and drinking-water safety: The significance of HPCs for water quality and human health. World Health Organization. 244 p.
2. Boe-Hansen R, Albrechtsen H-J, Arvin E, Jørgensen C (2002) Bulk water phase and biofilm growth in drinking water at low nutrient conditions. Water Res 36: 4477–4486. doi:10.1016/S0043-1354(02)00191-4.
3. Hammes F, Berger C, Köster O, Egli T (2010) Assessing biological stability of drinking water without disinfectant residuals in a full-scale water supply system. J Water Supply Res Technol 59: 31. doi:10.2166/aqua.2010.052.
4. Hammes F, Berney M, Wang Y, Vital M, Köster O, et al. (2008) Flow-cytometric total bacterial cell counts as a descriptive microbiological parameter for drinking water treatment processes. Water Res 42: 269–277. doi:10.1016/j.watres.2007.07.009.
5. Juhna T, Birzniece D, Larsson S, Zulenkovs D, Sharipo A, et al. (2007) Detection of Escherichia coli in biofilms from pipe samples and coupons in drinking water distribution networks. Appl Environ Microbiol 73: 7456–7464. doi:10.1128/AEM.00845-07.
6. Liu G, Lut MC, Verberk JQJC, Van Dijk JC (2013) A comparison of additional treatment processes to limit particle accumulation and microbial growth during drinking water distribution. Water Res 47: 2719–2728. doi:10.1016/j.watres.2013.02.035.
7. Van der Kooij D (2000) Biological stability: A multidimensional quality aspect of treated water. In: Belkin S, editor. Environmental Challenges. Springer Netherlands. pp. 25–34.
8. Vital M, Stucki D, Egli T, Hammes F (2010) Evaluating the growth potential of pathogenic bacteria in water. Appl Environ Microbiol 76: 6477–6484. doi:10.1128/AEM.00794-10.
9. Vital M, Füchslin HP, Hammes F, Egli T (2007) Growth of Vibrio cholerae O1 Ogawa Eltor in freshwater. Microbiology 153: 1993–2001. doi:10.1099/mic.0.2006/005173-0.
10. Vital M, Hammes F, Egli T (2008) Escherichia coli O157 can grow in natural freshwater at low carbon concentrations. Environ Microbiol 10: 2387–2396. doi:10.1111/j.1462-2920.2008.01664.x.
11. LeChevallier MW, Schulz W, Lee RG (1991) Bacterial nutrients in drinking water. Appl Environ Microbiol 57: 857–862.
12. Niquette P, Servais P, Savoir R (2001) Bacterial dynamics in the drinking water distribution system of Brussels. Water Res 35: 675–682.
13. Liu W, Wu H, Wang Z, Ong SL, Hu JY, et al. (2002) Investigation of assimilable organic carbon (AOC) and bacterial regrowth in drinking water distribution system. Water Res 36: 891–898.
14. Eichler S, Christen R, Höltje C, Westphal P, Bötel J, et al. (2006) Composition and dynamics of bacterial communities of a drinking water supply system as assessed by RNA- and DNA-based 16S rRNA gene fingerprinting. Appl Environ Microbiol 72: 1858–1872. doi:10.1128/AEM.72.3.1858-1872.2006.
15. Hong P-Y, Hwang C, Ling F, Andersen GL, LeChevallier MW, et al. (2010) Pyrosequencing analysis of bacterial biofilm communities in water meters of a drinking water distribution system. Appl Environ Microbiol 76: 5631–5635. doi:10.1128/AEM.00281-10.
16. Pinto AJ, Xi C, Raskin L (2012) Bacterial community structure in the drinking water microbiome Is governed by filtration processes. Environ Sci Technol 46: 8851–8859. doi:10.1021/es302042t.
17. Lautenschlager K, Hwang C, Liu W-T, Boon N, Köster O, et al. (2013) A microbiology-based multi-parametric approach towards assessing biological stability in drinking water distribution networks. Water Res 47: 3015–3025. doi:10.1016/j.watres.2013.03.002.
18. Rittmann BE, Snoeyink VL (1984) Achieving biologically stable drinking water. J - Am Water Works Assoc 76: 106–114.
19. Van der Wielen PWJJ, van der Kooij D (2010) Effect of water composition, distance and season on the adenosine triphosphate concentration in unchlorinated drinking water in the Netherlands. Water Res 44: 4860–4867. doi:10.1016/j.watres.2010.07.016.
20. Vital M, Dignum M, Magic-Knezev A, Ross P, Rietveld L, et al. (2012) Flow cytometry and adenosine tri-phosphate analysis: Alternative possibilities to evaluate major bacteriological changes in drinking water treatment and distribution systems. Water Res 46: 4665–4676. doi:10.1016/j.watres.2012.06.010.
21. LeChevallier MW, Au K-K (2004) Water treatment and pathogen control: Process efficiency in achieving safe drinking-water. IWA Publishing. 136 p.
22. Delahaye E, Welté B, Levi Y, Leblon G, Montiel A (2003) An ATP-based method for monitoring the microbiological drinking water quality in a distribution network. Water Res 37: 3689–3696. doi:10.1016/S0043-1354(03)00288-4.
23. Francisque A, Rodriguez MJ, Miranda-Moreno LF, Sadiq R, Proulx F (2009) Modeling of heterotrophic bacteria counts in a water distribution system. Water Res 43: 1075–1087. doi:10.1016/j.watres.2008.11.030.
24. Ramseier MK, von Gunten U, Freihofer P, Hammes F (2011) Kinetics of membrane damage to high (HNA) and low (LNA) nucleic acid bacterial clusters in drinking water by ozone, chlorine, chlorine dioxide, monochloramine, ferrate(VI), and permanganate. Water Res 45: 1490–1500. doi:10.1016/j.watres.2010.11.016.
25. Hammes F, Goldschmidt F, Vital M, Wang Y, Egli T (2010) Measurement and interpretation of microbial adenosine tri-phosphate (ATP) in aquatic environments. Water Res 44: 3915–3923. doi:10.1016/j.watres.2010.04.015.
26. LeChevallier MW, Welch NJ, Smith DB (1996) Full-scale studies of factors related to coliform regrowth in drinking water. Appl Environ Microbiol 62: 2201–2211.
27. Polanska M, Huysman K, van Keer C (2005) Investigation of assimilable organic carbon (AOC) in flemish drinking water. Water Res 39: 2259–2266. doi:10.1016/j.watres.2005.04.015.
28. Ramseier MK, Peter A, Traber J, von Gunten U (2011) Formation of assimilable organic carbon during oxidation of natural waters with ozone, chlorine dioxide, chlorine, permanganate, and ferrate. Water Res 45: 2002–2010. doi:10.1016/j.watres.2010.12.002.
29. Van der Kooij D (1990) Assimilable organic carbon (AOC) in drinking water. In: McFeters GA, editor. Drinking Water Microbiology. New York, NY: Springer New York. pp. 57–87.
30. Weinrich LA, Jjemba PK, Giraldo E, LeChevallier MW (2010) Implications of organic carbon in the deterioration of water quality in reclaimed water distribution systems. Water Res 44: 5367–5375. doi:10.1016/j.watres.2010.06.035.
31. Jjemba P (2010) Guidance document on the microbiological quality and Biostability of reclaimed water following storage and distribution. WateReuse Research Foundation.
32. Lehtola MJ, Laxander M, Miettinen IT, Hirvonen A, Vartiainen T, et al. (2006) The effects of changing water flow velocity on the formation of biofilms and water quality in pilot distribution system consisting of copper or polyethylene pipes. Water Res 40: 2151–2160. doi:10.1016/j.watres.2006.04.010.
33. Manuel CM, Nunes OC, Melo LF (2007) Dynamics of drinking water biofilm in flow/non-flow conditions. Water Res 41: 551–562. doi:10.1016/j.watres.2006.11.007.
34. Bucheli-Witschel M, Kötzsch S, Darr S, Widler R, Egli T (2012) A new method to assess the influence of migration from polymeric materials on the biostability of drinking water. Water Res 46: 4246–4260. doi:10.1016/j.watres.2012.05.008.
35. Douterelo I, Husband S, Boxall JB (2014) The bacteriological composition of biomass recovered by flushing an operational drinking water distribution system. Water Res 54: 100–114. doi:10.1016/j.watres.2014.01.049.
36. Rossman LA (2000) EPANET 2 Users manual. National Risk Management Research Laboratory. U.S. Environmental Protection Agency, Cincinatti, Ohio.
37. Rubulis J, Dejus S, Meksa R (2011) Online measurement usage for predicting water age from tracer tests to validate a hydraulic model American Society of Civil Engineers. pp. 1488–1497. doi:10.1061/41203(425)133.
38. Berney M, Vital M, Hülshoff I, Weilenmann H-U, Egli T, et al. (2008) Rapid, cultivation-independent assessment of microbial viability in drinking water. Water Res 42: 4010–4018. doi:10.1016/j.watres.2008.07.017.
39. Lautenschlager K, Boon N, Wang Y, Egli T, Hammes F (2010) Overnight stagnation of drinking water in household taps induces microbial growth and changes in community composition. Water Res 44: 4868–4877. doi:10.1016/j.watres.2010.07.032.

Acknowledgments

The authors thank Stefan Kötzsch for critical input, Arturs Briedis, Edgars Grundbergs and Kaspars Neilands for assistance in the sampling campaigns, and sampling/information support from Rigas Udens Ltd.

Author Contributions

Conceived and designed the experiments: FH MV JR TJ. Performed the experiments: FH MV JR. Analyzed the data: FH MV JR TJ AN. Contributed reagents/materials/analysis tools: FH TJ. Wrote the paper: FH MV JR TJ AN.

40. Antoun EN, Dyksen JE, Hiltebrand DJ (1999) Unidirectional flushing: A powerful tool. J - Am Water Works Assoc 91: 62–71.

41. Friedman M, Kirmeyer GJ, Antoun E (2002) Developing and implementing a distribution system flushing program. J - Am Water Works Assoc 94: 48–56.

42. Prévost M, Rompré A, Coallier J, Servais P, Laurent P, et al. (1998) Suspended bacterial biomass and activity in full-scale drinking water distribution systems: Impact of water treatment. Water Res 32: 1393–1406. doi:10.1016/S0043-1354(97)00388-6.

43. Berney M, Hammes F, Bosshard F, Weilenmann H-U, Egli T (2007) Assessment and interpretation of bacterial viability by using the LIVE/DEAD BacLight Kit in combination with flow cytometry. Appl Environ Microbiol 73: 3283–3290. doi:10.1128/AEM.02750-06.

44. Grégori G, Citterio S, Ghiani A, Labra M, Sgorbati S, et al. (2001) Resolution of viable and membrane-compromised bacteria in freshwater and marine waters based on analytical flow cytometry and nucleic acid double staining. Appl Environ Microbiol 67: 4662–4670. doi:10.1128/AEM.67.10.4662-4670.2001.

45. Lisle JT, Pyle BH, McFeters GA (1999) The use of multiple indices of physiological activity to access viability in chlorine disinfected Escherichia coli O157:H7. Lett Appl Microbiol 29: 42–47. doi:10.1046/j.1365-2672.1999.00572.x.

46. Shi L, Günther S, Hübschmann T, Wick LY, Harms H, et al. (2007) Limits of propidium iodide as a cell viability indicator for environmental bacteria. Cytometry A 71A: 592–598. doi:10.1002/cyto.a.20402.

47. Tsai Y-P (2005) Impact of flow velocity on the dynamic behaviour of biofilm bacteria. Biofouling 21: 267–277. doi:10.1080/08927010500398633.

48. Srinivasan S, Harrington GW, Xagoraraki I, Goel R (2008) Factors affecting bulk to total bacteria ratio in drinking water distribution systems. Water Res 42: 3393–3404. doi:10.1016/j.watres.2008.04.025.

49. Van der Kooij D (1992) Assimilable organic carbon as an indicator of bacterial regrowth. J - Am Water Works Assoc 84: 57–65.

50. Siebel E, Wang Y, Egli T, Hammes F (2008) Correlations between total cell concentration, total adenosine tri-phosphate concentration and heterotrophic plate counts during microbial monitoring of drinking water. Drink Water Eng Sci Discuss 1: 71–86. doi:10.5194/dwesd-1-71-2008.

51. Nocker A, Sossa KE, Camper AK (2007) Molecular monitoring of disinfection efficacy using propidium monoazide in combination with quantitative PCR. J Microbiol Methods 70: 252–260. doi:10.1016/j.mimet.2007.04.014.

52. Van der Kooij D, Vrouwenvelder JS, Veenendaal HR (2003) Elucidation and control of biofilm formation processes in water treatment and distribution using the Unified Biofilm Approach. Water Sci Technol J Int Assoc Water Pollut Res 47: 83–90.

53. Mezule L, Larsson S, Juhna T (2013) Application of DVC-FISH method in tracking Escherichia coli in drinking water distribution networks. Drink Water Eng Sci 6: 25–31. doi:10.5194/dwes-6-25-2013.

54. Cowan DA, Casanueva A (2007) Stability of ATP in Antarctic mineral soils. Polar Biol 30: 1599–1603. doi:10.1007/s00300-007-0324-9.

55. Venkateswaran K, Hattori N, La Duc MT, Kern R (2003) ATP as a biomarker of viable microorganisms in clean-room facilities. J Microbiol Methods 52: 367–377. doi:10.1016/S0167-7012(02)00192-6.

56. Azam F, Hodson RE (1977) Dissolved ATP in the sea and its utilisation by marine bacteria. Nature 267: 696–698. doi:10.1038/267696a0.

57. Mempin R, Tran H, Chen C, Gong H, Ho KK, et al. (2013) Release of extracellular ATP by bacteria during growth. BMC Microbiol 13: 301. doi:10.1186/1471-2180-13-301.

58. Riemann B (1979) The occurrence and ecological importance of dissolved ATP in fresh water. Freshw Biol 9: 481–490. doi:10.1111/j.1365-2427.1979.tb01532.x.

6

Macro-Invertebrate Decline in Surface Water Polluted with Imidacloprid: A Rebuttal and Some New Analyses

Martina G. Vijver[1]*, Paul J. van den Brink[2,3]

1 Institute of Environmental Sciences (CML), Leiden University, Leiden, The Netherlands, **2** Alterra, Wageningen University and Research centre, Wageningen, The Netherlands, **3** Wageningen University, Wageningen University and Research centre, Wageningen, The Netherlands

Abstract

Imidacloprid, the largest selling insecticide in the world, has received particular attention from scientists, policymakers and industries due to its potential toxicity to bees and aquatic organisms. The decline of aquatic macro-invertebrates due to imidacloprid concentrations in the Dutch surface waters was hypothesised in a recent paper by Van Dijk, Van Staalduinen and Van der Sluijs (PLOS ONE, May 2013). Although we do not disagree with imidacloprid's inherent toxicity to aquatic organisms, we have fundamental concerns regarding the way the data were analysed and interpreted. Here, we demonstrate that the underlying toxicity of imidacloprid in the field situation cannot be understood except in the context of other co-occurring pesticides. Although we agree with Van Dijk and co-workers that effects of imidacloprid can emerge between 13 and 67 ng/L we use a different line of evidence. We present an alternative approach to link imidacloprid concentrations and biological data. We analysed the national set of chemical monitoring data of the year 2009 to estimate the relative contribution of imidacloprid compared to other pesticides in relation to environmental quality target and chronic ecotoxicity threshold exceedances. Moreover, we assessed the relative impact of imidacloprid on the pesticide-induced potential affected fractions of the aquatic communities. We conclude that by choosing to test a starting hypothesis using insufficient data on chemistry and biology that are difficult to link, and by ignoring potential collinear effects of other pesticides present in Dutch surface waters Van Dijk and co-workers do not provide direct evidence that reduced taxon richness and abundance of macroinvertebrates can be attributed to the presence of imidacloprid only. Using a different line of evidence we expect ecological effects of imidacloprid at some of the exposure profiles measured in 2009 in the surface waters of the Netherlands.

Editor: Christopher Joseph Salice, Texas Tech University, United States of America

Funding: These authors have no support or funding to report.

Competing Interests: For transparency reasons, we mentioning the following: PvdB's chair was cofunded between 2008 and 2011 by the following pesticide producers, Bayer, which produces imidacloprid and Syngenta. We feel that this cofunding provides no compete of interest since we don't claim that imidacloprid poses less risks or toxicity than stated in the Van Dijk et al. (2013) as in the current paper we only criticized their methodology. This current work has not been funded. Sponsors thus had no role in study design, data collection and analysis, decision to publish, or preparation of the manuscript.

* E-mail: vijver@cml.leidenuniv.nl

Introduction

The Netherlands is one of the world's foremost agricultural producers, with 2/3 of the total land mass devoted to agriculture or horticulture. Land use is highly intensive in terms of output per hectare or head of livestock [2]. To achieve such high outputs a vast range of agricultural chemicals are used, including fertilizers, veterinary drugs, pesticides and biocides. Different pesticides are used depending on the crop that is grown on the land. There are several routes that pesticides may enter surface waters. Pesticides may be washed into ditches and rivers by rainfall; surface waters can be contaminated by direct overspray or via runoff and leaching from agricultural fields [3]. Emission to surface waters (and thus pesticide residue concentrations) is dictated by many factors such as distance of the crop from the ditch and the mode of application, weather conditions and so on.

Neonicotinoids are the first new class of insecticides to be introduced in the last 50 years. The neonicotinoid imidacloprid is currently one of the most widely used insecticides in the world [4]. Recently, imidacloprid has received much negative attention: The use of certain neonicotoids has been restricted in some countries due to evidence of an unacceptably high risk of toxicity to bees, but this restriction was not in effect in the Netherlands at the time of writing this paper. On April 29, 2013, the European Union passed a two-year ban on the use of three neonicotinoids: European law restricts the use of imidacloprid, clothianidin, and thiamethoxam on flowering plants for two years unless compelling evidence comes out that proves that the use of the chemicals is environmentally safe [5]. This ban is partially, restricted to some applications in specific crops and likely covers 15% of the total use of the three neonicotinoids in the Netherlands [6]. Temporary suspensions had previously been enacted in countries such as France, Germany, Switzerland and Italy. In March 2013, a review of 200 studies on neonicotinoids was published by Mineau and Palmer [7], calling for a ban on neonicotinoid use as seed treatments because of their toxicity to birds, aquatic invertebrates, and other wildlife. The EPA – USA is now re-evaluating the safety of neonicotinoids.

Van Dijk and co-workers [1] aimed to assess the specific relationship between imidacloprid residues in Dutch surface waters, and the abundance of non-target macro-invertebrate taxa.

As also stated by the authors, finding a statistical relationship between those two datasets does not necessarily reflect causality, because there could be other factors (e.g. other pesticide residues, other local habitat factors) which drive observed patterns of abundance. We have some fundamental criticisms on the way the data were analysed and the results were interpreted, and we feel that this can be challenged by existing data. Therefore as a response to the paper of Van Dijk et al [1], and by using additional data, we explore their two key assumptions: 1) residues of pesticides other than imidacloprid, that are collinear with imidacloprid exposure either do not exist or have negligible effects on macroinvertebrate abundance and 2) that imidacloprid concentrations can be extrapolated successfully over 160 days and at a 1 km^2 spatial scale.

Materials and methods

Data collection and treatment

Data on pesticides concentrations in surface water in the Netherlands were obtained from the Dutch Pesticides Atlas. [8]. This is an online tool from which Dutch monitoring data can be collected and processed into a graphic format. Here, data of all pesticide active ingredients and metabolites (n = 634) collected in 2009 were used, since this data set is contiguous with the data used by Van Dijk et al. [1]. Only one year was selected since it can be expected that the correlations between pesticide occurrences will be year-specific, so this correlation should also be assessed for each year specifically. The 2009 dataset covered 302111 individual measurement records of which 19693 measurements exceeded the reporting limit (LOR). The measurements were performed on 4816 samples obtained from 723 different locations. The sample by pesticide dataset is characterised by missing values (90% of entries) and below LOR values (9% of all entries). This is a result of the fact that every water manager has his own suite of pesticides that is sampled, measured and evaluated. The selection of this suite of pesticides is based on the crops and land-use in their region. This selection of pesticides to be monitored improves the efficiency of the monitoring efforts of the individual water managers but yields a data set that has missing values and with many < LOR values when the data of multiple water managers are combined into one. To obtain frequency distributions of the imidacloprid concentrations, data from 2010 and 2011 have also been used.

Environmental quality standards (EQS) of all pesticides were as follows: for imidacloprid the annual average-EQS value (AA-EQS) is 0.067 μg/L (database value set 2-6-2010), and the maximum allowable concentration (MAC-EQS) is 0.2 μg/L (database value set 2-6-2010) as specified by the European Water Framework Directive. In addition, in the Netherlands, the maximum permissible concentration (MPC) of 0.013 μg/L is an important additional criterion (database value set 8-10-2008).

For all samples in which a pesticide could not be detected or quantified, the database substitutes a value of lower than the LOR. The values of reporting limits vary across samples (unique location x time). In our calculations these measurements below LOR are set as zero. We chose to do so, as choosing any other value below LOR would be arbitrary. Moreover, if not taking zero as a value, any other chosen value will result in relatively high toxicity at intensively measured surface waters even if the pesticides are not applied in that area since all measurements results in a lowest value possible of being below the LOR. These types of assumptions are inherent when working with data sets based on monitoring efforts.

Collinearity of imidacloprid concentrations with concentrations of other pesticides

Collinearity refers to a linear relationship between two explanatory variables, meaning that one can be linearly predicted from the others with a non-trivial degree of accuracy. Collinearity was determined on the data set of 2009 measurements restricted to all samples with at least one measurement above the LOR. The reduced data contained measured values for 18% of the samples, of which 8% of the total were measurements above the LOR. In order to assess the correlation between the concentrations of different pesticides we needed a sample by pesticide matrix with as little missing values as possible. From this gappy database, the largest closed data sets were extracted using Principal Component Analysis [9]. For this, measured values in the database were coded as one and missing data by zero. After running the PCA, the species-by-substance matrix was sorted, based on the scores of the substances and samples on the first principal component. Using this approach, it was possible to extract closed data sets by extracting groups of samples with the same score on the first principal component. Four data sets could be extracted that contained more than 100 samples in which the same pesticides were measured. One data set did not include imidacloprid and was not taken into account. The remaining three matrices contained 114, 108 and 191 samples, 27, 51 and 54 pesticides, with 11, 11 and 13% of the measurements above the LOR for data set 1, 2 and 3, respectively. All sampling points of data set 1 were within the provinces of Utrecht and Gelderland while all sampling points of data set 2 and 3 were located in the province of South Holland.

The log((1000 * conc) +1) transformed pesticide concentration values were analysed with Principal Component Analysis (PCA) using the Canoco5 computer programme [10], (see Zafar et al. [11] for the rationale of the transformation]. The pesticide data were centred and standardised for each pesticide. The graphical pictures based on orthogonal coordinate systems describe optimal variance in a dataset. Points that are clustered near each other have a strong correlation. PCA [9] transforms data to a new coordinate system such that the greatest variance by any projection of the data comes to lie on the first coordinate (called the first principal component), the second greatest variance on the second coordinate [12].

Calculating multi substance PAF

The potential affected fraction (PAF) is a common way to express ecotoxicological risks [13]. Following this approach, measured pesticides concentrations were translated into PAF using the species sensitivity distribution (SSD) approach. Toxicity data for each pesticide was obtained from De Zwart [14], and based on acute median effect concentrations (EC50) as derived in the laboratory (database eTox, RIVM as described in [14]). The eTox database consists mainly of data entries from the ECOTOX EPA database. The SSD for imidacloprid is given in Figure 1, and includes 41 different species from 7 different taxonomic groups. Underlying data including references are given in Table S1 of the Supplementary Information. The full database used for the multi substance PAF (msPAF) calculations contained data of 496 different pesticides with 75 different modes of action. To quantify the ecological impacts due to imidacloprid concentrations amongst all other pesticide concentrations as measured in the surface waters, the msPAF was calculated. Firstly, all concentrations of individual pesticides measured over one month per location were aggregated using the maximum measured value. Secondly individual pesticide concentrations were compared to the toxicity data resulting in the PAF. Thirdly, pesticides were grouped based on their mode of action. The PAF's of the pesticides with a similar mode-of-action were added using a concentration addition equation. In this equation, each substance concentration is divided by its effect concentration, ECxa, i.e., the concentration of a that represents a standard effect expressed as EC50 for endpoint x. This gives: Emix (Cmix) = (Ca/ECxa) + (Cb/ECxb) + …… In which Emix(Cmix) is the summed ratio of the mixture components at the exposure concentration of each chemical (Cx). Fourthly, the different pesticides groups with dissimilar mode-of-action were added using a response addition equation. In response addition, the toxicity of the substances in the mixture can be predicted from the product of the fractional effects of the mixture components. This gives Emix (Cmix) = 1 − ((1 − E(Ca)) * (1 − E(Cb)) * …… In which Emix(Cmix) is the calculated effect of the mixture, Ca the exposure concentration of substance a, and E(ca) the effect of substance a at concentration Ca.

Both models for mixture toxicity are described in Hewlett and Plackett [15]. Chemicals with an unknown mode-of-action were treated according to a unique mode-of-action. As a result an msPAF value per month per monitoring location was derived. In this study we reported the maximum msPAF of the year 2009. The quantification of the relative contribution of imidacloprid on the total chemical pressure as expressed by msPAF was based on acute toxicity data as insufficient chronic toxicity data were available in the literature.

Pairwise combinations of samples taken within 1 km and 160 days

Datasets on imidacloprid concentrations and abundances of macroinvertebrates were linked to each other by Van Dijk and co-workers [1] by using the criteria ≤1 km distance and ≤ 160 days

of time difference. We performed pairwise comparisons of imidacloprid measurements to determine whether imidacloprid concentrations at sites that meet these criteria, matched successfully. Therefore, all imidacloprid measurements were extracted from the 2009 data set. All sampling sites were first ranked on their x coordinate and the difference in distance with the next sample was assessed (using Pythagoras theorem). All site combinations which yielded a difference less than 1 km were extracted. The same procedure was performed using a ranking based on the y-coordinate. The site combinations from both queries were combined. This procedure is not exhaustive since two sites that are not ranked next to each other can also be closer to 1 km from each other, but is likely to find most combinations. The imidacloprid concentrations of all samples taken at the paired sites were compared to each other when the samples were taken within 160 days. The result of the comparison were categorised into: 1) two measurements below the LOR, 2) one measurement below and one above the LOR (0% matching), 3) two measurements above the LOR, of which the number of sample pairs that matched 100% (based on one decimal) was also noted. The analysis resulted in 37 pairs of sites containing a total of 260 observations and 584 concentration measurement pairs being evaluated.

Time series of imidacloprid exposure

For each sampling site it was determined how often imidacloprid samples were analysed. For 34 sampling sites 10 or more samples were analysed, of which imidacloprid was not detected in any of the samples at 14 sites (41%), and in less than half of the samples at 28 sites (82%). The concentration dynamics of the

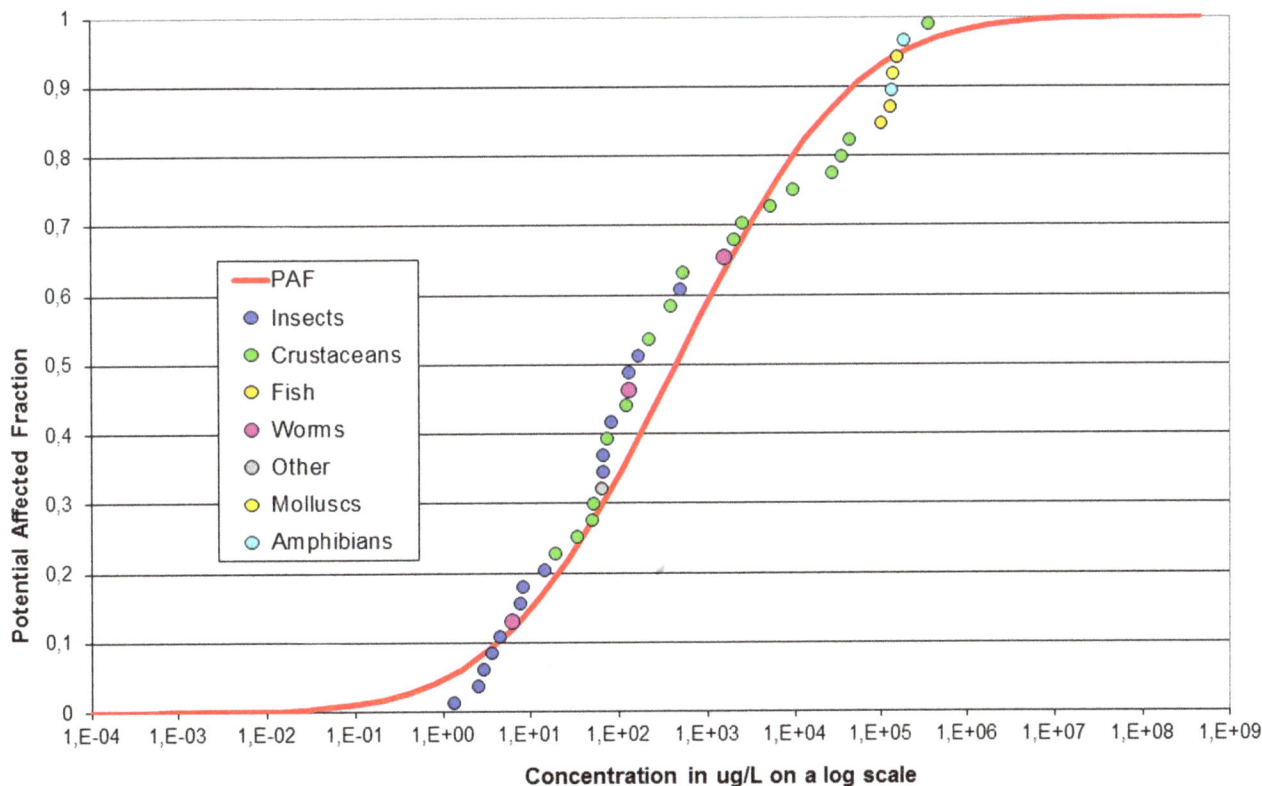

Figure 1. The Species Sensitivity Distribution of imidacloprid based on acute toxicity data. The data consist of 7 different taxomonic groups and 41 species. EPA database downloaded at Oct 23th 2013.

remaining 18% of the sites were plotted to evaluate whether chronic concentrations of imidacloprid may be expected.

Cumulative frequency of maximum imidacloprid concentrations

The measured maximum concentration of each site was compared with threshold concentrations based on the findings of Roessink et al. [16], i.e. the chronic EC10 of the mayfly species *Caenis horaria* and *Cloeon dipterum* ($\approx$0.03 µg/L) and the different environmental quality standards. In order to remove the within-site sample dependency, for each sampling site the maximum imidacloprid concentration was extracted. The analysis resulted in 225 negative measurements (below the LOR) and 226 positive measurements (above the LOR).

MPC exceedances of imidacloprid compared to other pesticides

Since only for a restricted number of pesticide AA-EQS and MAC-EQS values have been set in the WFD, we used the (Dutch) MPC standard to compare exceedance frequencies between pesticides. For this comparison, both the magnitude of exceedance as well as the frequency of exceedance was incorporated. Firstly, the exceedance of the MPC of an individual pesticide concentration was derived per measuring location. Secondly, the degree of standard exceedance was weighted according to the following classes: 0 ($\leq$MPC); 1 (> MPC and $\leq$ 2 x MPC), 2 (> 2 x MPC and $\leq$ 5 x MPC) and 5 (> 5 x MPC exceedance). Thirdly, the exceedance classes were summed over all measuring locations per year. Fourthly, pesticides were ranked on the basis of the weighted number of monitoring sites at which the MPC for the compound was exceeded, i.e. corrected for the number of monitoring sites by taking the percentage of sites that show an exceedance of the MPC. Compounds monitored at fewer than ten sites were ignored.

Results and Discussion

For many locations pesticide concentrations have been found to exceed the MPC in 2009 (see Fig. 2). Figure 2 shows that throughout the entire country more than one pesticide exceeds their respective quality standard, so this exceedance is not a common regionally problem. The maximum amount of pesticides exceeding their MPC in one sample is 35. From this it can be concluded that a single pesticide is not likely to drive solely the macro-invertebrate quality, rather all pesticides exceeding the quality standards should be considered.

Collinearity of imidacloprid concentrations with other pesticides

Figure 3A clearly shows that imidacloprid exposure is highly correlated with all chemicals placed on the right, lower side of the diagram, like carbendazim and DEET and to a lesser extend with the large group of chemicals which have a high loading with the horizontal axis, which explains almost double the amount of variance compared to the vertical axis. The results of the second data set (Fig. 3B) show that imidacloprid is placed in the centre of a large group of pesticides placed in the middle of the diagram, since it was measured only in a few samples (7% of the total). The results of the third data set shows a high occurrence of imidacloprid above the LOR (78% of all samples), with concentrations strongly collinear with those that have a high loading on the horizontal axis which explains almost triple the amount of variance of the vertical one (Fig. 3C). The results of the first and third data set show that the contribution of imidacloprid toxicity in surface waters cannot

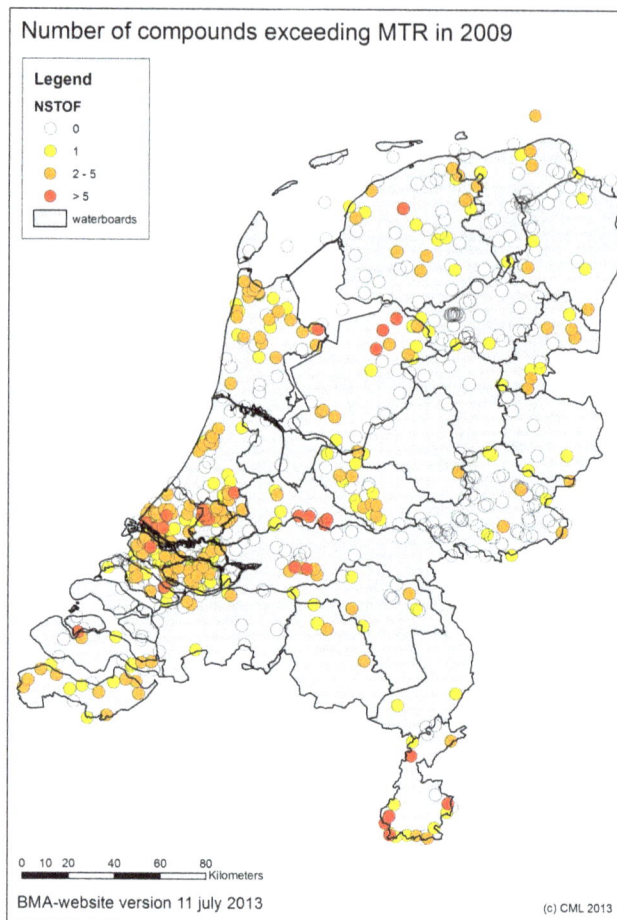

Figure 2. Number of pesticides exceeding the MPC in 2009. All monitoring locations in the Dutch surface waters with one (yellow); two till five pesticides concentrations (orange); and > five different pesticides (red) exceeding their MPC-values are depicted. Locations were measurements were performed but no exceedances were found are depicted in white.

easily be separated from the toxicity arising from other co-occurring pesticides, or indeed any other co-occurring chemical or physical stressing agent.

The correlations derived from the PCA-plots (Fig. 3) can also be explained from the fact that the active ingredient imidacloprid currently has several authorizations in 38 different products (database ctgb.nl [17], accessed 21-5-2013). The professional use ranges from the use in crops grown in glasshouses such as all different vegetables and in open systems for different bulbs of flowers, potatoes and sugarbeets. Imidacloprid is also registered for use in fruit trees including apple and pear trees. Generally, more than one pesticide is used to protect a specific crop from pest attack. Thus, depending on the land use type, imidacloprid is invariably emitted to surface waters in combination with other pesticides that are authorized to be used on those crops.

Imidacloprid contribution in the msPAF

The potentially affected fraction of the aquatic species by the measured pesticides is higher than 5% in 11 locations (reflecting 1.2 % of all monitoring sites) in the Netherlands in the year 2009. The maximum level that we determined based on the msPAF was 23% in the province of South-Holland. Imidacloprid contributed

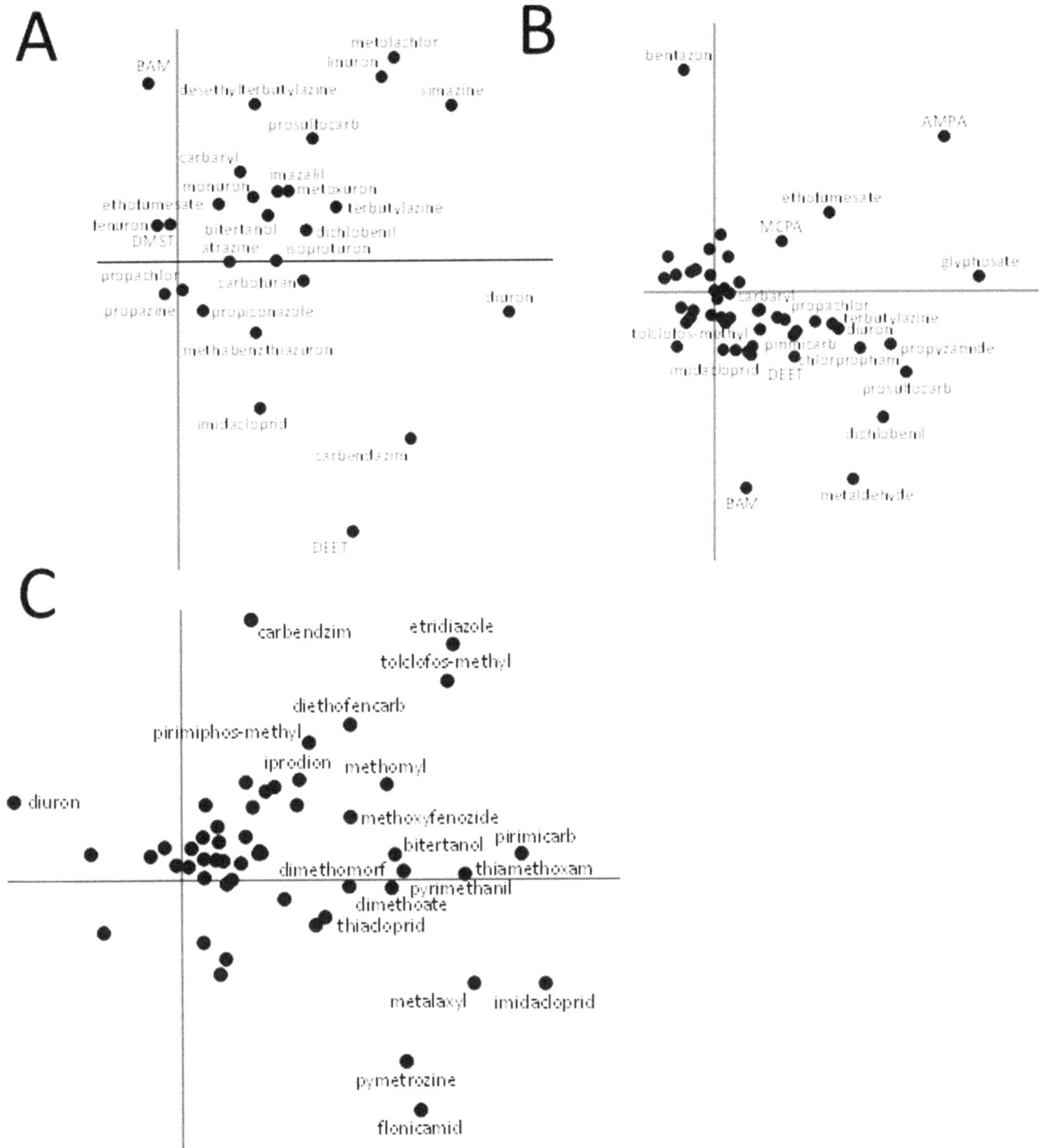

Figure 3. Results of the PCA analysis on data set 1 (A), 2 (B) and 3 (C). The PCA diagram of data set 1 displays 51% (33% on horizontal axis and 18% on vertical one) of the variation in chemical concentrations between the sites while 34% is displayed for data set 2 (21% on horizontal axis and 13% on vertical one) and 38% for data set 3 (28% on horizontal axis and 10% on vertical one).

in 8 out of 11 cases to this potential risk (Table 1). The relative contribution compared to other pesticides as measured at the same location at the same sampling time is rather modest and varied with a maximum of 21% at one location. Note that this calculation was based on acute toxicity data only, so likely is an underestimation of the potential risks that include both acute as chronic effects. From Table 1, it can be deduced that depending on

location, the contribution of specific individual active ingredients differs.

Pairwise combinations of samples

Imidacloprid measurements performed within a time window of 160 days which were taken at sites closer than 1 km from each other were compared. By this pairwise analysis we investigate if

Table 1. Contribution of imidacloprid to the msPAF at locations where msPAF > 5%.

x-coordinate	y-coordinate	Province	Total msPAF of measured pesticides (%)	Relative contribution of imidacloprid to the total msPAF of measured pesticides (%)
N 51 46 39.9	E 4 16 36.7	South Holland	22.53	0
N 52 1 29.6	E 4 30 24.7	South Holland	13.85	7.59
N 51 43 11.8	E 4 16 1.5	Zealand	12.48	0.002
N 51 52 33.5	E 4 10 26.2	South Holland	10.11	0
N 51 46 38.6	E 4 33 19.3	South Holland	9.91	0.009
N 51 45 0.4	E 4 25 46.2	South Holland	9.44	0
N 52 31 7.8	E 4 40 36.5	North Holland	9.25	0.014
N 51 57 10.2	E 4 15 8.8	South Holland	7.09	21.04
N 51 50 20	E 4 35 16.7	South Holland	6.61	0.001
N 51 21 52	E 4 2 10.1	Zealand	6.36	11.49
N 52 41 42.6	E 6 53 54.9	Drenthe	5.64	0.011

selected pairs of imidacloprid concentrations match with each other, and subsequently can be used to accurately link biological effect data and imidacloprid concentrations. Table 2 shows that in 39% of the comparisons there was no match in the presence of imidacloprid above the LOR, while only in 23% of the cases imidacloprid was present above the LOR in both samples. The remaining 38% of comparisons showed two measurements below the LOR. So when imidacloprid is found in at least one of the samples there is a large probability (62%) of not finding imidacloprid in the other site, which hampers the extrapolation of imidacloprid over a time window of 160 day and over a distance of 1 km (Table 2). We, therefore, conclude that the criteria used by Van Dijk et al. [1] to link chemical with biological observations result in a large probability (46%) of linking a site where imidacloprid was detected with a site, where the biological sample was taken, where actually no imidacloprid could be detected. The alternative, i.e. the first measurement being below the LOR and the second one above also has a relatively high probability (34%) (Table 2). Especially in a water-rich country such as the Netherlands, that has more than 350.000 km of ditch systems [18], it should be noted that sampling locations taken within 1 km, not necessarily have a hydrological connection with each other.

Imidacloprid dynamics

The concentration dynamics of imidacloprid (reflecting the concentrations of imidacloprid at the sampling locations with 10 or more samples taken in 2009 and with detection above the LOR in

at least 50% of those samples) are shown in Figure 4. In all but two (Fig. 4B and 4C) of these sampling sites the 28d, EC10 values for *C. horaria* and *C. dipterum* are exceeded for a period longer than 28 days, so at these sites chronic effects of imidacloprid exposure on mayflies can be expected. Also all standards are exceeded for some time in most of the sampling sites, with Fig. 4G showing the largest exeedence for a site near Boskoop in the province of South Holland. It should be noted that these 7 sites only constitute a small percentage (18%) of the total number of sites with 10 or more observations, so likely these exposure patterns represent the worst-cases of the exposure patterns at sites with 10 or more observations. Since we don't know whether there is a bias to measure imidacloprid more intensively at sites where exposure is expected we cannot extrapolate this to the whole population of sites.

Maximum concentrations of imidacloprid

Figure 5 shows the cumulative frequency of the all concentration measurements on the maximum level of imidacloprid for the years 2009, 2010 and 2011. The below LOR measurements are indicated at the 0.001 µg/L level and constituted 50, 53 and 55% of the maximum concentrations in 2009, 2010 and 2011, respectively. The results in Figure 4 show that peak concentrations of imidacloprid in the Dutch surface waters often exceeds the chronic effect concentrations of mayfly as determined in the chronic single species studies by Roessink et al. [16], as well as the three standards. In 2011 the MPC, 28d, EC10, AA-EQS and

Table 2. Result of the comparison of imidacloprid concentrations in samples taken in 2009 at sampling sites closer than 1 km and within 160 days.

Category	# sample pairs	% of total comparisons	% when 1st observation is above LOR	% when 1st observation is below LOR
Two below LOR	217	38		66
One below and above LOR	223	39	46	34
Two above LOR	134	23	54	
100% matching measurements	10	1.7		

LOR = analytical reporting limit.

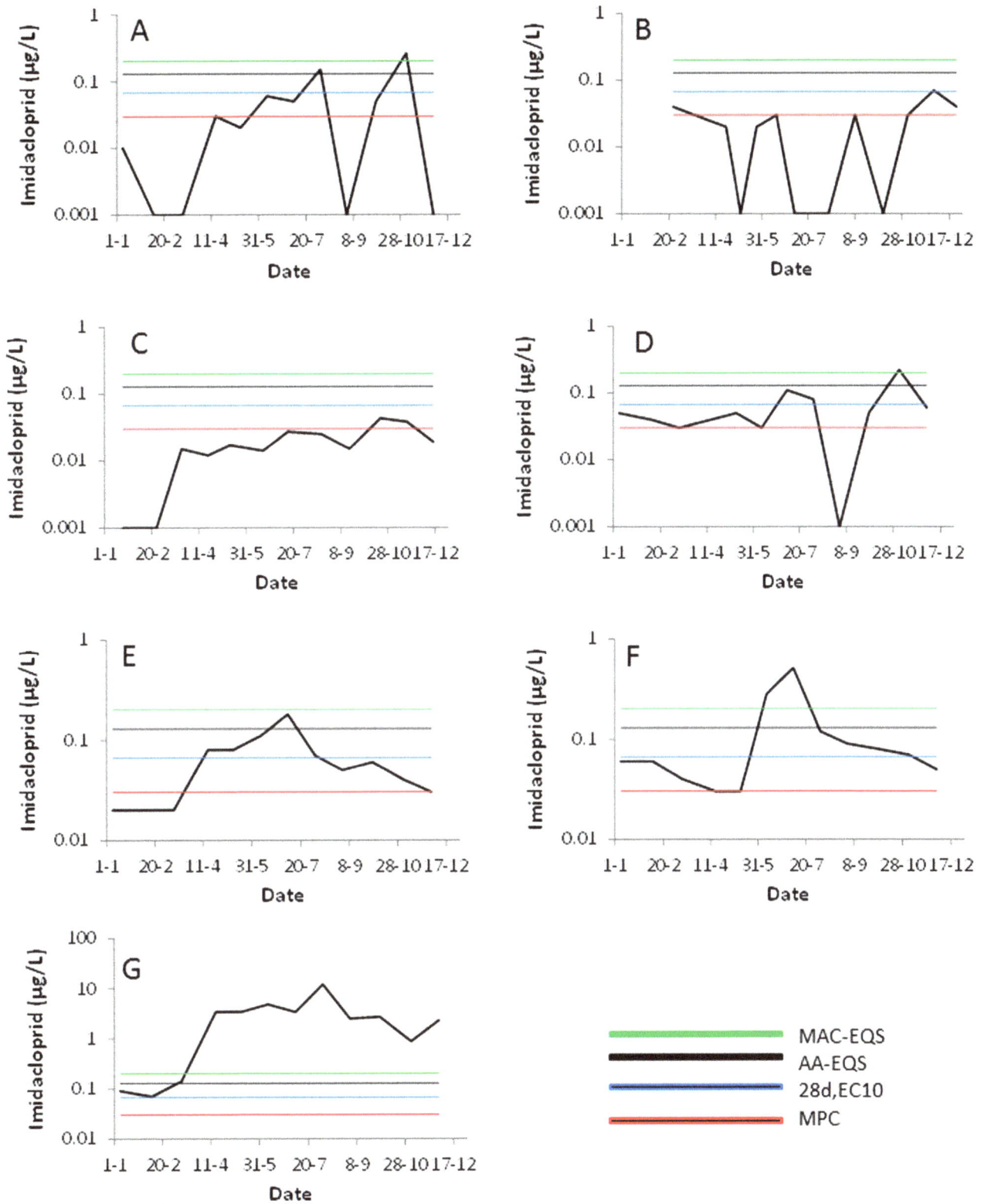

Figure 4. Concentration dynamics at the selected sampling sites (see text for procedure). The sampling sites 4A through 4G have X,Y coordinates of 108313,456412, 105888,455853, 103707,455196, 105927,453177, 170370,518957, 106781,503700 and 105079,453602, respectively. The horizontal lines denotes the MAC-EQS, the AA-EQS, the 28d, EC10 value for the mayflies C. horaria and C. dipterum (Roessink et al., 2013) and the MPC (top to bottom).

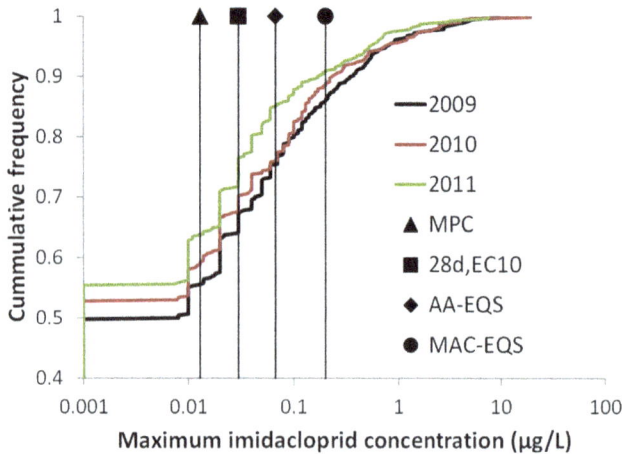

Figure 5. The cumulative frequency of the maximum imidacloprid concentrations of the sampling sites in 2009, 2010 and 2011, together with three standards and the 28d, EC10 of Cloeon dipterum and Caenis horaria.

MAC-EQS threshold values are exceeded by 36, 28, 15 and 9% of the maximum concentrations at the sampling sites, respectively. Since the Hazardous Concentration 5% based on 96h,EC10 values of 0.083 µg/L [16] corresponds more or less with the AA-EQS, acute effects of imidacloprid exposure cannot be excluded at a relatively large proportion of the sites ($\approx$15%). The maximum concentration is of course not a good predictor for the time weighted average concentration of 28d which should ideally be compared with the chronic threshold value of 0.03 µg/L. Still, when combining the results of the time-series (Fig. 3) and the exceedance of this threshold value by the maximum concentrations (Fig. 4) chronic effects of imidacloprid on insects like mayflies may be expected at a vast proportion of sites, with 28% being the most conservative estimate and 5% being the best guess. This 5% is calculated by multiplying the 28% chance of exceeding the threshold value by the maximum concentration and 15% chance of having above LOR measurements at more than 50% of the samples taken at a particular site where imidacloprid is measured at least 10 times. The comparison of the standards with the ecotoxicological threshold value for mayflies also suggests that the

MAC-EQS and AA-EQS are not fully protective for acute and chronic effects on insect taxa, respectively.

Exceedances of environmental quality standards

As stated in the Van Dijk et al [1] paper, in 2009 imidacloprid frequently exceeds quality standards for surface waters: 111 and 62 times for the AA-EQS and the MAC-EQS respectively [8,18]. In addition to the probability of exceeding a standard, also the magnitude of exceedance is important since it is likely that at higher magnitudes the ecological effects are more severe and maybe even last longer. Table 3 shows the compounds that exceeded the MPC most frequently in 2009, ranked according to degree of exceedance.

Imidacloprid was predicted to have a relatively large impact on the ecosystems compared to other pesticides, and gained third place in the Top 10 pesticides violating the environmental quality standards in respect to frequency and magnitude of exceedance. The number of measurements is high, as is also the number of locations from which the samples are taken. This means that monitoring is quite intensive for this compound, and surely covers many different surface waters belonging to different water managers and covering the geographical distribution of the different water types in the Netherlands. Although less intensively measured – a factor 5 to 10 – Table 3 also shows that other pesticides exceed the MPC more often. Thus although imidacloprid poses a significant ecological risk to surface waters in the Netherlands, it is not the only potential cause of degradation in macroinvertebrate abundance, as many other pesticides mentioned in Table 3 also exceed the MPC frequently (and in cases by orders of magnitude) and thus undoubtedly contribute to overall stress regime. It is a common flaw in ecological studies to selectively interpret individual causal agents within stressor regimes as the sole cause of observed phenomena, leading to erroneous conclusions.

Conclusion

Imidacloprid is one of several pesticides that can be detected in surface waters draining agricultural areas at levels frequently exceeding environmental quality standards. Despite this, we show here that key assumptions made by Van Dijk et al. [1] specifically relating to imidacloprid toxicity are not supported by observational data and, therefore, their assessment is unsuitable to determine threshold levels of effects. Specifically, the validity of

Table 3. Top10 pesticides exceeding the MPC in the Netherlands in the year 2009.

Pesticides name	No. of monitoring sites	% Exceedance	No. of measurements	% Exceedance
Captan	38	47	194	13
desethyl-terbuthylazin	63	37	299	10
Imidacloprid	451	44	2133	28
Triflumuron	24	21	142	4
Dicofol	24	17	142	3
Omethoaat	31	16	169	3
Foraat	51	14	313	2
Captafol	15	27	29	14
Fipronil	69	12	230	7
Pyraclostrobin	66	17	341	7

No. = number. The ranking of pesticides is based on frequency and magnitude of exceedances.

two assumptions: 1) that imidacloprid levels are not correlated with toxic levels of other pesticides residues and 2) that chemical exposure data can be extrapolated over a 1 km distance and 160 day time window are here shown to be highly questionable. The ecological status of field sites can be attributed to a complex suite of stressors resulting from a range of anthropogenic practices in the highly managed landscape of the Netherlands, of which pesticides are just one factor, and imidacloprid only one of many pesticides being applied, albeit an important one in terms of ecological risks. We therefore propose that any risk assessment should base the ecological threshold values not solely on field observations but also largely rely on the results of controlled experiments, since these types of experiments allow a full control of separating the imidacloprid stress from other stressors.

Supporting Information

Table S1 Acute toxicity values of imidacloprid (source eTox database, EPA database downloaded Oct 23th

References

1. Van Dijk TC, Van Staalduinen MA, Van der Sluijs JP (2013) Macro-invertebrate decline in surface water polluted with imidacloprid. PLOS ONE 8 (5) e62374.
2. Vijver MG, De Snoo GR (2012) Overview of the state-of-art of Dutch surface waters in the Netherlands considering pesticides. (chapter (9) In: The impact of pesticides, M. Jokanovic (ed.) AcedemyPublish.org, WY, USA. ISBN: 978-0-9835850-9-1.
3. Vijver MG, Van 't Zelfde M, Tamis WLM, Musters CJM, De Snoo GR (2008) Spatial and Temporal Analysis of Pesticides Concentrations in Surface Water: Pesticides Atlas. J Environ Sci Health Part B 43: 665–674.
4. Yamamoto I (1999) "Nicotine to Nicotinoids: 1962 to 1997". In Yamamoto, Izuru; Casida John. Nicotinoid Insecticides and the Nicotinic Acetylcholine Receptor. Tokyo: Springer-Verlag. pp. 3–27 ISBN: 443170213X.
5. McDonald-Gibson C (29 April 2013). *The Independent*. Retrieved 1 May 2013.
6. Van Vliet J, Vlaar LNC, Leendertse PC (2013) Toepassingen, gebruik en verbod van drie neonicotinoïden in de Nederlandse land en tuinbouw. CLM 825- 2013. Available: www.clm.nl. Accessed 2013 May 5.
7. Mineau P, Palmer C (2013) The impact of the nation's most widely used insecticides on birds. Neonicotinoid Insecticides and Birds. American Bird Conservancy. Available: http://www.abcbirds.org/abcprograms/policy/toxins/neonic_final.pdf.
8. Dutch pesticides atlas website. Available: http://www.bestrijdingsmiddelenatlas.nl, version 2.0. Institute of Environmental Sciences (CML) at Leiden University and Waterdienst of the Dutch Ministry of Infrastructure and Environment. Accessed 2013 Oct 23.

2013). Legend: Species selected for the toxicity test were given with their scientific name and with their species group. Toxicity data were given as log10 effect concentrations at which 50% of the organisms showed adverse effects. The scientific papers from which those data are collected are given.

Acknowledgments

The authors thank Donald Baird for his critical comments and language suggestions. We thank Dick de Zwart for providing the eTox database. All pesticides measurements compared to the different EU and MPC quality standards can be found and freely downloaded at www. bestrijdingsmiddelenatlas.nl [8].

Author Contributions

Conceived and designed the experiments: MGV PJB. Performed the experiments: MGV PJB. Analyzed the data: MGV PJB. Contributed reagents/materials/analysis tools: MGV PJB. Wrote the paper: MGV PJB.

9. Jolliffe IT (2002) Principal Component Analysis, Series: Springer Series in Statistics, 2nd ed., Springer, NY. ISBN 978-0-387-95442-4.
10. Ter Braak CJF, Šmilauer P (2012) Canoco reference manual and user's guide: software for ordination, version 5.0. Microcomputer Power, Ithaca, USA, 496 pp.
11. Zafar MI, Belgers JDM, Van Wijngaarden RPA, Matser A, Van den Brink PJ (2012) Ecological impacts of time-variable exposure regimes to the fungicide azoxystrobin on freshwater communities in outdoor microcosms. Ecotoxicol 21:1024–1038.
12. Van Wijngaarden RPA, Van den Brink PJ, Oude Voshaar JH, Leeuwangh P (1995) Ordination techniques for analyzing response of biological communities to toxic stress in experimental ecosystems. Ecotoxicol 4: 61–77.
13. Posthuma L, Suter GW II, Traas TP (eds) (2002) Species Sensitivity Distributions in Ecotoxicology. Lewis Publishers, Boca Raton, FL, USA.
14. De Zwart D (2005) Ecological effects of pesticide use in the Netherlands: Modeled and observed effects in the field ditch. Integrated Environmental Assessment and Management 1:123–134.
15. Hewlett PS, Plackett RL (1959) A unified theory for quantal responses to mixtures of drugs: non-interactive action. Biometrics 15:591–610.
16. Roessink I, Merga LB, Zweers HJ, Van den Brink PJ (2013) The neonicotenoid imidacloprid shows high chronic toxicity to mayfly nymphs. Environ Toxicol Chem 32: 1096 – 1100.
17. Statistics Netherlands. Available: http://www.statline.cbs.nl. Accessed 2013 May 21.
18. De Snoo GR, Vijver MG (eds) (2012) Bestrijdingsmiddelen en waterkwaliteit. Universiteit Leiden, 180 pp., ISBN: 978-90-5191-170-1.

7

Economic and Health Impacts Associated with a *Salmonella* Typhimurium Drinking Water Outbreak—Alamosa, CO, 2008

Elizabeth Ailes[1,2]**, Philip Budge**[2]**, Manjunath Shankar**[2]**, Sarah Collier**[1,2]**, William Brinton**[3]**, Alicia Cronquist**[4]**, Melissa Chen**[2]**, Andrew Thornton**[2]**, Michael J. Beach**[2]**, Joan M. Brunkard**[2]*

1 International Health Resources Consulting, Atlanta, Georgia, United States, 2 National Center for Emerging and Zoonotic Infectious Diseases, Centers for Disease Control and Prevention, Atlanta, Georgia, United States, 3 Regional Epidemiologist, San Luis Valley Public Health Emergency Preparedness and Response Program, Alamosa, Colorado, United States, 4 Disease Control and Environmental Epidemiology Division, Colorado Department of Public Health and Environment, Denver, Colorado, United States

Abstract

In 2008, a large *Salmonella* outbreak caused by contamination of the municipal drinking water supply occurred in Alamosa, Colorado. The objectives of this assessment were to determine the full economic costs associated with the outbreak and the long-term health impacts on the community of Alamosa. We conducted a postal survey of City of Alamosa (2008 population: 8,746) households and businesses, and conducted in-depth interviews with local, state, and nongovernmental agencies, and City of Alamosa healthcare facilities and schools to assess the economic and long-term health impacts of the outbreak. Twenty-one percent of household survey respondents (n = 369/1,732) reported diarrheal illness during the outbreak. Of those, 29% (n = 108) reported experiencing potential long-term health consequences. Most households (n = 699/771, 91%) reported municipal water as their main drinking water source at home before the outbreak; afterwards, only 30% (n = 233) drank unfiltered municipal tap water. The outbreak's estimated total cost to residents and businesses of Alamosa using a Monte Carlo simulation model (10,000 iterations) was approximately $1.5 million dollars (range: $196,677–$6,002,879), and rose to $2.6 million dollars (range: $1,123,471–$7,792,973) with the inclusion of outbreak response costs to local, state and nongovernmental agencies and City of Alamosa healthcare facilities and schools. This investigation documents the significant economic and health impacts associated with waterborne disease outbreaks and highlights the potential for loss of trust in public water systems following such outbreaks.

Editor: Martyn Kirk, The Australian National University, Australia

Funding: The Environmental Protection Agency Office of Science and Technology provided funding for this project but had no role in project design, data collection and analysis, decision to publish, or preparation of the manuscript.

Competing Interests: IHRC is a for profit staffing company. Government funds are put into an IHRC contract so that IHRC will supply trained individuals to accomplish the projects written into the contract. The individuals then work on site at CDC with other CDC staff members to accomplish project goals.

* E-mail: jbrunkard@cdc.gov

Introduction

Community-wide outbreaks associated with public drinking water systems are rare in the United States since drinking water regulations were implemented by the Environmental Protection Agency (EPA), beginning in 1974 with the Safe Drinking Water Act (SDWA) [1,2,3]. However, in 2008 a large community-wide outbreak occurred in Alamosa, Colorado caused by contamination of the town's unchlorinated municipal drinking water supply with *Salmonella* serotype *Typhimurium*.

Alamosa is a small municipality of approximately 8,800 residents situated between two mountain ranges in the San Luis Valley of south-central Colorado [4]. Prior to the outbreak, the City's municipal water was supplied by seven artesian wells and was not chlorinated [5]. On March 14, 2008, the Alamosa County Nursing Service was notified of three culture-confirmed cases of *S.* Typhimurium among residents of Alamosa, including two cases in infants. An epidemiologic investigation conducted by local and state public health authorities identified the city's municipal drinking water as the source of the outbreak [5]. From March 19–April 11, 2008, Alamosa water was deemed unsafe to drink and residents were under various water advisories. After April 11th, all areas of the water system had been hyperchlorinated and all drinking water restrictions lifted.

As a result of the outbreak, 434 cases, including 124 laboratory-confirmed cases, 20 hospitalizations, and one death were reported; a telephone survey conducted by the Colorado Department of Public Health and Environment (CDPHE) at the time of the outbreak indicated that an estimated 1,300 persons became ill (CDPHE, unpublished data). Anecdotal reports of subsequent complications due to *Salmonella* infections, such as anal abscesses and adverse pregnancy outcomes, were received by local public health authorities (W. Brinton, personal communication).

An extensive investigation conducted at the time of the outbreak involved multiple agencies. Water supply interruptions necessitated a large-scale response from local, state, and federal agencies, including the Colorado National Guard, and volunteer agencies. The economic burden to the community was thought to be significant due to business and school closures, missed work to care for ill family members, and the costs of obtaining potable water and other supplies. Because of the scope and extent of the outbreak and response, CDPHE and the local health department in Alamosa requested assistance from the Centers for Disease Control and Prevention (CDC) to assess the full economic and long-term health impacts on the community of Alamosa.

Methods

Ethics Statement

This data collection was judged by officials at CDC to be non-research public health practice, and therefore was not subject to Institutional Review Board (IRB) review. Nevertheless, written informed consent was obtained from all participants, and participants were given the option to refuse specific questions or to decline responding to the surveys.

Data Collection

We conducted a community-wide household survey to assess the health and economic impacts of the outbreak; a business survey to assess the costs incurred by businesses; school surveys to document closures and costs; a review of billing data from two local health care systems to assess direct healthcare expenditures; and interviews with local and state governmental and non-governmental agencies to document costs related to the emergency response efforts.

Household Survey

In October 2009, we sent or hand-delivered a survey to all households that received a water bill from the City of Alamosa (as of September 2009) and surveys were returned via postal mail to CDC. Survey questions covered topics such as residents' drinking water source before and after the outbreak, alternate water sources used during the outbreak, household illness during the outbreak, including potential long-term health consequences (e.g., joint, skin, urinary tract, eye, or other problems occurring within one month following diarrheal onset), and other demographic and household characteristics. Households were also asked to report economic costs associated with the outbreak, including costs associated with illness (e.g., over the counter medicine and out-of pocket costs for prescription medications, doctor's visits and hospitalization), caring for ill family members, securing alternate water sources (bottled water or water filters). To calculate indirect cost of illness, ill household members and caretakers were also asked to provide information on their occupation and daily wage (see Table S1 in File S1 for more information). Some questions were posed at the household-level (e.g., costs for purchase of bottled water) while others were reported for each member of the household (e.g. symptoms, occupation, and demographic characteristics). For our analysis, in order to be consistent with the case definition used during the outbreak, a case was defined as anyone who reported diarrhea (≥ 2 loose stools during a 24 hour period) during the outbreak. An affected household was any household with ≥ 1 person who experienced diarrhea during the outbreak.

Business Survey

To describe how the outbreak impacted local businesses financially, in October 2009 we sent a survey to all businesses inspected by CDPHE (N = 128), including retail food establish-

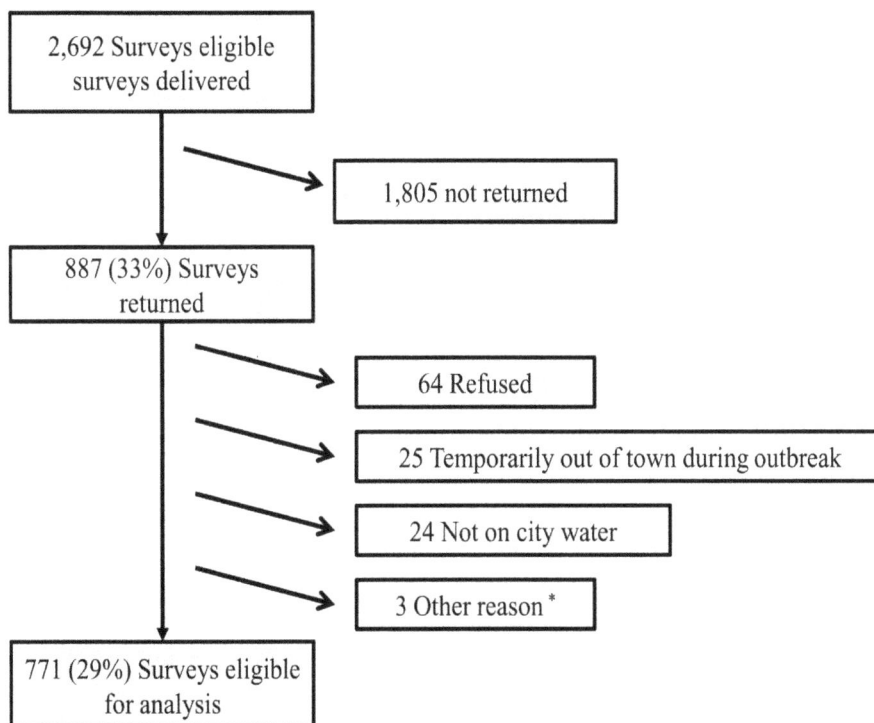

*Not living in Alamosa at time of outbreak (n=2) or respondent not $\geq$ 18 years of age (n=1)

Figure 1. Household survey response rate.

Table 1. Costs to City of Alamosa households associated with an outbreak of salmonellosis, Alamosa, Colorado 2008.

	Household Survey (n = 1,732 individuals, 771 households)					City of Alamosa (2008) (N = 8,746 persons, 3,302 households)	
			Cost($)				
			Reported in survey*			Simulation model†	
	n(%) incurred cost	n(%) reported cost*	Mean	Median	Total	Total	(range)
Outbreak-related expenses (n = 771 households)							
Bought bottled water	657(85%)	523(68%)	$87	$50	$45,530	$135,781	($30,498–$604,183)
Bought water filtration system	202(26%)						
Installation of filter	202(26%)	195(25%)	$180	$70	$35,084	$83,536	($7,760–$336,936)
Maintenance of filter	202(26%)	168(22%)	$121	$60	$20,343	$53,715	($3,974–$232,136)
Stay overnight somewhere else	123(16%)	90(12%)	$362	$233	$32,549	$113,266	($31,749–$403,950)
Subtotal					*$133,506*	*$386,298*	*($73,981–$1,577,205)*
Direct out-of-pocket health care costs (n = 242 households reporting at least one ill person with diarrhea)							
Bought items or received medical treatment	194(80%)	(0%)					
Bought nonprescription medicine	168(69%)	162(67%)	$40	$25	$6,483	$18,220	($4,617–$71,847)
Bought other things (e.g., Gatorade or diapers)	120(50%)	116(48%)	$44	$25	$5,102	$13,765	($4,127–$45,893)
Went to doctor or a clinic	75(31%)	64(26%)	$80	$43	$5,116	$12,653	($936–$54,660)
Received a prescription	32(43%)	28(44%)	$17	$13	$463	$1,850	($397–$5,193)
Had diagnostic tests (e.g., blood or stool test)	37(49%)	35(55%)	$96	$30	$3,355	$5,508	($408–$23,801)
Went to hospital/emergency room	31(13%)	21(9%)	$320	$40	$6,725	$10,603	($784–$45,825)
Subtotal					*$27,244*	*$62,599*	*($11,269–$247,219)*
Indirect cost of acute illness							
Ill persons (n = 369)							
Work full-time	156(42%)	87(24%)	$430	$300	$2,917	$183,644	($45,663–$480,473)
Work part-time	102(28%)	8(2%)	$215	$150	$37,889	$47,083	($11,716–$123,194)
Non-worker	111(30%)	(0%)	$0	$0	$0	$0	(n/a)
Caretakers (n = 106 caretakers)							
Paid caretakers	11(10%)	7(7%)	$178	$110	$1,245	$4,676	(n/a)
Unpaid caretaker (work full-time)	77(73%)	59(56%)	$913	$588	$54,165	$148,173	($31,958–$686,232)
Unpaid caretaker (work part-time)	15(14%)	1(1%)	$457	$294	$640	$14,434	($3,115–$66,844)
Unpaid caretaker (non-worker)	3(3%)	(0%)	$0	$0	$0	$0	(n/a)
Subtotal					*$96,856*	*$398,010*	*($97,218–$1,361,419)*
Total for Households in Alamosa					**$257,606**	**$846,907**	**($182,468–$3,185,843)**

*Respondents who incurred cost and reported an estimate of that cost (not every respondent who reported incurring a cost provided the specific cost estimate).
†Costs extrapolated to the City of Alamosa as: total individuals/households in City of Alamosa (from census)×% incurring costs (column 3 above)×cost distribution (Table S3 in File S1) using a Monte Carlo simulation model with 10,000 iterations. Total cost derived from median of 10,000 iterations and range represents the 5th to 95th percentiles of the 10,000 iterations of the Monte Carlo simulation model. See main text and Supporting Information (in File S1) for details.

ments (restaurants, hotels, nursing homes, and child care centers), and other businesses (N = 54) potentially affected by the water shortages (grocery stores, beauty salons, dentists, and animal clinics). Contact information was provided by CDPHE, or obtained through a telephone directory and internet searches. Businesses were asked how the outbreak affected their business, including whether the business had to close, lay off workers, and whether the business had to buy additional water or ice, lost or gained money overall, and if the business ever regained pre-outbreak levels. To encourage businesses to respond and protect confidentiality, we did not ask for the business name or address on the survey. Five of the 128 CDPHE-inspected businesses were located outside of Alamosa but had clientele likely to be comprised of Alamosa residents.

Interviews with Governmental and Non-governmental Agencies, Healthcare Facilities, and Schools

We interviewed via telephone or in-person staff from the City of Alamosa, Alamosa County Nursing Service, and CDPHE and the local chapter of the American Red Cross to ascertain estimates of the direct and indirect cost of the outbreak response to local and state governmental and non-governmental agencies. Respondents provided information on the cost of the response (e.g., lodging and meals for staff, truck rentals, etc.), the number of staff and their

Table 2. Potential long-term health consequences of *Salmonella* infection reported by survey respondents that were ill with diarrhea (≥2 loose stools during a 24-hour period) (n = 369) during an outbreak of salmonellosis, Alamosa, Colorado 2008.

Potential long-term health consequence	n (%)	Days after diarrhea began that problem started: mean (range)*	Duration (weeks)	
			Mean (range) time-limited duration of symptoms†	Symptoms still present at time of survey n(%)
Rash, itchiness or other skin problems	52 (14%)	5 days (0–30)	3 weeks (1–24)	11/52 (21%)
Arthritis, aching joints or other joint problems	51 (14%)	7 days (1–30)	3 weeks (0–16)	20/51 (39%)
Urinary tract problems (e.g., pain or burning during urination or a discharge)	32 (9%)	7 days (1–30)	4 weeks (1–30)	5/32 (16%)
Eye problems such as pain or redness	19 (5%)	4 days (1–7)	2.5 weeks (1–6)	2/19 (11%)
Abscess (skin, soft tissue, anal, etc.)	6 (2%)	5 days (2–14)	2 weeks (1–3)	1/6 (17%)
Other serious complications (e.g., bowel perforation or peritonitis, septic arthritis, or endocarditis)	7 (2%)	n/a‡	n/a‡	n/a‡

*As reported by 43/52 with skin problems, 36/51 with joint problems, 27/32 with urinary tract problems, 16/19 with eye problems, and 5/6 with abscesses.
†As reported 36/52 with skin problems, by 22/51 with joint problems, 15/32 with urinary tract problems, 13/19 with eye problems, and 4/6 with abscesses.
‡Questions were not asked.

aggregate labor hours spent responding to the outbreak. We also interviewed Alamosa health care providers, including a hospital, medical practices, nursing homes, and assisted living facilities to assess the outbreak impact and to request billing data to supplement cost estimates from the household surveys (see Supporting Information and Table S2 in File S1for more information). To determine the effects of the outbreak on educational institutions, we interviewed representatives from each of Alamosa's two public colleges, two private schools, and its public school district. School representatives were asked about the types of additional costs incurred because of the outbreak, including purchasing bottled water or hand sanitizer, paying for employee overtime, and costs for make-up days.

Analysis

All survey data were entered into a Microsoft Access 2007 database and descriptive analyses were conducted using SAS v. 9.2 (Cary, NC). We compared survey respondents' characteristics (sex, age, race/ethnicity and income) to the characteristics of the 2008 City of Alamosa population [6] using chi-square tests. All p-values were two-sided and the level of significance was 0.05.

For our cost estimates, we took a societal perspective and defined costs as expenses which would not be incurred if the outbreak had not occurred. Since almost all costs were incurred in 2008 and 2009, we did not apply a discount rate. No capital costs (materials with more than a 5 year useful life) were incurred. All the costs were recorded in 2008 U.S. dollars.

We built a Monte Carlo simulation model using @Risk software (Palisade Corporation, NY) to extrapolate the costs to the city of Alamosa. The model used the following formula (see Table S3 and Figure S1 in File S1 for more information):

Total cost = Number of individuals/households in Alamosa in 2008 (from census data)×Proportion of respondents who experienced a cost (from household survey) X.

Cost distribution of the given cost (from household survey).

We assumed that the proportion of respondents from the survey who experienced a given cost was the same proportion in the community who would have experienced the costs. We used the costs reported by all individuals/households in the survey to generate the cost distribution to fit the data for the model. The

details of these cost distributions are given in the Supporting Information (Table S3 in File S1). Using Monte Carlo simulation (10,000 iterations), we then extrapolated the costs to the city of Alamosa and the model's results are presented as the total cost (the median of the 10,000 iterations) and range (representing the 5th to 95th percentile of the 10,000 iterations). Additional methodological details about the model, methods for direct and indirect cost calculations, and extrapolation methods, are provided in the Supporting Information (see File S1).

Results

Household Survey

The community survey was distributed to all households that received municipal drinking water (N = 2,692). After excluding non-responders and ineligible responses (refusal, out of town during the outbreak, not on city water, or other reasons) 29% (n = 771) of households, representing 1,732 persons, returned surveys eligible for analysis (Figure 1). The median number of persons per household was two (range: 1–8), and the surveys analyzed represented approximately 20% of the City of Alamosa's 2008 population [4]. The majority (n = 458, 59%) of survey respondents were female; the median age of respondents was 57 years (range: 18–99). Half (n = 391, 51%) of the respondents identified themselves as white and 31% (n = 240) were of Hispanic ethnicity; 8% (n = 63) did not provide race or ethnicity information. Of the 620 (80%) households that reported income information, 14% (n = 89) made <$13,000, 18% (n = 111) $13,000–$25,000, 24% (n = 150) $25,000–$45,000, 21% (n = 130) $45,000–$75,000, and 23% (n = 140) >$75,000. Compared to the population of the City of Alamosa, our survey respondents tended to be older, more likely to be female (p<0.001), less likely to be of Hispanic ethnicity (p<0.001), and moderately more affluent (p<0.001) [6] (Table S4 in File S1).

Illness associated with the Outbreak. Approximately one-third (242/771, 31%) of households, and 21% (369/1732) of individual respondents, reported diarrheal illness during the outbreak. Fifty-seven percent (n = 187/329) of ill persons were female (including four who were pregnant); the median age of ill persons was 37 years (range: 0–98 years). By definition, all ill

Table 3. Number and percent of households that reported using various alternate water sources (n = 771 households)* during an outbreak of salmonellosis, Alamosa, Colorado 2008.

	Avoided activity	Bought bottled water	Used bottled/bulk water given out for free	Boiled tap water	Used water from outside Alamosa†	Used treated tap water†‡	Used un-boiled tap water	Used water from other source
Drinking	9(1%)	602(78%)	556(78%)	90(12%)	228(30%)	21(3%)	14(2%)	5(1%)
Cooking	33(4%)	465(60%)	531(60%)	233(30%)	238(31%)	28(4%)	42(5%)	9(1%)
Dishwashing	121(16%)	272(35%)	368(35%)	330(43%)	225(29%)	59(8%)	158(20%)	35(5%)
Brushing teeth	5(1%)	518(67%)	497(67%)	107(14%)	194(25%)	20(3%)	54(7%)	6(1%)
Showering/bathing	121(16%)	107(14%)	189(14%)	134(17%)	427(55%)	27(4%)	366(47%)	23(3%)

*Questions allowed for multiple options and therefore row totals do not sum to 100%.
†e.g., a friend's house, hotel, or artesian spring.
‡e.g., using chlorine or a filter.

persons experienced diarrhea out of which 30% (n = 110) reported bloody diarrhea. The median duration of illness was four days (range: 1–60 days). In total, survey respondents (n = 350) were sick for 2,341 person-days. Most (n = 194, 80%) of the Alamosa households with at least one ill person reported buying medicine or other items because of their illness (Table 1); 31% of households (n = 75) with at least one ill person sought care, and close to half of those who sought care reported receiving a prescription (n = 32, 43%) or having a diagnostic test performed (n = 37, 49%) (Table 1).

Twenty-nine percent (n = 108) of all ill persons reported experiencing ≥1 potential long-term health consequence of *Salmonella* infection, ranging from 14% reporting skin or joint problems to 2% reporting abscesses or more serious complications (e.g., bowel perforation, septic arthritis, or endocarditis) (Table 2). These symptoms began between four days (mean for eye problems) to seven days (mean for joint and urinary tract problems) after diarrhea onset. Among those whose symptoms had abated, symptoms lasted between two weeks (for abscesses) to four weeks (for urinary tract infections). However, 26% (n = 28) of those with long-term symptoms indicated that ≥1 symptom was still present at the time of the survey (18 months after the outbreak), ranging from 11% (for eye problems) to 39% (for joint problems) (Table 2).

Alternate water sources used during the outbreak. During the bottled water advisory, Do Not Use order, and boil water advisory, most households reported using bottled water (either purchased or donated) for drinking, cooking and brushing teeth (Table 3). Other water sources, such as boiled or treated tap water and water outside of Alamosa (e.g., at a hotel or friend's house), were used less frequently. Despite being told to avoid using tap water except for flushing toilets during the Do Not Use order, many households continued to use tap water for at least some potable and non-potable purposes (Table 4). Half of households (55%, n = 427) reported boiling water during the 24 days of the water emergency. Eighty-five percent of households (n = 657) bought bottled water during the outbreak and spent, on average, $87 (Table 1).

Drinking water impact. Most households (n = 699, 91%) reported municipal tap water as their main drinking water source at home prior to the outbreak (Figure 2). Eighteen months after the outbreak, the main drinking water sources among Alamosa survey respondents were: bottled water (38%, n = 292/771), municipal water (30%, n = 233), municipal water with a new filter installed (15%, n = 119), multiple (12%, n = 94), or other sources (4%, n = 28) (Figure 2). Taste, safety, and smell were the main reasons cited for switching from municipal tap water to bottled water or adding a new water filter after the outbreak (Table 5).

Household economic impacts. Outbreak-related costs for City of Alamosa residents totaled an estimated $846,907 (range: $182,468–$3,185,843) (Table 1). The estimated total cost associated with the purchase of supplemental bottled water, installing and maintaining new water filters, and paying to stay overnight elsewhere was $386,298. Direct costs associated with illness (e.g., purchasing over the counter medicines or other items, doctors' visits, prescription medications, diagnostic tests, and hospital stays) were estimated to total $62,599. Additional indirect costs of acute diarrheal illness and costs associated with paying for, or lost productivity due to, care of ill persons and children were estimated to total $398,010 (Table 1).

Business Survey

We distributed 177 surveys to establishments inside the City of Alamosa and 5 to businesses located outside the City of Alamosa that primarily served or employed Alamosa residents. Of the 182

Table 4. Number and percent of households that reported using *any* tap water for the following purposes during the various water advisories (n = 771 households) associated with an outbreak of salmonellosis, Alamosa, Colorado 2008.

	Bottled Water Advisory*	Do Not Use Order[†]	Boil Water Advisory[‡]
	(March 19–24, 2008)	(March 25–April 3, 2008)	(April 3–11, 2008)
Drinking	71(9%)	25(3%)	58(8%)
Cooking	212(28%)	73(9%)	233(30%)
Washing dishes	414(54%)	191(25%)	456(59%)
Brushing teeth	145(19%)	58(8%)	116(15%)
Showering/Bathing	516(67%)	256(33%)	498(65%)

*During the bottled water advisory, residents were told to use bottled water for drinking, cooking, brushing teeth, and dishwashing but that if no bottled water was available they could boil their water.
[†]During the Do Not Use order, while the distribution system was being hyperchlorinated, residents were told to only use their tap water for flushing toilets.
[‡]During the boil water advisory, residents were told to boil their water before using it for drinking, cooking, or brushing teeth.

surveys, 21 (12%) were undeliverable and 50 (50/161, 31%) were returned. The following surveys were excluded: three because the business was not open during the outbreak and one because it was a city jail. Forty-six (46/161, 29%) surveys were eligible for analysis: 41 surveys were included in the primary analysis and an additional five surveys were analyzed separately because the business was either located outside Alamosa or not connected to municipal water.

The 41 businesses located in Alamosa and connected to municipal water included retail stores (n = 8, 20%), restaurants or other food service establishments (n = 6, 15%), beauty salons or barber shops (n = 6, 15%), child care centers (n = 4, 10%), nursing homes or long-term care facilities (n = 3, 7%), and other types of businesses (n = 10, 24%); two (5%) businesses did not specify the type of establishment. One-third (14/41, 34%) of responding businesses closed during the outbreak (mean length of business closure: 8.4 days). Approximately half of businesses reported losing money due to the outbreak, with a median loss of $8,750 (range:

$400–$200,000) (Table 6). One business outside of Alamosa reported a total loss of $13,967. Four businesses (10%) never returned to pre-outbreak financial levels, including one business that closed permanently due to the outbreak. The total estimated cost of the outbreak extrapolated to the City of Alamosa businesses that are known to have used municipal water (n = 156) was $625,561 (range: $14,209–$2,817,036) (Table 6).

Governmental and Nongovernmental Agencies, Healthcare Facilities, and Schools

Outbreak response cost estimates to local, regional, and state governmental and volunteer organizations totaled $823,314 (Table 7). This estimate primarily included the governmental response costs for the National Guard, incident management teams, and personnel, and other costs (such as transportation, supplies, lodging, etc.) at state, county, and local levels. The volunteer response, coordinated primarily by the American Red Cross, lasted 21 days, involved 1,035 people who contributed over

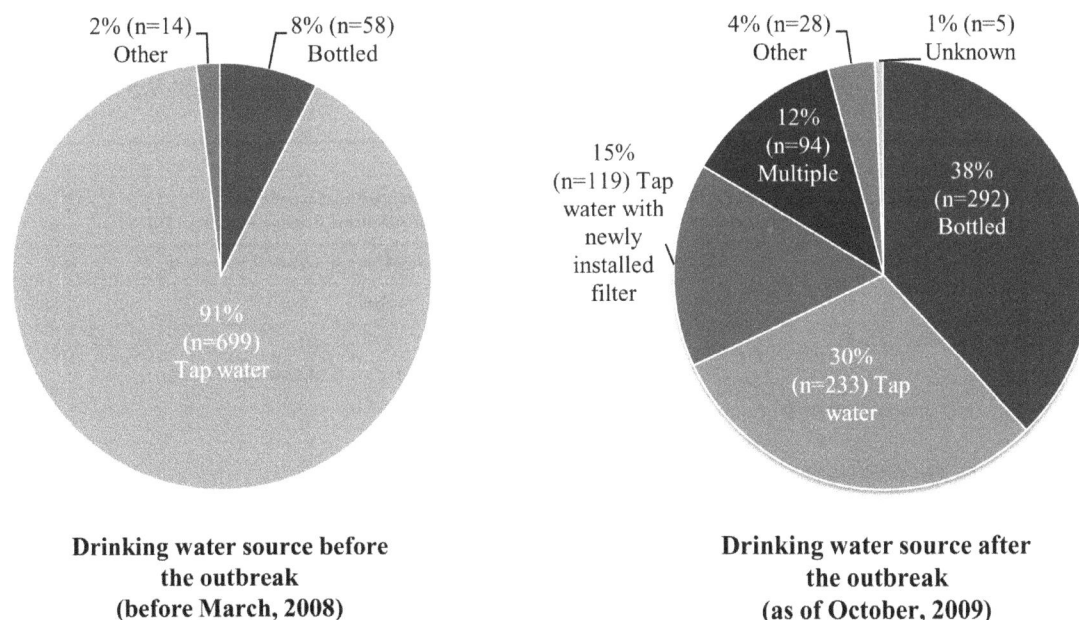

Drinking water source before the outbreak (before March, 2008)

Drinking water source after the outbreak (as of October, 2009)

Figure 2. Change in water source after the outbreak (n = 771 households) following an outbreak of salmonellosis, Alamosa, Colorado 2008.

Table 5. Reasons households cited for switching from tap water as a main drinking water source to bottled water as a main drinking water source (n = 249) or installing a new water filter (n = 114) after an outbreak of salmonellosis, Alamosa, Colorado 2008.

	Switched from tap to bottled water (n = 249)*	Added a new filter (n = 114)*
Taste	134(55%)	75(66%)
Safety	110(45%)	42(37%)
Smell	60(25%)	35(31%)
Color or other reason	29(12%)	16(14%)

*Categories are not mutually exclusive so percentages can sum to >100%.

5,289 hours to the effort, and totaled an estimated $80,710 (Table 7).

Most health care providers reported significant expenses in securing and providing clean water (for drinking, bathing, housekeeping and other uses) or disposable supplies. However, these costs could not be estimated because most could not retrospectively itemize these expenses. Only one local hospital was able to provide billing records for outbreak-related care it provided to 104 of the 124 laboratory-confirmed cases. The estimated total cost of health insurance payments for Alamosa City residents that sought health care was $244,985 (range: $65,615–$928,915) (Table 8). The five public and private schools and colleges were closed for, on average, 2 days (range: 0 to 5) due to the outbreak. However, anecdotally we learned that the outbreak occurred during spring break for most of the five public and private schools and colleges in Alamosa, which helped minimize the impact of school closures. Only two schools reported any substantial financial impact, which totaled $23,898 (Table 9). These costs were related to paying overtime and purchasing bottled water and other items.

Total Costs

The total estimated economic impact of the outbreak, including costs to City of Alamosa residents, businesses, schools, and healthcare facilities and the governmental and non-governmental

outbreak response was approximately $2.6 million (range: $1.1 million–$7.8 million dollars) (Table 9). The largest contributors (32.8%) to this cost were direct and indirect costs for City of Alamosa residents, followed by the cost of the outbreak response to governmental organizations (26.4%), and costs to Alamosa businesses (24.3%).

Discussion

Since passage of the Safe Drinking Water Act and its amendments by EPA [1,2,3], community-wide drinking water outbreaks in the U.S. are rare [7]. The *Salmonella* outbreak that occurred in Alamosa, Colorado in 2008 was one of the largest drinking water-associated outbreaks reported in the U.S. since the 1993 *Cryptosporidium* outbreak in Milwaukee, which sickened an estimated 400,000 people [8]. The estimated economic impact of the Alamosa outbreak totaled over $2.6 million. An unanticipated consequence of the outbreak was the loss of trust in the public water system after the outbreak.

Despite our comprehensive approach, this outbreak cost estimate is lower than previous epidemiologic studies of outbreaks in public water systems, perhaps due to our conservative methodological approach and the differences in the size of the affected population or duration of the outbreak. Harrington et al. estimated the economic impact of a 1984 waterborne outbreak of

Table 6. Outbreak-associated costs to City of Alamosa businesses associated with an outbreak of salmonellosis, Alamosa, Colorado 2008.

	Business Survey (n = 46)			Extrapolated to Sample of City of Alamosa Businesses (N = 156)	
				Estimated using simulation model*	
	n(%)	Mean	Median	Total (range)	
Businesses inside Alamosa and on city water (n = 41)					
Lost money	22(54%)	$35,306	$8,750	**$625,561**	($14,209–$2,817,036)
No change	17(41%)				
Did better because of the outbreak	2(5%)				
Businesses outside Alamosa or not on city water (n = 5)					
Lost money	1(20%)	$13,967	$13,967	**$13,967**	n/a
No change	3(60%)				
Did better because of the outbreak	1(20%)				
Total				**639,528**	($28,176–$2,831,003)

*Costs extrapolated to the City of Alamosa as: total businesses on municipal water (N = 156) ×% incurring costs (column 3 above) ×cost distribution (Table S3 in File S1) using a Monte Carlo simulation model with 10,000 iterations. Total cost derived from median of 10,000 iterations and range represents the 5th to 95th percentiles of the 10,000 iterations of the Monte Carlo simulation model. See main text and Supporting Information (in File S1) for details.

Table 7. Outbreak response costs associated with an outbreak of salmonellosis, Alamosa, Colorado 2008.

	Personnel	Other*	Total
Governmental organizations			
Federal Government	$19,040[†]		$19,040
State of Colorado	$215,925	$316,449[‡]	$532,374
Alamosa County	$52,817	$7,582	$60,399
City of Alamosa	$50,872	$19,023	$69,895
Non-governmental organizations			
Volunteer organizations	$40,135[#]	$40,575	$80,710
Other organizations	$60,896[†]		$60,896
Total	**$439,685**	**$383,629**	**$823,314**

*Includes transportation, supplies, lodging, etc.
[†]As captured by City of Alamosa record-keeping.
[‡]Includes expenses covered by the state disaster fund (e.g., National Guard, incident management teams); these were all included in the "Other" category because personnel costs were not reported separately from other expenses.
[#]Includes $7,920 for staff overtime (3 persons and a total of 750 hours of overtime) and $32,215 in estimated indirect costs associated with volunteer time (4,589 volunteer hours estimated at Colorado minimum wage rate of $7.02/hour).

giardiasis in a Pennsylvania county at $18.2–133.3 million (in 2008 dollars) and did not include the cost of the outbreak response or the impact on local businesses [9]. The outbreak occurred in a community of 25,000 households, resulted in 370 cases of giardiasis and a boil water advisory that lasted at least 99 days (270 days for half of those affected). Corso et al. estimated that the massive 1993 cryptosporidiosis outbreak in Milwaukee, Wisconsin (population ~ 1.6 million) affecting 403,000 individuals cost an estimated $143.4 million (in 2008 dollars) in direct healthcare expenditures and productivity losses [10]; the cost of using an alternate water source during the outbreak, costs to local businesses, and the cost of the outbreak response were not

included. Despite these differences, the limited number of cost analyses for waterborne outbreaks underscores the need to conduct these analyses for future outbreak investigations and the utility of including longer follow-up investigations to capture long-term costs.

In our assessment, 31% of households and 21% of survey respondents became ill during the outbreak. Approximately one-third of those who became sick reported a potential long-term health consequence following their diarrheal illness and, of those, 26% were still experiencing symptoms 18 months after the outbreak. Because all symptoms were based on self-report, and may have been coincidental to, rather than caused by the *Salmonella* outbreak, we included only symptoms that began within 30 days of diarrheal illness. In addition, similar frequencies of post-*Salmonella* infection joint pain [11,12] and other symptoms [13] have been observed in previous studies, including a Canadian study that found that such symptoms can persist for three years post-infection [14]. Although they were rare, we also found that 2% of cases experienced a more serious complication of infection such as bowel perforation, peritonitis, septic arthritis, or endocarditis, complications from *Salmonella* infection that have been reported elsewhere [15]. Unfortunately, we were unable to estimate the direct and indirect costs associated with these long-term sequelae. However, the time and costs associated with these are likely to have been substantial. For instance, patients with arthritis and other rheumatologic conditions had average annual medical care expenditures of $1,891 and earned $1,590 less than individuals without these conditions in 2003 [16]. Additionally, urinary tract infections cost an estimated $1.6 billion per year in 1994 in direct and indirect costs [17].

Over 90% of households reported that municipal water was their main drinking water source at home prior to the outbreak. After the outbreak, 38% of respondents mainly drank bottled water and only 30% of households continued to primarily drink tap water; an additional 15% purchased a new filter or filtration system. The purchase of bottled water and installation and maintenance of filters cost City of Alamosa residents approximately $273,000 during the outbreak. Almost half (45%) of survey

Table 8. Health insurance payments for Alamosa City residents that sought healthcare during an outbreak of salmonellosis, Alamosa, Colorado 2008.

	Household Survey (n = 1,732)	Hospital A Cost Estimates (n = 104 culture-confirmed cases and 139 separate healthcare visits)			Extrapolated to City of Alamosa (N = 8,746)	Cost ($)[‡]
	n(%)	n(%)	Mean	Median	N(%)[†]	Total (range)
Ill Persons	369(21%)				1,423(16%)	
Sought care	107(29%)				413(29%)	
Clinic/doctors' office*	76(71%)	67(48%)	$129	$93	293(71%)	$26,275 ($6,430–$107,324)
Emergency department	22(21%)	67(48%)	$693	$390	85(21%)	$36,600 ($7,351–$182,111)
Hospitalized	9(8%)	5(4%)	$7,011	$3,159	35(8%)	$182,110 ($51,834–$639,480)
Total		139				$244,985 ($65,615–$928,915)

*Because data were obtained from the hospital only, clinic/doctor's office visit costs only include laboratory but not physicians' fees.
[†]We have removed the background rate of diarrhea in the population (5%) to get the percent of illness due to outbreak (21%−5% = 16%). The number of ill persons was the denominator for subsequent proportions who incurred the costs (e.g., 29%,71%, 21% and 8%).
[‡]Costs extrapolated to the City of Alamosa as: total population (N = 8,746)×% incurring costs (column 3 above)×cost distribution (Table S3 in File S1) using a Monte Carlo simulation model with 10,000 iterations. Total cost derived from median of 10,000 iterations and range represents the 5th to 95th percentiles of the 10,000 iterations of the Monte Carlo simulation model. See main text and Supporting Information (in File S1) for details.

Table 9. Total costs associated with an outbreak of salmonellosis, Alamosa, Colorado 2008.

	Total	(range)	% of total
City of Alamosa households	**$846,907**	**($182,468–$3,185,843)**	**32.8%**
Outbreak-related expenses	$386,298	($73,981–$1,577,205)	
Direct out-of-pocket health care costs	$62,599	($11,269–$247,219)	
Indirect costs of acute illness and caretaking	$398,010	($97,218–$1,361,419)	
Alamosa businesses	**$625,561**	**($14,209–$2,817,036)**	**24.3%**
Businesses outside the City of Alamosa	**$13,967**	**n/a**	**0.5%**
Governmental organizations	**$681,708**	**n/a**	**26.4%**
Federal Government	$19,040	n/a	
State of Colorado	$532,374	n/a	
Alamosa County	$60,399	n/a	
City of Alamosa	$69,895	n/a	
Non-governmental organizations	**$141,606**	**n/a**	**5.5%**
Volunteer organizations	$80,710	n/a	
Other	$60,896	n/a	
Health insurance payments	**$244,985**	**($65,615–$928,915)**	**9.5%**
School and colleges	**$23,898**	**n/a**	**0.9%**
Grand Total	**$2,578,632**	**($1,123,471–$7,792,973)**	**100%**

Details provided in Tables 1, 6 & 8. Extrapolation done using Monte Carlo simulation model. See main text and Supporting Information in File S1 for details.

respondents cited safety concerns as a reason for switching from tap to bottled water. This lack of trust was also apparent in survey participants' comments, such as: "I will never again fully trust the system or drink any tap water without some concern…" and "I still don't feel safe drinking or cooking with the city water… I have spent a lot of money buying bottled water."

The economic impact of the outbreak on the sample of businesses was one of the largest expenses, totaling $626,000 and accounting for 24% of the total outbreak costs. Approximately half of businesses that responded indicated that they lost money and approximately one-third had to close temporarily during the outbreak. Only 60% reported ever returning to pre-outbreak financial levels, including one that noted that "it took 2–3 months to get back to previous levels." Because the survey was sent 18 months after the outbreak, it could have failed to reach businesses that might have been forced to close because of the outbreak. Household survey responses corroborated this; one respondent noted that "we couldn't pay our mortgage [and] lost our restaurant. We now both work for someone else for not as much pay. We had our restaurant for 25 years."

This assessment was subject to several limitations. First, our outbreak cost estimate is likely an underestimate. It does not include health care costs for individuals who sought care outside of Alamosa (either because some ill individuals may not have responded to the survey or because we were only able to obtain hospital-associated costs from one local hospital). We also were unable to assign an estimate for the one death associated with the outbreak. Alamosa is the geographic and commercial center of the San Luis Valley, and many people from surrounding areas work and dine in Alamosa but the survey did not capture business-related costs or health impact for people who live outside of Alamosa or for businesses that either did not receive a survey or did not respond. Additionally, outbreak response costs incurred by the federal government and by local organizations or municipalities outside the City of Alamosa that contributed to the outbreak response are likely incomplete. Household survey respondents also

mentioned various costs not covered in the questionnaire, such as the cost of gas, disposable plates/utensils, or pet care associated with the outbreak. Additionally, direct and indirect costs associated with the long-term health consequences of *Salmonella* infection were not assessed. Outbreak-associated costs were also limited to those incurred during the outbreak, even though it is likely that many costs have been incurred since the outbreak.

Second, we assumed that the survey respondents were a representative sample of the City of Alamosa population, yet our survey respondents differed by age, sex, ethnicity, and socioeconomic status [6]. In addition, our survey was conducted 18 months after the outbreak, and persons who responded to the survey may have been more likely to be sick during the outbreak and therefore to respond to the survey, although the attack rate estimated from our household survey is similar to the estimate found in a survey of Alamosa residents immediately after the outbreak (CDPHE, unpublished data). Third, our indirect cost estimates were based on assumptions (see Supporting Information in File S1 for details) about the value of caretakers' and ill people's time. However, the wages reported in our survey were similar to that reported for the Colorado non-Metropolitan Statistical Area that includes Alamosa [18]. Finally, our relatively low response rate (~ 30%), although similar to that of other mailed surveys [19], may mean that the results of the household and business surveys are not reflective of the experiences of all City of Alamosa households and businesses.

The likely source of the outbreak was determined to be animal contamination of a storage tank that had numerous cracks and entry points [5]. This outbreak highlights the critical importance of robust inspection of public drinking water storage facilities, identification of system deficiencies during required sanitary surveys, and maintaining staffing and resources for adequate follow-up for any deficiencies identified. Although it is now being chlorinated, the City of Alamosa's water prior to the outbreak was derived from an unchlorinated ground water source. The recently promulgated Ground Water Rule (GWR) [20,21] requires most community water systems to complete initial sanitary surveys by

2012. Once fully implemented, it should help reduce the risk for similar outbreaks in the future. Nevertheless, a deficiency in the distribution system (i.e., storage tank contamination) was the primary cause of this outbreak. Maintaining the integrity of the nation's drinking water systems is a fundamental safeguard to protecting public health and preventing economic damage from waterborne disease outbreaks and should be a top public policy imperative.

Acknowledgments

The authors gratefully acknowledge the contributions of all of those who aided in the outbreak response and in this investigation, particularly the Alamosa County Nursing Service and the American Red Cross.

Disclaimer

The findings and conclusions in this report are those of the authors and do not necessarily represent the views of the Centers for Disease Control and Prevention.

Author Contributions

Conceived and designed the experiments: EA PB SC WB AC MB JB. Performed the experiments: EA PB SC WB AT MC MB JB. Analyzed the data: EA PB MS SC AT. Wrote the paper: EA PB MS SC MB JB.

References

1. Environmental Protection Agency (1975) Water programs: national interim primary drinking water regulations. Federal Register. 59566–59574.
2. Pontius F, Roberson JA (1994) The current regulatory agenda: an update. Major changes to USEPA's current regulatory agenda are anticipated when the SDWA is reauthorized. Journal of American Water Works Association 86: –63.
3. Pontius F (1997) Implementing the 1996 SDWA amendments. Journal of American Water Works Association 89: 18–36.
4. U.S. Census Bureau 2008 Population Estimates, Alamosa City Coloardo. Available: http://www.census.gov/popest/data/cities/totals/2009/tables/SUB-EST2009-04-08.csv.
5. Falco R, Williams SI (2009) Waterborne Salmonella Outbreak in Alamosa, Colorado March and April 2008: Outbreak Identification, Response, and Investigation. Available at: http://www.cdphe.state.co.us/wq/drinkingwater/pdf/AlamosaInvestRpt.pdf Accessed 9/3/2010.
6. U.S. Census Bureau 2005–2009 American Community Survey 5-Year Estimates: Alamosa City, CO. Available: http://factfinder2.census.gov/bkmk/table/1.0/en/ACS/09_5YR/DP5YR5/1600000US0801090.
7. Craun GF, Brunkard JM, Yoder JS, Roberts VA, Carpenter J, et al. (2010) Causes of outbreaks associated with drinking water in the United States from 1971 to 2006. Clin Microbiol Rev 23: 507–528.
8. Mac Kenzie WR, Hoxie NJ, Proctor ME, Gradus MS, Blair KA, et al. (1994) A massive outbreak in Milwaukee of cryptosporidium infection transmitted through the public water supply. N Engl J Med 331: 161–167.
9. Harrington W, Krupnick AJ, Spofford WO (1989) The economic losses of a waterborne disease outbreak. Journal of Urban Economics 25: 116–137.
10. Corso PS, Kramer MH, Blair KA, Addiss DG, Davis JP, et al. (2003) Cost of illness in the 1993 waterborne Cryptosporidium outbreak, Milwaukee, Wisconsin. Emerg Infect Dis 9: 426–431.
11. Samuel MP, Zwillich SH, Thomson GT, Alfa M, Orr KB, et al. (1995) Fast food arthritis–a clinico-pathologic study of post-Salmonella reactive arthritis. J Rheumatol 22: 1947–1952.
12. Townes JM, Deodhar AA, Laine ES, Smith K, Krug HE, et al. (2008) Reactive arthritis following culture-confirmed infections with bacterial enteric pathogens in Minnesota and Oregon: a population-based study. Ann Rheum Dis 67: 1689–1696.
13. Doorduyn Y, Van Pelt W, Siezen CL, Van Der Horst F, Van Duynhoven YT, et al. (2008) Novel insight in the association between salmonellosis or campylobacteriosis and chronic illness, and the role of host genetics in susceptibility to these diseases. Epidemiol Infect 136: 1225–1234.
14. Buxton JA, Fyfe M, Berger S, Cox MB, Northcott KA (2002) Reactive arthritis and other sequelae following sporadic Salmonella typhimurium infection in British Columbia, Canada: a case control study. J Rheumatol 29: 2154–2158.
15. Cohen JI, Bartlett JA, Corey GR (1987) Extra-intestinal manifestations of salmonella infections. Medicine (Baltimore) 66: 349–388.
16. Yelin E, Murphy L, Cisternas MG, Foreman AJ, Pasta DJ, et al. (2007) Medical care expenditures and earnings losses among persons with arthritis and other rheumatic conditions in 2003, and comparisons with 1997. Arthritis Rheum 56: 1397–1407.
17. Foxman B, Barlow R, D'Arcy H, Gillespie B, Sobel JD (2000) Urinary tract infection: self-reported incidence and associated costs. Ann Epidemiol 10: 509–515.
18. Bureau of Labor Statistics. May 2009 Metropolitan and Nonmetropolitan Area Occupational Employment and Wage Estimates: Eastern and Southern Colorado Monmetropolitan Area. Available: http://www.colorado.gov/cs/Satellite/CDLE-LaborLaws/CDLE/1251566749488.
19. Link MW, Battaglia MP, Frankel MR, Osborn L, Mokdad AH (2006) Address-based versus random-digit-dial surveys: comparison of key health and risk indicators. Am J Epidemiol 164: 1019–1025.
20. Environmental Protection Agency (2000) National primary drinking water regulations: ground water rule. 40 CFR Parts 141 and 141: Federal Register. 30194–30274.
21. Environmental Protection Agency (2006) National primary drinking water regulations: ground water rule. 40 CFR Parts 141 and 141: Federal Register. 65573–65660.

8

Perceptional and Socio-Demographic Factors Associated with Household Drinking Water Management Strategies in Rural Puerto Rico

Meha Jain[1]*, Yili Lim[1], Javier A. Arce-Nazario[2], María Uriarte[1]

1 Department of Ecology, Evolution and Environmental Biology, Columbia University, New York, New York, United States of America, 2 Department of Biology, University of Puerto Rico in Cayey, Cayey, Puerto Rico, United States of America

Abstract

Identifying which factors influence household water management can help policy makers target interventions to improve drinking water quality for communities that may not receive adequate water quality at the tap. We assessed which perceptional and socio-demographic factors are associated with household drinking water management strategies in rural Puerto Rico. Specifically, we examined which factors were associated with household decisions to boil or filter tap water before drinking, or to obtain drinking water from multiple sources. We find that households differ in their management strategies depending on the institution that distributes water (i.e. government PRASA vs community-managed non-PRASA), perceptions of institutional efficacy, and perceptions of water quality. Specifically, households in PRASA communities are more likely to boil and filter their tap water due to perceptions of low water quality. Households in non-PRASA communities are more likely to procure water from multiple sources due to perceptions of institutional inefficacy. Based on informal discussions with community members, we suggest that water quality may be improved if PRASA systems improve the taste and odor of tap water, possibly by allowing for dechlorination prior to distribution, and if non-PRASA systems reduce the turbidity of water at the tap, possibly by increasing the degree of chlorination and filtering prior to distribution. Future studies should examine objective water quality standards to identify whether current management strategies are effective at improving water quality prior to consumption.

Editor: Joan Muela Ribera, Universitat Rovira i Virgili, Spain

Funding: Funding was provided by awards to MU from National Science Foundation DEB 0620910 and the Crosscutting Initiatives program, Earth Institute, Columbia University. This study was partially supported by Award Number P20MD006144 from the National Center on Minority Health and Health Disparities. The content is solely the responsibility of the authors and does not necessarily represent the official views of the National Center on Minority Health and Health Disparities or the National Institutes of Health. This work was also supported by the National Science Foundation Graduate Research Fellowship under Grant No. 11-44155 awarded to MJ. The funders had no role in study design, data collection and analysis, decision to publish, or preparation of the manuscript.

Competing Interests: The authors have declared that no competing interests exist.

* E-mail: mj2415@columbia.edu

Introduction

Over 700 million people across the globe do not have access to clean drinking water, leading to high levels of chronic waterborne illnesses [1–3]. This is particularly problematic in rural communities that do not receive adequately treated water from government facilities and may not have access to appropriate technologies to treat water locally [4,5]. Scientists and policymakers have long considered the best ways to improve access to potable water, yet identifying the most effective ways to manage drinking water is difficult given that it is typically managed by multiple public and private agencies [6–8]. Drinking water is often extracted and treated at different spatial scales (e.g. regional, watershed, and household level), resulting in management by various stakeholders that act at each of these scales (e.g. governmental, private, and household sectors; [9,10]. Given the complexity of drinking water management, policy makers and agencies (e.g. World Health Organization) over the past decade have increasingly recognized the importance of household water management, particularly in regions where government and community water treatment facilities are ineffective [11,12].

Households play an important role in determining the water quality experienced by individuals, as households are the last point of management prior to consumption [4,12].

To target the most successful interventions, it is important to understand the socio-cultural context of current household water management decisions [13]; by understanding how households manage their drinking water and why, policymakers can more effectively target intervention strategies to improve water quality prior to consumption. Though most households in a given community face the same water quality at the tap, some may treat their water prior to consumption while others may not [14,15]. This variation in household water management is influenced by a variety of factors, including knowledge of water treatment practices prior to distribution, perceptions of water quality at the tap, and socio-demographic characteristics of the decision-maker [14,16,17]. For example, previous studies have found that households are more likely to treat their tap water when they believe that government or community treatment facilities are ineffective [18,19], or when they believe that water quality is low at the tap [15]. While previous studies have examined the importance of these factors individually, few studies have considered these multiple drivers within the same analysis. Doing so is important because it identifies which factors are the most influential for household decision-making. This knowledge can then be used to identify and target interventions that are in line with current household perceptions, which has been shown to result in a greater rate of intervention uptake and success [20].

Our study assesses which factors most strongly influence household water management decisions, specifically whether households filter or boil their tap water prior to consumption or whether they obtain drinking water from multiple sources, in rural Puerto Rico. It is important to understand household water management in this region because previous studies have suggested that broader water management institutions do not always provide adequate water quality at the tap, particularly in rural, mountainous regions that are far from government treatment facilities [21]. There are two broad categories of institutions that manage drinking water for the island's four million people: government-managed Puerto Rico Aqueduct and Sewer Authority (PRASA) systems (which serve approximately 3.8 million people), and private and community non-PRASA systems (which serve approximately 400 communities, or up to 250,000 people), which are found primarily in mountainous regions that are too far to be connected to PRASA treatment facilities [21,22]. While the non-PRASA category encompasses a range of management strategies, given decentralized management where each community typically develops their own management plan, it is widely believed that non-PRASA communities in general are exposed to low water quality at the tap due to ineffective management of water prior to distribution. The Puerto Rico Department of Health (PRDOH) considers non-PRASA systems to be a health threat since they typically do not comply with federal water quality standards [23]. This is because about fifty percent of non-PRASA systems obtain water from surface sources and there is little or no monitoring of water quality in these communities [24]. Previous studies estimate that 30% of non-PRASA systems lack any water treatment infrastructure [22], and water is not treated consistently even when water infrastructure exists [22,25]. PRASA systems on the other hand typically filter and chlorinate water at treatment facilities before distribution and provide water quality assessments required by the U.S. Federal Potable Water Standards. Despite centralized management, PRASA systems are often plagued by water shortages and high rates of sediment loading and turbidity, which can result in non-compliances with the US Environmental Protection Agency (EPA) water quality standards [26]. This is because many filtration plants, particularly in mountainous regions, are not equipped to handle water filtration during periods of heavy rainfall [23], which is especially problematic given Puerto Rico's high frequency of tropical storms [27].

Given the possibility of inadequate water treatment by non-PRASA and PRASA facilities, some households have developed management strategies that are thought to improve drinking water quality prior to consumption. These strategies include filtering or boiling tap water or obtaining water from alternate sources like private wells and local markets. In this study, we assessed which perceptional factors that have been postulated to be important in previous literature are most associated with households that undertake water management strategies in rural, mountainous Puerto Rico [15,18,19]. Specifically, we predict the following in order of importance:

(1) households will have different management techniques depending on whether water is provided by government (PRASA) or community (non-PRASA) institutions likely due to differences in water quality at the tap;

(2) households that have problems with institutional water management prior to distribution are more likely to treat water;

(3) households are more likely to treat water if they perceive that water from the tap is of low quality;

(4) households that have less knowledge about how their water is treated prior to distribution are more likely to treat their water.

We quantify the relative importance of these various factors for household decision-making to better guide future water quality assessments and interventions in rural Puerto Rico. While our results are specific to Puerto Rico, we argue that our methodology can also be implemented in other regions to better understand the drivers of household water management and more effectively target interventions to those households vulnerable to low water quality.

Methods

Study site

Data were collected in eight different community sectors within the Cayey Mountain range in Puerto Rico from June to August of 2009. Our study focused on communities in this region because they are thought to be at high risk for low water quality given that they are rural and found in mountainous terrain, which makes them difficult to connect to PRASA treatment facilities. We specifically focused on villages found in Cayey and Patillas municipalities (Figure 1), which contain a large number of non-PRASA communities. Both municipalities are similar in socio-economic and development status. The median household income was $10,923 in Cayey and $9,375 in Patillas in 2000, which were lower than the island average of $13,189 [28]. We selected PRASA and non-PRASA communities that were adjacent to one another in each of the two municipalities. This was possible when we interviewed communities at the boundary where PRASA systems stopped serving communities with piped government water. This paired sampling design reduced possible confounding effects from socio-economic and geographic factors and allowed us to better assess whether households make different decisions based on if PRASA or non-PRASA institutions manage their water. Initial communities (n = 2) were selected based on where our field team had previous experience and knew PRASA and non-PRASA communities were adjacent to one other. We then used a snowball technique and visited additional communities (n = 6) that were suggested to us by the initial community contact [29]. While the communities that we selected for sampling were not entirely selected at random given this snowball technique, we believe that they are representative of the broader region given that each of our four pairs of PRASA and non-PRASA communities were spread across a wide geographic area in the Cayey mountain range (up to 15 km between our four sites).

Data collection

We surveyed 218 respondents across the eight community sectors considered in our study. Each community sector ranged in size from 50 to 200 households, but to ensure comparability we selected adjacent PRASA and non-PRASA communities that were approximately the same size. We aimed to interview 20 to 30 households in each community, and selected survey households at random distributed equally throughout each community. A summary of the number of survey respondents in PRASA verus

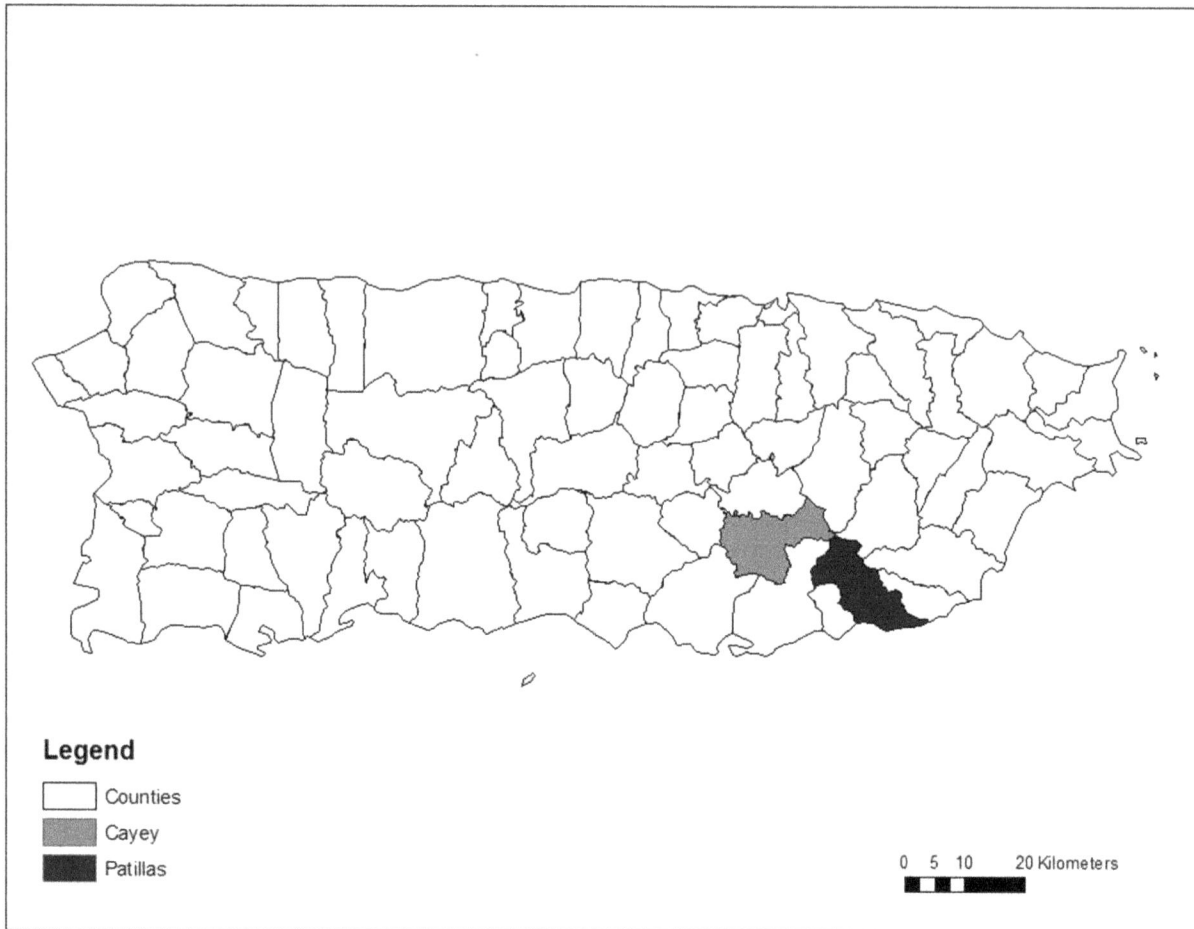

Figure 1. Map of Study Region in Puerto Rico. Municipalities where surveys were conducted are highlighted in gray. We did not list specific communities that we visited to keep the communities we surveyed anonymous.

non-PRASA communities is given in the supplementary information (Table S1). We then spoke to the household member who answered the door and identified which member of the household was in charge of household water management decisions. If that family member was home, we then conducted the oral structured survey with that family member. If the family member in charge of water management decisions was not at home we skipped that household and did not include it in our survey sample.

Ethics statement. Surveys were approved by the Columbia University Institutional Review Board under protocol number IRB-AAAE0079 and informed consent was written. Surveys were conducted in Spanish by local research assistants. We asked all respondents if we could audio record their interviews in order to keep a record of responses and to assist in confirming written responses and only did so if the interviewee gave permission. Our survey instrument contained questions related to whether households undertake any drinking water management prior to consumption, the respondent's perceptions of institutional water management and water quality at the tap, and socio-demographic information for the respondent. Details about each question are listed below, and all data collected were self-reported.

We asked respondents how they managed their drinking water sources prior to consumption, which serves as the dependent variable in our analyses. We grouped responses into two different types of strategies that households may undertake to cope with

inadequate water quality. One coping strategy is to increase *the number of drinking water sources* used in the household. Households may diversify sources of drinking water by purchasing bottled water or obtaining drinking water from a personal well. The second coping strategy considered in this study is if households *treat tap water before drinking*. If households believe that their tap water is of inadequate quality, they may filter or boil it before drinking.

We also collected data on the following variables that have been suggested to be important for household water management decisions in previous studies. These variables serve as covariates in our statistical models and we discuss specific data that were collected for each variable of interest. As outlined in the introduction, we believe that management institution type, problems with institutional water management, perceptions of water quality, and knowledge of water treatment will influence household decisions to manage drinking water.

Management institution. We considered *the type of institution that manages water* (i.e. *PRASA or non-PRASA*) as a fixed effect because the way that specific institutions manage water may influence household decision-making. This may occur if institutions influence the behavior of households via uniform rules and norms [30]. Institutions may also affect household decision-making if they expose all households in a given community to the same quality of resource. Previous studies have shown that mismanagement of water treatment by institutions may negatively impact water quality experienced by all households within the distribution system [14].

Problems with institutional management. As a broad measure of whether households believe that institutions effectively manage water, which has been shown to be important in the previous literature [18,19], we asked households whether they *have problems with how their water is managed* by PRASA or non-PRASA operators. We predict that respondents who have more problems with institutional management are more likely to treat tap water since they may believe that their water was inadequately treated before distribution.

Perceptions of water quality at the tap. Even though all households in a given community are exposed to the same water quality at the tap, varying perceptions may lead to heterogeneous behavior among decision-makers. Previous studies have shown that perceptions of water quality are strong drivers of household water management decisions [15]. To assess water quality perceptions, we asked respondents to rank the *quality of their tap water* on a scale of 1 to 4, where 1 equals poor water quality and 4 equals excellent water quality. We predict that households that believe they have poor water quality are more likely to develop coping strategies.

Knowledge of institutional management. Given that previous studies have suggested that increased knowledge of institutional management practices influences individual decision-making [31], we asked respondents whether they *knew how their water was treated* before it is piped to their homes. We predict that households that have less knowledge of how their water was treated by management institutions are more likely to treat water given that they may not trust that their water was treated prior to distribution. Previous studies have suggested a link between increased knowledge, transparency, and trust [32,33].

Socio-economic and demographic variables. Various socio-economic and demographic factors, such as income, age, and gender of the decision-maker, can influence household decisions [34,35]. We considered the age and gender of the respondent as controls in our analysis, but did not include income in our final models because only half of our interviewees responded to this question. Income data were collected as self-reported annual income for the household in $10,000 US increments (e.g. $10,000–$20,000, $20,000–$30,000, etc.). However, to test whether income may be important for water management decisions in our region, we ran our statistical models on the subset of data with income. We found that the income variable was never significant ($p > 0.05$), suggesting that it is not a significant driver of water management decisions in this region. Furthermore, since we are interested in quantifying the relative importance of various perceptional and socio-demographic factors for decision-making, excluding income from the analysis should not impact our results; instead, it would at most reduce the amount of variance explained by our models.

Statistical analyses

We conducted three sets of analyses to identify how water management and the drivers of water treatment decisions varied across households in our study. First, we used ANOVA to compare institution types for our two dependent variables of interest: the number of water sources and water treatment. We also compared the distribution of our covariates between institution type using ANOVA analyses. These simple comparisons illustrate whether there were significant differences in coping strategies, perceptions, and socio-demographic factors between households in PRASA and non-PRASA communities.

In a second set of analyses, we used separate logistic regressions to assess the effects of all covariates (Table 1) on the two response variables of interest. To assess whether these covariates have different effects on household decision-making in PRASA and non-PRASA communities, we included interactions between management institution (i.e. PRASA, non-PRASA) and the other covariates. To avoid parameter tradeoffs and clarify interpretation of the results, we dropped covariates that had a correlation > 0.4. Based on this criterion, we dropped gender from our analysis. We then conducted stepwise variable selection using AIC_c to select the best model [36]. To facilitate the interpretation of effect magnitudes among covariates, all continuous predictors were standardized by subtracting their mean and dividing by twice their standard deviation [37]. Goodness of fit was calculated using the universal goodness of fit le Cessie and Houwelingen test [38] in the Design package (Version 2.3-0) in R Project Software (R Statistical computing 2012, Version 2.14.1 was used for all analyses).

Finally, to assess the relative importance of each variable, we dropped each variable one at a time from the best logistic regression model and compared the AIC_c from the resulting model with the AIC_c from the best model. Variables that contributed most to model fit, and therefore were the most important in our analysis, had the largest change in AIC_c between the best model and the model with the variable in question dropped [39].

Results

ANOVA results

Several variables differed between PRASA and non-PRASA households (Table 2). Considering water management strategies, non-PRASA households were significantly more likely to obtain water from multiple sources, whereas PRASA households were significantly more likely to treat their tap water before drinking. This simple analysis suggests that households in PRASA and non-PRASA communities mitigate perceived low water quality in different ways. Considering perceptional variables, Non-PRASA households were significantly more likely to know how their institutions managed drinking water prior to distribution and non-PRASA households were also more likely to report higher water quality than PRASA households (Table 2).

Logistic regression models

The most important predictor of household decisions to obtain water from multiple sources was the institution that manages water (e.g. PRASA vs non-PRASA; Table 3, Figure 2A). Respondents in non-PRASA communities were more likely to obtain water from multiple sources than those from PRASA communities. Using the le Cessie and Houwelingen goodness of fit test, there is not a significant difference between observed and predicted values from the model suggesting good model fit ($z = 0.78$, $sd = 0.19$, $p = 0.44$).

The best predictors of household decisions to treat tap water before drinking were the institution that manages water, perceptions of water quality, and the interaction between the institution that manages water and problems with institutional management (Table 3, Figure 2B). PRASA households were significantly more likely to treat their water before drinking than non-PRASA households. Households that reported lower water quality were also more likely to treat their tap water, regardless of water management institution. Finally, the significant interaction between the institution that manages water and whether a household reported problems with institutional management suggests that non-PRASA households that had problems with institutional management were more likely to treat tap water before drinking than PRASA households. Le Cessie and

Table 1. Description and hypothesized relationship for each of the variables considered in our statistical models.

Variable	Variable Code	Description	Hypothesis
Number of Water Sources	Num Source	Number of drinking water sources (0= one source, 1= multiple sources)	Dependent Variable
Treat Water	Treat Water	Whether a household filtered or boiled tap water before drinking (0= No, 1= Yes)	Dependent Variable
Institution Type	Water System	Which water system the household receives water from (i.e. PRASA =0, Non-PRASA =1)	+
Knowledge of Treatment	Treatment Knowledge	Identified if individual had knowledge of how institution (PRASA or Non-PRASA) treated water before it arrives at the tap (i.e. No =0, Yes =1)	-
Reported Problems with Institutional Management	Problems	Whether the respondent reported problems with the way institutions manage water (i.e. No =0, Yes =1)	+
Perceptions of Water Quality	Water Quality	Self-reported quality of drinking water from the tap (i.e. poor =1, fair =2, good =3, excellent =4)	-
Demographic Data	Age	Age	Control
Gender	Gender	Gender (0= Male, 1= Female)	Control

Variable, coding method, description, and the hypothesized relationship with the likelihood of adopting coping strategies for all covariates considered in both statistical models. A positive relationship indicates that the variable would lead to increased coping, as defined by a higher likelihood of treating water and obtaining water from multiple sources.

Houwelingen goodness of fit test indicated a good fit between predicted and observed data ($z = -1.31$, sd $= 0.14$, p $= 0.19$).

Variable importance

To understand the relative importance of each covariate considered in our logistic models (Table 1), we conducted a full model logistic regression and assessed the importance of each factor based on its contribution to model fit as measured by the change in AIC_c when that variable was dropped from the full model. In the model that predicted which households were more likely to obtain water from multiple sources, we found that the institution that manages water contributed most to model fit (Figure 3A). This suggests that whether households were from PRASA or non-PRASA communities was the most important variable for predicting whether households obtain water from multiple sources. The remainder of the variables in the model contributed little to model fit.

For the model that identified whether households treat or do not treat water, the institution that manages water was also the best predictor (Figure 3B). This suggests that whether households are

from PRASA or non-PRASA communities was the most important variable to explain whether households treat or do not treat their water. Perceptions of water quality also contributed significantly to model fit (Figure 3B) suggesting that this variable is also important.

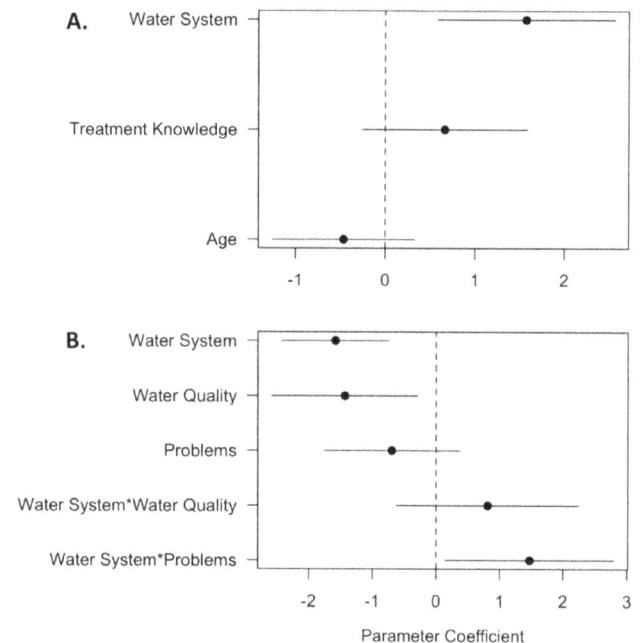

Table 2. Comparison of each variable considered in our statistical models by institution type (PRASA vs non-PRASA).

Variable	Mean value by Institution		ANOVA results		
	PRASA	Non-PRASA	d, f	F	P
Number of Water Sources	0.06	0.26	1, 187	14.28	<0.001*
Treat Water	0.71	0.42	1, 187	16.26	<0.001*
Treatment Knowledge	0.49	0.73	1, 187	11.74	<0.001*
Problems	0.38	0.49	1, 187	2.47	0.12
Water Quality	2.41	2.98	1, 187	21.69	<0.001*
Age	53.20	50.25	1, 187	1.35	0.25

Mean value by institution (i.e. PRASA, Non-PRASA) and ANOVA results (degrees of freedom, F-statistic, p-value) are reported for each variable. * indicates p<0.05.

Figure 2. Parameter Estimate Plot of All Variables Considered in the Two Models that Predict Household Water Management Strategies. Standard errors are plotted as black lines. The variable is significant if standard error bars do not cross the zero axis. For the number of water sources (A), institution type is significant (p<0.005). For whether households treat water (B), institution type (p<0.001), perceptions of water quality (p<0.05), and the interaction between institution type and if households have a problem with institutional management (p<0.05) are significant.

Table 3. Results for each statistical model predicting which factors are associated with household water management strategies.

Response Variable	Covariates considered in logit model	Parameter Coefficient (Standard Error)	p value	N	GOF (p value)
Number of Water Sources	Water System	1.57 (0.52)	<0.005*	189	0.44
Number of Water Sources	Treatment Knowledge	0.66 (0.48)	0.17	189	0.44
Number of Water Sources	Age	−0.47 (0.42)	0.27	189	0.44
Treat Water	Water System	−1.58 (0.44)	<0.001*	189	0.19
Treat Water	Water Quality	−1.43 (0.60)	0.02*	189	0.19
Treat Water	Problems	−0.69 (0.56)	0.23	189	0.19
Treat Water	Water System*Water Quality	0.81 (0.75)	0.28	189	0.19
Treat Water	Water System*Problems	1.47 (0.70)	0.04*	189	0.19

Variables considered, parameter coefficients with standard error, p values, sample size, and goodness of fit for both of the full models including interaction terms. The first model predicts whether households obtain water from one or more sources, and the second model predicts whether households treat or do not treat their water. Significance of at least 5% is highlighted with a *.

Discussion

Policy-makers and agencies have increasingly recognized the importance of household water management for potable water provisioning given that households are the last point of management prior to consumption [13]. By understanding which factors most influence household water management, policy makers can better identify and target intervention strategies that improve access to clean drinking water. In this study, we examined household water management in rural Puerto Rico. It is important to understand household level management in these communities given that both government PRASA and community non-PRASA water treatment may be ineffective at providing clean water at the

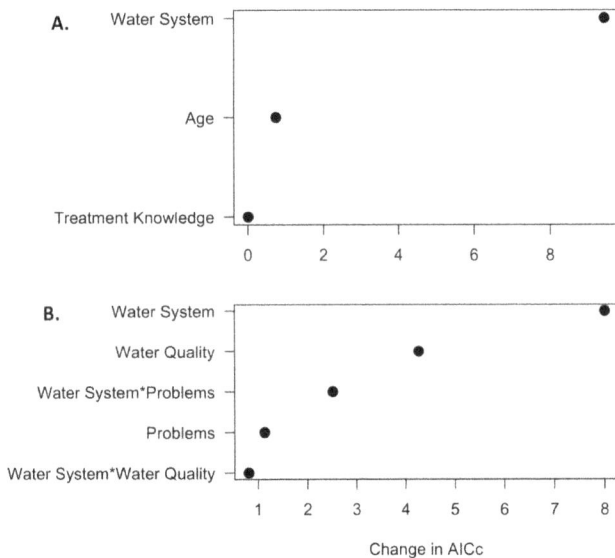

Figure 3. Importance of Each Covariate for Model Fit in the Two Models that Predict Household Water Management Strategies. Change in AIC$_c$ for each of the covariates considered in the full logit model for the number of drinking water sources (A) and whether households treat or do not treat water (B). Larger changes in AIC$_c$ values suggest that the variable contributed more to overall model fit. In both analyses (A and B), the institutional variable Water System (i.e. PRASA, non-PRASA) is the variable that contributes most to overall model fit. In the analysis of whether households treat water (B), water quality perceptions were also an important variable.

tap. Specifically, we analyzed (1) whether households obtained water from multiple sources or filtered or boiled tap water before drinking, and (2) which perceptional and socio-demographic factors were most associated with these management decisions. Our analysis suggests that three of our four initial predictions are correct: households manage water differently based on whether they are in PRASA or non-PRASA communities, households are more likely to treat water if they have problems with institutional management, and households are more likely to treat water if they believe that their tap water is of low quality (Figure 2). The fourth factor we predicted to be important in our analysis, whether households had knowledge of how water was treated prior to distribution, was not significant in our analyses.

The institution that manages water (i.e. PRASA vs non-PRASA) was the strongest driver of household drinking water management (Figure 3). PRASA households were more likely to filter or boil their water before drinking, whereas non-PRASA households were more likely to obtain water from multiple sources (Figure 2A). Differences in management strategies between PRASA and non-PRASA communities may be due to differences in perceptions of low water quality, possibly because of differences in water quality at the tap [14]. In PRASA communities, our informal discussions with community members indicate that perceptions of low water quality are due to the bad taste and odor of tap water, which community members attribute to over-chlorination. PRASA treatment facilities typically add chlorine to water prior to distribution, which has been associated with a reduction in bacteria such as *Escherichia Coli* (E. Coli) [40,41]. However, based on our informal interviews with community members across our survey area, it is possible that PRASA systems are over-chlorinating water in this region; these anecdotal claims are bolstered by objective water quality measures collected by the government for the *barrios* (sub-districts) considered in our study, which show periods when chlorine levels are higher than those recommended by the EPA (> 4.0 ppm, Fig. S1) [24,42,43]. Thus, in PRASA communities, families may filter or boil their tap water in order to improve the smell and taste of water prior to consumption. In non-PRASA communities, discussions with community members suggest that perceptions of low water quality are due to turbidity, which community members attribute to the lack of treatment by non-PRASA institutions. Based on discussions with community members and the operators of non-PRASA systems, it appears as if water was not regularly treated (e.g. via chlorine addition or filters) in storage tanks prior to distribution, which resulted in increased water turbidity at the tap. Households mitigated this perceived low water quality by obtaining water from other sources, like store-bought bottled water or filtered water from friends and relatives in PRASA communities.

Second, households are more likely to treat water if they believe that their water was ineffectively managed by treatment facilities prior to distribution. This corroborates previous studies that show that households and communities increase water management efforts if they believe that government or private agencies ineffectively manage water prior to distribution [18,19]. This result is only significant for non-PRASA communities (Figure 2B), suggesting that perceptions of institutional effectiveness drive decisions to treat water only in non-PRASA households. Institutional perceptions may play a stronger role in non-PRASA relative to PRASA communities because institutional management of drinking water is decentralized; given decentralized management, households in non-PRASA communities often play a stronger role in community-level water management than do households in PRASA communities, where water management is centralized within government agencies. Informal discussions with non-PRASA community members support this interpretation: non-PRASA households state that they feel a strong connection to water management institutions due to increased knowledge of treatment practices (Table 2) and the ability to participate in water management by speaking with local water operators or attending community meetings.

Finally, we found that perceptions of water quality were significant predictors of whether households were more likely to treat their water via filtering and boiling (Figure 2B). These results corroborate previous studies that find that households are more likely to manage their water if they perceive that their tap water is of low quality [15]. It is important to note that we only examined water quality perceptions and not objective water quality metrics at the household level, and it is unclear how well these two measures correlate with one another. If these two measures are not related, this could lead to water management decisions that result in low drinking water quality. For example, households may perceive that their water is of good quality, resulting in no treatment at the tap, when in reality objective water quality measures show that water treatment is required prior to consumption. Future studies should measure objective water quality standards in this region both before and after household treatment of drinking water to determine whether households are accurately perceiving low water quality and treating water effectively.

Based on the three main findings outlined above, we have several recommendations to improve water quality management in this region. First, we argue that both PRASA and non-PRASA institutions would likely improve water quality if they took household perceptions into account and understood how households manage water after it is distributed to the tap. Specifically, PRASA systems may improve water quality if they take steps to improve the taste and odor of tap water. If this low water quality is caused by over-chlorination as many people in PRASA communities believe, these systems should reduce the amount of chlorine used or let chlorinated water sit in storage tanks to allow for dechlorination prior to distribution while controlling for environmental variables that may increase chlorination byproducts [44]. Non-PRASA systems, on the other hand, may benefit by reducing the amount of turbidity at the tap, possibly by filtering water prior to distribution; this, and chlorination, may reduce perceived low water quality at the household scale. Second, objective water quality assessments should be coupled with these household level survey results to focus intervention strategies on the most

vulnerable populations, particularly those households that have low water quality but do not treat their water or that treat their water ineffectively. For example, PRASA households perceive low water quality due to bad taste and odor possibly caused by over-chlorination, however, one of the main strategies to mitigate this problem is filtering tap water. Yet to dechlorinate water, expensive active carbon filters are required [45] and these filters were typically not used in this region, suggesting that household strategies to filter water may be ineffective at reducing chlorine content. Finally, given that perceptions of institutional effectiveness appear to influence household management decisions, particularly in non-PRASA communities, we argue that these agencies should strengthen perceptions of institutional effectiveness by increasing the involvement of local community members in water management decisions. If community members have an increased say in how water is managed prior to distribution, it is likely that there will be improved water management given that household-level concerns about water quality are more likely to be addressed [46,47].

It is important to note that this study examined household perceptions of water quality and management, and it is possible that these perceptions are inaccurate when compared to objective measures. For example, most PRASA households believed that the bad taste and odor of tap water were caused by over-chlorination at treatment plants prior to distribution, but it is possible that the bad taste and odor were caused by other factors, like the addition of air or exposure to old pipes during the distribution process [48,49]. Future work should quantify objective water quality and assess whether current management strategies are effective at improving water quality prior to consumption. Second, we conducted our analyses based on survey data collected for over 200 people who live in the Cayey Mountain range. It is possible that our results would differ if we increased the scope of this study, particularly to other regions in Puerto Rico that may have different management strategies in PRASA and non-PRASA systems. Future studies should conduct similar perceptional studies across the island to better identify how universal the findings of this study are. Finally, it is important to note that we used the broad category of non-PRASA to encompass a wide range of institutions. Given that non-PRASA management is decentralized and individual communities are making water management decisions, it is possible that each non-PRASA system managed water slightly differently prior to distribution. We argue, however, that the coarse institutional categorization of non-PRASA is important particularly for policy given that the government uses this coarse categorization in water quality and compliance monitoring [23]. Future work should examine the heterogeneity in water management across non-PRASA systems to identify whether certain management strategies result in different outcomes for water quality and management at the household scale.

In conclusion, this study highlights the importance of social surveys and decision-making analyses to better identify how households currently manage drinking water and which factors influence household management decisions. Our results suggest that both community-level properties, like the type of institution that manages water prior to distribution, and household-level factors, like water quality perceptions, are important for predicting household-level water management behavior. By understanding household perceptions of both water quality and treatment of water prior to distribution, policy-makers can better identify and target intervention strategies that are tailored to current household decision-making. This is important given that previous studies have suggested that policies have a higher chance of uptake and success if they are created considering the local context [20].

Supporting Information

Figure S1 Free chlorine levels in ppm in PRASA and non-PRASA communities across our survey area. Data for PRASA communities were obtained from government databases collected at the barrio level, and data for non-PRASA communities were collected by our field team across several of our study communities of interest. These data suggest that free chlorine levels are typically lower in non-PRASA communities than PRASA communities, and several PRASA measurements have free chlorine levels higher than those recommended by the EPA (4.0 ppm, dotted horizontal line). This suggests that there may be over-chlorination in some PRASA communities.

Table S1 Number of interviewees in Non-PRASA and PRASA communities in our two study municipalities. We

do not provide specific names of the communities or sectors surveyed in order to keep anonymity of our participants.

Acknowledgments

We would like to thank our field assistants Natalia Rodriguez and Yazmin Rivera, who conducted the household interviews and helped enter data during the summer of 2009, and Derek Berezdivin, who also helped enter data in 2010.

Author Contributions

Conceived and designed the experiments: MJ YL MU. Performed the experiments: MJ YL. Analyzed the data: MJ. Wrote the paper: MJ MU JA YL.

References

1. Teunis P, Medema GJ, Kruidenier L, Havelaar AH (1997) Assessment of the risk of infection by Cryptosporidium or Giardia in drinking water from a surface water source. Water Research.
2. Hellard ME, Sinclair MI, Forbes AB, Fairley CK (2001) A randomized, blinded, controlled trial investigating the gastrointestinal health effects of drinking water quality. Environ Health Perspect 109: 773–778.
3. Prüss A, Kay D, Fewtrell L, Bartram J (2002) Estimating the burden of disease from water, sanitation, and hygiene at a global level. Environ Health Perspect 110: 537–542.
4. Trevett AF, Carter RC, Tyrrel SF (2004) Water quality deterioration: A study of household drinking water quality in rural Honduras. International Journal of Environmental Health Research 14: 273–283.
5. Hunter PR, Toro GIR, Minnigh HA (2010) Impact on diarrhoeal illness of a community educational intervention to improve drinking water quality in rural communities in Puerto Rico. Bmc Public Health 10,
6. Cash DW, Adger WN, Berkes F, Garden P, Lebel L, et al. (2006) Scale and cross-scale dynamics: Governance and information in a multilevel world. Ecology and Society 11.
7. Berkes F (2006) From community-based resource management to complex systems: The scale issue and marine commons. Ecology and Society 11.
8. Sarker A, Ross H, Shrestha KK (2008) A common-pool resource approach for water quality management: An Australian case study. Ecological Economics 68: 461–471.
9. Lebel L, Garden P, Imamura M (2005) The politics of scale, position, and place in the governance of water resources in the Mekong region. Ecology and Society 10.
10. Saravanan VS (2008) A systems approach to unravel complex water management institutions. Ecological Complexity 5: 202–215.
11. Mintz E, Bartram J, Lochery P, Wegelin M (2001) Not Just a Drop in the Bucket: Expanding Access to Point-of-Use Water Treatment Systems. American Journal of Public Health 91: 1565–1570.
12. Clasen TF, Cairncross S (2004) Editorial: Household water management: refining the dominant paradigm. Tropical Medicine & International Health 9: 187–191.
13. Sobsey MD (2002) Managing water in the home: accelerated health gains from improved water supply. World Health Organization.
14. Gartin M, Crona B, Wutich A, Westerhoff P (2010) Urban Ethnohydrology: Cultural Knowledge of Water Quality and Water Management in a Desert City. Ecology and Society 15: 36.
15. Hu Z, Morton LW, Mahler RL (2011) Bottled Water: United States Consumers and Their Perceptions of Water Quality. Int J Env Res Pub He 8: 565–578.
16. Fielding KS, Russell S, Spinks A, Mankad A (2012) Determinants of household water conservation: The role of demographic, infrastructure, behavior, and psychosocial variables. Water Resour Res 48.
17. Sabau G, Haghiri M (2008) Household willingness-to-engage in water quality projects in western Newfoundland and Labrador: a demand-side management approach. Water and Environment Journal 22: 168–176.
18. Zérah M (2000) Water, unreliable supply in Delhi. Manohar Publishers.
19. Katuwal H, Bohara AK (2011) Coping with poor water supplies: empirical evidence from Kathmandu, Nepal. J Water Health 9: 143–158.
20. Jehu-Appiah C, Aryeetey G, Agyepong I, Spaan E, Baltussen R (2012) Household perceptions and their implications for enrolment in the National Health Insurance Scheme in Ghana. Health Policy and Planning 27: 222–233.
21. Molina-Rivera W (1998) Estimated water use in Puerto Rico, 1995. Washington D.C.: US. Geological Survey Open-File Report.
22. Guerrero-Preston R, Norat J, Rodriguez M, Santiago L, Suarez E (2008) Determinants of compliance with drinking water standards in rural Puerto Rico between 1996 and 2000: a multilevel approach. P R Health Sci J 27: 229–235.
23. Quinones F (2005) PRASA has ample water supplies. Water Industry News.
24. Environmental Protection Agency (2010) Puerto Rico Aqueduct and Sewer Authority (PRASA) Pollutant Discharge Settlement. Environmental Protection Agency.

25. Toro GR, Minnigh HA (2004) Regulation and Financing of Potable Water Systems in Puerto Rico: A Study in Failure in Governance. AWRA Dunde: Scotland.
26. de Cardenas SC (2011) Does private management lead to improvement of water services? Lessons learned form the experience of Bolivia and Puerto Rico. University of Iowa.
27. Boose E, Serrano M, Foster D (2004) Landscape and regional impacts of hurricanes in Puerto Rico. Ecol Monogr 74: 335–352.
28. US Bureau of the Census (2000) Census of population: social and economic characteristics. Washington D.C. USA: Department of Commerce, Economics, and Statistics Administration.
29. Biernacki P, Waldorf D (1981) Snowball Sampling: Problems and Techniques of Chain Referral Sampling. Sociological Methods & Research 10: 141–163.
30. North DC (1991) Institutions. The Journal of Economic Perspectives 5: 97–112.
31. Makutsa P, Nzaku K, Ogutu P, Barasa P (2001) Challenges in implementing a point-of-use water quality intervention in rural Kenya. American Journal of Public Health 91: 1571–1573.
32. Peters RG, Covello VT, McCallum DB (1997) The determinants of trust and credibility in environmental risk communication: an empirical study. Risk Analysis 17: 43–54.
33. Palanski ME, Kahai SS, Yammarino FJ (2011) Team Virtues and Performance: An Examination of Transparency, Behavioral Integrity, and Trust. J Bus Ethics 99: 201–216.
34. Jorgensen B, Graymore M, O'Toole K (2009) Household water use behavior: An integrated model. Journal of Environmental Management 91: 227–236.
35. Below TB, Mutabazi KD, Kirschke D, Franke C, Sieber S, et al. (2012) Can farmers' adaptation to climate change be explained by socio-economic household-level variables? Global Environmental Change 22: 223–235.
36. Hurvich C, Tsai C (1989) Regression and Time-Series Model Selection in Small Samples. Biometrika 76: 297–307.
37. Gelman A, Hill J (2007) Data Analysis Using Regression And Multilevel/Hierarchical Models. Cambridge Univ Pr.
38. le Cessie S, van Houwelingen J (1991) A Goodness-of-Fit Test for Binary Regression-Models, Based on Smoothing Methods. Biometrics 47: 1267–1282.
39. Burnham KP, Anderson DR (2002) Model Selection and Multimodel Inference. 2nd ed. New York: Springer.
40. Payment P, Trudel M, Plante R (1985) Elimination of viruses and indicator bacteria at each step of treatment during preparation of drinking water at seven water treatment plants. Appl Environ Microbiol 49: 1418–1428.
41. Arnold BF, Colford JM (2007) Treating water with chlorine at point-of-use to improve water quality and reduce child diarrhea in developing countries: a systematic review and meta-analysis. Am J Trop Med Hyg 76: 354–364.
42. Autoridad de Acueductos y Alcantarillados (2001-2010) Water Quality Reports.
43. Puerto Rico Department of Public Health (1996-2009) Water Quality Reports.
44. Chowdhury S, Champagne P, McLellan PJ (2009) Source of the Total Environment. Science of the Total Environment, The 407: 4189–4206.
45. Worley JL (2000) Evaluation of Dechlorinating Agents and Disposable Containers for Odor Testing of Drinking Water. Virginia Polytechnic Institute and State University.
46. Olsson P, Folke C, Berkes F (2004) Adaptive Comanagement for Building Resilience in Social-Ecological Systems. Environmental Management 34: 75–90.
47. Larson AM, Soto F (2008) Decentralization of Natural Resource Governance Regimes. Annu Rev Env Resour 33: 213–239.
48. Sangodoyin AY (1993) Water quality in pipe distribution systems. Environmental Management and Health.
49. Young WF, Horth H, Crane R, Odgen T, Arnott M (1996) Taste and Odour Threshold Concetrations of Potential Potable Water Contaminants. Water Research 30: 331–340.

9

Environmental Fate of Soil Applied Neonicotinoid Insecticides in an Irrigated Potato Agroecosystem

Anders S. Huseth[1], Russell L. Groves[2]*

1 Department of Entomology, Cornell University, New York State Agricultural Experiment Station, Geneva, New York, United States of America, 2 Department of Entomology, University of Wisconsin-Madison, Madison, Wisconsin, United States of America

Abstract

Since 1995, neonicotinoid insecticides have been a critical component of arthropod management in potato, *Solanum tuberosum* L. Recent detections of neonicotinoids in groundwater have generated questions about the sources of these contaminants and the relative contribution from commodities in U.S. agriculture. Delivery of neonicotinoids to crops typically occurs as a seed or in-furrow treatment to manage early season insect herbivores. Applied in this way, these insecticides become systemically mobile in the plant and provide control of key pest species. An outcome of this project links these soil insecticide application strategies in crop plants with neonicotinoid contamination of water leaching from the application zone. In 2011 and 2012, our objectives were to document the temporal patterns of neonicotinoid leachate below the planting furrow following common insecticide delivery methods in potato. Leaching loss of thiamethoxam from potato was measured using pan lysimeters from three at-plant treatments and one foliar application treatment. Insecticide concentration in leachate was assessed for six consecutive months using liquid chromatography-tandem mass spectrometry. Findings from this study suggest leaching of neonicotinoids from potato may be greater following crop harvest in comparison to other times during the growing season. Furthermore, this study documented recycling of neonicotinoid insecticides from contaminated groundwater back onto the crop via high capacity irrigation wells. These results document interactions between cultivated potato, different neonicotinoid delivery methods, and the potential for subsurface water contamination via leaching.

Editor: Christopher J. Salice, Texas Tech University, United States of America

Funding: This research was supported by the Wisconsin Potato Industry Board and the National Potato Council's State Cooperative Research Program FY11-13. The funders had no role in study design, data collection and analysis, decision to publish, or preparation of the manuscript.

Competing Interests: RLG has received research funding, not related to this project, from Bayer CropScience, DuPont, Syngenta, and Valent U.S.A.

* E-mail: groves@entomology.wisc.edu

Introduction

The neonicotinoid group of insecticides is among the most broadly adopted, conventional management tools for insect pests of annual and perennial cropping systems [1]. Benefits of the neonicotinoid group of compounds include flexibility of application, diversity of active ingredients, and broad spectrum activity [2]. Moreover, growers have readily adopted neonicotinoids for two specific reasons: first, these compounds are fully systemic in plants after soil application and second, several new generic formulations have recently become available which have incentivized their continued use in many crops [1–3]. Since 2001, the United States Environmental Protection Agency (EPA) has classified several neonicotinoids as either conventional, reduced-risk pesticides, or as organophosphate alternatives [4],[5]. EPA certification often requires replacement of older, broad-spectrum pesticides with newer, more specific products for management of key economic pests. Critical attributes of replacement insecticides include documented reductions in human and environmental risk when compared to older, broad-spectrum pesticides [5]. Despite acceptance of neonicotinoid insecticides as reduced-risk by growers and regulatory agencies, nearly two decades of wide-spread, repetitive use has resulted in several insecticide resistance

issues, impacts on native and domestic pollinators, and unanticipated environmental impacts [6–9].

The environmental fate of several neonicotinoid active ingredients have been previously assessed. Previous studies focused on degradation and movement processes in soil, leachate, and runoff [10–15]. The leaching potential of the neonicotinoids into groundwater, as well as persistence in the plant canopy, is related to properties of the chemicals and delivery method of the compound to the crop (Fig. S1)[12],[15],[16]. Soil application (e.g., seed treatment or in-furrow) has been adopted as the principal form of insecticide delivery in potato production as it provides the longest interval of pest control, while also reducing non-target impacts, and limits exposure to workers when compared to foliar application methods. Since 1995, soil-applied neonicotinoids (i.e., clothianidin, imidacloprid, thiamethoxam) have been the most common pest management strategy used to control infestation of Colorado potato beetle, *Leptinotarsa decemlineata* Say; potato leafhopper, *Empoasca fabae* Harris; green peach aphid, *Myzus persicae* Sulzer; and potato aphid, *Macrosiphium euphorbiae* Thomas. The now widespread and extensive use of these systemic neonicotinoid insecticides, coupled with the recent detection of thiamethoxam in groundwater [17],[18], supports the hypothesis that potato pest management may contribute a portion of the documented neonicotinoid contaminants reported in

Wisconsin, USA. Furthermore, we hypothesized that neonicotinoid insecticides applied to potato are most vulnerable to leaching in the spring season when the root system of the plant has yet to fully exploit all of the active ingredient applied directly in the seed furrow. Large rain events at this time could drive insecticide leaching from potato and subsequent groundwater contamination at large scales. In this study, we examined how neonicotinoid concentrations in leachate were altered in response to different insecticide delivery methods using potatoes grown under commercial production practices. We also report the patterns of historic neonicotinoid insecticide detections in groundwater using water quality surveys collected by the Wisconsin Department of Agriculture, Trade and Consumer Protection-Environmental Quality Section (WI DATCP-EQ). Second, using potato as a model system, we analyzed leachate captured below different seed treatments, soil-applications, and foliar delivery treatments for thiamethoxam using liquid chromatography-tandem mass spectrometry (LC/MS/MS) over two consecutive field seasons. In this experiment, thiamethoxam was chosen as one representative insecticide in a broader group of water-soluble neonicotinoids. Moreover, this active ingredient represented the majority of positive neonicotinoid detections in groundwater monitoring surveys conducted by the WI DATCP-EQ [17], [18]. Third, using identical quantitative methods, we measured thiamethoxam concentration in irrigation water collected from operating, high-capacity irrigation wells at two time points in each sampling year. And finally, we characterize irrigation use and production trends of crops that may contribute to neonicotinoid detection in groundwater. Results of this study increase our understanding about the influence of insecticide delivery method on the neonicotinoid insecticides leaching from potato into the surrounding environment.

Materials and Methods

Ethics Statement

No specific permits were required for the field study described here. Access to field sites was granted by the private landholder to conduct leaching experiments. No specific permissions were needed to present publically available records provided by Wisconsin Department of Agriculture, Trade and Consumer Protection or Wisconsin Department of Natural Resources. Field studies did not involve any endangered or protected species.

Groundwater Contamination

Permanent groundwater monitoring wells, maintained by the WI DATCP-EQ, were used to measure neonicotinoid contamination of subsurface water resources as one component of an ongoing study documenting agrochemical (e.g., insecticides, herbicides, nutrients) impact on groundwater quality. Beginning in 2006, analytical water quality assessments for neonicotinoid contamination were conducted by the Wisconsin Department of Agriculture Trade and Consumer Protection-Bureau of Laboratory Services. Concentrations of acetamiprid, clothianidin, dinotefuran, imidacloprid, and thiamethoxam were monitored in 20–30 different monitoring well locations from 2006–2012. Presented are positive detections of those insecticides in different monitoring wells from 2006–2012 [17],[18]. Data provided by WI DATCP-EQ characterize the temporal and spatial profile of thiamethoxam and other neonicotinoid detections that occurred between 2008–2012. These data are presented in summary as a foundation for following objectives (Table 1).

Experimental Site and Design

In 2011 and 2012, leaching experiments were conducted 6 km west of Coloma, Wisconsin. Experiments were planted in two different fields approximately 0.5 km apart on 20 May 2011 and 11 May 2012. The soil at both sites consisted of Richford loamy sand (sandy, mixed, mesic, Typic Udipsamments) [19]. Soil composition was 7% clay, 82% sand, and 11% silt. Organic matter was 0.53 percent by weight. Study sites soils had a high infiltration rate (Hydrological Soil Group A), a high saturated hydraulic conductivity (K_{sat}) at 28 micrometers per second, and an available water capacity rating of 0.1 cm per cm [19]. No restrictive layer that would impede water movement through the soil has been documented [19]. Study site soil was formed in the bed of glacial Lake Wisconsin from parent material of glacial till overlain by glacial outwash [20]. Upper soil horizons (A and B) are sand with minimal structure. Subsurface soil (C horizon) had no structure. Irrigation pivots in sample fields withdrew water at a depth of 37 m and the water table depth (static water level) was approximately 6 m for both sites [21].

A randomized complete block design with four insecticide delivery treatments and an untreated control was established using the potato cultivar, 'Russet Burbank'. Plots were 0.067 ha in size and planted at a rate of one seed piece per 0.3 m with 0.76 m spacing between rows. Each year, experiments were nested within a different ~32 ha commercial potato field, and maintained under commercial management practices by the producer (e.g., nutrient application timing, chemical usage, tillage practices, etc.), with the exception of insecticide inputs. The decision to locate these experiments in commercial fields was, in part, based upon access to a center pivot irrigation system to best duplicate water inputs used to produce commercial potato in Wisconsin. All other inputs and production strategies (e.g. tillage, fumigation, fertility, and disease management) were conducted by the producer with equipment and products in a manner consistent with the best management practices for potato production in Wisconsin. Prior to planting in each season, a tension plate lysimeter (25.4×25.4×25.4 cm) was buried at a depth of 75 cm below the soil surface. Lysimeters were constructed of stainless steel with a porous stainless steel plate affixed to the top to allow water to flow into the collection basin over each sampling interval. Experimental blocks were connected with 9.5 mm copper tubing to a primary manifold and equipped with a vacuum gauge. A predefined, fixed suction was maintained under regulated vacuum at 107±17 kPa (15.5±2.5 lb per in^2) with a twin diaphragm vacuum pump (model UN035.3 TTP, KnF, Trenton, NJ) connected to a 76 L portable air tank. Each treatment block was equipped with a data-logging rain gauge (Spectrum Technologies, Inc. model # 3554WD1) recording daily water inputs at a five minute interval. Data was offloaded with Specware 9 Basic software (Spectrum Technologies, Inc., Plainfield, IL, USA) and aggregated into daily irrigation or rain event totals using the *aggregate* and *dcast* function in R (package: reshape2, [22]). Irrigation event records were obtained from the grower to identify days and estimated inputs of water application throughout the growing season.

Insecticides and Application

Thiamethoxam treatments (Platinum 75SG, 75% thiamethoxam per formulated unit, Syngenta, Greensboro, NC) were selected to represent a common, soil-applied insecticide in potato. A second formulation of thiamethoxam was selected to represent a common pre-plant insecticide seed treatment in potato (Cruiser 5FS, 47.6% thiamethoxam per formulated unit, Syngenta, Greensboro, NC). Each insecticide formulation is used to manage early season infestations of Colorado potato beetle, potato

Table 1. Positive (means±SD) neonicotinoid detections in groundwater from 2008–2012, State of Wisconsin Department of Agriculture Trade and Consumer Protection.

Year	County	Area potato (ha)[a]	Row crops (ha)[b]	Percent potato[c]	Well ID	N positive samples	Insecticide concentration (µg/L)[d]		
							clothianidin	imidacloprid	thiamethoxam
2008	Adams	2,617	21,385	10.9	6	2	-	-	4.34 (4.97)
	Grant	0	47,827	0.0	10	1	-	-	1.25
	Iowa	18	25,795	0.1	11,12,13	9	-	-	1.50 (0.67)
	Richland	29	9,582	0.3	16	1	-	-	0.69
	Sauk	30	31,931	0.1	17	2	-	-	2.41 (1.32)
	Waushara	2,630	29,447	8.2	20	2	-	-	0.67 (0.05)
2009	Adams	3,989	24,894	13.8	6	2	-	-	5.31 (5.12)
	Dane	22	101,527	0.0	9	1	-	-	1.61
	Iowa	343	33,375	1.0	11,12	3	-	-	1.31 (0.68)
	Richland	87	14,402	0.6	16	1	-	-	1.26
	Sauk	328	40,571	0.8	17	2	-	-	3.00 (0.94)
2010	Adams	4,188	24,871	14.4	6	4	3.43	-	2.97 (2.04)
	Brown	1	39,322	0.0	7	1	-	-	0.52
	Dane	34	110,979	0.0	8,9	4	0.54 (0.24)	0.54	1.08
	Grant	49	74,566	0.1	10	1	0.73	-	-
	Iowa	356	38,840	0.9	11,12,13	7	-	-	1.25 (1.02)
	Sauk	188	45,309	0.4	17	5	0.41	-	1.81 (0.88)
	Waushara	4,184	33,576	11.1	19,20	2	-	2.77 (0.81)	-
2011	Adams	4,066	27,693	12.8	2,5,6	9	0.63 (0.36)	0.33	0.63 (0.26)
	Brown	7	38,309	0.0	7	1	-	-	0.21
	Dane	33	107,214	0.0	8	2	0.62 (0.19)	-	-
	Grant	13	75,436	0.0	10	1	0.30	-	-
	Iowa	47	40,138	0.1	12	4	-	0.34 (0.09)	0.88 (0.23)
	Portage	7,364	45,324	14.0	15	1	-	-	0.32
	Sauk	213	46,686	0.5	17,18	5	0.54 (0.10)	-	1.92 (0.43)
	Waushara	4,536	36,676	11.0	19,20,21,23	23	0.25 (0.03)	0.78 (0.69)	1.40 (0.56)
2012	Adams	4,263	27,037	13.6	1,3,4,6	6	0.52 (0.30)	0.51 (0.26)	0.27
	Dane	11	115,501	0.0	8	1	0.67	-	-
	Grant	4	72,920	0.0	10	1	0.26	-	-
	Iowa	369	40,764	0.9	12	2	0.24	0.28	0.44
	Juneau	907	28,542	3.1	14	2	0.42 (0.18)	-	0.20
	Portage	7,622	46,337	14.1	15	2	-	0.47	0.47
	Waushara	5,904	38,999	13.1	21,22,23	13	-	0.68 (0.88)	1.51 (0.72)
	summary			N=23		67	25	30	68

Table 1. Cont.

Year	County	Area potato (ha)[a]	Row crops (ha)[b]	Percent potato[c]	Well ID	N positive samples	Insecticide concentration (μg/L)[d]		
							clothianidin	imidacloprid	thiamethoxam
						Average	0.62 (0.63)	0.79 (0.83)	1.59 (1.51)
						Range	0.21–3.34	0.26–3.34	0.20–8.93

[a] Acreage estimates generated from USDA National Agricultural Statistics Service – Cropland Data Layer, 2008–2012 [26].
[b] Row crops class is the sum of the following crop areas (ha): maize, soy, small grains, wheat, peas, sweet corn, and miscellaneous vegetables and fruits.
[c] Percent potato calculated as the potato area grown annually divided by total arable row crop acreage (other row crops + potato).
[d] Positive neonicotinoid detections extracted from long-term, groundwater wells maintained by the WI-DATCP-EQ Program.

leafhopper, and colonizing aphid in Wisconsin potato crops. Commercially formulated insecticides were applied at maximum labeled rates for in-furrow (140 g thiamethoxam ha^{-1}) and seed treatment (112 g thiamethoxam ha^{-1} at planting density of 1,793 kg seed ha^{-1}) for potato [23]. A calibrated CO_2 pressurized, backpack sprayer with a single nozzle boom was used to deliver an application volume of 94 liters per hectare at 207 kPa through a single, extended range, flat-fan nozzle (TeeJet XR80015VS, Spraying Systems, Wheaton, IL) for in-furrow applications. Spray applications were directed onto seed pieces in the furrow at a speed of one meter per second and furrows were immediately closed following application. Seed treatments were applied using a calibrated CO_2 pressurized backpack sprayer with a single nozzle boom delivering an application volume of 102.2 L per hectare at 207 kPa through a single, extended range, flat-fan nozzle (TeeJet XR80015VS, Spraying Systems, Wheaton, IL) was used for delivery of thiamethoxam in water (130 mL) directly to suberized, cut seed pieces (23 kg) 24 hours prior to planting. Seed treatments were allowed to dry in the absence of light at 20°C during that pre-plant period. A novel soil application method, impregnated copolymer granules, was included as another treatment in an attempt to stabilize applied insecticide in the soil. Polyacrylamide horticultural copolymer granules (JCD-024SM, JRM Chemical, Cleveland, OH) were impregnated at an application rate of 16 kg per hectare. The polyacrylamide treatment was included as a novel delivery method to stabilize insecticide in the rooting zone and possibly reduce leaching in the early season. Thiamethoxam (0.834 g, Platinum 75SG) was initially diluted in 250 mL of deionized water and 100 μL of blue food coloring was incorporated into solution to ensure uniform mixing (brilliant blue FCF). Insecticide solutions were mixed with 75 g polyacrylamide then stirred until the liquid was absorbed and a uniform color was observed. Impregnated granules were vacuum dried in the absence of light for 24 hours at 20°C. Treated granules were divided into even quantities per row and evenly distributed into the four treatment rows for each polyacrylamide plot. A single untreated flanking row was planted between plots. All soil-applied insecticides were applied on 20 May 2011 and 11 May 2012 at the time of planting.

Two foliar applications of thiamethoxam (Actara 25WG, 25% thiamethoxam per formulated unit, Syngenta, Greensboro, NC) sprayed on the same plot were included as a fourth delivery treatment. Two successive neonicotinoid applications are recommended for foliar control of pests in potato [23]. Foliar thiamethoxam was applied using a calibrated CO_2 pressurized backpack sprayer delivering an application volume of 187.1 liters per hectare at 207 kPa through four, extended range flat-fan nozzles (TeeJet XR80015VS, Spraying Systems, Wheaton, IL) spaced at 45.2 cm. The first foliar application was followed approximately seven days later with a second equivalent rate of thiamethoxam to total the season-long maximum labeled rate (105 g thiamethoxam ha^{-1}) [23] and were timed to coincide with the appearance of 1st and 2nd instar larvae of native populations of *L. decemlineata*. Foliar applications of thiamethoxam were applied on 28 June and 5 July in 2011 and 15 and 22 June in 2012. Although total amounts of active ingredient differ by formulation, these rates are identical to registered label recommendations [23] and reflect the maximum amount of active ingredient used on an average hectare of cultivated potato. Specific chemical properties of formulated thiamethoxam that affect solubility and leaching potential in soil can be found in Gupta et al. [15] and the references therein (Fig. S1).

Chemical Extraction and Quantification

Lysimeter leachate was sampled twice monthly beginning on June 1 of each year and concluding in October of 2011 and November of 2012. Total leachate volume was recorded for each plot. A 500 mL subsample was taken from each plot into a 0.5 L glass vessel and immediately placed on ice and refrigerated at 4–6°C in the laboratory prior to analysis. Samples were homogenized into a 400 mL monthly (i.e., two samples per month) sample as percent volume per volume dependent on total catch measured in the field. Neonicotinoid residues from monthly water samples were extracted using automated solid phase extraction (AutoTrace SPE workstation, Zymark, Hopkinton, MA) with LiChrolut EN SPE columns (Merk KGaA, Darmstadt, Germany). If visual inspection of sample found excessive sediment contamination, samples were filtered through a 0.45 μm filter prior to extraction. Columns were conditioned prior to extraction with 3 mL of methanol (MeOH) and 3 mL of water. 210 mL of sample were loaded onto columns and rinsed with 10 mL of water then dried under flowing nitrogen for 15 minutes (N-evap, Organomation, Berlin, MA). Samples were eluted using a 50% ethyl acetate (EtOAc) and 50% methanol solution to collect a 2 mL sample fraction. Sample extract fractions were analyzed using a Waters 2690 HPLC/Micromass Quattro LC/MS/MS (Waters Corporation, Milford, MA). All thiamethoxam residues were identified, quantified, and confirmed using LC/MS/MS by the Wisconsin Department of Agriculture Trade and Consumer Protection-Bureau of Laboratory Services. The method detection limit (MDL) of the extraction procedure was $0.2\ \mu g\ L^{-1}$. Specific conditions for all quantitative procedures follow WI-DATCP Standard Operating Procedure #1009 developed from Seccia et al. [24] and references therein.

Irrigation Use and Crop Area

To determine the extent of irrigated agriculture present within the watershed, we utilized current high capacity well pumping data and irrigated agriculture estimates derived from digital imagery. Publically available operator reporting data for high capacity agricultural pivots were obtained from the Wisconsin Department of Natural Resources Bureau of Drinking Water and Groundwater. Records included location information and pumping volume for the year 2012. High capacity wells service several irrigated fields and often these fields are further divided into individual crop management units each with unique irrigation requirements. We digitized the area watered by all identifiable center pivot, linear move, and traveling gun irrigation systems using digital aerial photography to measure the total number of management units present within the greater Central Wisconsin Water Management Unit watershed [25] (ArcGIS version 10.1, Redlands, CA). Fields were subdivided into management units using the consistent divisions in crop types with a sequence of National Agricultural Statistics Service Cropland Data Layer (NASS-CDL) [26] thematic data and aerial photography images [25] from 2010–2012.

To determine agronomic trends in the Central Sands vegetable production region of Wisconsin, we used a combination of publically available land use data and current neonicotinoid registration information. A geospatial watershed management boundary layer delineated by the Wisconsin Department of Natural Resources [27] was used to generally define the spatial extent where agriculture could be contributing to the detection of neonicotinoid insecticides in subsurface water. The Central Wisconsin Water Management Unit extent was used to estimate annual crop composition using the NASS-CDL [26] from 2006–2012 using ArcGIS. From these data, we selected major crops that

frequently receive either seed or in-furrow soil-applied neonicotinoid insecticide treatments. Application rates were identical for several similar crops (e.g. soybean and green bean), and so, we chose to aggregate crops based on insecticide rate and crop type into three primary groups: maize, beans, and potato [23],[28–30]. These crop groups comprise the majority of production area in the Central Wisconsin Water Management Unit extent. To our knowledge, limited information exists documenting the proportion of different soil-applied neonicotinoid active ingredients that are used on a per crop basis in the Central Wisconsin Water Management Unit. Based on this level of uncertainty, we chose not to extend tabulated crop areas to a direct calculation or estimate of neonicotinoid active ingredients applied.

Data Analysis

To determine the impact of different insecticide delivery treatments on thiamethoxam leachate detected over time, we reported the mean concentration over a period of several months. All lysimeter analyses included samples where neonicotinoid insecticides were not detected (i.e., zero detections). All data manipulation and statistical analyses of leachate concentrations were performed in R, version 2.15.2 [31] using the base distribution package. Functions used in the analysis are available in the base package of R unless otherwise noted. Observed concentration for time points in each year were subjected to a repeated-measures analysis of variance (ANOVA) using a linear mixed-effects model to determine significant delivery (i.e. treatment), date, and delivery×date effects ($P<0.05$). Because the agronomic conditions differed between years and given that our comparison of interest was at the insecticide delivery treatment level, insecticide concentrations were analyzed separately for each year. Mixed-effects models (i.e., repeated-measures analysis of variance) were fit using the *lme* function (package nlme, [32]). Empirical autocorrelation plots from unstructured correlation model residuals were examined using the *ACF* function (package nlme, [32]). Correlation among within-group error terms were structured and examined in three ways: first, unstructured correlation, second, with compound symmetry using the function *corCompSymm* and third, with autoregressive order one covariance using the function *corAR1* (package nlme, [32][33]). Since models were not nested, fits of unstructured, compound symmetry, and autoregression order one covariance were compared using Akaike's information criterion statistic with the function *anova* (test = "F"). Data were transformed with natural logarithms before analysis to satisfy assumptions of normality, however untransformed means are graphically presented. In 2012, a single lysimeter in the polyacrylamide treatment of the leachate study malfunctioned and these observations were dropped from subsequent analyses leading to an unbalanced replicate number for that treatment (N = 3) in 2012. Water input data collected from tipping bucket samplers were averaged across block by day and aggregated as cumulative water inputs using the *cumsum* function. All summary statistics and model estimates were extracted using *aggregate*, *summary*, and *anova* functions.

Results and Discussion

Groundwater Detections

Neonicotinoid insecticides were detected at 23 different well monitoring well locations by WI-DATCP-EQ surveys between the years 2008 and 2012 (Table 1). These annual surveys, administered by WI-DATCP-EQ, occur at sensitive geologic or hydrogeologic locations that are at high risk of non-point source agrochemical leaching. Specifically, two agriculturally intensive

A

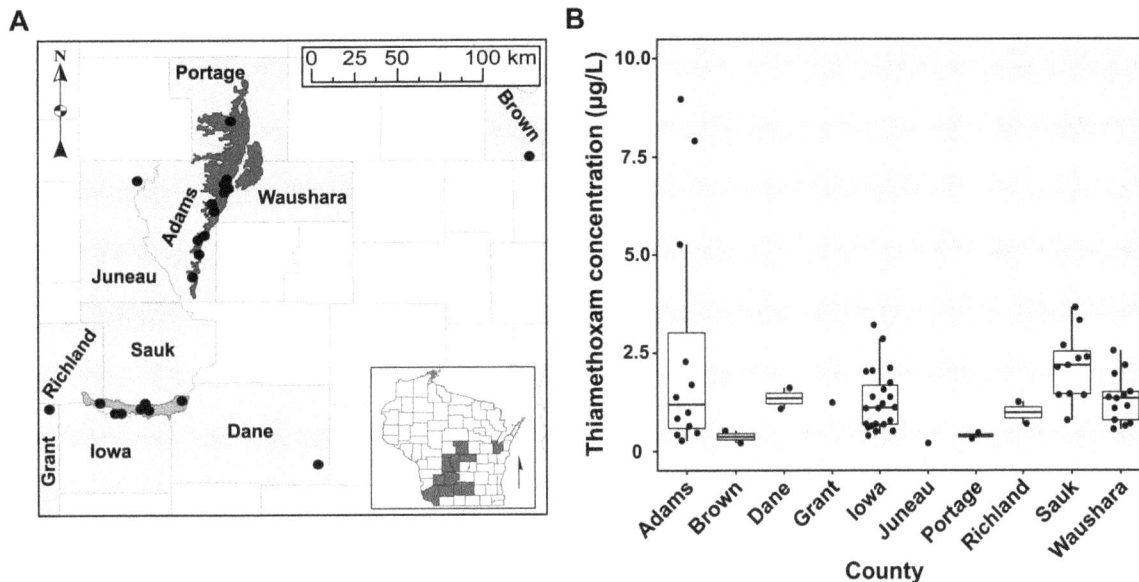

B

Figure 1. Positive thiamethoxam residue detections in groundwater 2008–2012. Points in the map (A) correspond to positive detection locations. Dark grey shaded region indicates the Central Sands potato production region. Light grey delimits the Lower Wisconsin River potato production region. Positive detections were obtained from established agrochemical monitoring wells collected by the Wisconsin Department of Agriculture, Trade and Consumer Protection (DATCP)-Environmental Quality division in collaboration with the Wisconsin DATCP Bureau of Laboratory Services. Boxplots (B) indicate average concentration detected from 2008–2012. Points show individual measured concentrations.

production regions of the state, the Central Sands and Lower Wisconsin River valley, are classified as high-risk areas for groundwater contamination and are frequently monitored for the presence of common agrochemicals (Fig. 1A). These regions have well-drained, sandy soils and easily accessible groundwater for irrigation that has driven agricultural intensification focused on vegetable production. Commercial potato is a key component in the agricultural production sequence, but is also rotated with many other specialty crops such as: carrots, onions, peas, pepper, processing cucumber, sweet corn, and snap beans. Unfortunately, the unique soil and water characteristics supporting a profitable

specialty crop production system are also particularly vulnerable to groundwater contamination with water-soluble agricultural products [34–36]. Regulatory exceedences of nitrates and herbicide products (e.g. triazines, triazinones, and chloroacetamide) have been commonplace for several years [34–37], but recent detections of neonicotinoid contaminants have created new groundwater quality concerns. Beginning in the spring of 2008, two wells had detections of 1.25 and 1.47 μg L^{-1} thiamethoxam in Grant and Sauk Counties, WI (Fig. 1B, Table 1). Subsequent sampling later that season identified six additional locations for a total of 17 independent positive thiamethoxam detections that year. Since

A   **B**

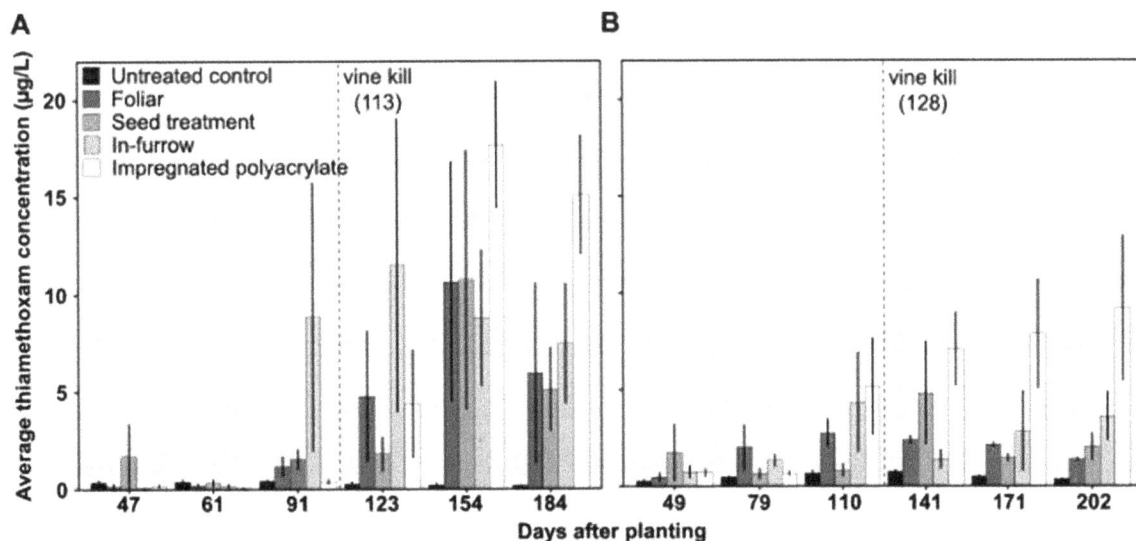

Figure 2. Thiamethoxam concentration in leachate from potato. Average thiamethoxam ($\pm$SD) recovered from in-furrow and foliar treatments in (A) 2011 an (B) 2012. Dotted lines indicate the date that the producer applied vine desiccant prior to harvest. Lysimeter studies continued in undisturbed soil following vine kill.

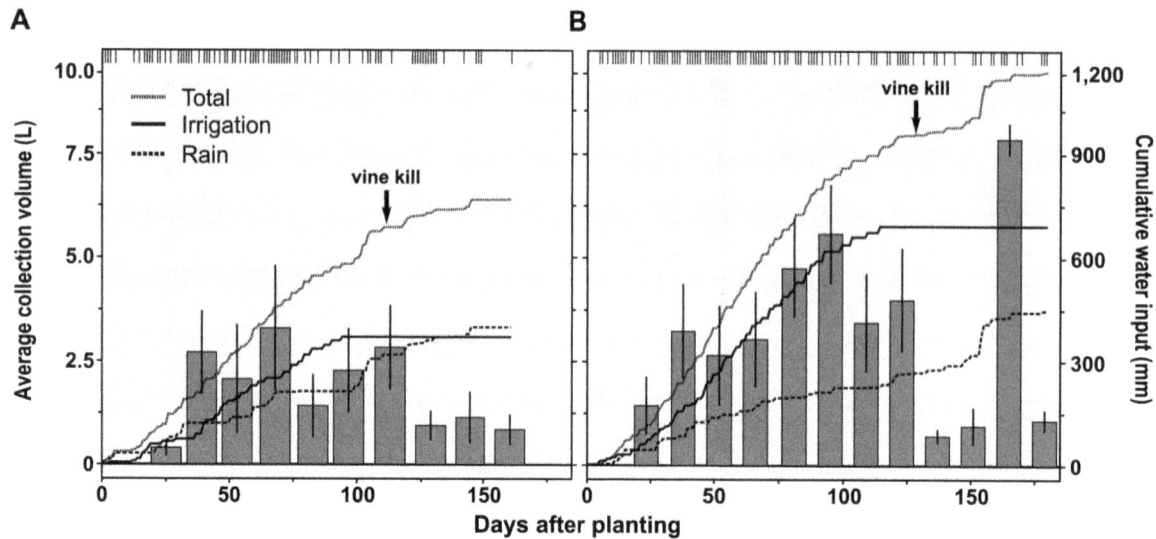

Figure 3. Water input volumes, 2011 and 2012. Water inputs and leachate volume collected in lysimeter studies in (A) 2011 and (B) 2012. Lines indicate cumulative water measured in tipping bucket rain gauges installed in plots each season. Bar plots indicate average leachate volume ($\pm$SD) collected in lysimeters on a bi-monthly sampling frequency. Hash marks at the top of each figure indicate days that overhead irrigation or rainfall occurred in each season.

these early detections, the WI-DATCP-EQ [17],[18] has repeatedly detected thiamethoxam, imidacloprid, and clothianadin residues at 23 different monitoring well locations over a five-year period (Table 1). Although the sampling effort was not uniformly distributed within the state, neonicotinoid detections often correspond to areas where intensive irrigated agricultural production occurs (Fig. 1A). As an indication of specialty crop production intensity, we used county-level potato abundance to better describe trends in historical neonicotinoid detections. Observed frequency and magnitude of neonicotinoid detections did not consistently correspond to potato abundance (Table 1). Although the contribution of potato production to the observed detections was not clear, regulatory agencies have continued to pursue this interaction by sampling where potato occurs at a high density, specifically the Central Sands and Wisconsin River Valley. Groundwater sampling strategies have provided a useful timeline of non-point source agrochemical pollution events in subsurface water resources. Identifying the origin of pollutants in the state is complicated by the diversity of neonicotinoid registrations, application methods and formulations; currently Wisconsin has 164 different registrations for field, forage, tree fruit, vegetable, turf, and ornamentals crops (6 acetamiprid, 18 clothianadin, 4 dinotefuran, 108 imidacloprid, 1 thiacloprid, 26 thiamethoxam) [38].

Neonicotinoid Losses and Concentrations in Leachate

The neonicotinoid insecticide thiamethoxam was included in field experiments to investigate the potential for leaching losses associated with different types of pesticide delivery. Specifically, formulations of thiamethoxam were applied as foliar and as at plant systemic treatments in commercial potato over two years and at two different irrigated fields. We hypothesized that thiamethoxam would be most vulnerable to leaching early in the season when plants were small and episodic heavy rains can be common. Interestingly, we observed the greatest insecticide losses following vine-killing operations which occurred more than 100 days after planting (Fig. 2). Detections of thiamethoxam in lysimeters varied between insecticide delivery treatments through time in 2011 (delivery$\times$date interaction, $F = 2.1$; d.f. $= 20,88$; $P = 0.0131$) and again in 2012 (delivery$\times$date interaction, $F = 1.8$; d.f. $= 20,87$; $P = 0.0384$). Moreover, the impregnated polyacrylamide delivery produced the greatest amount of thiamethoxam leachate late in each growing season (Fig. 2) when compared with other types of insecticide delivery.

Early season rainfall was not exceptionally heavy in either year of this experiment (Fig. 3). The accumulation of leachate detections in lysimeters likely is reflected by the steady application of irrigation water and rainfall. One clear exception to this pattern occurred in 2012 at 155–156 days after planting when 89 mm of

Table 2. Neonicotinoid concentration from irrigation water, 2011 and 2012.

| Date | Days after planting | Insecticide concentration (μg/L)[a] | |
		clothianidin	thiamethoxam
28 June 2011	39	-	0.310
1 September 2011	114	-	0.327
10 July 2012	60	-	0.533
15 August 2012	96	0.225	0.580

[a]Samples obtained from irrigation pivots while under operation in potato fields containing lysimeter experiments.

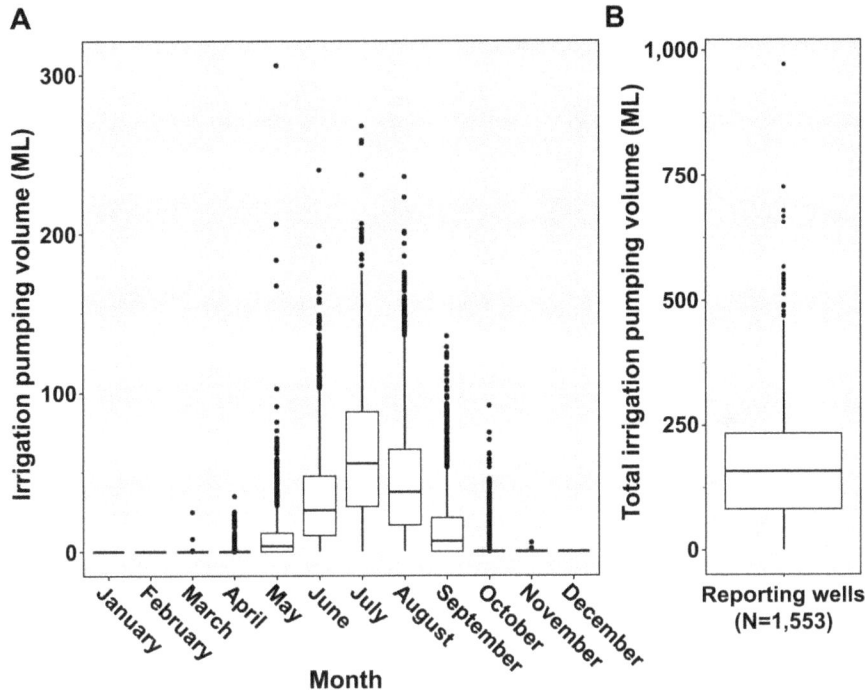

Figure 4. Reported irrigation inputs in the Central Wisconsin River Water Management Unit. Average reported agricultural pumping (megaliters, ML) in the Central Wisconsin River Water Management Unit for 2012. Monthly pumping records were reported by growers to the Wisconsin Department of Natural Resources Bureau of Drinking Water and Groundwater. Upper and lower whiskers extend to the values that are within 1.5*Inter-quartile range beyond the first (25%) and third (75%) percentiles. Data beyond the end of whiskers indicate outlier values and have been plotted as points.

rain fell within a 24-hour period. Peak detections of thiamethoxam in 2012 began to trend upward following this rain event, however the timing of similar detections across treatments in 2011 occurred at about the same time. One additional explanation may be that increased levels of pesticide losses are associated with plant death or senescence. In each year of this study, the largest proportion of pesticide detections in leachate occurred after vine killing with herbicide in the potato crop. Vine killing in commercial potato production is a common practice designed to aid the tubers in developing a periderm. Perhaps the rapid loss in root function following plant death permits excess pesticide to be solubilized and washed through the soil profile more quickly in root channels. In both seasons of this study, however, large episodic rain events did not occur early in the growing season. These results do appear, however, to document low to moderate levels of leaching losses that occur throughout the season even when the crop is managed at nominal evapo-transpirative need.

Untreated control plots also yielded low-level detections of thiamethoxam throughout both seasons. To better understand these insecticide detections in control plots, we sampled water directly from the center pivot irrigation system providing irrigation directly to the potato crop. Samples were taken while the systems were operational from lateral spigots mounted on the well casings. In both years, samples revealed low concentrations of thiamethoxam present in the groundwater at two time points in each sample season (Table 2) from which irrigation water was being drawn. Clothianidin was also present at a single time point in 2012 (Table 2). These positive detections of low-dose thiamethoxam were obviously being unintentionally applied directly to the crop through irrigation and this information is new to the producers in the Central Sands of Wisconsin. Although systemic neonicotinoids have recently been detected from surface water runoff and catch

basins associated with irrigated orchards [10], [39], to our knowledge no other study has documented the occurrence of neonicotinoids in subsurface groundwater being recycled through operating irrigation wells. Currently, the known exposure pathways for insecticide residues are most often associated with direct application or systemic movement of insecticides in floral structure and guttation water [8],[9],[40].

The implications for non-target effects resulting from these groundwater contaminants is currently unknown, but could be important considering the scale of irrigation ongoing in the Central Sands potato agroecosystem in Wisconsin (Fig. S2). Using a combination of aerial photography and NASS Cropland Data Layers, we identified 2,530 different irrigated field units distributed within the Central Wisconsin River Water Management Unit (Fig. S2). In all, 71,864 hectares of irrigated cropland were identified within the extent of the water management unit. Average irrigated field unit size was 28.4±17.7 hectares (min. 1, max 138). Irrigation use patterns demonstrated clear increases in the summer months of the 2012 growing season (Fig. 4). Average annual pumping volume reported to the Wisconsin Department of Natural Resources in 2012 was 170.6±115.6 megaliters (ML) of irrigation water (min. 0.00001, max 972.1) distributed over 1,553 reporting wells. Peak pumping volumes occurred in the month of July, averaging 61±43.3 ML (min. 0, max 286.4). The timing of peak pumping correspond with crop demands for and reproductive phases of common open and closed pollination crops grown in the region.

While considerable attention has been focused on the positive attributes of the neonicotinoids [1–3], an increasing body of research suggests substantial negative impacts not only in terms of pest resistance development (e.g., Colorado potato beetle), but also impacts on non-target organisms and surrounding ecosystems

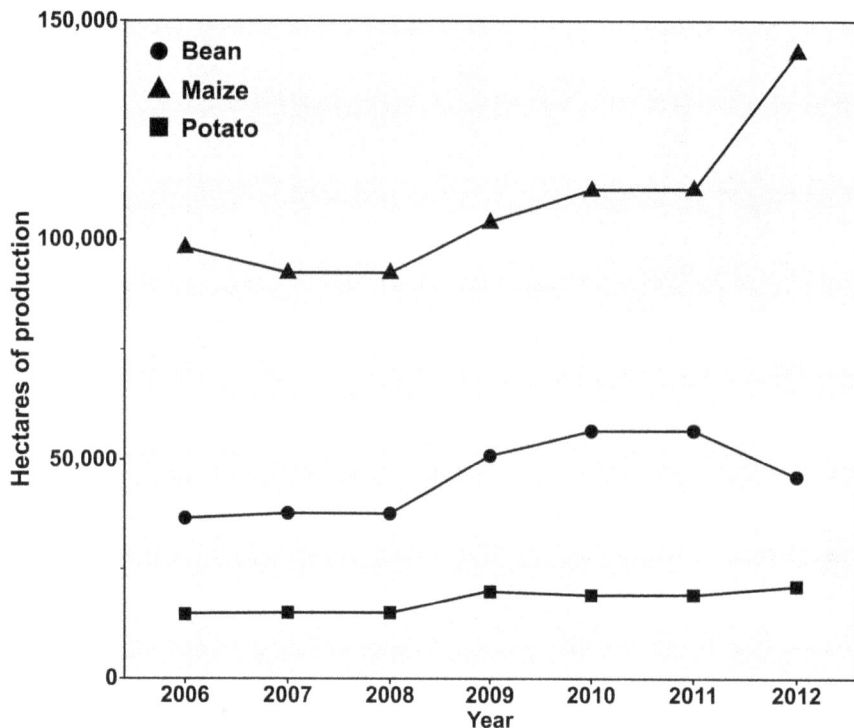

Figure 5. Crop area grown in the Central Wisconsin River Water Management Unit. Cropping trends in the Central Wisconsin River Water Management Unit from 2006–2012. Crop groups are often planted with a soil-applied neonicotinoid insecticide for insect pest management. Crop totals within the water management unit were tabulated from annual USDA-NASS Cropland Data Layers [26].

[8],[10],[41–44]. Recent studies have documented the negative influence of neonicotinoids on pollinator population health (both native and managed) which, in turn, created substantial concern about the long-term sustainability of these pesticides in agriculture [7],[11],[43],[45–49]. Exposures to pollinators reportedly occur through chronic, sub-lethal contact with low concentrations of neonicotinoid residues in pollen, nectar, waxes, and guttation drops of common crop plants [50–53]. Gill et al. [43] and Whitehorn et al. [54] found that low concentrations (≤ 10 µg L^{-1}) of imidacloprid significantly reduced colony-level health in bumblebee (*Bombus terrestris* L.). Imidacloprid residues measured by those authors are consistent with insecticide concentrations found in nectar and pollen of flowering crops, further supporting the direct crop-pollinator toxicological pathway hypothesis [47],[52],[54],[55]. Though they have received much less attention, many closed pollination crops also provide resources for pollinators (e.g., pollen, water)[56],[57]. These crops also rely on neonicotinoids and may have currently undescribed risks for non-target organisms through indirect contaminant pathways in the agroecosystem [51],[58].

Possible exposure related to a high frequency of irrigation could drive the exposure of non-target arthropods to low concentrations of neonicotinoid insecticides in irrigation water. Although such impacts have yet to be documented directly, new comprehensive reviews of neonicotinoid environmental impacts have demonstrated numerous unanticipated impacts occurring at the ecosystem scale [9],[58]. In the Wisconsin agroecosystem, neonicotinoids are used on a large proportion of crops grown with irrigation [28],[29]. Trends in production show increased maize production over the past six years in the Central Wisconsin River Water Management Unit (Fig. 5). As a result of common neonicotinoid seed treatment on maize, accelerating production may partially explain the increased frequency of neonicotinoid detection in

groundwater. Unfortunately, little crop-specific pesticide information exists for individual neonicotinoids at the watershed scale [26]. Although measurement of specific contributions of crops to measured insecticide contamination is currently not available, this study demonstrates a research approach to better understand leaching from different application methods. Improved understanding of crops and insecticide delivery that results in greater risk of insecticide leaching will inform targets to reduce aquifer contamination and recirculation of soil-applied insecticides. Area-wide application of neonicotinoid insecticides through irrigation water applications may have considerable unanticipated or undocumented environmental impacts for non-target organisms through chronic low-dose exposure to insecticides.

Conclusions

To gain a better understanding of the seasonal cycle of neonicotinoids moving from the potato system, this study used an experimental approach to document the leaching potential of common neonicotinoid application methods. Results presented here benefit both potato producers and regulators by identifying trends in leachate losses for these commonly used, water-soluble insecticides. Lysimeter experiments documented loss of thiamethoxam following the application of vine desiccants at the conclusion of the potato production season. Leachate losses did vary among the different delivery methods over time indicating some variability in the patterns of pesticide leachate throughout the season. Quantification of crops commonly using neonicotinoid soil applications in the Central Wisconsin Water Management Unit highlights the need to research leaching potential from soil-applied neonicotinoids in other commodities. Documentation of several neonicotinoids in irrigation water suggests a new candidate pathway for non-target environmental impacts of insecticides.

Supporting Information

Figure S1 Chemical structures and properties of common neonicotinoid insecticides. Chemical structures were drawn using ChemDraw (version 13, Perkin Elmer Inc., Waltham, MA). Properties of each active ingredient were accessed from the National Center for Biotechnology Information PubChem online interface. Available: https://pubchem.ncbi.nlm.nih.gov/. Accessed 2014 Mar 20.

Figure S2 Irrigated field locations in the Central Wisconsin River Water Management Unit. Distribution of fields irrigated with high capacity wells (n = 2530) in the Central Wisconsin River Water Management Unit [27]. Points indicate locations of individual irrigation units identified from aerial photography using ArcGIS.

References

1. Jeschke P, Nauen R, Schindler M, Elbert A (2010) Overview of the status and global strategy for neonicotinoids. J Agric Food Chem 59: 2897–2908.
2. Elbert A, Haas M, Springer B, Thielert W, Nauen R (2008) Applied aspects of neonicotinoid uses in crop protection. Pest Manag Sci 64: 1099–1105.
3. Jeschke P, Nauen R (2008) Neonicotinoids–from zero to hero in insecticide chemistry. Pest Manag Sci 64: 1084–1098.
4. United States Environmental Protection Agency (2003) Imidacloprid; pesticide tolerances. Fed Regist 68: 35303–35315.
5. United States Environmental Protection Agency (2012) What is the conventional reduced risk pesticide program? Available: http://www.epa.gov/opprd001/workplan/reducedrisk.html. Accessed 2012 Oct 17.
6. Szendrei Z, Grafius E, Byrne A, Ziegler A (2012) Resistance to neonicotinoid insecticides in field populations of the Colorado potato beetle (Coleoptera: Chrysomelidae). Pest Manag Sci 68: 941–946.
7. Cresswell JE, Desneux N, vanEngelsdorp D (2012) Dietary traces of neonicotinoid pesticides as a cause of population declines in honey bees: An evaluation by Hill's epidemiological criteria. Pest Manag Sci 68: 819–827.
8. Blacquiere T, Smagghe G, Van Gestel CA, Mommaerts V (2012) Neonicotinoids in bees: A review on concentrations, side-effects and risk assessment. Ecotoxicology 21: 973–992.
9. Goulson D (2013) An overview of the environmental risks posed by neonicotinoid insecticides. J Appl Ecol 50: 977–987.
10. Starner K, Goh KS (2012) Detections of the neonicotinoid insecticide imidacloprid in surface waters of three agricultural regions of California, USA, 2010–2011. Bull Environ Contam Toxicol 88: 316–321.
11. Miranda GR, Raetano CG, Silva E, Daam MA, Cerejeira MJ (2011) Environmental fate of neonicotinoids and classification of their potential risks to hypogean, epygean, and surface water ecosystems in Brazil. Human and Ecological Risk Assessment: An International Journal 17: 981–995.
12. Gupta S, Gajbhiye V, Agnihotri N (2002) Leaching behavior of imidacloprid formulations in soil. Bull Environ Contam Toxicol 68: 502–508.
13. Papiernik SK, Koskinen WC, Cox L, Rice PJ, Clay SA, et al. (2006) Sorption-desorption of imidacloprid and its metabolites in soil and vadose zone materials. J Agric Food Chem 54: 8163–8170.
14. Chiovarou ED, Siewicki TC (2008) Comparison of storm intensity and application timing on modeled transport and fate of six contaminants. Sci Total Environ 389: 87–100.
15. Gupta S, Gajbhiye V, Gupta R (2008) Soil dissipation and leaching behavior of a neonicotinoid insecticide thiamethoxam. Bull Environ Contam Toxicol 80: 431–437.
16. Juraske R, Castells F, Vijay A, Muñoz P, Antón A (2009) Uptake and persistence of pesticides in plants: measurements and model estimates for imidacloprid after foliar and soil application. J Hazard Mater 165: 683–689.
17. Wisconsin Department of Agriculture, Trade and Consumer Protection (2010) Fifteen years of the DATCP exceedence well survey. WI-DATCP, Madison, WI.
18. Wisconsin Department of Agriculture, Trade and Consumer Protection (2011) Agrichemical Management Bureau annual report – 2011. Available: http://datcp.wi.gov/Environment/Water_Quality/ACM_Annual_Report/. Accessed 2012 Jul 10.
19. United States Department of Agriculture - Natural Resources Conservation Soil Service (2013) Web Soil Survey. USDA-NRCS, Washington, DC. Available: http://websoilsurvey.sc.egov.usda.gov. Accessed 2014 Jan 8.
20. Cooley ET, Lowery B, Kelling KA, Speth PE, Madison FW, et al. (2009) Surfactant use to improve soil water distribution and reduce nitrate leaching in potatoes. Soil Sci 174: 321–329.
21. Wisconsin Department of Natural Resources (2013) DNR drinking water system: high capacity wells. WI DNR, Madison, WI. Available: http://dnr.wi.gov/topic/wells/highcapacity.html. Accessed 2013 Aug 22.
22. Wickham H (2007) Reshaping data with the reshape package. Journal of Statistical Software 21: 1–20. Available: http://www.jstatsoft.org/v21/i12/. Accessed 2011 Jan 15.
23. Bussan A, Colquhoun J, Cullen E, Davis V, Gevens A, et al. (2012) Commercial vegetable production in Wisconsin. Publication A3422. University of Wisconsin-Extension, Madison WI.
24. Seccia S, Fidente P, Barbini DA, Morrica P (2005) Multiresidue determination of nicotinoid insecticide residues in drinking water by liquid chromatography with electrospray ionization mass spectrometry. Anal Chim Acta 553: 21–26.
25. United States Department of Agriculture - National Agricultural Imagery Program (2010) Wisconsin NAIP. USDA-NAIP, Washington, DC. Available: http://datagateway.nrcs.usda.gov/. Accessed 2011 Jan 15.
26. United States Department of Agriculture - National Agricultural Statistics Service Cropland Data Layer (2012) Wisconsin Cropland data layer. USDA-NASS, Washington, DC. Available: http://nassgeodata.gmu.edu/CropScape/. Accessed 2013 May 10.
27. Wisconsin Department of Natural Resources (2002) Wisconsin DNR 2003 watersheds. Wisconsin DNR, Madison, WI. Available: http://dnr.wi.gov/maps/gis/documents/dnr_watersheds.pdf. Accessed 2013 Jun 12.
28. Thelin GP, Stone WW (2013) Estimation of annual agricultural pesticide use for counties of the conterminous United States, 1992–2009. US Department of the Interior, US Geological Survey.
29. Stone WW (2013) Estimated annual agricultural pesticide use for counties of conterminous United States, 1992–2009. U.S. Geological Survey Data Series 752, 1-p. pamphlet, 14 tables.
30. Cullen EM, Davis VM, Jensen B, Nice GRW, Renz M (2013) Pest management in Wisconsin field crops. Publication A3646.University of Wisconsin-Extension, Madison WI. Available: http://learningstore.uwex.edu/pdf/A3646.PDF as of 08/18/2013. Accessed 2013 Aug 23.
31. Team R Core (2011) R: A language and environment for statistical computing (Version 2.15.2). Vienna, Austria: R foundation for statistical computing; 2012. Available: http://cran.r-project.org. Accessed 2012 Jun 15.
32. Pinheiro J, Bates D, DebRoy S, Sarkar D (2007) Linear and nonlinear mixed effects models. R package version 3.1–108.
33. Pinheiro J, Bates D (2000) Mixed-effects models in S and S-PLUS. New York: Springer.
34. Mossbarger Jr W, Yost R (1989) Effects of irrigated agriculture on groundwater quality in corn belt and lake states. J Irrig Drain Eng 115: 773–790.
35. Kraft GJ, Stites W, Mechenich D (1999) Impacts of irrigated vegetable agriculture on a humid North-Central US sand plain aquifer. Ground Water 37: 572–580.
36. Saad DA (2008) Agriculture-related trends in groundwater quality of the glacial deposits aquifer, central Wisconsin. J Environ Qual 37: 209–225.
37. Postle JK, Rheineck BD, Allen PE, Baldock JO, Cook CJ, et al. (2004) Chloroacetanilide herbicide metabolites in Wisconsin groundwater: 2001 survey results. Environ Sci Technol 38: 5339–5343.
38. Agrian Inc. (2013) Advanced product search. Available: http://www.agrian.com/labelcenter/results.cfm. Accessed 2013 Mar 21.
39. Hladik ML, Calhoun DL (2012) Analysis of the herbicide diuron, three diuron degradates, and six neonicotinoid insecticides in water–Method details and application to two Georgia streams: U.S. Geological Survey Scientific Investigations Report 2012–5206.
40. Hopwood J, Vaughan M, Shepherd M, Biddinger D, Mader E, et al. (2012) Are neonicotinoids killing bees? A review of research into the effects of neonicotinoid insecticides on bees, with recommendations for action. Xerces Society for Invertebrate Conservation, USA.
41. Casida JE (2012) The greening of pesticide–environment interactions: Some personal observations. Environ Health Perspect 120: 487–493.

Acknowledgments

We thank the cooperating growers for generously allowing us to conduct lysimeter studies on their farm. We thank Amy DeBaker, Rick Graham, Jeff Postle, Wendy Sax, Stan Senger, and Steve Sobek of Wisconsin DATCP for their support of this project. We thank Dave Johnson and Robert Smail of Wisconsin DNR-Water Bureau for providing 2012 irrigation use data. We thank Birl Lowery and Mack Naber for input on lysimeter design and installation. We thank Scott Chapman, Ken Frost, and David Lowenstein for their help installing lysimeters. We thank Claudio Gratton, George Kennedy, Jessica Petersen, and Wesley Stone for their insightful comments on earlier versions of this manuscript. We thank the Wisconsin Potato and Vegetable Growers Association for continued support of our research efforts.

Author Contributions

Conceived and designed the experiments: ASH RLG. Performed the experiments: ASH. Analyzed the data: ASH. Wrote the paper: ASH RLG

42. Krupke CH, Hunt GJ, Eitzer BD, Andino G, Given K (2012) Multiple routes of pesticide exposure for honey bees living near agricultural fields. PLoS ONE 7: e29268.

43. Gill RJ, Ramos-Rodriguez O, Raine NE (2012) Combined pesticide exposure severely affects individual-and colony-level traits in bees. Nature 491: 105–108.

44. Seagraves MP, Lundgren JG (2012) Effects of neonicotinoid seed treatments on soybean aphid and its natural enemies. J Pest Sci 85: 125–132.

45. Cresswell JE, Page CJ, Uygun MB, Holmbergh M, Li Y, et al. (2012) Differential sensitivity of honey bees and bumble bees to a dietary insecticide (imidacloprid). Zoology 115: 365–371.

46. Henry M, Beguin M, Requier F, Rollin O, Odoux J, et al. (2012) A common pesticide decreases foraging success and survival in honey bees. Science 336: 348–350.

47. Stoner KA, Eitzer BD (2012) Movement of soil-applied imidacloprid and thiamethoxam into nectar and pollen of squash (*Cucurbita pepo*). PLoS ONE 7: e39114.

48. Tapparo A, Marton D, Giorio C, Zanella A, Soldà L, et al. (2012) Assessment of the environmental exposure of honeybees to particulate matter containing neonicotinoid insecticides coming from corn coated seeds. Environ Sci Technol 46: 2592–2599.

49. Tomé HVV, Martins GF, Lima MAP, Campos LAO, Guedes RNC (2012) Imidacloprid-induced impairment of mushroom bodies and behavior of the native stingless bee *Melipona quadrifasciata anthidioides*. PloS ONE 7: e38406.

50. Chauzat M, Faucon J, Martel A, Lachaize J, Cougoule N, et al. (2006) A survey of pesticide residues in pollen loads collected by honey bees in France. J Econ Entomol 99: 253–262.

51. Girolami V, Mazzon L, Squartini A, Mori N, Marzaro M, et al. (2009) Translocation of neonicotinoid insecticides from coated seeds to seedling guttation drops: A novel way of intoxication for bees. J Econ Entomol 102: 1808–1815.

52. Laurent FM, Rathahao E (2003) Distribution of (14C) imidacloprid in sunflowers (*Helianthus annuus* L.) following seed treatment. J Agric Food Chem 51: 8005–8010.

53. Mullin CA, Frazier M, Frazier JL, Ashcraft S, Simonds R, et al. (2010) High levels of miticides and agrochemicals in North American apiaries: Implications for honey bee health. PLoS ONE 5: e9754.

54. Whitehorn PR, O'Conner S, Wackers FL, Goulson D (2012) Neonicotinoid pesticide reduces bumble bee colony growth and queen production. Science 336: 351–352.

55. Dively GP, Kamel A (2012) Insecticide residues in pollen and nectar of a cucurbit crop and their potential exposure to pollinators. J Agric Food Chem 60: 4449–4456.

56. Free JB (1993) Insect pollination of crops. London: Academic Press.

57. Klein A, Vaissière BE, Cane JH, Steffan-Dewenter I, Cunningham SA, et al. (2007) Importance of pollinators in changing landscapes for world crops. Proc R Soc B 274: 303–313.

58. Sánchez-Bayo F, Tennekes HA, Goka K (2013) Impact of systemic insecticides on organisms and ecosystems. *In* Stanislav T, editor, Insecticides - Development of Safer and More Effective Technologies. Rijeka: InTech. 367–416.

10

Impact of the Provision of Safe Drinking Water on School Absence Rates in Cambodia: A Quasi-Experimental Study

Paul R. Hunter[1]*, Helen Risebro[1], Marie Yen[2], Hélène Lefebvre[2], Chay Lo[3], Philippe Hartemann[4], Christophe Longuet[5], François Jaquenoud[2]

1 Norwich School of Medicine, University of East Anglia, Norwich, United Kingdom, **2** 1001 fontaines pour demain, Caluire et Cuire, France, **3** Teuk Saat 1001, Phnom Penh, Cambodia, **4** Département Environnement et Santé Publique, Faculté de médecine de Nancy - Université de Lorraine, Nancy, France, **5** Fondation Mérieux, Lyon, France

Abstract

Background: Education is one of the most important drivers behind helping people in developing countries lift themselves out of poverty. However, even when schooling is available absenteeism rates can be high. Recently interest has focussed on whether or not WASH interventions can help reduce absenteeism in developing countries. However, none has focused exclusively on the role of drinking water provision. We report a study of the association between absenteeism and provision of treated water in containers into schools.

Methods and Findings: We undertook a quasi-experimental longitudinal study of absenteeism rates in 8 schools, 4 of which received one 20 L container of treated drinking water per day. The water had been treated by filtration and ultraviolet disinfection. Weekly absenteeism rates were compared across all schools using negative binomial model in generalized estimating equations. There was a strong association with provision of free water and reduced absenteeism (Incidence rate ratio = 0.39 (95% Confidence Intervals 0.27–0.56)). However there was also a strong association with season (wet versus dry) and a significant interaction between receiving free water and season. In one of the intervention schools it was discovered that the water supplier was not fulfilling his contract and was not delivering sufficient water each week. In this school we showed a significant association between the number of water containers delivered each week and absenteeism (IRR = 0.98 95%CI 0.96–1.00).

Conclusion: There appears to be a strong association between providing free safe drinking water and reduced absenteeism, though only in the dry season. The mechanism for this association is not clear but may in part be due to improved hydration leading to improved school experience for the children.

Editor: C. Mary Schooling, CUNY, United States of America

Funding: The study received financial support as charitable donations from Fondation Ensemble, Danone Group, Agence Française de Développement, Fondation Mérieux, and Fondation Avenir Finance. The funders had no role in study design, data collection and analysis, decision to publish, or preparation of the manuscript.

Competing Interests: In 2010 PRH received travelling expenses and a small honorarium from Danone to give a talk at a science meeting organised by the company.

* E-mail: Paul.Hunter@uea.ac.uk

Introduction

The receipt of a good quality education is one of the most important factors in enabling children to fulfil potential in later life and reduce poverty [1]. Increased educational attainment is also associated with substantial health gains especially on child health in future generations including reduction in child mortality [2,3]. Important gains in child health may be associated even with future mothers improved access to primary education alone [3]. The importance of access to education is reflected within the Millennium Development Goals of the commitment to ensure that all children can complete a course of primary education [4]. In an earlier review of studies from developing countries, the author pointed out that time spent learning being linked to educational achievement is one of the most consistent findings [5]. However, as pointed out by Abadzi [1], instructional time available to children in many developing countries is often markedly reduced. Indeed Abadzi concluded that "assumptions

about the pro-poor poverty alleviation effect of education may be unrealistic", and that additional public investment may fail to mitigate poverty, unless it improves instructional delivery [1]. There are many reasons for this reduced educational contact time in low income countries, some of which are institutional such as teacher absenteeism, frequent school closures, etc [1]. However, even when schools are open, pupil absenteeism can be high [1]. Clearly reducing student absenteeism is important to improving educational attainment and consequent poverty alleviation.

Recent interest has turned towards the potential role of improving water and sanitation provision in schools as a tool towards improving children's health and educational achievements. In a recent systematic review, the authors identified 41 studies that reported on the impact of water, hygiene and sanitation interventions on health and educational outcomes of which 8 were concerned with absenteeism [6]. Most of these studies were from developed nations. Although there was some indication of links between water and sanitation prevision and absenteeism, this was strongest around sanitation provision and absenteeism in menstruating girls. No strong conclusions could be made around the importance of drinking water. In probably the largest study of its type Freeman et al. reported on a large cluster randomised trial in Kenya on the impact of water, sanitation and hygiene interventions on absenteeism [7]. This study included over 6000 pupils from 135 schools in three study arms: a control with no intervention, a second study arm with hygiene promotion and chlorine for drinking water treatment and a third study arm with sanitation improvement in addition to the hygiene promotion and water treatment. The authors found no significant difference between any of the study arms, unless they did further sub-group analysis. Furthermore, by restricting their drinking water intervention to provision of near-to-use chlorination, Freeman et al. only tested the impact of drinking water disinfection and did not undertake an adequate assessment of the value of provision of drinking water per se in schools on educational outcomes [7]. Furthermore, serious doubts have been raised about the health value of household chlorination of drinking water as blinded studies have repeatedly found no benefit [8,9]. The available evidence of the benefits or otherwise of drinking water interventions targeted at the level of the school in developing countries is, therefore, weak. We report a quasi-experimental study of the impact of provision of treated water in containers to schools on recorded absenteeism.

Methods

This study was approved by the University of East Anglia Faculty of Health ethical committee and the Cambodian National Ethics Committee for Health Research. Given that we did not introduce any intervention, only summaries of routinely collected data were obtained and no person specific data was collected, there was no requirement for informed consent to be obtained.

The intervention being investigated was delivery of treated water in containers to schools. These water containers were provided free of charge to schools and funded by "1001 fontaines pour demain" (1001F), a non-governmental and not-for-profit organization based in Caluire, France. 1001F has been working in Cambodia since 2005. The basic model is to identify local entrepreneurs and financially support them to build a local plant to bottle filtered and ultraviolet disinfected water in cleaned and disinfected 20 L containers. Most of these containers are then sold to local customers. During and after start-up 1001F technical staff provide training and an ongoing quality assurance scheme. Funding for 1001F is mainly from private donors, though it has also received financial support from French Embassies in the countries where it works. A video highlighting 1001F's work can be seen at the following link: http://fr.youtube.com/watch?v = 8bykbVECVrE. In some of the villages, 1001F paid the entrepreneur to provide free water to the village school. Each participating school was provided with 1001F water in containers to be placed in the classroom so that each child could take water whenever they wished. For those schools participating in the scheme one 20 L bottle of water was delivered to each class each day. Given that the average class size was 38 children, this equates to approximately 0.53 L per child per day. The overall cost of the scheme was US$1.4 per child per year.

In this study, we obtained absenteeism data from the four schools where 1001F were providing free water. In addition we obtained this same data from four schools not in receipt of free water. In a related community study of childhood diarrhoea we were conducting a longitudinal study of childhood diarrhoea and water use in 25 villages. These villages had been chosen at random

from all villages with an established 1001F presence or through a process of propensity score matching, the details of which is described elsewhere [10]. Four schools from these 25 villages were in receipt of the free school water scheme and willing and able to provide absenteeism data. Four control schools were chosen from the other 25 villages based on number of registered students present and the proportion of students under 14 years closest to those values of the intervention schools. The head teacher was then approached and invited to participate.

Data collection was based on routinely collected absenteeism data provided to the study team by the head teacher. Data was provided from the week beginning 4th December 2011 to 31st May 2012. This period spread over two school terms one of which was in the dry season and the other the wet season.

Data analysis was done using STATA version 11. Absenteeism rates per week were calculated as the number of days absent/ (5 ×children registered). Random effects negative binomial regression analyses were done using a generalized linear model with a random intercept for school. The outcome variable was the number of days lost in each week from absenteeism and the number of children enrolled in the school was the exposure variable. The predictor variables were whether or not the school received water and season. Interaction terms were included for intervention and season.

In one school it was discovered that the number of water containers delivered fell short of the contracted amount. A further regression analysis was done for this school with days missed in the week being the dependent variable. The actual number of water containers delivered in the week and days missed at all the other schools combined were predictor variables. The analysis was restricted to the dry season and excluded holiday weeks.

Results

Table 1 shows certain key characteristics of to the villages where each of the eight schools were based. It can be seen that across most characteristics the intervention and control schools were generally very similar. The main exception is that very few people in intervention villages have access to improved water or sanitation compared to the control villages. This is not too surprising as the 1001F had primarily targeted its intervention at schools in areas where it was known that the local community had poor access to improved drinking water. Also of note was that rather more of the populations of the intervention villages were reported as being migrants. The predominant source of drinking water in the control schools was whatever the children brought in from their home. In one control school (C2) children also had access to a hand pump and jar in a pagoda about 100 m from the school and in another (C3) there was a rainwater harvesting tank for which children were reported to have some use.

Data was collected for 26 consecutive weeks. Three schools were closed during week 18, all schools were closed during week 19, and all but one in week 20. The dry season was taken to include all the weeks before the break in week 19 and the wet season in weeks subsequent to this holiday. Across all eight schools this represented 60,194 child weeks of follow-up. The overall absenteeism rate was 5.57%. Figure 1 shows the absenteeism rate for each school by week. The most obvious finding was the dramatic increase in absenteeism during the wet season, towards the end of the study period. This was not surprising given the fact that in many villages, children would be kept off school at this time to help in the fields.

Table 2 shows the results of the negative binomial regression analysis comparing absenteeism rates using whether or not the

Table 1. Characteristics of the villages in which the schools were based.

School code	C1	C2	C3	C4	I1	I2	I3	I4
Number of children in school	379	438	267	450	587	271	954	174
Number of households in village	318	320	405	191	331	256	227	192
Population	1634	1706	1951	942	1378	1126	1075	873
% population male	49	50	48	51	51	50	49	49
% population <5 yrs old in village	8	10	11	13	8	12	12	10
% population 5 to 14 yrs old	20	27	25	26	19	28	25	24
% population with lower secondary education or greater	96	96	95	98	92	98	99	98
% adult female literacy rate	84	58	72	34	66	50	73	64
% working in primary sector (mainly agricultural)	68	96	81	87	85	94	66	94
% population migrants	23	4	32	20	57	48	36	65
% with access to improved water	48	72	1	63	5	0	2	2
% with access to sanitation	84	65	28	19	14	6	16	38

School codes prefixed with I received the free water and those prefixed by C did not. Data from The National Institute of Statistics, Ministry of Planning, Royal Government of Cambodia (http://www.nis.gov.kh/index.php/online-statistics/resultonline).

school received free 1001F water and season as predictor variables. In addition we investigated the interaction between season and receipt of 1001F water. It can be seen that absenteeism was less than half in the intervention schools compared to those who did not receive 1001F water. Given the significant interaction term the association between having 1001F water and reduced absenteeism was restricted to the dry season with no such association in the wet season (as was also clear in figure 1).

At the end of the study period it became clear that one of the suppliers was not fully fulfilling their contract as they did not have sufficient capacity to provide water to the school and to their paying customers. Although container water was provided this fell short of the contracted amount. The remaining three schools received all their assigned supplies. Table 3 shows the results of the regression analysis of absenteeism in the school with incomplete water delivery adjusted for within week absenteeism in other participating schools. There was a significant association between the number of containers of water delivered in the week and reported absenteeism. For every extra container delivered there was a 2.9% reduction in absenteeism (95% confidence intervals (CI) 0.5 to 5.1%). The association was also tested between absenteeism and delivery in the previous week. Absenteeism was not associated with the number of containers delivered in the previous week.

Discussion

In this study we have shown lower absenteeism in schools receiving free containers of 1001F water. However, this association was only seen in the dry season and not in the wet season. There were also strong seasonal effects as absenteeism in several of the schools increased dramatically during the wet season, irrespective of water delivery. We were informed that this increase in absenteeism during the early wet season was partly because children were frequently kept off school to help in the fields. We have, furthermore, shown that in one school where delivery of water containers fell short of the contracted amount, absenteeism rates were associated with the number of containers delivered in the week. As far as we are aware this is the first study to show that provision of adequate safe drinking water in school can affect attendance in a developing country.

Clearly one has to be cautious when interpreting the results of an observational study like this. Nevertheless, taking both analyses together, this gives a fairly strong indication that provision of safe palatable drinking water is indeed associated with reduced absenteeism. Firstly although this study was not blinded and so potentially open to some form of reporting bias, school absenteeism rates are not subjective and so our results should not be as at risk of reporting bias that has affected many other studies of water and health in low income countries [8]. We cannot, of course, exclude bias in the way the classroom teacher records the daily attendance register or in how the school compiles absenteeism data from the class registers. However, any such bias is far less likely when based on register records than may be expected by asking children to recall their absence history during interview as was done in the only other study of school absence and WASH [7]. Secondly, although it is plausible that selection of schools for the intervention may have led to a degree of bias, it is difficult to see how this would have affected the association found between number of containers delivered and absenteeism in the school with incomplete contract fulfilment. Of particular note here was that the intervention schools were generally in areas with poor domestic access to improved drinking water supplies and sanitation. If inadequate drinking water and sanitation does

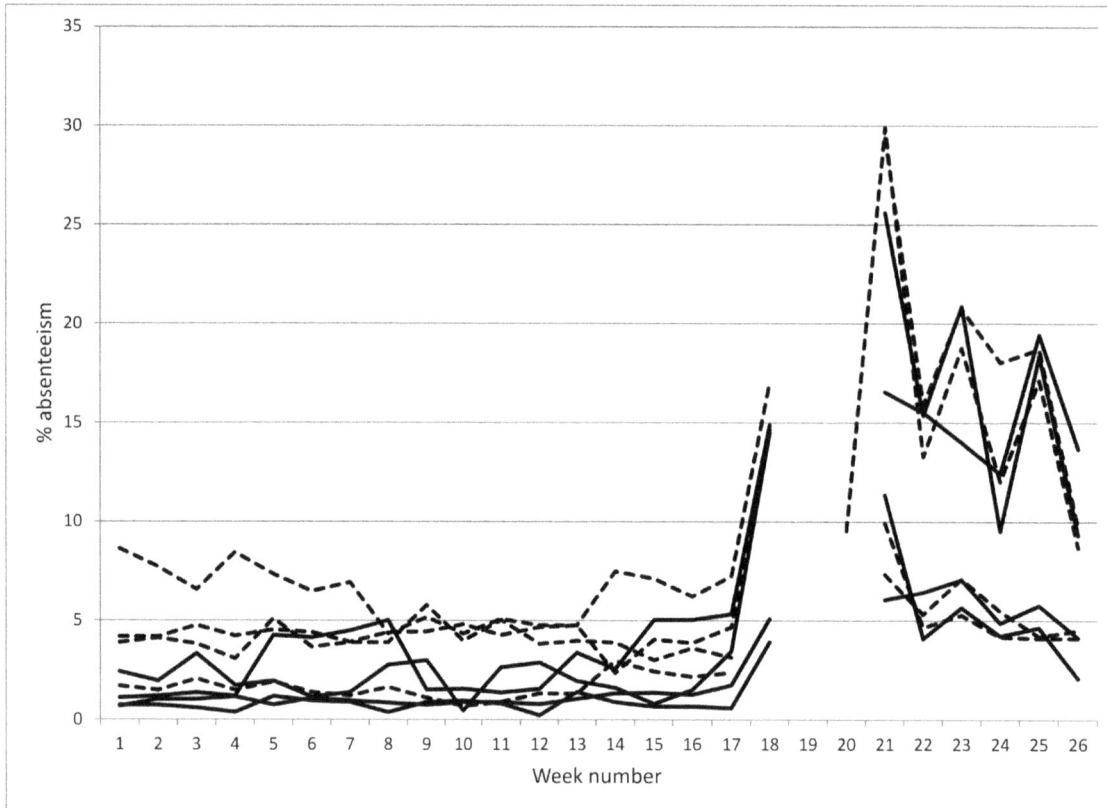

Figure 1. Absenteeism rate by school and week. Solid line shows rates for intervention schools and broken line for control schools.

impact of school absenteeism, then if anything this source of bias would be expected to increase absenteeism in the intervention schools rather than reduce it. It is not clear what effect if any the greater number of migrants in some villages would have on absenteeism in school. We would however suggest that further randomised studies are required before a more definitive conclusion can be made.

This leaves the question of what was the mechanism between water supply provision and absenteeism. In this study we were not able to collect any data on the reasons for the absenteeism. Given the fact that the association was between absenteeism rates and water delivery in the same week and not the previous week, we are not suggesting that this association was primarily due to a reduction in waterborne infectious disease. A possible explanation in our view may be that by providing readily available palatable and safe water in the classroom, children are more likely to drink during the school day and so not become dehydrated. Even mild

dehydration in vulnerable groups such as young children has been suggested as being associated with various adverse health effects [11]. Furthermore, in a recent study from a hot dry region of Italy, the authors showed that supplementary drinking water was associated with improved cognition and an improved subjective sense of vigour [12]. This Italian study is in line with similar findings from several previous researchers [13,14]. What this suggests therefore is that provision of supplementary water sufficiently improves the child's general wellbeing as well as the learning and experience of the school day as he/she is better hydrated. Consequently they are more likely to attend school the following day if they had felt good at school the previous day.

Even if, as we suspect, the main reason for the reduced absenteeism in the intervention group is due to improved hydration rather than a reduction in waterborne disease, this should not be taken as an indication that the provision of water of any quality would be acceptable. The link between contaminated

Table 2. Risk factors for absenteeism rates in schools.

Predictor		Incidence Rate Ratio	Lower 95% CI	Upper 95% CI	P
Receives 1001F Water	No	1			<0.001
	Yes	0.386	0.266	0.560	
Season	Dry	1			<0.001
	Wet	2.618	2.279	3.008	
Season-1001F water interaction	Yes-Wet	1.991	1.603	2.472	<0.001

Table 3. Risk factors for absenteeism in school with incomplete delivery of water containers during the dry season.

Predictor		Incidence Rate Ratio	Lower 95% CI	Upper 95% CI	P
Water delivered in week	/container	0.971	0.949	0.995	0.016
Absenteeism days in other schools	/days missed in week	1.000	0.992	1.007	0.976

drinking water and disease risk is well accepted and it is clear that the main risk falls on young children [15]. Any scheme to increase drinking water provision in the classroom that does not ensure that that water is safe to drink is likely to put the children at risk of waterborne disease. However, providing safe water in the school environment does not necessarily mean children will drink it. Indeed taste appears to be a major determinant affecting whether or not people continue to use safe drinking water sources [16,17]. Chlorination of drinking water is associated with poorer taste for many people [18,19]. On the other hand filtration can be associated with improved taste [20]. The fact that 1001F water uses filtration and Ultraviolet disinfection but not chlorination would mean that it would have better taste qualities than other safe water sources and so may be more likely to be used by children.

In conclusion, we have shown a significant association between provision of supplementary water in the classroom and reduced absenteeism rates. With the delivery mechanism in this study the cost per child is modest, but the potential benefit to children's education and subsequent life potential could be extremely large. There is a great need for further research in this area, especially randomised control trials and studies aimed at determining the biological mechanisms behind this reduction in absenteeism.

Acknowledgments

We would like to thank Indochina Research of Phnom Penh, Cambodia who undertook all the field work.

Author Contributions

Conceived and designed the experiments: PRH HR FJ HL MY PH C. Longuet. Performed the experiments: MY HL C. Lo. Analyzed the data: PRH FJ. Wrote the paper: PRH HR MY HL C. Longuet PH C. Lo FJ.

References

1. Abadzi H (2009). Instructional Time Loss in Developing Countries: Concepts, Measurement, and Implications. World Bank Res Obser 24: 267–290.
2. Gakidou E, Cowling K, Lozano R, Murray CJL (2010) Increased educational attainment and its effect on child mortality in 175 countries between 1970 and 2009: a systematic analysis. Lancet 376: 959–974.
3. Basu AM, Stephenson R (2005) Low levels of maternal education and the proximate determinants of childhood mortality: a little learning is not a dangerous thing. Soc Sci Med 60: 2011–2023.
4. United Nations (UN) (2011). The Millennium Development Goals Report 2011; United Nations: New York, NY, USA.
5. Fuller B (1987) What School Factors Raise Achievement in the Third World? Review of Educat Res 57: 255–292.
6. Jasper C, Le T-T, Bartram J (2012) Water and Sanitation in Schools: A Systematic Review of the Health and Educational Outcomes. Int. J. Environ. Res. Public Health 9: 2772–2787.
7. Freeman MC, Greene LE, Dreibelbis R, Saboori S, Muga R, et al. (2012) Assessing the impact of a school-based water treatment, hygiene and sanitation programme on pupil absence in Nyanza Province, Kenya: a cluster-randomized trial. Trop Med Int Health 17: 380–391.
8. Schmidt W-P, Cairncross S (2009) Household water treatment in poor populations: Is there enough evidence for scaling up now? Environm Sci Technol 43: 986–992.
9. Hunter PR (2009) House-hold water treatment in developing countries comparing different intervention types using meta-regression. Environm Sci Technol 43: 8991–8997.
10. Hunter PR, Risebro H, Yen M, Lefebvre H, Lo C, et al. (2013) Water source and diarrhoeal disease risk in children under 5 years old in Cambodia: a prospective diary based study. BMC Public Health 13: 1145.
11. Maughan RJ (2012) Hydration, morbidity, and mortality in vulnerable populations. Nutr Rev 70 (Suppl. 2): S152–S155.
12. Fadda R, Rapinett G, Grathwohl D, Parisi M, Fanari R, et al. (2012)Effects of drinking supplementary water at school on cognitive performance in children. Appetite 59: 730–737.
13. Bar-David Y, Urkin J, Kozminsky E (2005) The effect of voluntary dehydration on cognitive functions of elementary school children. Acta Paediatrica 94: 1667–1673.
14. Benton D, Burgess N (2009) The effect of the consumption of water on the memory and attention of children. Appetite 53: 143–146.
15. Risebro H, Breton L, Aird H, Hooper A, Hunter PR (2012) Risk of infectious intestinal disease in consumers drinking from private water supplies: A prospective cohort study. PLOS One 7(8): e42762.
16. Huber AC, Mosler HJ (2013) Determining behavioral factors for interventions to increase safe water consumption: a cross-sectional field study in rural Ethiopia. Int J Environ Health Res 23: 96–107.
17. Tamas A, Mosler HJ (2011) Why Do People Stop Treating Contaminated Drinking Water With Solar Water Disinfection (SODIS)? Health Educ Behav 38: 357–366.
18. Heiner JD, Simmons EA, Hile DC, Wedmore IS (2011) A Blinded, Randomized, Palatability Study Comparing Variations of 2 Popular Field Water Disinfection Tablets. Wild Environ Med 22: 329–332.
19. Doria MD, Pidgeon N, Hunter PR (2009) Perceptions of drinking water quality and risk and its effect on behaviour: A cross-national study. Sci Total Environ 407: 5455–5464.
20. Ngai TKK, Shrestha RR, Dangol B, Maharjan M, Murcott SE (2007) Design for sustainable development - Household drinking water filter for arsenic and pathogen treatment in Nepal. J Environ Sci Health A 42: 1879–1888.

11

Enteric Pathogens in Stored Drinking Water and on Caregiver's Hands in Tanzanian Households with and without Reported Cases of Child Diarrhea

Mia Catharine Mattioli[1], Alexandria B. Boehm[1], Jennifer Davis[1,2], Angela R. Harris[1], Mwifadhi Mrisho[3], Amy J. Pickering[1,2]*

1 Environmental and Water Studies, Department of Civil and Environmental Engineering, Stanford University, Stanford, California, United States of America, 2 Woods Institute for the Environment, Stanford University, Stanford, California, United States of America, 3 Ifakara Health Institute, Bagamoyo Research and Training Unit, Bagamoyo, Tanzania

Abstract

Background: Diarrhea is one of the leading causes of mortality in young children. Diarrheal pathogens are transmitted via the fecal-oral route, and for children the majority of this transmission is thought to occur within the home. However, very few studies have documented enteric pathogens within households of low-income countries.

Methods and Findings: The presence of molecular markers for three enteric viruses (enterovirus, adenovirus, and rotavirus), seven *Escherichia coli* virulence genes (ECVG), and human-specific *Bacteroidales* was assessed in hand rinses and household stored drinking water in Bagamoyo, Tanzania. Using a matched case-control study design, we examined the relationship between contamination of hands and water with these markers and child diarrhea. We found that the presence of ECVG in household stored water was associated with a significant decrease in the odds of a child within the home having diarrhea (OR = 0.51; 95% confidence interval 0.27–0.93). We also evaluated water management and hygiene behaviors. Recent hand contact with water or food was positively associated with detection of enteric pathogen markers on hands, as was relatively lower volumes of water reportedly used for daily hand washing. Enteropathogen markers in stored drinking water were more likely found among households in which the markers were also detected on hands, as well as in households with unimproved water supply and sanitation infrastructure.

Conclusions: The prevalence of enteric pathogen genes and the human-specific *Bacteroidales* fecal marker in stored water and on hands suggests extensive environmental contamination within homes both with and without reported child diarrhea. Better stored water quality among households with diarrhea indicates caregivers with sick children may be more likely to ensure safe drinking water in the home. Interventions to increase the quantity of water available for hand washing, and to improve food hygiene, may reduce exposure to enteric pathogens in the domestic environment.

Editor: Lionel G. Filion, University of Ottawa, Canada

Funding: This study was supported by the National Science Foundation (SES-0827384) and by the Stanford University Shah Research Fellowship. The funders had no role in study design, data collection and analysis, decision to publish, or preparation of the manuscript.

Competing Interests: The authors have declared that no competing interests exist.

* E-mail: amyjanel@stanford.edu

Introduction

Over 6.5 million children died in 2012 before reaching their fifth birthday [1]. Globally, almost 10% of these deaths are attributed to diarrhea, and the highest rates of child mortality occur in sub-Saharan Africa [1]. The east African country of Tanzania is a nation that continues to struggle with the burden of childhood diarrhea. In 2010, diarrheal diseases were responsible for almost 9% of all deaths of Tanzanian children under the age of five, just behind malaria (11%) and pneumonia (15%) [2]. Diarrhea-causing pathogens are transmitted via the fecal-oral route and, in low-income countries like Tanzania, it has been suggested that up to 88% of all child diarrhea cases can be attributed to inadequate sanitation, unsafe water, and/or insufficient hygiene [3]. The large diarrheal burden in Tanzania is consistent with the fact that only half of the Tanzanian population has access to improved drinking water sources, and only 10% has access to improved sanitation [4].

Fecal contamination is traditionally monitored using fecal indicator bacteria (FIB) characteristic of animal and human feces (*e.g.*, *Escherichia coli*, enterococci, and fecal coliforms). The concentration of these organisms is regularly used to evaluate health risk associated with different exposure pathways, such as recreational water use and drinking water [5]. For young children, exposure to feces is thought to occur primarily within the home [6], and previous research in low-income countries has documented high levels of FIB in household stored drinking water [7–9] and on hands of mothers [10–13].

Several studies have investigated the behavioral determinants of FIB levels in stored water and on hands in order to identify the most promising household-level interventions for preventing childhood diarrhea. Safe storage containers and point-of-use (POU) treatments have been found to significantly reduce FIB contamination of household stored drinking water [14,15]. Other studies investigating hand contamination found that increasing the frequency of hand washing with soap reduces FIB contamination on hands, whereas some household chores increase hand contamination [16,17].

Several studies have found associations between FIB on hands and/or in stored drinking water and diarrheal illness in children [18–20]. For example, one study in the Philippines found that children drinking water with high levels of *E. coli* had significantly higher rates of diarrhea than those drinking less contaminated water [19], and a recent study in Tanzania found FIB contamination on hands to be significantly associated with diarrheal illness among household members [17]. However, other research suggests that FIB are inadequate indicators of the health risks associated with fecal contamination in the home [21–23]. For example, in Ecuador Levy *et al.* found childhood diarrhea to be associated with levels of *E. coli* in drinking water, but not with the concentration of enterococci or somatic coliphage [24]. Similarly, a meta-analysis by Gundry *et al.* was unable to determine any clear relationship between FIB levels in household drinking water and incidence of child diarrhea [22].

One possible reason for the equivocal findings regarding the relationship between FIB and child diarrhea is that FIB are present in both human and non-human animal feces [5]. In addition, FIB have historically been used to assess water quality in countries with temperate climates. They have subsequently been found to occur naturally and even proliferate in water [25,26], soil [27–29], and sands [30], particularly in tropical environments [31,32]. As such, FIB in water and on hands are often not strongly associated with the presence of human enteric pathogens [33–36], and there is very limited research on the association between indicators and pathogens in South America, Asia, and Africa [37]. In particular, viruses such as human enterovirus, adenoviruses, and rotavirus show limited association with FIB [33,37–43]. Given the low infectious dose of these viruses, as well as their significant contribution to child diarrhea globally, the limited utility of FIB to indicate viral presence in water and on hands is of particular concern [44,45].

For these reasons, in low-income countries with widespread fecal contamination, FIB are likely not adequate indicators for understanding the relationship between water management and hygiene behaviors and diarrheal illness among young children. Molecular fecal markers such as human-specific *Bacteroidales* that are unique to human fecal sources [46], as well as molecular markers of enteric pathogens, may be better indicators of contamination. In turn, such indicators may be more useful for identifying behavioral interventions that can prevent childhood diarrhea. However, few studies have reported the prevalence of enteric pathogens or human-specific fecal contamination in the household environment of low-income countries [16,33,47].

In a recent study conducted in Bagamoyo, Tanzania, Mattioli *et al.* analyzed source water, stored drinking water, and hand rinse samples from 93 households for molecular markers of enteric viruses and *E. coli* virulence genes [33]. The authors found a significant association between viral markers on hands and in stored drinking water, as well as between *E. coli* virulence genes in stored and source waters. FIB were found not to be good predictors of viral markers in water or on hands, although turbidity was associated with viral markers on hands. However, Mattioli *et al.* did not examine the association between microbial

contamination and either child health or water management and hygiene behaviors among sampled households.

The present study builds on Mattioli *et al.* [33] by using data collected from 223 households in Bagamoyo, Tanzania to investigate whether the presence of enteropathogen and human-specific *Bacteroidales* genes in water and on hands is associated with reported cases of child diarrhea. Of the 223 households, 213 households are distinct from those described previously [33]. In addition, we combine the microbial data collected uniquely for this study with data from the 93 households described in Mattioli *et al.* [33] in order to evaluate the association of water management and hygiene behaviors with enteropathogen and human-specific *Bacteroidales* genes in stored drinking water and hand rinse samples.

Methods

Setting

The study was conducted in Bagamoyo, Tanzania (06°28'S, 38°55'E), approximately 70 km north of Dar es Salaam. Households–defined as groups of people that sleep and eat together in a dwelling on a regular basis–that included at least one child under five years of age were enrolled in the study. The data used in the present study were collected from a subset of 1219 households surveyed during the baseline phase of a household water and hygiene behavioral intervention trial from March-May 2010 [33].

Data Collection

In each participating household, a hand rinse sample was taken from the respondent (adult female caregiver). Hand rinse sampling involves the participant placing her hands, one at a time, into a sterile sample bag containing 350 ml of sterile distilled water, a sampling method used successfully in a number of previous studies [13,16,17,33]. A sample of stored water that was intended for drinking and cooking was also collected from each household. Enumerators documented whether or not the water storage container was covered, then asked each respondent to extract water in the manner she usually would, pouring it into a 1.63 liter sterile sample bag (VWR, Radnor, PA). Enumerators noted whether the respondent's hand touched the water during extraction, as well as the extraction method used (*e.g.*, decanting, filling a cup or bowl). Enumerators inquired whether and how the water had been treated, and for how long the water had been stored prior to sampling. Finally, respondents were asked to identify the water source from which the stored water had been collected. Water samples were tested for chlorine using a dip chlorine strip (Hach Co., Loveland, CO); because chlorine was never detected, there was no need to add sodium thiosulfate to neutralize residual.

Along with water and hand rinse samples, enumerators conducted interviews with the female caretaker of the youngest child in the household. Information was collected regarding household water management and hygiene behaviors, water supply and sanitation services, household socioeconomic and demographic characteristics, and illness status of household members. Tanzanian enumerators participated in a 4-week training that included instruction on survey content and administration, electronic data collection, and sterile sampling technique. The survey instrument underwent multiple iterations and pre-tests.

Ethics Statement

Participants were informed in the local language (Kiswahili) of all study procedures and the time required for participation. Written informed consent was obtained from the mother or primary female adult caretaker of the under-five children in the household. When younger caregivers were interviewed (15–17), an adult household member was present. The Tanzanian Commission for Science and Technology, the Tanzanian National Institute for Medical Research (NIMRI) Ethics Sub-Committee, the Ifakara Health Institute Institutional Review Board (IRB), and Stanford University's IRB (IRB Protocol #17971) approved the consent procedures and study protocol.

Laboratory Analysis

All water and hand rinse samples were stored in a cooler on ice and transported to a local laboratory for microbial analysis by membrane filtration within six hours of collection. The turbidity of the water and hand rinse samples was measured using a LaMotte 2020e Turbidity Meter (LaMotte Company, Chestertown, MD). The fecal indicator bacteria, *E. coli* and enterococci, were enumerated following USEPA Methods 1604 and 1600, respectively [48,49]. $MgCl_2$ was added to the water and hand rinse filters to facilitate capture of viral particles [50], and the samples were subsequently passed through a 0.45 µm-pore size membrane filter as outlined in Mattioli *et al.* [33]. The filters were then treated with RNAlater (Qiagen, Germantown, MD) to stabilize RNA/DNA [51] and stored at −80°C. *E. coli* membrane filters with *E. coli* biomass (from EPA method 1604) were removed from agar after counting and were treated with RNAlater and stored at −80°C. Filters were stored for up to 5 months at −80°C until being transported back to Stanford University (Stanford, CA, USA) for molecular processing. Details on field and lab blanks, turbidity measurements, filtration volumes, culture assay detection limits, and sample transport were previously described by Mattioli *et al.* [33].

The presence or absence of the nucleic acids from three enteric viruses (enterovirus, adenovirus, and rotavirus), the human-specific *Bacteroidales* marker, as well as seven *E. coli* virulence genes was measured in all samples. These pathogens were chosen for analysis because rotavirus, pathogenic *E. coli*, and *Shigella* spp. are believed to be major viral and bacterial etiologies of childhood diarrhea [52,53]; in addition, enteroviruses and adenoviruses are recognized as important etiological agents of gastroenteritis for children in the developing world [45,54].

Presence of *E. coli* virulence genes in preserved *E. coli* biomass was determined using multiplex polymerase chain reactions (PCR) [33]. This method determines the presence of diarrheagenic *E. coli* virulence genes that are commonly found in *Shigella* spp., as well as five different pathotypes of *E. coli* including enteroinvasive *E. coli* (EIEC), enteropathogenic *E. coli* (EPEC), enteroaggregative *E. coli* (EAEC), enterotoxigenic *E. coli* (ETEC), and enterohemorrhagic *E. coli* (EHEC) [47]. These virulence genes include *stx1* and *stx2* (present in EHEC), *eaeA* (present in EHEC and EPEC), *STIb* and *LTI* (present in ETEC), *ipaH* (present in EIEC and *Shigella* spp.), and *aggR* (present in EAEC) [55].

The three enteric viruses (enterovirus [56], rotavirus [57], and adenovirus [58]) and the human-specific *Bacteroidales* fecal marker (BacHum) [59] were detected using end-point PCR (BacHum) or reverse transcriptase-PCR (viruses) with a hydrolysis probe. Details on nucleic acid extraction, molecular detection assays, assay detection limits, and inhibition analyses can be found in Mattioli *et al.* [33].

Statistical Analysis

Data were analyzed using SAS Enterprise Guide version 4.3 (SAS Institute Inc., Cary, NC). The term 'enteric virus' is defined as the presence or absence of at least one of the three enteric viruses measured. The acronym ECVG is defined as the presence or absence of at least one of the seven *E. coli* virulence genes measured, and the term BacHum is defined as the presence or absence of the human-specific *Bacteroidales* fecal marker. Results are considered statistically significant at a level of $p \leq 0.05$.

A matched case-control study design was used to evaluate the relationship between cases of diarrhea in children under five and the presence of molecular markers of enteric pathogens on hands of mothers and in household stored water. Out of the 1219 households surveyed in the larger household water and hygiene behavioral intervention trial, the total number of children classified as sick with GI illness was 113 (among 112 unique households). These children served as our 'cases' in the case-control study. A case of gastrointestinal illness (GI) was defined as a child having three or more loose/watery stools per 24 hour period, blood in the stool, and/or vomiting using a two-day recall period. Reported GI illness was thus nearly contemporaneous with the sampling and interview. Healthy children (no reported symptoms) in the same household as a GI case were excluded from being controls. Children presenting with non-GI symptoms (*e.g.*, coughing, congestion) and their siblings were also excluded from being controls.

Cases were matched to controls post sample collection by one-to-one propensity score matching (PSM) with no replacement using STATA (version 11; Stata Corporation, College Station, TX) [60]. PSM was employed to reduce the chance of bias resulting from systematic differences between cases and controls and to improve effect estimation efficiency [61]. The following variables were included in the model to generate the propensity scores for matching: child age; the number of families in the housing unit; the number of times mother reported washing hands with soap the day prior to interview; whether the household was located in an urban or rural community; whether the mother reported that the youngest child uses a latrine regularly; whether the household has an on-plot water source; if the child's palms were observed to have visible dirt; whether the mother works outside the home; the number of liters of water collected *per capita* per day; and whether the household's main drinking water source type was a borewell or tap. An equal number of children (among 111 unique households) were selected by PSM as matched controls.

Conditional logistic regression was used to calculate matched odds ratios (OR) representing the association between childhood diarrhea and the presence of contamination on the primary caregiver's hands and in the household's stored drinking water (case-control analysis). Exact p-values and confidence intervals (CI) were calculated. As a robustness check and to control for potential confounders in the case-control analysis, bivariate analyses of household demographics and water, sanitation, and hygiene behaviors/characteristics between case and control households were performed. The bivariate analyses showed which factors were independently associated with child case status. The student's T-test was used to compare mean values between case and control households. Pooled variances were used unless the Equality of Variances Folded F-Statistic was significant ($p<0.05$), in which case a Satterthwaite Test of unequal variance was used. The chi-square test was used to test for differences in proportions of binary variables between case and control households; when N <5 a Fisher's exact test was used. An ANOVA test was used to evaluate differences in proportions between case and control households for categorical variables, and the Wilcoxon-Mann-Whitney Test was used for non-parametric comparisons for non-normally distributed continuous variables. Variables with a p-value ≤0.20 in the bivariate analysis were included in a multivariate logistic regression model of diarrhea. Model reduction was performed by backwards selection until all remaining variables were found to be significant ($p \leq 0.05$). Adjusted odds ratios (AOR) and 95% confidence intervals were calculated using conditional logistic regression, adjusting for those variables independently associated with diarrhea [62]. The possibility of overmatching with PSM was also assessed and is described in the Supporting Information (SI).

Multivariate logistic regression was used to model the presence of contamination (separate models were estimated for ECVG, enteric virus, and BacHum) on hands and in household stored water as a function of hand hygiene and water management behaviors, respectively. Each separate model included data from 306 unique households. Two hundred twenty-three of these households were part of the case-control analysis presented above. The additional 83 households were drawn from the dataset published by Mattioli et al. [33].

The hand hygiene behaviors examined in the model of hand contamination include reported liters per capita per day used for hand washing; whether the respondent reported washing her hands with soap within one hour prior to having her hand rinse taken; and the respondent's reported activity immediately prior to having the hand rinse sample taken (i.e., washing clothes, dishes, or child; hand washing; food preparation/eating/serving; or other activity such as gardening/farming and sweeping; versus sitting).

The water management behaviors examined in the model of stored drinking water contamination include whether the respondent reported actively treating her stored water (i.e., boiling, chlorinating, filtering, or solar water disinfection); the reported length of time the sampled water had been stored in the home; whether the observed extraction method of stored water for sampling by the respondent was "risky" (i.e., dipping a short-handled cup, mug, or bowl); whether the stored drinking water was reportedly collected from an improved source; the average reported amount of time that household members spend fetching water per day; and whether the bacterial or viral gene(s) being modeled was present in the respondent's hand rinse sample. The models also include control variables for the presence of an infant in the household, household monthly expenditure per capita, and whether the household uses a facility with improved sanitation infrastructure. Other control variables considered but that did not contribute significantly to any of the models include whether the household used a private or shared latrine; the number of children under five in the household; whether the primary caregiver worked outside of the home, and whether the primary caregiver was able to read and write.

For the purposes of this manuscript, improved sanitation infrastructure is defined as a toilet or latrine with a cement slab, septic tank, or flush tank into a piped sewer system or pit latrine. This definition differs from the WHO/UNICEF Joint Monitoring program definition [63] in which only privately owned sanitation facilities are considered improved. This definition allowed us to examine for the effects of private versus shared sanitation and improved versus unimproved sanitation infrastructure separately in our models.

Results

Matched Case-control Analysis

Among the 113 children classified as sick with GI illness within the 48 hours prior to interview, 80 children were reported to have 3 or more loose/watery stools per 24 hour period, 6 were reported to have blood in the stool, and 53 were reported to have vomited. Variables used in the PSM method, as well as household demographics and water, sanitation, and hygiene characteristics stratified by case versus control status are presented in Table S1.

At least one of the three enteric viruses measured was found on 21% of mother's hands (47/222) and in 3% of stored water samples (7/216). Rotavirus was the most frequently detected virus both on hands (10%) and in stored water (2%). Enterovirus was detected in 8% of hand samples, but was never detected in household stored water; adenovirus was detected in 5% of hand rinse samples and in 1% of stored water samples. BacHum was found in 39% of hand rinse samples (86/222) and in 14% of stored water samples (30/216).

More than half (59%) of all households in the study had at least one E. coli virulence gene (ECVG) detected in their stored water, and 41% of respondents had ECVG detected on their hands. The presence of individual virulence genes among all water and hand rinse samples processed for the case-control analysis is reported in Table S2. Across case and control households, the mean E. coli and enterococci concentration in household stored water was 1.5 (SD 1.0) and 0.5 (SD 0.6) log CFU/100 ml (N = 219), respectively. The mean E. coli and enterococci concentration in hand rinse samples was 2.5 (SD 1.0) and 2.7 (SD 1.0) log CFU/2 hands (N = 223), respectively. The concentration of FIB on hands and in stored drinking water stratified by case status can be found in Table S2. There was no significant difference between case and control households in the concentration of E. coli or enterococci found on mother's hands or in stored drinking.

Table 1 presents the prevalence of molecular markers of enteric viruses and ECVG in household stored water and hand rinse samples, stratified by case and control households, as well as the odds ratios from the matched case-control analysis. The presence of ECVG in a household's stored water was associated with a 2-fold decrease in the odds of at least one child under five years of age in the household reporting symptoms of diarrhea (OR = 0.51 [95% CI 0.27–0.93]; $p = 0.03$). Similarly, the presence of the E. coli virulence gene, Lt1, on the primary caregiver's hands (suggesting the presence of the E. coli pathotype enterotoxigenic E. coli (ETEC) [55]) was also associated with a significant decrease in the odds of a child in the household having diarrhea (OR = 0.25 [95% CI 0.05, 0.93]; $p = 0.04$).

There were no significant associations between virus detection in water or hand rinses and diarrhea case status in children. The presence of each enteric virus on hands and in stored water was consistently associated with increased odds of a child in the household having diarrhea (Table 1), although none achieved statistical significance. The presence of at least one enteropathogen molecular marker (enteric virus or ECVG) on hands or in water was also not significantly associated with diarrhea.

The presence of BacHum in either stored water or hand rinse samples was not associated with the odds of a child having diarrhea at the time of visit. Similarly, the presence of FIB (E. coli or enterococci) on hands or in stored water was not associated with cases of diarrhea. Higher concentrations of FIB in stored water (categorical variable: 1 to <11, 11 to 100, or >100 CFU/100 mL, Table 1) were also not associated with the odds of a child having diarrhea.

Case and control households were found to be similar with respect to the variables used in the PSM (Table S1, all $p > 0.05$). Cases and controls had no statistically different characteristics (all $p > 0.05$), with one exception: a larger percentage of control households had their stored drinking water covered (98% versus 89%, $p < 0.01$). The results of the bivariate analyses performed as a robustness check and to control for other potential confounders in the case-control analysis can be found in the SI (Results S1 and Table S1). The revised odds ratios after adjusting for the variables

Table 1. Prevalence of *E. coli* virulence genes (ECVG), enteric virus genes, human-specific *Bacteroidales* genes, and FIB detected in household stored drinking water and hand rinse samples of respondents with at least one child younger than five years old that were either sick with diarrhea (cases) versus matched healthy children under five years of age (controls).

	HANDS						STORED WATER					
	Case (%)	Control (%)	OR	95% CI[c]		P	Case (%)	Control (%)	OR	95% CI[c]		P
ECVG [a]	40.0	44.1	0.86	0.48	1.53	0.68	52.7	67.9	0.51	0.27	0.93	0.03[†]
ipaH	20.0	31.5	0.50	0.24	1.01	0.05	23.2	33.9	0.58	0.31	1.07	0.09
aggR	13.6	15.3	0.92	0.39	2.19	1.00	21.4	32.1	0.54	0.26	1.07	0.08
Lt1	4.5	13.5	0.25	0.05	0.93	0.04[†]	11.6	17.0	0.71	0.31	1.57	0.46
STIb	0.0	0.9	1.00*	0.00	19.00	1.00	1.8	1.8	1.00	0.07	13.80	1.00
eaeA	4.5	5.4	0.83	0.20	3.28	1.00	17.0	15.2	1.15	0.51	2.64	0.85
stx1	9.1	17.1	0.53	0.22	1.19	0.14	20.5	30.4	0.54	0.26	1.07	0.08
stx2	0.0	0.9	1.00*	0.00	19.00	1.00	0.0	2.7	0.26*	0.00	1.71	0.25
Enteric Virus [b]	24.8	17.0	1.69	0.82	3.66	0.18	4.6	1.8	2.50	0.41	26.25	0.45
Rotavirus	12.4	8.0	1.56	0.63	4.07	0.40	3.7	0.0	5.29*	0.90	∞	0.13
Adenovirus	6.2	4.5	1.50	0.36	7.23	0.75	0.9	1.8	0.50	0.01	9.61	1.00
Enterovirus	8.8	6.3	1.60	0.46	6.22	0.58	0.0	0.0				
At least 1 enteric virus or ECVG	55.8	54.0	1.12	0.63	1.97	0.79	54.0	66.4	0.59	0.32	1.08	0.09
Human *Bacteroidales*	33.6	43.8	0.64	0.35	1.13	0.13	11.1	16.2	0.67	0.27	1.59	0.42
Escherichia coli[¥]	81.4	74.3	1.44	0.76	2.80	0.29	81.4	87.6	0.63	0.28	1.37	0.28
1 to <11 CFU/100 mL							13.3	20.4	0.48	0.16	1.39	0.21
11 to 100 CFU/100 mL							31.9	33.6	0.70	0.25	1.85	0.57
>100 CFU/100 mL							36.3	33.6	0.79	0.29	2.04	0.75
Enterococcus[¥]	84.1	87.6	0.71	0.28	1.73	0.54	79.6	89.4	0.48	0.20	1.06	0.07
1 to <11 CFU/100 mL							18.6	15.9	0.60	0.20	1.75	0.43
11 to 100 CFU/100 mL							31.0	38.1	0.47	0.19	1.10	0.09
>100 CFU/100 mL							30.1	35.4	0.46	0.17	1.16	0.11

The study consisted of 112 unique case households (containing 113 case children) and 111 unique households with only healthy children (containing 113 matched, control children).
[a]At least one of the seven pathogenic *E. coli* virulence genes (ECVG) measured present.
[b]At least one of the three enteric viruses measured (rotavirus, adenovirus, enterovirus) present.
[c]CI, confidence interval.
[¥]Presence/Absence of CFU per 2 hands; Presence/Absence or within specified range of CFU/100 mL stored drinking water with 0 CFU/100 mL as the reference group.
*Indicates a median unbiased estimate.
[†]Statistically significant (p≤0.05).

significantly associated with case status are presented in Table S3. Since there were no substantive differences between the adjusted case-control analyses and the unadjusted analyses, the case-control results were considered robust. The incorporation of numerous matching variables in the PSM model may have overmatched households, causing the case-control effect estimates to be biased downward [61,64,65]. Therefore, the potential effect of over-matching on our results was assessed. The results of the overmatching analysis (see SI and Tables S4, S5, S6, and S7) show that matching case and control children on several control variables did not affect the results.

Behavioral Determinants of Enteric Pathogen and Fecal Markers

We modeled ECVG, enteric virus, and BacHum presence in hand rinse samples and household stored water as a function of hygiene and water management behaviors. Household socioeconomic and demographic characteristics, as well as water supply, sanitation, and hygiene practices of all 306 households used in the models can be found in Table S8. Among the households

modeled, twenty-five percent of households included an infant (<1 yr) present at the time of visit, with an average of 1.3 (SD 0.5) children under the age of five. The age of the primary caregivers in our study ranged from 16 to 68 years (median 25 yr). Households reported spending an average of $17 US (SD $9) per month per person. Only 15% of households had an on-plot water source, but 85% reported using an improved source as their main source of drinking water according to the WHO/UNICEF Joint Monitoring program definition [63] (*e.g.*, tap, borewell, or rainwater). A minority of households (16%) reported to have treated their stored water. At the time of sampling, mean storage time was 33 hours (SD 29), and 94% of stored water containers were observed to be covered. Thirty-eight percent of households had access to a sanitation facility with improved sanitation infrastructure, and 51% percent of households reported having a private sanitation facility.

The models of ECVG, enteric virus, and BacHum presence in hand rinse samples and household stored water are presented in Tables 2 and 3, respectively. We found a limited number of significant determinants of hand contamination. The liters *per capita* used per day for hand washing (natural-log transformed) was the only statistically significant explanatory hygiene behavior in the model of enteric virus presence on hands ($p<0.01$). Results indicate that doubling the quantity of water used *per capita* per day for hand washing was associated with a 2-fold decrease in the odds of a respondent having an enteric virus on her hands. In the model of ECVG presence on hands, the odds of detecting ECVG was approximately three times greater for those respondents who reported handling food or washing (both $p<0.01$) prior to sampling versus those reporting sitting. No hand hygiene behaviors were significant in the model of BacHum. It is important to note that our model explains only a small portion of the variation in microbial prevalence on hands, highlighting the unexplained variability in hand contamination within these communities.

Contamination of stored water was associated with several water management variables. Stored drinking water collected from an improved source was associated with a 2.7-fold reduction in the odds of detecting ECVG contamination ($p = 0.02$) and a 5.6-fold decrease in the odds of detecting BacHum ($p<0.01$). Conversely, the odds of enteric virus and BacHum being present in a household's stored water was 32.7 and 3.6 times greater for households where enteric viruses or BacHum were also present on the hands of the respondent (both $p<0.01$). Also, the use of a facility with improved sanitation infrastructure was associated with a 1.7-fold decrease in the odds of detecting ECVG in the household's stored drinking water ($p = 0.04$).

Discussion

The overall prevalence of enteropathogen genes and the human-specific *Bacteroidales* molecular marker in stored drinking water and hand rinse samples analyzed for this study (Table S2) was similar to that found in Bagamoyo, Tanzania by Mattioli *et al.* [33]. Other studies in Tanzania have also documented ECVG, enteric virus, and *Bacteroidales* molecular markers in soil, on household surfaces, on produce, and on hands in Tanzania [16,47].

Unexpectedly, our study found that the presence of ECVG in stored water and on hands was associated with decreased odds of a child under the age of five having reported diarrhea–indicating a higher prevalence of ECVG among control households. One possible explanation for the observed negative association between ECVG presence and diarrhea is that an episode of child diarrhea in the household might have triggered efforts by the primary caregiver to improve stored drinking water quality, such as treatment or collection from an improved source. A greater percentage of respondents caring for children with diarrhea reported boiling their stored drinking water (16%) than those in control households (9%), although the difference did not reach statistical significance (p = 0.1). Notably, a prior study in Tanzania also found microbial water quality to be cleaner among households with child diarrhea [17].

Asymptomatic pathogen shedding among healthy control children could help explain the absence of association between the pathogens detected in the household environment and child diarrhea. A recent study to identify the etiology of child diarrhea in Tanzania found that 52% of healthy controls were infected with an enteric pathogen [53]. Asymptomatic shedding could be the result of subclinical infections, persistent shedding after illness symptoms have subsided, ingestion of pathogens below an infectious dose, or immunity developed by those children living in households with persistent pathogen contamination [66–69].

Several reported water management and hygiene behaviors were found to be significantly associated with molecular indicators

of pathogen contamination in the household. For example, modeling the presence of enteric virus marker on hands suggests that the volume of water used for hand washing may be important in reducing enteric virus transmission. Currently, the World Health Organization does not recommend a minimum volume of water per person per day specifically for hand washing, with the assumption that this volume is dependent on level of service and water fetching distance [70]. Households in our study reported using an average of 1.8 L (SD = 1.3) per person per day for hand washing and had an average of 5.4 people (SD = 2.3) per household (Table S8). Therefore, for a household in our study to reduce their risk of viral hand contamination by 2-fold, each household would have to access an average of 19.3 extra L of water (3.6 L per person) per day for hand washing. This would be roughly equivalent to a household collecting one extra 20-L jerry can of water per day.

Respondents reporting activities involving washing soiled household items (such as clothes or dishes), and those who reported handling food immediately prior to sampling, were more likely to have ECVG detected on their hands. This may suggest, as proposed by others [71], that water used for washing might serve as a transport mechanism for bacterial pathogens within the home. Previous studies found Tanzanian fresh market produce to be contaminated with ECVG [47] and have detected increased levels of FIB on hands after handling produce or preparing food [16]. This may imply that bacterial pathogens could also be transferred from produce to hands and would explain the association between ECVG on hands and recent food handling observed in this study.

Stored drinking water collected from households using improved water sources was less likely to be contaminated with ECVG and BacHum. This result adds nuance to published evidence of significant post-supply FIB contamination of stored water, particularly stored water collected from improved sources [8,14,72,73]. Together this suggests that while stored water may become re-contaminated with FIB after collection, improvements to water sources may still provide safer water at the point-of-use by preventing bacterial pathogens and human feces from entering drinking water prior to collection. Households with improved sanitation infrastructure also had reduced odds of ECVG detection in their stored water, suggesting that, like water sources, improvements to sanitation infrastructure (*e.g.*, the addition of a concrete slab, septic tank, or flush tank) may reduce the risk of pathogenic bacteria entering the household's stored drinking water.

Stored water collected from the household of a respondent with BacHum detected on her hands was more likely to be contaminated with BacHum, implying that hands may be a source of human fecal contamination in stored drinking water. However, due to the cross-sectional study design, we cannot confirm the directionality of contamination. Interestingly, reported water treatment, storage time, use of a "risky" extraction method, and water fetching time were not found to be associated with markers of pathogen or human fecal presence in stored water. This result stands in contrast to other research which found higher FIB levels in drinking water of households that performed risky extraction methods [14] and lower FIB levels in households that actively treated stored water [15] or were served by on-site water sources [74]. However the results are consistent with previous research in Tanzania identifying fecal contamination on hands as the strongest predictor of fecal contamination levels in stored water [17]. Thus, the effects of safe water management behaviors may be offset or muted by contamination from hand contact.

Viral marker prevalence may have been too low in this study to detect significant associations with diarrhea or water management

Table 2. Binary logistic regression model of *E. coli* virulence genes (ECVG), enteric virus genes, and human-specific *Bacteroidales* genes presence in hand rinse samples as a function of hygiene behaviors.

HANDS

Variable	ECVG, N = 256*						Enteric Virus, N = 258*						Human-Specific *Bacteroidales*, N = 258*					
	pseudo R² = 0.09, Max-rescaled R² = 0.12						pseudo R² = 0.06, Max-Rescaled R² = 0.10						pseudo R² = 0.03, Max-Rescaled R² = 0.04					
	Likelihood Ratio: χ2 = 24.31, p<0.01						Likelihood Ratio: χ² = 16.60, p=0.06						Likelihood Ratio: χ² = 6.93, p=0.64					
	β	SE	P	OR	95% CI		β	SE	P	OR	95% CI		β	SE	P	OR	95% CI	
Intercept	−0.87	0.33	0.01				−1.43	0.37	0.00				−0.57	0.32	0.07			
Liters per capita per day used for hand washing[a]	−0.01	0.27	0.98	0.99	0.59	1.68	−1.00	0.34	0.00†	0.37	0.19	0.72	−0.15	0.26	0.56	0.86	0.51	1.43
Respondent washed hands within 1 h prior to sampling[b]	−0.08	0.30	0.77	0.92	0.51	1.64	−0.67	0.38	0.08	0.51	0.24	1.09	−0.13	0.29	0.66	0.88	0.50	1.56
Washing (clothes, dishes, hands, child) vs. Sitting[b,c]	1.04	0.41	0.01†	2.84	1.28	6.32	−0.22	0.56	0.69	0.80	0.27	2.39	0.05	0.41	0.91	1.05	0.47	2.34
Hand Washing vs. Sitting[b,c]	0.90	1.43	0.53	2.46	0.15	40.77	1.41	1.44	0.32	4.12	0.25	68.73	−13.76	894.20	0.99	<0.01	<0.01	>999.9
Food Preparation vs. Sitting[b,c]	1.23	0.34	0.00†	3.41	1.74	6.69	0.10	0.41	0.81	1.11	0.49	2.48	0.55	0.33	0.10	1.73	0.90	3.34
Other vs. Sitting[b,c]	−0.25	0.57	0.66	0.78	0.25	2.39	0.31	0.58	0.59	1.37	0.44	4.24	−0.38	0.56	0.49	0.68	0.23	2.04
Use of facility with improved sanitation infrastructure[b]	−0.30	0.29	0.30	0.74	0.43	1.30	0.16	0.34	0.64	1.17	0.60	2.26	0.29	0.28	0.30	1.34	0.78	2.30
Infant (<1 yr) present in household[b]	0.49	0.30	0.11	1.63	0.90	2.96	−0.27	0.37	0.47	0.77	0.37	1.58	−0.14	0.30	0.63	0.87	0.48	1.56
Regular monthly expenditures per person per 1000 TZS[d]	0.06	0.10	0.53	1.06	0.88	1.28	0.22	0.11	0.04	1.24	1.01	1.53	0.01	0.09	0.95	1.01	0.84	1.20

[a] Ln-transformed.
[b] Binary variables with values of 0 and 1.
[c] Refers to the reported activity prior to the respondent having their hand rinse sample taken.
[d] TZS Tanzanian Shillings.
*N <306 because sample was lost or survey response not collected.
† Statistically significant (p≤0.05).

Table 3. Binary logistic regression model of *E. coli* virulence genes (ECVG), enteric virus genes, and human-specific *Bacteroidales* gene presence in household stored water as a function of water management behaviors.

	STORED WATER																	
	ECVG, N=276*						Enteric Virus, N=267*						Human-Specific *Bacteroidales*, N=267*					
	pseudo R²=0.06, Max-Rescaled R²=0.09						pseudo R²=0.11, Max-Rescaled R²=0.45						pseudo R²=0.13 Max-Rescaled R²=0.24					
	Likelihood Ratio: χ^2=17.83, p=0.04						Likelihood Ratio: χ^2=32.17, p<0.01						Likelihood Ratio: χ^2=35.57 p<0.01					
Variable	β	SE	p	OR	95% CI		β	SE	p	OR	95% CI		β	SE	p	OR	95% CI	
Intercept	1.34	0.89	0.13				-25.02	418.8	0.95				-3.89	1.38	0.00			
Stored water reportedly treated[b,c]	-0.11	0.37	0.77	0.90	0.43	1.87	0.40	1.29	0.76	1.49	0.12	18.76	0.02	0.60	0.98	1.02	0.32	3.26
Time water stored in household (h)[a]	0.09	0.11	0.45	1.09	0.87	1.37	-0.41	0.36	0.25	0.66	0.33	1.34	0.47	0.25	0.06	1.60	0.99	2.60
Observed extraction method of stored water was risky[b,d]	-0.04	0.38	0.92	0.96	0.46	2.03	12.16	285.8	0.97	1.90×10^6	<0.01	>999.9	-0.17	0.56	0.77	0.85	0.28	2.56
Stored water reportedly collected from improved source[b,e]	-1.01	0.41	0.01†	0.36	0.16	0.82	11.01	306.1	0.97	6.02×10^4	<0.01	>999.9	-1.69	0.45	0.00†	0.18	0.08	0.45
Water Fetching Time Per Day (min)[a]	0.03	0.15	0.87	1.03	0.77	1.37	-0.15	0.42	0.72	0.86	0.38	1.96	0.37	0.25	0.14	1.44	0.89	2.35
Modeled contamination present on hands of primary caregiver[b]	0.26	0.26	0.33	1.30	0.77	2.17	3.49	1.11	0.00†	32.7	3.68	290.42	1.28	0.43	0.00†	3.60	1.55	8.38
Use of facility with improved sanitation infrastructure[b]	-0.55	0.27	0.04†	0.58	0.34	0.98	-0.70	0.87	0.42	0.50	0.09	2.75	0.41	0.44	0.35	1.51	0.64	3.59
Infant (<1 yr) present in household[b]	-0.07	0.30	0.80	0.93	0.52	1.66	-11.96	243.5	0.96	0.00	<0.01	>999.9	-0.35	0.52	0.50	0.71	0.25	1.96
Regular monthly expenditures per person per 1000 TZS[f]	-0.08	0.09	0.37	0.92	0.77	1.10	-0.42	0.35	0.23	0.66	0.33	1.30	-0.05	0.14	0.75	0.95	0.72	1.26

[a]Ln-transformed.
[b]Binary variables with values of 0 and 1.
[c]Boiling, chlorinating, filtering, or SODIS (versus no treatment including settling).
[d]Cup, mug, or bowl (versus pouring, long handled dipper, or spigot).
[e]Borewell, rainwater, or tap (versus shallow well, cart/tanker, surface water, or vendor).
[f]TZS Tanzanian Shillings.
*N <306 because sample was lost or survey response not collected.
†Statistically significant (p≤0.05).

behaviors given our sample size. Also, the lower limit of detection of the viral assays ranged from 10^0 to 10^2 genomic units per 100 ml stored water or per two hands. Thus, health relevant concentrations below our detection limit could have been present but been undetected in our study [75]. Despite these limitations, viral prevalence trended toward an increase in the odds of a child having diarrhea in the household, and therefore should be prioritized for further research. Our findings suggest that research focused on child exposure to enteric pathogens in the household environment should include rotavirus, as it was the most prevalent virus found in both hand rinses and stored water of sample households. Rotavirus was also recently reported to be one of the most important etiological agents of childhood diarrhea in a multi-country, prospective case-control study [76].

The results presented herein should be interpreted in consideration of several study limitations. The cross-sectional design precludes determining the direction of effect between child diarrhea, behaviors, and exposure. Also, previous research has shown that microbial contamination of hands and drinking water can vary significantly over time [14,16], and our cross sectional study design does not allow us to consider this variability. In addition, only nucleic acids of enteric pathogens were detected in this study; our methods do not characterize the infectivity or viability of the pathogens targeted. Diarrhea and many of the water management and hygiene behaviors used in our models were self-reported by the respondent; self-reported data has been found to introduce inaccuracy and bias into estimates of behavior [77]. Finally, we were unable to measure all potential enteric pathogens (*e.g.*, protozoa) [76] in this study, and it is possible that these unmeasured pathogens may have had a positive association with diarrhea.

Few studies have looked at enteric pathogen and human fecal molecular markers in household environments of low-income communities [16,33,47]; this study contributes new knowledge by examining the association between hygiene and water management behaviors and the presence of these markers. The identification of behaviors associated with molecular markers of enteric pathogens and human fecal contamination on hands and in water in Bagamoyo can be used to inform the development of more efficacious interventions aimed at reducing the burden of childhood diarrhea in other low-income communities. For example, our work suggests that increasing the quantity of water available for hand washing may reduce enteric virus transmission from hands. In addition, improvements in food hygiene practices and sanitation infrastructure may help alleviate pathogenic *E. coli* contamination within the home, while improved source water may prevent human fecal and pathogenic *E. coli* contamination of a household's stored drinking water. To our knowledge, this is the first study to examine the association between child diarrhea and molecular markers of enteric pathogens in household stored water and hand rinses. Our results warrant further investigation into why stored drinking water was less frequently contaminated with bacterial pathogens in households with sick children. In combination, our analyses highlight the need to better understand the relative contribution and interdependence of household exposure routes to the burden of child diarrhea.

Supporting Information

Table S1 Child-level descriptive statistics from the case-control analysis. The table includes variables used in the PSM, as well as household (HH) demographics, and water, sanitation, and hygiene characteristics by case and control status.

Table S2 ECVG, enteric virus gene, Human *Bacteroidales* gene, and FIB (*Escherichia coli* and *Enterococcus*) prevalence for households in the case-control study.

Table S3 Adjusted, matched case-control analysis results.

Table S4 Unmatched household case-control analysis results using original controls.

Table S5 Unmatched household case-control analysis results using additional controls.

Table S6 Prevalence (%) of ECVG, enteric virus genes, and Human *Bacteroidales* genes in additional control households. The additional households, along with the original (case-control analysis) control households, were used in the robustness checks.

Table S7 Robustness checks logistic regression results. These results were used to evaluate whether the propensity score is significantly associated with pathogen presence in the original (case-control analysis) control households.

Table S8 Household-level descriptive statistics from the water management and hygiene behavior models. Household-level demographics, and water, sanitation, and hygiene characteristics of the households used in the logistic regression model of contamination as a function of water management and hygiene behaviors.

Methods S1 Detailed description of methods used for the overmatching analysis.

Results S1 Results of robustness checks and overmatching analysis.

Acknowledgments

The authors acknowledge Rebecca Gilsdorf, Michael Harris, Debbie Lee, Emily Viau, and Maggie Montgomery for their support in the field and the laboratory. We acknowledge our collaborators Salim Abdulla and Omar Juma at the Ifakara Health Institute in Bagamoyo, Tanzania. This project would not have been possible without the Tanzanian lab and field teams and participating households.

Author Contributions

Conceived and designed the experiments: MCM ABB JD ARH MM AJP. Performed the experiments: MCM ARH AJP. Analyzed the data: MCM AJP. Contributed reagents/materials/analysis tools: ABB JD MM. Wrote the paper: MCM ABB JD ARH MM AJP.

References

1. Walker CL, Rudan I, Liu L, Nair H, Theodoratou E, et al. (2013) Global burden of childhood pneumonia and diarrhoea. Lancet 381: 1405–1416.
2. Liu L, Johnson HL, Cousens S, Perin J, Scott S, et al. (2012) Global, regional, and national causes of child mortality: an updated systematic analysis for 2010 with time trends since 2000. Lancet 379: 2151–2161.
3. Prüss-Üstün A, Bos R, Gore F, Bartram J (2008) Safer water, better health: costs, benefits and sustainability of interventions to protect and promote health. Geneva: World Health Organization.
4. Organization WH, UNICeF (2012) Progress on sanitation and drinking water: 2012 update. Geneva: World Health Organization.
5. Dufour AP, Ballentine P (1986) Ambient Water Quality Criteria for Bacteria-1986 (Bacteriological ambient water quality criteria for marine and fresh recreational waters). Washington, DC: U.S. Environmental Protection Agency.
6. Cairncross AM (1990) Health Impacts in Developing Countries: New Evidence and New Prospects. Water and Environment Journal 4: 571–575.
7. Trevett AF, Carter RC, Tyrrel SF (2005) The importance of domestic water quality management in the context of faecal-oral disease transmission. J Water Health 3: 259–270.
8. Wright J, Gundry S, Conroy R (2004) Household drinking water in developing countries: a systematic review of microbiological contamination between source and point-of-use. Trop Med Int Health 9: 106–117.
9. Jensen PK, Ensink JH, Jayasinghe G, van der Hoek W, Cairncross S, et al. (2002) Domestic transmission routes of pathogens: the problem of in-house contamination of drinking water during storage in developing countries. Trop Med Int Health 7: 604–609.
10. Luby SP, Agboatwalla M, Raza A, Sobel J, Mintz ED, et al. (2001) Microbiologic effectiveness of hand washing with soap in an urban squatter settlement, Karachi, Pakistan. Epidemiol Infect 127: 237–244.
11. Luby SP, Agboatwalla M, Billhimer W, Hoekstra RM (2007) Field trial of a low cost method to evaluate hand cleanliness. Trop Med Int Health 12: 765–771.
12. Pinfold JV (1990) Faecal contamination of water and fingertip-rinses as a method for evaluating the effect of low-cost water supply and sanitation activities on faeco-oral disease transmission. II. A hygiene intervention study in rural north-east Thailand. Epidemiol Infect 105: 377–389.
13. Davis J, Pickering AJ, Rogers K, Mamuya S, Boehm AB (2011) The effects of informational interventions on household water management, hygiene behaviors, stored drinking water quality, and hand contamination in peri-urban Tanzania. Am J Trop Med Hyg 84: 184–191.
14. Levy K, Nelson KL, Hubbard A, Eisenberg JN (2008) Following the water: a controlled study of drinking water storage in northern coastal Ecuador. Environ Health Perspect 116: 1533–1540.
15. Sobsey MD, Stauber CE, Casanova LM, Brown JM, Elliott MA (2008) Point of use household drinking water filtration: A practical, effective solution for providing sustained access to safe drinking water in the developing world. Environ Sci Technol 42: 4261–4267.
16. Pickering AJ, Julian TR, Mamuya S, Boehm AB, Davis J (2011) Bacterial hand contamination among Tanzanian mothers varies temporally and following household activities. Trop Med Int Health 16: 233–239.
17. Pickering AJ, Davis J, Walters SP, Horak HM, Keymer DP, et al. (2010) Hands, water, and health: fecal contamination in Tanzanian communities with improved, non-networked water supplies. Environ Sci Technol 44: 3267–3272.
18. Brown JM, Proum S, Sobsey MD (2008) Escherichia coli in household drinking water and diarrheal disease risk: evidence from Cambodia. Water Sci Technol 58: 757–763.
19. Moe CL, Sobsey MD, Samsa GP, Mesolo V (1991) Bacterial indicators of risk of diarrhoeal disease from drinking-water in the Philippines. Bull World Health Organ 69: 305–317.
20. VanDerslice J, Briscoe J (1995) Environmental interventions in developing countries: interactions and their implications. Am J Epidemiol 141: 135–144.
21. Jensen PK, Jayasinghe G, van der Hoek W, Cairncross S, Dalsgaard A (2004) Is there an association between bacteriological drinking water quality and childhood diarrhoea in developing countries? Trop Med Int Health 9: 1210–1215.
22. Gundry S, Wright J, Conroy R (2004) A systematic review of the health outcomes related to household water quality in developing countries. J Water Health 2: 1–13.
23. Khush RS, Arnold BF, Srikanth P, Sudharsanam S, Ramaswamy P, et al. (2013) H2S as an indicator of water supply vulnerability and health risk in low-resource settings: a prospective cohort study. Am J Trop Med Hyg 89: 251–259.
24. Levy K, Nelson KL, Hubbard A, Eisenberg JN (2012) Rethinking indicators of microbial drinking water quality for health studies in tropical developing countries: case study in northern coastal Ecuador. Am J Trop Med Hyg 86: 499–507.
25. Power ML, Littlefield-Wyer J, Gordon DM, Veal DA, Slade MB (2005) Phenotypic and genotypic characterization of encapsulated Escherichia coli isolated from blooms in two Australian lakes. Environ Microbiol 7: 631–640.
26. Viau EJ, Goodwin KD, Yamahara KM, Layton BA, Sassoubre LM, et al. (2011) Bacterial pathogens in Hawaiian coastal streams–associations with fecal indicators, land cover, and water quality. Water Res 45: 3279–3290.
27. Byappanahalli MN, Whitman RL, Shively DA, Sadowsky MJ, Ishii S (2006) Population structure, persistence, and seasonality of autochthonous Escherichia coli in temperate, coastal forest soil from a Great Lakes watershed. Environ Microbiol 8: 504–513.
28. Anderson KL, Whitlock JE, Harwood VJ (2005) Persistence and differential survival of fecal indicator bacteria in subtropical waters and sediments. Appl Environ Microbiol 71: 3041–3048.
29. Byappanahalli MN, Fujioka RS (1998) Evidence that tropical soil environment can support the growth of Escherichia coli. Water Science and Technology 38: 171–174.
30. Yamahara KM, Walters SP, Boehm AB (2009) Growth of enterococci in unaltered, unseeded beach sands subjected to tidal wetting. Appl Environ Microbiol 75: 1517–1524.
31. Fujioka RS, Tenno K, Kansako S (1988) Naturally occurring fecal coliforms and fecal streptococci in Hawaii's freshwater streams. Toxicity Assessment 3: 613–630.
32. Ferguson AS, Mailloux BJ, Ahmed KM, van Geen A, McKay LD, et al. (2011) Hand-pumps as reservoirs for microbial contamination of well water. J Water Health 9: 708–717.
33. Mattioli MC, Pickering AJ, Gilsdorf RJ, Davis J, Boehm AB (2013) Hands and water as vectors of diarrheal pathogens in Bagamoyo, Tanzania. Environ Sci Technol 47: 355–363.
34. Lemarchand K, Lebaron P (2003) Occurrence of Salmonella spp. and Cryptosporidium spp. in a French coastal watershed: relationship with fecal indicators. FEMS Microbiol Lett 218: 203–209.
35. Horman A, Rimhanen-Finne R, Maunula L, von Bonsdorff CH, Torvela N, et al. (2004) Campylobacter spp., Giardia spp., Cryptosporidium spp., noroviruses, and indicator organisms in surface water in southwestern Finland, 2000–2001. Appl Environ Microbiol 70: 87–95.
36. Bonadonna L, Briancesco R, Ottaviani M, Veschetti E (2002) Occurrence of Cryptosporidium oocysts in sewage effluents and correlation with microbial, chemical and physical water variables. Environ Monit Assess 75: 241–252.
37. Wu J, Long SC, Das D, Dorner SM (2011) Are microbial indicators and pathogens correlated? A statistical analysis of 40 years of research. J Water Health 9: 265–278.
38. Jiang S, Noble R, Chu W (2001) Human adenoviruses and coliphages in urban runoff-impacted coastal waters of Southern California. Appl Environ Microbiol 67: 179–184.
39. Noble RT, Fuhrman JA (2001) Enteroviruses detected by reverse transcriptase polymerase chain reaction from the coastal waters of Santa Monica Bay, California: low correlation to bacterial indicator levels. Hydrobiologia 460: 175–184.
40. Harwood VJ, Levine AD, Scott TM, Chivukula V, Lukasik J, et al. (2005) Validity of the indicator organism paradigm for pathogen reduction in reclaimed water and public health protection. Appl Environ Microbiol 71: 3163–3170.
41. Pusch D, Oh DY, Wolf S, Dumke R, Schroter-Bobsin U, et al. (2005) Detection of enteric viruses and bacterial indicators in German environmental waters. Arch Virol 150: 929–947.
42. Baggi F, Demarta A, Peduzzi R (2001) Persistence of viral pathogens and bacteriophages during sewage treatment: lack of correlation with indicator bacteria. Res Microbiol 152: 743–751.
43. Ferguson AS, Layton AC, Mailloux BJ, Culligan PJ, Williams DE, et al. (2012) Comparison of fecal indicators with pathogenic bacteria and rotavirus in groundwater. Sci Total Environ 431: 314–322.
44. Fong TT, Lipp EK (2005) Enteric viruses of humans and animals in aquatic environments: health risks, detection, and potential water quality assessment tools. Microbiol Mol Biol Rev 69: 357–371.
45. Ramani S, Kang G (2009) Viruses causing childhood diarrhoea in the developing world. Curr Opin Infect Dis 22: 477–482.
46. Bernhard AE, Field KG (2000) A PCR assay To discriminate human and ruminant feces on the basis of host differences in Bacteroides-Prevotella genes encoding 16S rRNA. Appl Environ Microbiol 66: 4571–4574.
47. Pickering AJ, Julian TR, Marks SJ, Mattioli MC, Boehm AB, et al. (2012) Fecal contamination and diarrheal pathogens on surfaces and in soils among Tanzanian households with and without improved sanitation. Environ Sci Technol.
48. USEPA (2002) Method 1604: Total coliforms and Escherichia coli in water by membrane filtration using a simultaneous detection technique (MI Medium). Washington, D.C.: United States Environmental Protection Agency, Office of Water.
49. USEPA (2009) Method 1600: Enterococci in water by membrane filtration using membrane-Enterococcus Indoxyl-D-Glucoside Agar (mEI). Washington, D.C.: United States Environmental Protection Agency, Office of Water.
50. Victoria M, Guimaraes F, Fumian T, Ferreira F, Vieira C, et al. (2009) Evaluation of an adsorption-elution method for detection of astrovirus and norovirus in environmental waters. J Virol Methods 156: 73–76.
51. Keating DT, Malizia AP, Sadlier D, Hurson C, Wood AE, et al. (2008) Lung tissue storage: optimizing conditions for future use in molecular research. Exp Lung Res 34: 455–466.
52. Ashbolt NJ (2004) Microbial contamination of drinking water and disease outcomes in developing regions. Toxicology 198: 229–238.
53. Gascon J, Vargas M, Schellenberg D, Urassa H, Casals C, et al. (2000) Diarrhea in children under 5 years of age from Ifakara, Tanzania: a case-control study. J Clin Microbiol 38: 4459–4462.

54. Silva PA, Stark K, Mockenhaupt FP, Reither K, Weitzel T, et al. (2008) Molecular characterization of enteric viral agents from children in northern region of Ghana. J Med Virol 80: 1790–1798.

55. Kaper JB, Nataro JP, Mobley HL (2004) Pathogenic *Escherichia coli*. Nat Rev Microbiol 2: 123–140.

56. Walters SP, Yamahara KM, Boehm AB (2009) Persistence of nucleic acid markers of health-relevant organisms in seawater microcosms: implications for their use in assessing risk in recreational waters. Water Res 43: 4929–4939.

57. Jothikumar N, Kang G, Hill VR (2009) Broadly reactive TaqMan assay for real-time RT-PCR detection of rotavirus in clinical and environmental samples. JIN2@cdc.gov. J Virol Methods 155: 126–131.

58. Jothikumar N, Cromeans TL, Hill VR, Lu X, Sobsey MD, et al. (2005) Quantitative real-time PCR assays for detection of human adenoviruses and identification of serotypes 40 and 41. Appl Environ Microbiol 71: 3131–3136.

59. Kildare BJ, Leutenegger CM, McSwain BS, Bambic DG, Rajal VB, et al. (2007) 16S rRNA-based assays for quantitative detection of universal, human-, cow-, and dog-specific fecal *Bacteroidales*: a Bayesian approach. Water Res 41: 3701–3715.

60. Dehejia RH, Wahba S (2002) Propensity score-matching methods for nonexperimental causal studies. Review of Economics and statistics 84: 151–161.

61. Wacholder S, Silverman DT, McLaughlin JK, Mandel JS (1992) Selection of controls in case-control studies. III. Design options. Am J Epidemiol 135: 1042–1050.

62. Howard CM, Handzel T, Hill VR, Grytdal SP, Blanton C, et al. (2010) Novel risk factors associated with hepatitis E virus infection in a large outbreak in northern Uganda: results from a case-control study and environmental analysis. Am J Trop Med Hyg 83: 1170–1173.

63. WHO-UNICEF (2012) Progress on Drinking Water and Sanitation: 2012 Update. Geneva: World Health Organization.

64. Day NE, Byar DP, Green SB (1980) Overadjustment in case-control studies. Am J Epidemiol 112: 696–706.

65. Breslow NE, Day NE, Cancer IAfRo (1980) Statistical methods in cancer research. vol. 1: The analysis of case-control studies.

66. Levine MM, Robins-Browne RM (2012) Factors that explain excretion of enteric pathogens by persons without diarrhea. Clinical infectious diseases : an official publication of the Infectious Diseases Society of America 55 Suppl 4: S303–311.

67. VanDerslice J, Briscoe J (1993) All coliforms are not created equal: A comparison of the effects of water source and in-house water contamination on infantile diarrheal disease. Water Resources Research 29: 1983–1995.

68. Black RE, Merson MH, Rowe B, Taylor PR, Alim AA, et al. (1981) Enterotoxigenic *Escherichia coli* diarrhoea: acquired immunity and transmission in an endemic area. Bulletin of the World Health Organization 59: 263.

69. Valentiner-Branth P, Steinsland H, Fischer TK, Perch M, Scheutz F, et al. (2003) Cohort study of Guinean children: incidence, pathogenicity, conferred protection, and attributable risk for enteropathogens during the first 2 years of life. J Clin Microbiol 41: 4238–4245.

70. Howard G, Bartram J (2003) Domestic water quantity, service level, and health. Geneva: World Health Organization.

71. Thompson J (2001) Drawers of Water II: 30 years of change in domestic water use & environmental health in east Africa. Summary. Iied.

72. Kremer M, Leino J, Miguel E, Zwane AP (2011) Spring cleaning: Rural water impacts, valuation, and property rights institutions*. The Quarterly Journal of Economics 126: 145–205.

73. Gasana J, Morin J, Ndikuyeze A, Kamoso P (2002) Impact of water supply and sanitation on diarrheal morbidity among young children in the socioeconomic and cultural context of Rwanda (Africa). Environ Res 90: 76–88.

74. Brown J, Hien VT, McMahan L, Jenkins MW, Thie L, et al. (2012) Relative benefits of on-plot water supply over other 'improved' sources in rural Vietnam. Trop Med Int Health 18: 65–74.

75. Haas CN, Rose JB, Gerba C, Regli S (1993) Risk assessment of virus in drinking water. Risk Anal 13: 545–552.

76. Kotloff KL, Nataro JP, Blackwelder WC, Nasrin D, Farag TH, et al. (2013) Burden and aetiology of diarrhoeal disease in infants and young children in developing countries (the Global Enteric Multicenter Study, GEMS): a prospective, case-control study. The Lancet.

77. Curtis V, Cousens S, Mertens T, Traore E, Kanki B, et al. (1993) Structured observations of hygiene behaviours in Burkina Faso: validity, variability, and utility. Bull World Health Organ 71: 23–32.

12

Novel Microbiological and Spatial Statistical Methods to Improve Strength of Epidemiological Evidence in a Community-Wide Waterborne Outbreak

Katri Jalava[1][*][¤], Hanna Rintala[2], Jukka Ollgren[1], Leena Maunula[3], Vicente Gomez-Alvarez[4], Joana Revez[3], Marja Palander[1], Jenni Antikainen[5], Ari Kauppinen[6], Pia Räsänen[1], Sallamaari Siponen[1], Outi Nyholm[1], Aino Kyyhkynen[1], Sirpa Hakkarainen[2], Juhani Merentie[2], Martti Pärnänen[2], Raisa Loginov[5], Hodon Ryu[4], Markku Kuusi[1], Anja Siitonen[1], Ilkka Miettinen[6], Jorge W. Santo Domingo[4], Marja-Liisa Hänninen[3], Tarja Pitkänen[6]

1 Department of Infectious Disease Surveillance and Control, National Institute for Health and Welfare, Helsinki, Finland, **2** Siilinjärvi municipality, Finland, **3** Department of Food Hygiene and Environmental Health, University of Helsinki, Helskinki, Finland, **4** Office of Research and Development, United States Environmental Protection Agency, Cincinnati, Ohio, United States of America, **5** Laboratory HUSLAB, Helsinki University Hospital, Helsinki, Finland, **6** Department of Environmental Health, National Institute for Health and Welfare, Kuopio, Finland

Abstract

Failures in the drinking water distribution system cause gastrointestinal outbreaks with multiple pathogens. A water distribution pipe breakage caused a community-wide waterborne outbreak in Vuorela, Finland, July 2012. We investigated this outbreak with advanced epidemiological and microbiological methods. A total of 473/2931 inhabitants (16%) responded to a web-based questionnaire. Water and patient samples were subjected to analysis of multiple microbial targets, molecular typing and microbial community analysis. Spatial analysis on the water distribution network was done and we applied a spatial logistic regression model. The course of the illness was mild. Drinking untreated tap water from the defined outbreak area was significantly associated with illness (RR 5.6, 95% CI 1.9–16.4) increasing in a dose response manner. The closer a person lived to the water distribution breakage point, the higher the risk of becoming ill. Sapovirus, enterovirus, single *Campylobacter jejuni* and EHEC O157:H7 findings as well as virulence genes for EPEC, EAEC and EHEC pathogroups were detected by molecular or culture methods from the faecal samples of the patients. EPEC, EAEC and EHEC virulence genes and faecal indicator bacteria were also detected in water samples. Microbial community sequencing of contaminated tap water revealed abundance of *Arcobacter* species. The polyphasic approach improved the understanding of the source of the infections, and aided to define the extent and magnitude of this outbreak.

Editor: Martyn Kirk, The Australian National University, Australia

Funding: No external funding execpt routine governmental work funding was obtained for this work. The funders had no role in study design, data collection and analysis, decision to publish, or preparation of the manuscript.

Competing Interests: The authors have declared that no competing interests exist.

* Email: katri.jalava@thl.fi

¤ Current address: Environmental Health Department, City of Tampere, Finland

Introduction

Community-wide waterborne outbreaks are characterized by a large number of exposed people with high attack rates [1–6]. Waterborne outbreaks are frequently associated with large number of symptomatic cases in a point source manner. Such outbreaks may be caused by a failure in the drinking water distribution system [1,3,4] or water treatment breakthrough of contaminating agents due to heavy rainfall or other excess weather conditions [2,5]. The water distribution system can be potentially contaminated with multiple pathogens during a relatively short period of time as the result of intrusion from surface or waste water [7]. Indeed, waterborne outbreaks with multiple causative organisms, e.g. *Campylobacter* spp., norovirus-like organisms, *Shigella* and enterohaemorrhagic *Escherichia coli* (EHEC) have been described [1,3,4,6,8,9]. In particular, when norovirus and

sapovirus types are implicated in large scale waterborne outbreaks, this strongly indicates drinking water distribution system contamination by a human faecal sources [10,11]. Sapovirus usually causes sporadic infections [12] but has been isolated from cases of waterborne outbreaks [12,13].

Waterborne outbreaks may be classified according to the level of evidence indicating that the drinking water was the cause of the outbreak. Evidence may be found by microbiological and/or epidemiological studies and the level of evidence can be assessed according to standardized criteria [14]. During the years 1998–2009 there have been 3–10 waterborne outbreaks in Finland annually and the outbreaks have typically been detected in small community groundwater plants with fewer than 500 consumers [7]. The implicated technical failures for the groundwater contamination in Finland have been flooding and surface run-off caused by heavy rains or rapid melting of snow. Also intrusion of contaminated water and cross-connections in the water distribution system play important role as a cause of Finnish waterborne outbreaks [15,16]. Most common causative microbes have been norovirus and *Campylobacter* [7].

In waterborne outbreak investigations, the delay between environmental investigation and the original presence of the pathogen within a water body has often hampered the detection of causative microbiological agents [15]. Special methods suitable for concentration of large water volumes, such as the use of large diameter membrane filters or ultrafiltration apparatus have been developed to increase the sensitivity of microbial detection in dilute water samples [17–19]. For the identification of a faecal contamination source, various detection methods targeting specific pathogens and microbial water quality indicators [15] and faecal source host specific molecular assays are available [20]. The use of RNA-based methods may not only help detecting but also identify active and thus potentially infective pathogens, providing a better estimate of public health risks than current DNA-based methods [21]. Additionally, next generation sequencing (NGS) can provide a path towards detecting multiple microbial taxa within complex microbial communities including rare members [17]. The high number of reads per sample in NGS applications also enables the use of these techniques for source tracking and identification of the transmission route of an outbreak [22].

Epidemiological cohort or case control studies may be applied in drinking water outbreaks due to the point source nature of the contamination and a well-defined population [3,11,23]. Web-based questionnaires are now increasingly used in outbreak investigations [24], especially in large and/or widely spread outbreaks [25]. Additionally, novel spatial methods have shown potential to define the source and location of the outbreak [26–28]. Modelling may be used to describe the person-to-person transmission or infection spreading from the environment [28]. Spatial variation in disease incidence has been studied in detail [27]. Also distance from water tanks has been evaluated by statistical methods with respect to diarrhoeal incidence [26].

The outbreak

The main water pipe was accidentally broken on 4th July, 2012 during road construction work in Vuorela, a community of 3000 inhabitants within the municipality of Siilinjärvi in Eastern Finland. The pipe breakage caused the contents of the upper drinking water storage reservoir to leak into the road construction pit. The pipe breakage was fixed within 14 hours, flushed and quality of the water was shown to fulfill the hygienic quality criteria. On the 16th July, the local environmental health authorities were informed by the health care centre of an excess number of patients with gastrointestinal symptoms. Given the recent pipe break in the area, a waterborne outbreak was suspected. The local outbreak control team of health and environmental authorities and waterworks personnel was activated, national outbreak awareness team was informed and consulted. On the following day, the results from tap water samples revealed faecal contamination of the water further confirming the waterborne nature of the outbreak. An immediate boil water notice was issued on 17th July and collection of patient and drinking water samples was initiated. The water distribution pipeline was subsequently flushed, the water storage reservoir was cleaned and the whole distribution system was disinfected with chlorine. The boil water notice was cancelled when the outbreak was declared over on 3rd August 2012.

The aim of our study was to reveal the role of contaminated water as a cause of a community wide outbreak detected in a small Finnish municipality during July 2012. We used a polyphasic approach carrying out advanced epidemiological, microbiological and environmental investigation to verify the source and scale of the outbreak. We included advanced and novel statistical, epidemiological and molecular microbiology methods, not previously applied to outbreak investigations according to our knowledge.

Materials and Methods

This study was part of public health response. According to Finnish legislation, no ethical approval is needed in this type of response. This study and related sampling and modelling were part of an official waterborne outbreak investigation, which is on the responsibility of municipal health and public health authorities (many of the coauthors are representatives of respective authorities).

Epidemiological investigation

Data collection. The contaminated part of the water distribution system provides drinking water for the area of Vuorela and Toivala in the municipality of Siilinjärvi. The total population of this area was 5934 and the exposed population who were served by the contaminated water distribution system in the defined outbreak area was 2931 persons (source population). The age, sex and living coordinates for the population were obtained from the National Population Register. A case was defined as a person staying or living in the Vuorela area during July 2012 with diarrhoea or two of the following symptoms: nausea, vomiting, stomach ache or fever. We excluded persons who were absent from the outbreak area during the whole study period and those who travelled abroad during July 2012. Based on the geographical coordinates we also determined the number of households or blocks of houses with unique water delivery points in the affected area.

Study design. A retrospective cohort study was conducted. A web based questionnaire was designed to define the extent and cause of the outbreak. The exposed population was informed by the local newspaper and press releases on the municipality website to participate in the study. All inhabitants of the Vuorela and Toivala area were invited as well as those visiting or working in this area. Data was collected between 19th July and 1st August, 2012, and any person living or spending time during the study period in the defined outbreak area was eligible to participate in the study. Study participants were asked about their basic demographic characteristics, clinical symptoms (from 4th to 20th July, 2012) and habits of consumption of tap water in the defined outbreak area (from 1st to 30th July, 2012).

Data analysis. We used a commercial web based questionnaire from Webropol (www.webropol.fi). The data was analysed as univariate factors calculating the risk ratios for all risk factors asked and frequencies of the illness as a retrospective cohort sample using R [29]. Subsequently, a binary log and logistic regression models with case status as the outcome variable and those explanatory variables that were significant in the univariate analysis were included, the analysis were performed in R. Furthermore, as we had the information of all the inhabitants, we compared the age and sex between the non-cases and the source population using a standard Wald's statistic for calculating the confidence intervals for the observed difference in percentages between the groups.

Spatial analysis and regression. For the spatial analysis, only those persons with address information available from the National Population Register and who replied to the cohort study were included. We obtained information on name, address, date of birth and living location co-ordinates. Sampling locations in the water distribution pipeline were plotted on a schematic map obtained from the local construction office (Figure 1). The shortest distance via the water pipe to the pipe breakage was determined from the obtained digital map using an R package gdistance. The distance data was subsequently allocated to each person living in the area. We also included a spatial correlation variable for the model to explain the possible transmission of the infection within households and closely living contacts and due to the other possibly unmeasured spatially correlated factors. We evaluated the model by residual diagnostics, residuals were plotted against the predicted values and Cook's diagnostic values were calculated [30]. We subsequently categorised the distance to four equal groups and compared the proportion of ill persons in each. The statistical calculations were done in R using both routine and special spatial packages [29]. The code for calculating the spatial distance via the water pipe in R which is presented in (File S1).

Clinical microbiology investigation

Microbiological analysis of patient samples. A total of 25 patients were sampled between 17th July and 2nd August, 2012. The age of the sampled patients had a median of 43 years (range 5–85 years). The potential causative agents were tested broadly, each from 5–21 samples aiming of qualitative detection of possible pathogens (Table 1). The initial samples were analysed in the local clinical laboratory with routine tests for common pathogens. As no major pathogens were detected, more samples were collected and analysed in national reference or specialized laboratories. A part of the patient samples were tested for enteric bacterial pathogens using standard clinical microbiology culture methods and PCR for the target genes of *Campylobacter jejuni/coli* (*rimR, gyrB*), *Salmonella* spp. (*invA*), *Shigella* spp./enteroinvasive *E. coli* (*ipaH, invE*) and *Yersinia* spp. (*virF, rumB*) [31]. The presence of diarrheagenic *E. coli* (i.e. EHEC, EPEC, ETEC, EIEC/*Shigella* spp. and EAEC) virulence genes (*pic, bfpB, invE, hlyA, elt, ent, escV, eae, ipaH, aggR, stx₁, stx₂, estlb, estla* and *astA*) were

determined using multiplex PCR and qPCR techniques from mixed preliminary cultures grown on CLED (cystine lactose electrolyte deficient) agar medium plates [31,32]. The *ipaH* and *invG* genes are specific for both *Shigella* spp. and EIEC and the tests do not distinguish between these two organisms. If PCR for EHEC genes was positive, the specific colony was picked out when possible from the mixed culture plate for a single PCR of the *stx* genes. Electron microscopy was performed to detect enteric viruses, such as noro-, rota-, adeno-, entero-, sapo- and astroviruses. In addition, for norovirus reverse transcription (RT)-qPCR [33,34] was performed. For sapovirus analysis, two different PCR protocols were used [35,36] and nucleic acid sequences were determined from the amplicons of the polymerase region [37]. Samples were also tested for enteric parasites *Giardia* and *Cryptosporidium* using enzyme immunoassay method for antigen detection [38]. Seven frozen faecal samples were tested in a retrospective testing scenario for the presence of *Arcobacter* spp. by culture method as described previously [39] and the plates were inspected regularly for up to 3 weeks [40]. A species-specific multiplex-PCR was performed for detection of *A. buzleri, A. cryaerophilus* and *A. skirrowii* as described previously [41]. The samples were also tested using an additional *Arcobacter* genus specific PCR method [42].

Environmental investigation

Environmental sampling and analysis. The drinking water in the distribution network of Vuorela and Toivala area is UV-disinfected groundwater produced in the nearby Jäläniemi waterworks. The Siilinjärvi municipality owns and operates this public drinking waterworks and its distribution network. The employees of the Siilinjärvi municipality, including health protection authorities and personnel of the waterworks were responsible of environmental sampling and delivery of samples to expert laboratories. The water quality management routines include testing for coliform bacteria, heterotrophic plate counts and *Clostridium perfringens* at the waterworks and around the

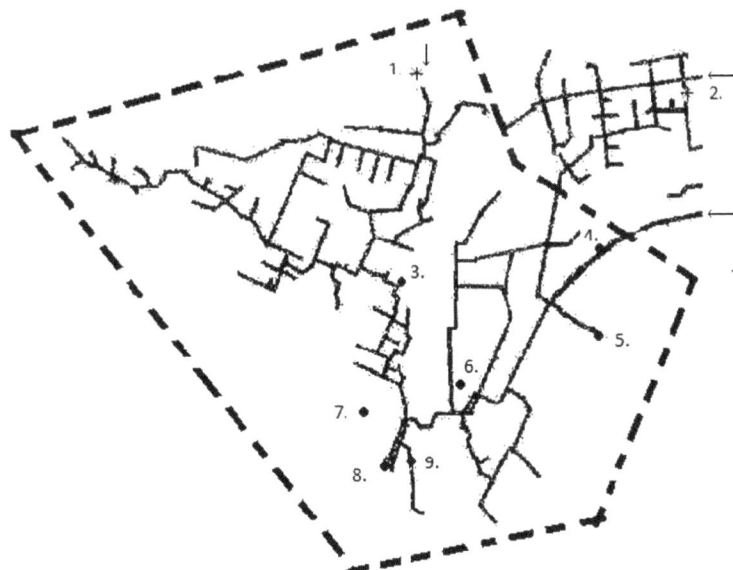

Figure 1. Schematic map of the water pipe of a defined outbreak area in Vuorela, July 2012. The outbreak (boil water notice) area is indicated by dashed line (- - -), the water sampling points (1–9) are coded as (●) with a positive culture finding and (*) with a negative finding. Arrows indicate the inflow points of the water from the water plant (outside the figure). Points 1,2,3,6,7 and 9 are tap water sampling locations, point 4 is the water pipe line breakage point (surface water), point 5 represent drinking water from the upper water storage reservoir and point 8 is the municipal effluent sampling location.

Table 1. Microbiological results of faecal samples from symptomatic patients of a waterborne outbreak in Vuorela, July 2012.

Microbial pathogen and methods used	Number of patients tested positive/number of tested
Campylobacter jejuni/coli[1,2]	1[5]/21
Salmonella spp.[1,2]	0/21
Shigella spp.[1,2]	0/21
Yersinia spp.[1,2]	0/21
E. coli	
- EHEC[1,2]	2[6]/12
- EPEC[2]	6/12
- ETEC[2]	0/12
- EIEC/*Shigella*[1,2]	0/12
- EAEC[2]	2/12
Arcobacter spp.[1,2]	0/7
Norovirus[2,3]	0[2]/12, 0[3]/17
Adenovirus[3]	0/17
Enterovirus[3]	1/17
Sapovirus[2,3]	5[2,7]/12, 3[3]/17
Astrovirus[2,3]	0[2]/7, 0[3]/17
Giardia[4]	0/19
Cryptosporidium[4]	0/19

Methods used were cultivation[1], PCR/RT-PCR[2] (polymerase chain reaction/reverse transcription-PCR), EM[3] (electron microscopy) and/or EIA[4] (Enzyme Immunoassay). [5]*Campylobacter jejuni* and [6]EHEC O157:H7 (from one sample) and [7]sapovirus GII.P3 were isolated from the samples.

water distribution network according to European regulation [43]. The water quality results were acceptable prior to the outbreak with the last results dating back to 7th June and 2nd July, 2012. After accidental pipe breakage at 4th July, the pipeline was fixed within 14 hours, then flushed and sampled from two points (locations before the breakage point outside the schematic map and after the breakage point, point 6 in Figure 1) in 5th July, 2012 for indicator bacteria testing. The water quality was shown to comply the legal microbiological quality criteria (i.e. absence of coliform bacteria, *E. coli* and intestinal enterococci with no abnormal change in heterotrophic plate counts, Table 2).

A total of 65 water samples were collected between 5th July and 29th August (Figure 1). The majority of samples (n = 54) were tap water samples collected from various parts of the distribution system (1–10 sampling events/location) (Table 2). Also a raw water sample from a water production well, two biofilm samples from water meters and a sample from the community wastewater influent were collected. Surface water from a rainwater collection well and ditches at the pipeline breakage location (n = 5) were sampled as potential contamination sources. Both small (500–2000 mL) and large (10 L) scale water samples were analysed for pathogens. On-site large volume sampling (more than 100 L) using tangential flow ultrafiltration was conducted 17th July at one of the distribution system sites where high counts of faecal bacteria were detected at the previous day small scale sampling (point 7 represented in Figure 1). The ultrafiltration was conducted without sodium polyphosphate using a semi-automated system [17,18] and the bovine-serum pre-treated ultrafilter was eluted after sampling with 9:1 retentate/elution solution and 500 ml elution solution [17,44].

Culture methods were used to analyse *E. coli* and coliform bacteria [45,46], intestinal enterococci and heterotrophic plate counts [46], and vegetative cells and spores of *Clostridium perfringens* [47] from water samples. The presence/absence of

faecal bacterial pathogens, thermotolerant *Campylobacter* spp. and *Salmonella* spp., was determined using culture-based selective enrichment methods [46,48], and EHEC was analysed using serotype O157 specific immunomagnetic separation coupled with the selective culture enrichment [49]. Enteric viruses were concentrated as previously described [50], except using Sartolon polyamide membranes (diameter 47 mm, pore size 45 μm; Sartorius, Goettingen, Germany). Norovirus, rotavirus and adenovirus were analysed using previously described RT-qPCR and qPCR methods [48,50,51]. For sapovirus analysis, the faecal protocol was applied to water as described earlier. Protozoan parasites *Giardia* and *Cryptosporidium* were determined from a tap water concentrate using immunomagnetic separation and epifluorescence microscopy according to the international standard method [52] and from water meter biofilm samples using molecular techniques [53]. The concentration of free and total chlorine in the drinking water distribution was measured on-site using the Palintest Micro 1000 photometer and in the laboratory using ISO 7393 standard method [54].

Further identification of pathogens. *E. coli* isolates from seven tap water samples (n = 19) and DNA extracts (n = 4) were screened for presence of virulence genes of EHEC, EPEC, ETEC, EIEC and EAEC [31,32]. In addition, the same virulence genes were tested from mixed cultures (tap water and rainwater sample) collected from membranes during the *E. coli*/coliform analysis. PFGE-profiles of *C. jejuni* isolates (n = 6) from patient and environmental samples were produced using *Kpn*I and *Sma*I restriction enzymes [55].

Extracted nucleic acids from total (DNA) and active bacterial fraction (RNA) in water samples were used as template to amplify faecal source identifiers and the bacterial 16S rRNA gene for microbial source tracking (MST) and NGS applications, respectively. MST assays were performed as previously described [21]. The samples were obtained from the upper storage reservoir water

Table 2. The counts of water quality indicator bacteria (range of MPN or CFU/100 mL), occurrence of faecal pathogens and chlorine concentrations in the water samples taken in July and August, 2012 in Vuorela, Finland.

Date (number of samples)	E. coli	Coliform bacteria	Entero-coccus	Faecal pathogens	Chlorine (mg/l)
Drinking water distribution system samples from the clean area					
5–19 Jul (12)	0	0–1	0	Not detected/3–6 samples[1]	ND, 0.4–1.8[2]
Drinking water distribution system samples from the boil water notice area					
5 Jul (1)	0	0	ND	ND	ND
16 Jul (4)	0–150	0–150	0–17	Arcobacter spp./1–4 samples[3]	ND
17 Jul (4)	0–21	0–34	0–2	EHEC, EPEC and EAEC virulence genes[4]	ND
18–20 Jul (10)	0[5]	0	0	Not detected/2–4 samples[1]	<0.1–1.6
23–30 Jul (13)	ND[6]	ND	ND	Not detected/3 samples[7]	<0.1–2.0
1–29 Aug (9)	ND[6]	ND	ND	ND	0.4–1.9
Water storage reservoir					
17 July (1)	110	190	15	Norovirus and adenovirus[8]	ND
21 July (1)	0	0	ND	Not detected[9]	ND
Biofilms from the water meters removed from the boil water notice area					
1 Aug (2)	ND	ND	ND	Arcobacter spp.[10]	ND
Raw water at the groundwater abstraction plant					
9 August (1)	0	0	ND	Not detected[11]	ND
Surface water samples (contaminant source)					
23 Jul (1)	86	450	44	Campylobacter jejuni 10–100 cfu/l, EPEC virulence genes[12]	ND
29 Aug (5)	0–94	ND	5–80	ND	ND
Community wastewater					
25 Jul (1)	ND	ND	ND	Sapovirus	ND

ND; not determined.

[1] A portion of the samples were selected for *Salmonella*, *Campylobacter*, enterohaemorragic *E. coli* (EHEC) culture analyses and for norovirus analysis.

[2] Measured from three locations at 19 July.

[3] Samples were tested for noro-, adeno-, rota- and sapoviruses, *Campylobacter* and *E. coli* virulence genes. *Arcobacter* was tested from DNA extracts and genus specific PCR was positive in one sample (point 9 in Figure 1).

[4] *E. coli* virulence genes were detected after ultrafiltration from one sampling location (point 5 in Figure 1). *Salmonella*, *Campylobacter*, EHEC (culture method), noro-, rota- and sapovirus, *Giardia* and *Cryptosporidium* were not detected (1–4 samples tested/method).

[5] One colony of *Clostridium perfringens* was found from 1 000 mL of tap water sample taken from the most contaminated area.

[6] *Clostridium perfringens* was analyzed and not detected from 5 000 mL samples.

[7] Samples were tested for sapovirus.

[8] Sample was tested for noro-, adeno-, rota- and sapoviruses, *Salmonella*, *Campylobacter* and *E. coli* virulence genes. *Clostridium perfringens* was detected (10 CFU/L).

[9] Sample was tested for *Campylobacter* and noroviruses. *Clostridium perfringens* was detected (2 CFU/L).

[10] Sample tested for *Campylobacter*, *Arcobacter*, *Giardia* and *Cryptosporidium*.

[11] Sample tested for *Campylobacter*.

[12] Sample tested for *Campylobacter*, *E. coli* virulence genes, noro- and adenoviruses, *Giardia* and *Cryptosporidium*. *Clostridium perfringens* was detected (40 CFU/L).

before and after cleaning (point 5 in Figure 1) and from a tap water sample collected during the contamination episode (point 7 in Figure 1). The MST assays included the analysis of faecal bacterial groups *Bacteroidales* spp. (GenBac3 assay [56]) and human-specific *Bacteroidales* (HF183 assay [57]).

For Illumina MiSeq NGS, we utilized barcoded primers 515F and 806R to produce 250 bp pair-ended sequences [58]. Reads were processed and analyzed using the software MOTHUR v1.30.2 ([59]; http://www.mothur.org) as described previously [60]. Briefly, fastq files with forward and reverse reads were used to form contigs. The reads that met the following criteria were excluded from further analysis: the length was no greater than 255 bp; contained ambiguous bases (N) or homopolymers greater than 7 bases; identified as chimera; or classified as Chloroplasts or Mitochondria. Reads were aligned and sorted with >97% similarity into operational taxonomic units (OTUs). Taxonomic classification was obtained using the tool Classifier in the

Ribosomal Database Project II release 10.28 [61]. The raw reads were deposited in the NCBI Sequence Read Archive (SRA) under accession number SRP041117. Using MEGA v5.2 [62], a phylogenetic tree based on the aligned 16S rRNA gene sequences (~255 bp) was constructed with the Maximum Likelihood (ML) method using Tamura-Nei model [63] with 1,000 bootstrap replicates. The tree was used to infer the phylogenetic relationship among sequences classified as *Arcobacter* obtained in this study. *Sulfurospirillum deleyianum* (NR_074378) and *Campylobacter fetus* (L04314) were used as outgroup. Three water samples and two water meter samples were subsequently analysed for *Arcobacter* spp. using genus specific PCR [42] and species specific multiplex-PCR [41] from the DNA extracted as previously described [50].

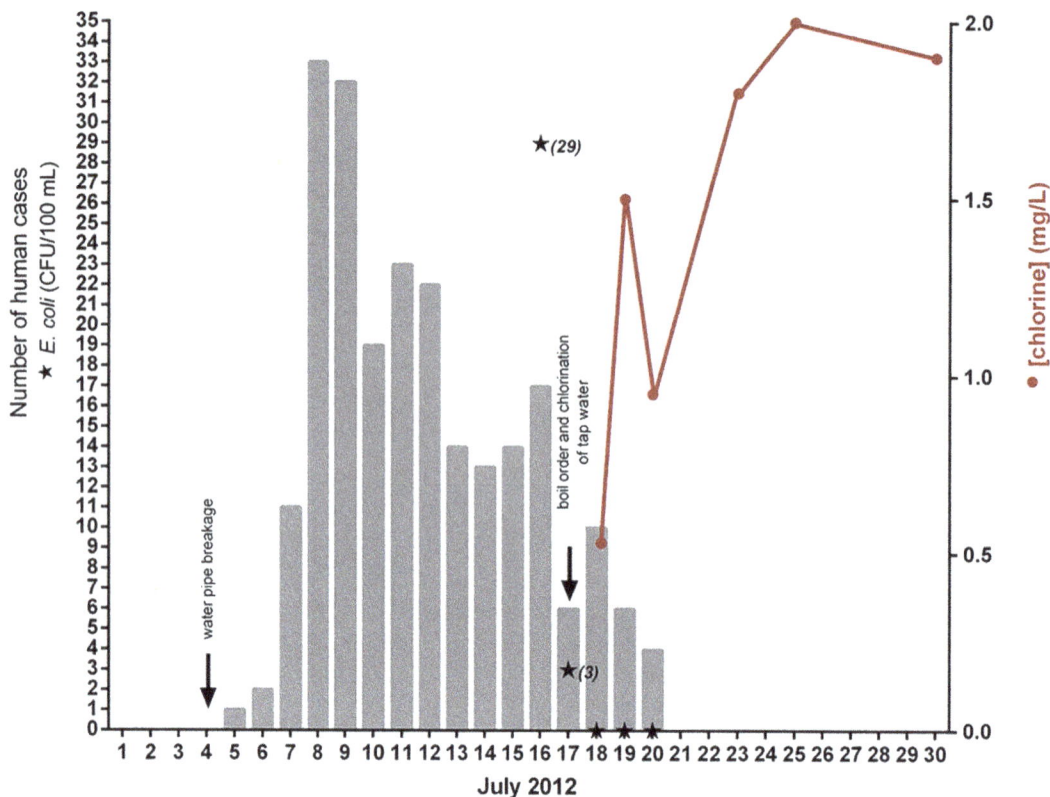

Figure 2. Epidemic curve of a waterborne outbreak in Vuorela, July 2012 based on the reported onset date of illness of the cases, and *E. coli* bacteria counts and chlorine levels in the point 7 (See Fig. 1) of the water distribution network.

Results

Epidemiological investigation

Descriptive analysis. Of all 2931 inhabitants (source population) of the defined outbreak area, 473 (16%) persons participated in the study (study population). We excluded 19 persons absent from the outbreak area during the whole study period and 23 due to travelling abroad. In total, we identified 225 cases and 206 healthy persons from the cohort study according to the case definition. However, for the individual risk factors data contained a few missing values. During the four week period after the water distribution breakage, an estimated 800 excess persons visited either a nurse or GP in the local health care centre. The outbreak curve presented in Figure 2 implicates a point source outbreak.

Demographic characteristics. Of the 431 respondents, 33% (135/408) were male. The mean age was 41.2 years (range 1–80 years). As there were few children among the respondent (10 in total, 3 among non-cases), we excluded the children from

comparison analysis. The proportion of male (33%, 60/182) among the non-cases in the study population was lower to that of the source population (48%, 1074/2252). This difference for the male proportions of 14.5% (7.4%–21.6%) was statistically significant. Persons were slightly older among the source population compared to the study population, Table 3. Also, there were 50.2% (232/462) households (unique water delivery points) with at least one person under 18 years of age in the source population compared to 68.5% (126/184) among the study population. This difference was significant, CI for the difference was 18.3% (10.2%–26.4%). Therefore, households with children were more likely to respond to the study.

The course of the illness was mild, only one person was admitted to hospital. The main symptoms were stomach ache 88% (199/225), nausea 85% (191/225) and diarrhoea 82% (185/225). The length of the illness had a median of 3 days (range 1–30), 16% (35/225) of the cases sought for medical assistance and one patient was hospitalized (37-year old female with no underlying medical condition). Absence from work due to outbreak illness was

Table 3. Comparison of the study population to source population with respect to age groups in a waterborne outbreak in Vuorela, July 2012.

Age group	Study population (non-cases)% of population (number of persons)	Source population % of population (number of persons)	% difference (95% confidence intervals)
20–39 years	40% (78/195)	29% (655/2252)	11% (3.9%–18.1%)
≥40 years	60% (117/195)	71% (1597/2252)	−11% (−18.1%−−3.9%)

Table 4. The Univariate and multivariate results for individual risk factors and the generalized additive model risk ratios with the spatial term of a waterborne outbreak in Vuorela, July 2012.

	Explanatory variable	Risk ratio or univariate log regression exp (β-values), (95% confidence intervals) for individual risk factors	Multivariable generalized additive logit model, exp (β-values), (p-value)
Personal characteristic	Age (continuous in years)	0.99 (0.98–0.99)	0.975 (0.0061)
Drinking at home	Tap water	2.2 (1.2–4.1)	5.90 (0.0037)
	Water from own well	1.0 (0.57–1.87)	n/a
	Bottled water	0.86 (0.69–1.07)	n/a
	Boiled water	0.69 (0.53–0.90)	n/a
Drinking water in Vuorela (outside home)	Tap water	1.6 (1.2–2.0)	n/a
	Water from own well	0.83 (0.56–1.22)	n/a
	Bottled water	0.90 (0.72–1.12)	n/a
	Boiled water	0.80 (0.56–1.15)	n/a
Spatial variables	Distance from the breakage by waterpipe (metres)	0.99950 (0.99930–0.99969)	0.998 (0.060)
	Spatial variable (coordinates)	n/a	n/a (0.002)

reported by 31% (133/431) of the study participants, the total number of working days lost because of illness was 398. Taking care of a sick child was not included.

Univariate analysis of the cohort. Of all possible cases occurring in the area, 225 cases responded to the study in an area with a population of 2931 persons. Of the water related risk factors, drinking untreated tap water in the outbreak area had a risk ratio of 5.6 (95% CI 1.9–16.4), also drinking untreated tap water at home RR 2.2 (95% CI 1.2–4.1) and outside home RR 1.6 (95% CI 1.2–2.0) in the outbreak area were associated with illness, Table 4. Drinking boiled water at home was a protective factor, RR 0.69 (95% CI 0.53–0.90), drinking well water or bottled water were not significant either at home or outside home. It was observed that the risk ratio increased as the number of glasses of water consumed increased. This further strengthened the role of the contaminated tap water as a cause of the outbreak in a dose response manner, table 5.

Multivariate and spatial analysis of the cohort. Of those fulfilling eligibility criteria, 20 participants had no address information, 81 lived outside the defined boil water notice area and 17 could not be found in the national population register, leaving 313 persons for the spatial analysis. After these exclusions, 154 cases and 159 non-cases were used for the spatial analysis.

Age, drinking tap water at home and distance from the breakage point were significant in the multivariable model, Table 4. Younger persons were more likely to become ill (p = 0.0061), risk also increased if tap water was consumed at home (p = 0.0037). Distance from the leakage point was inversely associated with becoming a case (p = 0.060). Also spatial variable was significant in the generalized additive model (p = 0.002), this is likely to reflect the person to person transmission within households or neighbourhoods (Table 4). The distance via the water pipe was shorter for the cases compared with the non-cases (Table 4). This indicates that the closer one lived to the water breakage point, the more contaminated the drinking water was and therefore the likelihood of the illness was higher. The diagnostics of the model for randomized quantile residuals were normally distributed to suggest correct specification of the model. Between the categorised groups, the proportion of ill persons was higher among those living closer to water breakage point, Table 6.

Clinical microbiology investigation

Patient faecal samples. The microbiological analysis of faecal samples from the patients identified several pathogens (Table 1). The pathogens identified included sapovirus (detected from 5 patients of 12 tested, 2 sapoviruses were genotyped as

Table 5. The dose-response between the illness and the amount of water consumed at home in a waterborne outbreak in Vuorela, July 2012.

Number of glasses of water consumed at home per day	Cases/total number of persons in the implicated group (%)	Risk ratio (95% confidence intervals)	p-value
0	6/28	reference	
1–3	49/114 (43%)	2.01 (1.06–4.81)	0.065
4–6	112/193 (58%)	2.71 (1.48–6.40)	0.0066
7–9	37/48 (77%)	3.60 (1.94–8.54)	0.00054
10 or more	9/19 (47%)	2.21 (0.96–5.68)	0.068

Those not drinking water at home served as a control group.

Table 6. Categorized distance and proportion of cases within those groups in a waterborne outbreak in Vuorela, July 2012.

Distance categories	% (Cases/total)
Distance 1 (<2332)	72.4% (63/87)
Distance 2 (2332–2713)	45.5% (35/77)
Distance 3 (2713–3202)	43.5% (37/85)
Distance 4 (>3202)	29.7% (19/64)

GII.P3), enterovirus (detected from one patient out of 17 tested) and *C. jejuni* (isolated from one patient out of 21 tested). Two sapovirus nucleic acid sequences were submitted to Genbank with accession numbers KJ200380 and KJ200381. In addition, specific virulence genes of pathogenic *E. coli* were detected, including genes from EHEC (2/12), EPEC (6/12) and EAEC (2/12). In subsequent culturing of the virulence gene positive samples, one *E. coli* O157:H7 positive culture was found. No suspect *Arcobacter* spp. were detected by culture methods or by PCR.

Environmental investigation

Faecal microbes in the drinking water distribution. Faecal indicator bacteria were detected in four out of 11 tap water samples taken from different sampling points of the distribution on 16th and 17th July, 2012 (Table 2). The bacterial counts from the drinking water reservoir sample (point 5 in Figure 1) taken at the same time were high for total coliform bacteria, *E. coli* and enterococci. In addition, low numbers of adenoviruses and noroviruses (genogroup II) were detected in the storage reservoir sample. Water samples tested negative for sapovirus, although the community waste water influent was shown to contain sapovirus. Specific virulence genes of EHEC, EPEC and EAEC were detected from the large volume tap water sample (point 7 in Figure 1).

Cause of the water contamination. Technical investigations revealed that the water storage reservoir had been rapidly filled with contaminated water after the pipe breakage repair. The microbial contaminants unintentionally funnelled to the storage reservoir could remain viable as no disinfection was used in the distribution system (Table 2). Subsequently the contaminated water was introduced to the Vuorela and Toivala distribution system due to the water usage. The water level in the storage reservoir varies depending on water usage and pumping of the fresh water from the groundwater abstraction plant into the reservoir. The water quality at the groundwater abstraction plant was tested acceptable and free of microbial contaminants shortly after the breakage (Table 2). De-contamination of the reservoir was ensured by emptying and then mechanical cleaning, washing and chlorination of the inner surfaces of the storage reservoir. Indicator bacteria were no longer detected from the water distribution system after flushing and successful chlorination (Table 2). However, it took approximately a week before *C. perfringens* spores were absent in the water samples from the contaminated distribution system and chlorine levels reached the minimum target concentration (0.5 mg/l) at the most distant parts of the distribution.

Microbial source tracking. GenBac3 and HF183 assays showed the presence of *Bacteroides* spp. and human-specific *Bacteroidales*, respectively, in the contaminated tap water and in the storage reservoir (Table S1). There were more rRNA markers present compared to the rRNA gene (DNA) markers (rRNA:rDNA ratios were higher) in the storage reservoir sample than in the contaminated tap water sample. After cleaning of the storage reservoir, the GenBac3 and HF183 marker concentrations declined to below the quantification and detection limits, respectively (Table S1). A reactive surface water sample from the rainwater well was collected one week after the outbreak notification to evaluate the potential environmental sources near the pipe breakage location (Table 2). *C. jejuni* and EPEC virulence genes were identified from the surface water sample. Also indicator bacteria; coliforms, *E. coli* and intestinal enterococci were abundant. Five *C. jejuni* isolates from this suspected contamination source had identical *Kpn*I and *Sma*I profiles in PFGE, but different from those of the *C. jejuni* patient isolate (data not shown).

Bacterial communities in water. Community analysis based on the bacterial 16S rRNA region showed a higher diversity for contaminated sampling sites, while upper storage reservoir after the cleaning showed a lower diversity (Table S2). It is noteworthy that the community diversity measured as a total number of OTUs in the community was higher within the DNA reads than within the RNA reads (Table S2). Taxonomic analysis of contaminated water samples indicated a high presence of the class Beta-proteobacteria including the family *Comamonadaceae* and genus *Zoogloea*, and Gamma-proteobacteria including affiliated family *Methylococcaceae* and genera *Methylobacter* and *Pseudomonas* (Figure 3, A and B; Table S3).

Interestingly, also a significant share of reads generated from the contaminated water samples affiliated with Epsilon-proteobacteria genus *Arcobacter* (Table S3 and Table S4). Further sequence analysis revealed close relationship to *A. cryaerophilus* group 1A LMG9865 (OTU00012 in Figure 4). However, the corresponding tap water sample was negative for the species-specific *Arcobacter* multiplex-PCR (for *A. butzleri*, *A. cryaerophilus*, *A. skirrowii*) and the storage reservoir samples were no longer available for species-specific testing. *Arcobacter* genus specific PCR was positive for a tap water sample and a water meter sample collected from another location (point 9 in Figure 1) but these samples remained negative by culture. Although the sequences from both samples (Genbank with accession number KJ196910) showed high sequence similarity (>98%) with *A. cryaerophilus* (uncultured), *A. skirrowii* and *A. cibarius*, no species attribution can be made.

After the cleaning and chlorination of the upper storage reservoir, the relative abundance of Beta- and Gamma-proteobacteria decreased and the abundance of *Sphingobacteria* including genus *Pedobacter* increased (Figure 3, C; Table S3). The reads associated *Bacteroides*, *Escherichia* and *Clostridia* in the contaminated water samples were absent from the storage reservoir sample after cleaning and chlorination (Table S4).

Discussion

This study describes a large-scale municipality area wide outbreak due to breakage of drinking water distribution pipeline during a road construction work. The illness in the community became apparent only two weeks after the incident but the cause of the illness was obvious. This was confirmed using a polyphasic approach applying microbiological, epidemiological and statistical methods. Multiple causative microorganisms were isolated from patient samples. No major definite causative pathogen was identified but sapovirus was most frequently detected in patient samples. Also various types of virulent *E. coli*, *C. jejuni* and *Arcobacter* spp. might have played a role in the onset of gastrointestinal symptoms. All these organisms have been previously associated with waterborne transmission [5,9,12,13,23]. By

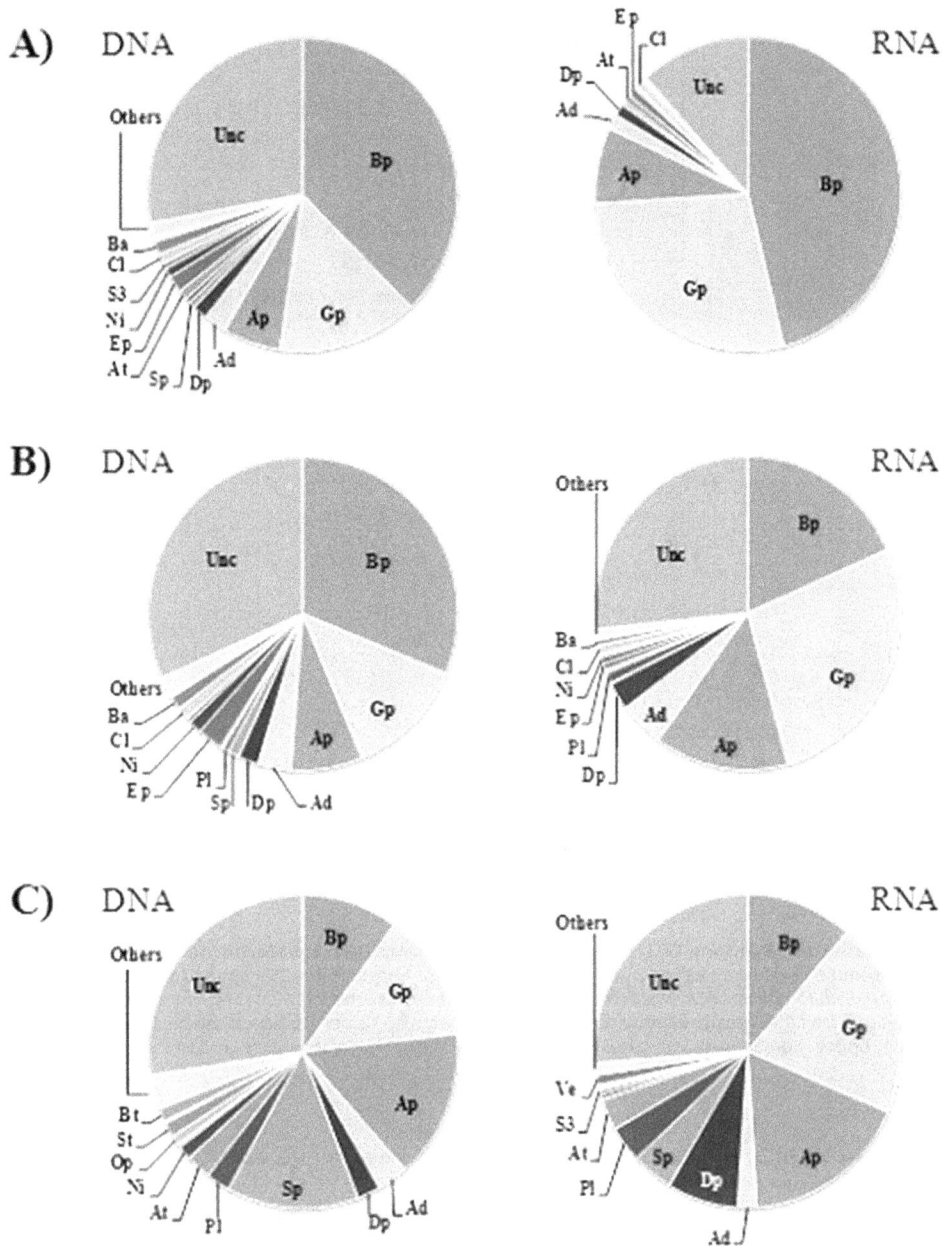

Figure 3. Distribution of the Bacteria domain as determined by taxonomic identification of partial 16S rRNA gene sequencing (at class level) in a waterborne outbreak in Vuorela, July 2012. Charts and tables represent the cumulative distribution of total DNA and RNA extracted from samples of A) the upper storage reservoir before cleaning (point 5 in Figure 1), B) tap water during contamination (point 7 in Figure 1) and C) the upper storage reservoir after cleaning. Legend: Beta-proteobacteria (Bp); Gamma-proteobacteria (Gp); Alpha-proteobacteria (Ap); Acidobacteria (Ad); Delta-proteobacteria (Dp); Sphingobacteria (Sp); Planctomycetacia (Pl); Actinobacteria (At); Epsilon-proteobacteria (Ep); Nitrospira (Ni); Verrucomicrobia Subdivision3 (S3); Clostridia (Cl); Bacteroidia (Ba); Opitutae (Op); Spartobacteria (St); Verrucomicrobiae (Ve); Bacteroidetes (Bt); Others (classes each representing <1%); unclassified (Unc).

applying the spatial and microbiological analysis, we could identify the contamination route for this outbreak. The distance to the water breakage point was inversely associated with illness both by epidemiological and spatial methods. We further confirmed the role of the contaminated water as a vehicle by novel spatial analysis. The spatial method provided here could be used to compare the likelihoods between possible candidate point source locations. Effective control measures and rapid and continuous

communication with the media were essential to ensure appropriate public health measures.

The overall clinical illness in most patients was mild with only 16% accessing medical assistance and only one patient being hospitalized. Majority of the cases had gastrointestinal symptoms with mild course and relatively short length. Overall, the number of cases was likely to be an underestimate for the whole population. The case definition used in the present study was relatively sensitive due to the mild overall nature of the illness. The

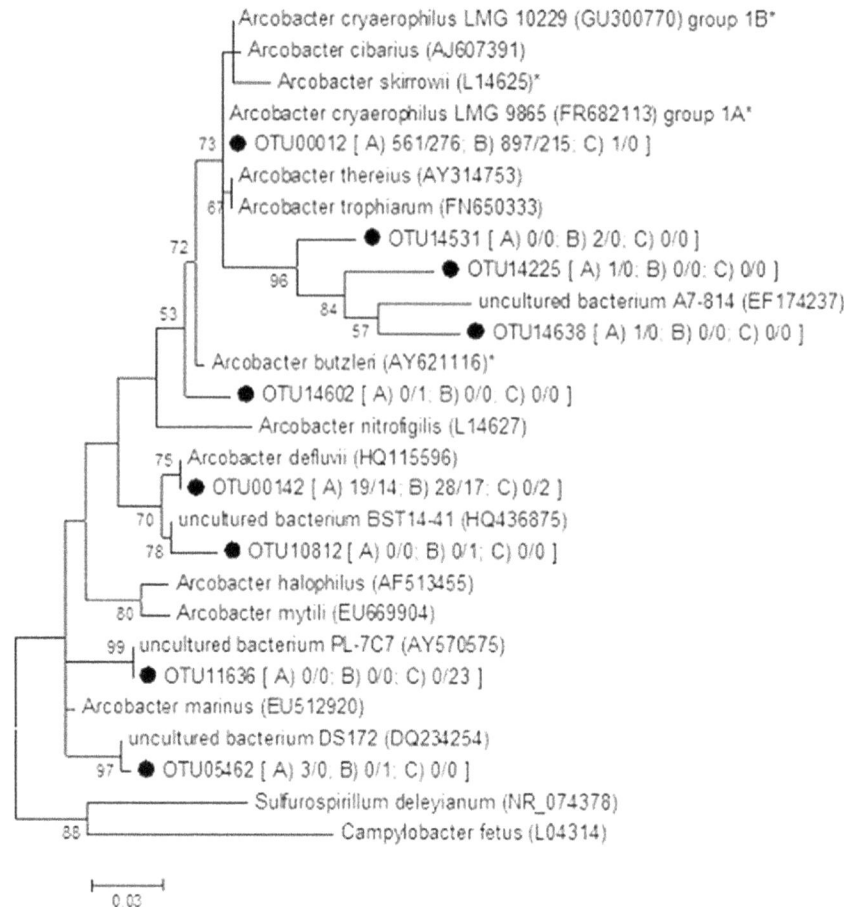

Figure 4. Phylogenetic relationships among OTUs (●) of the genus *Arcobacter* in a waterborne outbreak in Vuorela, July 2012. The tree was inferred from a maximum likelihood analysis of aligned 16S rRNA gene sequence (≈255 bp) and nodes with a bootstrap value ≥50% of 1 000 replicates are identified. *Sulfurospirillum deleyianum* (NR_074378) and *Campylobacter fetus* (L04314) were used as outgroup. Number in bracket represents the total amount of DNA/RNA reads identified in samples from A) the upper storage reservoir before cleaning, B) tap water during contamination and C) the upper storage reservoir after cleaning. *A. butzleri, A. cryaerophilus* and *A. skirrowii* have been associated with gastrointestinal diseases [76].

contaminated water consumed whether at or outside home in the defined outbreak area was associated with illness. We further showed by epidemiological methods that drinking contaminated tap water was associated with illness in a dose-response manner. The effect levelled off at the highest dose, often observed in food or waterborne outbreaks [64,65], this has been observed also in previous studies [1,3,9]. Furthermore, drinking boiled water was a protective factor indicating the effectiveness of the implemented control measures.

It is challenging to estimate the true attack rate in the present outbreak. It is quite likely that people showing symptoms responded more actively in the study compared to non-cases. We had a unique opportunity to compare the study population to the source population and found that women responded more actively to the study. This has been found also previously [66–68]. As we also found slightly more younger people among the non-cases compared to the source population, therefore, the observed effect of age should be interpreted with caution. The crude attack rate obtained from the questionnaire study (225/431, 52%) is within the same range to what has been observed in similar outbreaks, namely between 31–88% [2–4]. The true number of ill persons is difficult to estimate, but the educated evaluation of the local general practitioner (800 cases) is a fair estimate.

By applying the spatial and microbiological analysis, we could identify the contamination route for this outbreak. The contaminated water from the breakage initially filled the water storage reservoir and subsequently this water was distributed to the community over a number of days. The drinking water storage reservoirs have contributed to the transmission of waterborne infections also elsewhere, usually due to the improper maintenance and structure of the reservoirs [46,69,70]. In our case, the exact cause of the contamination was pipeline breakage at the road construction site. The storage reservoir prolonged the contaminant transport time to the water consumers. The likelihood of illness was higher closer to the water breakage point as measured by water pipeline length. Also the higher amount of water consumed in the boil water notice area increased the likelihood of illness. Additionally, younger persons (i.e. parents of small children) might have been more likely to become infected presumably due to the fact that they had more contact to young infants.

The source of the illness was obvious from the beginning of the outbreak, but we initially searched for one major pathogen as a causative agent yet aware that this type of outbreak is often caused by multiple pathogens. The sapovirus finding was novel and has rarely been detected in previous Finnish waterborne outbreaks [13]. According to recent reports, sapoviruses can be commonly found in environmental waters [10,71,72] highlighting the importance of including sapovirus detection into the outbreak investigations. Sapoviruses cause acute gastroenteritis primarily among young children [36], and indeed, young age was suspected to be one risk factor in this study. This result, together with the presence of sapoviruses in communal wastewater influent, support the assumption that sapovirus was commonly circulating in the community. We did not detect sapoviruses in the drinking water samples, presumably due to delayed sampling and low viral concentration in the drinking water. We also found that adults had symptomatic sapovirus infections. Adults have been found to have sapovirus infections in other studies [12,73,74]. The occurrence of human infecting sapoviruses among the patient cases together with the detection of human-specific genetic marker HF183, adenoviruses and noroviruses from the contaminated drinking water suggests that human faecal material was present at the pipe breakage site. No leakage was identified in a waste-water pipe line located in the same construction pit as the drinking water pipeline. Instead, it was concluded that surface water runoff from the nearby recreational area was the potential source of the faecal microbes. The abundant pathogenic *E. coli* findings from patient and water samples were not surprising bearing in mind the nature of the contamination. One positive *C. jejuni* finding in a patient may have had sporadic origin as the environmental isolate was of different type.

We used a novel NGS approach to study the microbial communities in the tap water distributed in the contaminated water pipe network. To our knowledge, this was the first time that high-throughput sequencing was employed for detection of potential causative bacterial agents in a waterborne outbreak case. By using the novel microbial community analysis for the water samples, we could show an abundance of *Arcobacter* spp. in the drinking water distribution during the contamination. Further analysis found high sequence similarity with potential pathogenic *Arcobacter* spp. (*A. cryaerophilus*, *A. skirrowii* and *A. cibarius*). Although the abundance of reads of *Arcobacter* correlated with the quality of water (increase in contaminated waters), the resolution of *Arcobacter* species lineage was not possible due to ambigious nature of the 16S rRNA sequences [75]. Since *Arcobacter* spp. have been suspected as a cause of gastroenteritis in humans [5,76], we aimed to isolate these from the remaining patient samples. No *Arcobacter* spp. were isolated perhaps due to the small quantity, long storage time (more than one year) or storage conditions of the samples. Nevertheless, considering that there is no gold standard for the isolation of these putative pathogens, their presence cannot be ruled out.

The NGS approach also provided information on the bacterial communities in the non-chlorinated and faecally contaminated drinking water. Until now, the analysis of indigeneous microbial community composition in non-chlorinated drinking water distribution systems has been scarce [77]. The detected change in the storage reservoir bacterial community structure after the cleaning and chlorination proves the effectiveness of these measures. Moreover, our result showing higher community diversity within the DNA reads than within the RNA reads suggest that DNA-based methods may not effectively discriminate between active and dormant populations. However, more work is needed to better characterize the bacterial community changes in the different drinking water distribution systems. For example, understanding the role of drinking water retention time in the network [77] and the effect of drinking water treatment processes such as filtration on the bacterial communities in the distribution [78] would facilitate the sustainable management of the microbial water quality in the distribution networks.

We further confirmed the role of the contaminated water as a vehicle by novel spatial analysis. By calculating the distance for each household to the water breakage point via the water pipe, we could show that the probability of illness decreased by the increasing length of the water pipe. However, it should be noted that non-symptomatic persons in the same households and neighbourhoods participated more actively in the cohort study. We used all cases in the model, including potential secondary cases. As the spatial term describes just secondary spread and was significant in the spatial logistic regression model, it suggests that there were indeed a number of secondary cases among the case patients. In addition to distance from the breakage point, age was inversely associated with illness and consumption of water in the Vuorela area was positively associated in a spatial logistic regression.

The study was limited by a slow response due to notification delay as it took a relatively long time before the outbreak was detected and confirmed. Therefore, the causative agents could not be isolated from the water samples. The microbial community analysis also warrants further studies. We used 16S rRNA as a target of the NGS studies. In the future, with careful design of pathogen specific primers for sequencing purposes, it might be possible to abandon the requirement of culture isolation for genotyping purposes. In successful NGS applications, the huge number of reads per sample potentially enables the identification phylogenetic relatedness between the causative pathogenic strains in different samples [79]. Subsequently, this might allow the use of NGS approach for source tracking and understanding of the transmission route of an outbreak.

Conclusions

We used novel and existing statistical, microbiological and spatial methods to characterize a community wide waterborne outbreak. These methods may be applied to wide range of food and waterborne outbreak investigations and beyond. In particular, microbial community analysis in combination with traditional culture and PCR based methods aided to clarify the potential causative agents. In addition, the cohort study was carried out as a rapid web based application added with a novel spatial method to show the statistical association to the suspected point source leakage of the water pipe network. We also confirmed previous observations that women and younger people were more likely to respond to this type of study. In the present outbreak the event associated with the outbreak was fairly obvious, but in many outbreak situations the source cannot be easily identified [3]. Failures in the water distribution networks are common causes of waterborne outbreaks [2]. The method provided here could be used to compare the likelihoods between possible candidate point source locations.

Supporting Information

Table S1   Numbers of the GenBac3 and HF183 markers ($\log_{10}$ copies 100 mL^{-1}) in TaqMan rRNA-targeted RT-qPCR (rRNA) and rRNA gene-targeted qPCR (rDNA) assays.

Table S2   Community diversity estimates ($\pm$ CI) of the domain *Bacteria*.

Table S3   Taxonomic affiliation of the most abundant *Bacteria* domain representatives.

Table S4 Abundance of reads associated with faecal bacteria and/or disease agents detected in the Vuorela and Toivala drinking water distribution system.

File S1 R code for calculating the shortest direct distance and the distance via the water pipe between each inhabitant location and water pipe breakage point.

Acknowledgments

We thank Anne-Mari Rissanen, Paula Muona and Mirja Rissanen and all the other personnel at the water works, in the local analytical laboratories and municipality of Siilinjärvi for their help in the outbreak investigation. We thank Pekka Pere from the University of Helsinki in helping with the statistical analysis. The opinions expressed in the paper are those of the authors and do not necessarily reflect the official positions and policies of the US Environmental Protection Agency or any other public authorities mentioned in this study. Any mention of product or trade names does not constitute recommendation for use by the US Environmental Protection Agency or any other public authorities mentioned in this study.

Author Contributions

Conceived and designed the experiments: KJ JO LM VGA JR H. Rintala AS JWSD MLH TP. Performed the experiments: KJ H. Rintala LM VGA JR M. Palander JA A. Kauppinen PR SS ON A. Kyyhkynen SH JM M. Pärnänen RL H. Ryu TP. Analyzed the data: KJ JO LM VGA JR A. Kauppinen A. Kyyhkynen SH M. Pärnänen MK AS JWSD MLH TP. Contributed reagents/materials/analysis tools: KJ JO LM VGA JR M. Palander JA A. Kyyhkynen SH JM M. Pärnänen RL H. Ryu MK AS IM JWSD MLH TP. Wrote the paper: KJ JO LM VGA JR A. Kauppinen MK AS JWSD MLH TP.

References

1. Maurer AM, Sturchler D (2000) A waterborne outbreak of small round structured virus, *Campylobacter* and *Shigella* co-infections in la Neuveville, Switzerland, 1998. Epidemiol Infect 125: 325–332.
2. Laursen E, Mygind O, Rasmussen B, Ronne T (1994) Gastroenteritis: A waterborne outbreak affecting 1600 people in a small Danish town. J Epidemiol Community Health 48: 453–458.
3. Jakopanec I, Borgen K, Vold L, Lund H, Forseth T, et al. (2008) A large waterborne outbreak of campylobacteriosis in Norway: The need to focus on distribution system safety. BMC Infect Dis 8: 128-2334-8-128.
4. Laine J, Huovinen E, Virtanen MJ, Snellman M, Lumio J, et al. (2011) An extensive gastroenteritis outbreak after drinking-water contamination by sewage effluent, Finland. Epidemiol Infect: 1–9.
5. Fong TT, Mansfield LS, Wilson DL, Schwab DJ, Molloy SL, et al. (2007) Massive microbiological groundwater contamination associated with a waterborne outbreak in Lake Erie, South Bass Island, Ohio. Environ Health Perspect 115: 856–864. 10.1289/ehp.9430.
6. Bopp DJ, Sauders BD, Waring AL, Ackelsberg J, Dumas N, et al. (2003) Detection, isolation, and molecular subtyping of *Escherichia coli* O157:H7 and *Campylobacter jejuni* associated with a large waterborne outbreak. J Clin Microbiol 41: 174–180.
7. Zacheus O, Miettinen IT (2011) Increased information on waterborne outbreaks through efficient notification system enforces actions towards safe drinking water. J Water Health 9: 763–772. 10.2166/wh.2011.021.
8. Dev VJ, Main M, Gould I (1991) Waterborne outbreak of *Escherichia coli* O157. Lancet 337: 1412.
9. Gubbels SM, Kuhn KG, Larsson JT, Adelhardt M, Engberg J, et al. (2012) A waterborne outbreak with a single clone of *Campylobacter jejuni* in the Danish town of Koge in May 2010. Scand J Infect Dis 44: 586–594. 10.3109/00365548.2012.655773.
10. Nenonen NP, Hannoun C, Larsson CU, Bergstrom T (2012) Marked genomic diversity of norovirus genogroup I strains in a waterborne outbreak. Appl Environ Microbiol 78: 1846–1852. 10.1128/AEM.07350-11.
11. Parshionikar SU, Willian-True S, Fout GS, Robbins DE, Seys SA, et al. (2003) Waterborne outbreak of gastroenteritis associated with a norovirus. Appl Environ Microbiol 69: 5263–5268.
12. Svraka S, Vennema H, van der Veer B, Hedlund KO, Thorhagen M, et al. (2010) Epidemiology and genotype analysis of emerging sapovirus-associated infections across Europe. J Clin Microbiol 48: 2191–2198. 10.1128/JCM.02427-09.
13. Rasanen S, Lappalainen S, Kaikkonen S, Hamalainen M, Salminen M, et al. (2010) Mixed viral infections causing acute gastroenteritis in children in a waterborne outbreak. Epidemiol Infect 138: 1227–1234. 10.1017/S0950268809991671.
14. Tillett HE, de Louvois J, Wall PG (1998) Surveillance of outbreaks of waterborne infectious disease: Categorizing levels of evidence. Epidemiol Infect 120: 37–42.
15. Pitkänen T (2013) Review of *Campylobacter* spp. in drinking and environmental waters. J Microbiol Methods 95: 39–47. 10.1016/j.mimet.2013.06.008.
16. Miettinen IT, Zacheus O, von Bonsdorff CH, Vartiainen T (2001) Waterborne epidemics in Finland in 1998–1999. Water Sci Technol 43: 67–71.
17. Hill VR, Kahler AM, Jothikumar N, Johnson TB, Hahn D, et al. (2007) Multistate evaluation of an ultrafiltration-based procedure for simultaneous recovery of enteric microbes in 100-liter tap water samples. Appl Environ Microbiol 73: 4218–4225. 10.1128/AEM.02713-06.
18. Rhodes ER, Hamilton DW, See MJ, Wymer L (2011) Evaluation of hollow-fiber ultrafiltration primary concentration of pathogens and secondary concentration of viruses from water. J Virol Methods 176: 38–45. 10.1016/j.jviromet.2011.05.031.

19. Hijnen WA, Biraud D, Cornelissen ER, van der Kooij D (2009) Threshold concentration of easily assimilable organic carton in feedwater for biofouling of spiral-wound membranes. Environ Sci Technol 43: 4890–4895.
20. Toledo-Hernandez C, Ryu H, Gonzalez-Nieves J, Huertas E, Toranzos GA, et al. (2013) Tracking the primary sources of fecal pollution in a tropical watershed in a one-year study. Appl Environ Microbiol 79: 1689–1696. 10.1128/AEM.03070-12.
21. Pitkänen T, Ryu H, Elk M, Hokajarvi AM, Siponen S, et al. (2013) Detection of fecal bacteria and source tracking identifiers in environmental waters using rRNA-based RT-qPCR and rDNA-based qPCR assays. Environ Sci Technol 47: 13611–13620. 10.1021/es403489b.
22. Di Bella JM, Bao Y, Gloor GB, Burton JP, Reid G (2013) High throughput sequencing methods and analysis for microbiome research. J Microbiol Methods. 10.1016/j.mimet.2013.08.011.
23. Riera-Montes M, Brus Sjolander K, Allestam G, Hallin E, Hedlund KO, et al. (2011) Waterborne norovirus outbreak in a municipal drinking-water supply in Sweden. Epidemiol Infect 139: 1928–1935. 10.1017/S0950268810003146.
24. de Jong B, Ancker C (2008) Web-based questionnaires - a tool used in a *Campylobacter* outbreak investigation in Stockholm, Sweden, October 2007. Euro Surveill 13: 18847.
25. Oxenford CJ, Black AP, Bell RJ, Munnoch SA, Irwin MJ, et al. (2005) Investigation of a multi-state outbreak of *Salmonella hvittingfoss* using a web-based case reporting form. Commun Dis Intell Q Rep 29: 379–381.
26. Bessong PO, Odiyo JO, Musekene JN, Tessema A (2009) Spatial distribution of diarrhoea and microbial quality of domestic water during an outbreak of diarrhoea in the Tshikuwi community in Venda, South Africa. J Health Popul Nutr 27: 652–659.
27. Dangendorf F, Herbst S, Reintjes R, Kistemann T (2002) Spatial patterns of diarrhoeal illnesses with regard to water supply structures–a GIS analysis. Int J Hyg Environ Health 205: 183–191. 10.1078/1438-4639-00151.
28. Tuite AR, Tien J, Eisenberg M, Earn DJ, Ma J, et al. (2011) Cholera epidemic in haiti, 2010: Using a transmission model to explain spatial spread of disease and identify optimal control interventions. Ann Intern Med 154: 593–601. 10.1059/0003-4819-154-9-201105030-00334.
29. R_Core_Team R (2013) R: A language and environment for statistical computing. Vienna, Austria.
30. Hosmer DWJ, Lemeshow S, Strudivant RX (2013) Applied logistic regression. 3rd edition, Wiley.
31. Antikainen J, Kantele A, Pakkanen SH, Laaveri T, Riutta J, et al. (2013) A quantitative polymerase chain reaction assay for rapid detection of 9 pathogens directly from stools of travelers with diarrhea. Clin Gastroenterol Hepatol 11: 1300–1307.e3. 10.1016/j.cgh.2013.03.037.
32. Antikainen J, Tarkka E, Haukka K, Siitonen A, Vaara M, et al. (2009) New 16-plex PCR method for rapid detection of diarrheagenic *Escherichia coli* directly from stool samples. Eur J Clin Microbiol Infect Dis 28: 899–908. 10.1007/s10096-009-0720-x.
33. Summa M, von Bonsdorff CH, Maunula L (2012) Pet dogs–a transmission route for human noroviruses? J Clin Virol 53: 244–247. 10.1016/j.jcv.2011.12.014.
34. Ronnqvist M, Ratto M, Tuominen P, Salo S, Maunula L (2013) Swabs as a tool for monitoring the presence of norovirus on environmental surfaces in the food industry. J Food Prot 76: 1421–1428. 10.4315/0362-028X,JFP-12-371.
35. Oka T, Katayama K, Hansman GS, Kageyama T, Ogawa S, et al. (2006) Detection of human sapovirus by real-time reverse transcription-polymerase chain reaction. J Med Virol 78: 1347–1353. 10.1002/jmv.20699.
36. van Maarseveen NM, Wessels E, de Brouwer CS, Vossen AC, Claas EC (2010) Diagnosis of viral gastroenteritis by simultaneous detection of adenovirus group F, astrovirus, rotavirus group A, norovirus genogroups I and II, and sapovirus in two internally controlled multiplex real-time PCR assays. J Clin Virol 49: 205–210. 10.1016/j.jcv.2010.07.019.

37. Reuter G, Zimsek-Mijovski J, Poljsak-Prijatelj M, Di Bartolo I, Ruggeri FM, et al. (2010) Incidence, diversity, and molecular epidemiology of sapoviruses in swine across Europe. J Clin Microbiol 48: 363–368. 10.1128/JCM.01279-09.

38. Rimhanen-Finne R, Hanninen ML, Vuento R, Laine J, Jokiranta TS, et al. (2010) Contaminated water caused the first outbreak of giardiasis in Finland, 2007: A descriptive study. Scand J Infect Dis 42: 613–619. 10.3109/00365541003774608.

39. Revez J, Huuskonen M, Ruusunen M, Lindstrom M, Hanninen ML (2013) Arcobacter species and their pulsed-field gel electrophoresis genotypes in Finnish raw milk during summer 2011. J Food Prot 76: 1630–1632. 10.4315/0362-028X.JFP-13-083.

40. Merga JY, Leatherbarrow AJ, Winstanley C, Bennett M, Hart CA, et al. (2011) Comparison of Arcobacter isolation methods, and diversity of Arcobacter spp. in Cheshire, United Kingdom. Appl Environ Microbiol 77: 1646–1650. 10.1128/AEM.01964-10.

41. Houf K, Tutenel A, De Zutter L, Van Hoof J, Vandamme P (2000) Development of a multiplex PCR assay for the simultaneous detection and identification of Arcobacter butzleri, Arcobacter cryaerophilus and Arcobacter skirrowii. FEMS Microbiol Lett 193: 89–94.

42. Harmon KM, Wesley IV (1996) Identification of Arcobacter isolates by PCR. Lett Appl Microbiol 23: 241–244.

43. European U (1998) Council directive 98/83/EC of 3rd November on the quality of water intended for human consumption. Official Journal of European Communities L330: 32–54.

44. Polaczyk AL, Narayanan J, Cromeans TL, Hahn D, Roberts JM, et al. (2008) Ultrafiltration-based techniques for rapid and simultaneous concentration of multiple microbe classes from 100-L tap water samples. J Microbiol Methods 73: 92–99. 10.1016/j.mimet.2008.02.014.

45. ISO_9308-2 A (2012) Water quality - enumeration of Escherichia coli and coliform bacteria – part 2: Most probable number method. International Organization for Standardization, Geneva, Switzerland.

46. Pitkanen T, Miettinen IT, Nakari UM, Takkinen J, Nieminen K, et al. (2008) Faecal contamination of a municipal drinking water distribution system in association with Campylobacter jejuni infections. J Water Health 6: 365–376.

47. ISO/CD 6461-2 (2002) Water quality - Detection and enumeration of the spores of sulfite-reducing anaerobes (clostridia) – part 2: Method by membrane filtration. Revised committee draft. International Organization for Standardization, Geneva, Switzerland.

48. Pradhan SK, Kauppinen A, Martikainen K, Pitkänen T, Kusnetsov J, et al. (2013) Microbial reduction in waste water treatment using Fe^{3+} and Al^{3+} coagulants and PAA disinfectant. J Water Health 11: 581–589.

49. NMKL 164 A (2005) Escherichia coli O157. Detection in food and feeding stuffs, Nordic committee on food analysis.

50. Kauppinen A, Ikonen J, Pursiainen A, Pitkänen T, Miettinen IT (2012) Decontamination of a drinking water pipeline system contaminated with adenovirus and Escherichia coli utilizing peracetic acid and chlorine. J Water Health 10: 406–418. 10.2166/wh.2012.003.

51. Kauppinen A, Martikainen K, Matikka V, Veijalainen AM, Pitkänen T, et al. (2014) Sand filters for removal of microbes and nutrients from wastewater during a one-year pilot study in a cold temperate climate. J Environ Manage 133: 206–213. 10.1016/j.jenvman.2013.12.008.

52. ISO_15553 (2006) Water quality - isolation and identification of Cryptosporidium oocysts and Giardia cysts from water. International Organization for Standardization, Geneva, Switzerland.

53. Guy RA, Payment P, Krull UJ, Horgen PA (2003) Real-time PCR for quantification of Giardia and Cryptosporidium in environmental water samples and sewage. Appl Environ Microbiol 69: 5178–5185.

54. ISO 7393 (1985) Water quality – Determination of free chlorine and total chlorine. International Organization for Standardization, Geneva, Switzerland.

55. Nakari UM, Hakkinen M, Siitonen A (2011) Identification of persistent subtypes of Campylobacter jejuni by pulsed-field gel electrophoresis in Finland. Foodborne Pathog Dis 8: 1143–1145. 10.1089/fpd.2011.0882.

56. Siefring S, Varma M, Atikovic E, Wymer L, Haugland RA (2008) Improved real-time PCR assays for the detection of fecal indicator bacteria in surface waters with different instrument and reagent systems. J Water Health 6: 225–237. 10.2166/wh.2008.022.

57. Kildare BJ, Leutenegger CM, McSwain BS, Bambic DG, Rajal VB, et al. (2007) 16S rRNA-based assays for quantitative detection of universal, human-, cow-, and dog-specific fecal Bacteroidales: A bayesian approach. Water Res 41: 3701–3715. 10.1016/j.watres.2007.06.037.

58. Caporaso JG, Lauber CL, Walters WA, Berg-Lyons D, Lozupone CA, et al. (2011) Global patterns of 16S rRNA diversity at a depth of millions of sequences per sample. Proc Natl Acad Sci U S A 108 Suppl 1: 4516–4522. 10.1073/pnas.1000080107.

59. Schloss PD, Westcott SL, Ryabin T, Hall JR, Hartmann M, et al. (2009) Introducing mothur: Open-source, platform-independent, community-supported software for describing and comparing microbial communities. Appl Environ Microbiol 75: 7537–7541. 10.1128/AEM.01541-09.

60. Kozich JJ, Westcott SL, Baxter NT, Highlander SK, Schloss PD (2013) Development of a dual-index sequencing strategy and curation pipeline for analyzing amplicon sequence data on the MiSeq illumina sequencing platform. Appl Environ Microbiol 79: 5112–5120. 10.1128/AEM.01043-13.

61. Cole JR, Wang Q, Cardenas E, Fish J, Chai B, et al. (2009) The ribosomal database project: Improved alignments and new tools for rRNA analysis. Nucleic Acids Res 37: D141–5. 10.1093/nar/gkn879.

62. Tamura K, Peterson D, Peterson N, Stecher G, Nei M, et al. (2011) MEGA5: Molecular evolutionary genetics analysis using maximum likelihood, evolutionary distance, and maximum parsimony methods. Mol Biol Evol 28: 2731–2739. 10.1093/molbev/msr121.

63. Tamura K, Nei M, Kumar S (2004) Prospects for inferring very large phylogenies by using the neighbor-joining method. Proc Natl Acad Sci U S A 101: 11030–11035. 10.1073/pnas.0404206101.

64. ter Waarbeek HL, Dukers-Muijrers NH, Vennema H, Hoebe CJ (2010) Waterborne gastroenteritis outbreak at a scouting camp caused by two norovirus genogroups: GI and GII. J Clin Virol 47: 268–272. 10.1016/j.jcv.2009.12.002 [doi].

65. Teunis P, Van den Brandhof W, Nauta M, Wagenaar J, Van den Kerkhof H, et al. (2005) A reconsideration of the Campylobacter dose-response relation. Epidemiol Infect 133: 583–592.

66. Volken T (2013) Second-stage non-response in the swiss health survey: Determinants and bias in outcomes. BMC Public Health 13: 167-2458-13-167. 10.1186/1471-2458-13-167 [doi].

67. Johnson TP, Wislar JS (2012) Response rates and nonresponse errors in surveys. JAMA 307: 1805–1806. 10.1001/jama.2012.3532 [doi].

68. Martikainen P, Laaksonen M, Piha K, Lallukka T (2007) Does survey non-response bias the association between occupational social class and health? Scand J Public Health 35: 212–215. 10.1080/14034940600996563.

69. Miettinen IT, Pitkänen T, Nakari UM, Hakkinen M, Wermudsen K, et al. (2006) 5th nordic drinking water conference, 8.-10. June, 2006.

70. Richardson G, Thomas DR, Smith RM, Nehaul L, Ribeiro CD, et al. (2007) A community outbreak of Campylobacter jejuni infection from a chlorinated public water supply. Epidemiol Infect 135: 1151–1158. 10.1017/S0950268807007960.

71. Kitajima M, Oka T, Haramoto E, Katayama H, Takeda N, et al. (2010) Detection and genetic analysis of human sapoviruses in river water in Japan. Appl Environ Microbiol 76: 2461–2467. 10.1128/AEM.02739-09.

72. Murray TY, Mans J, Taylor MB (2013) First detection of human sapoviruses in river water in South Africa. Water Sci Technol 67: 2776–2783. 10.2166/wst.2013.203.

73. Mikula C, Springer B, Reichart S, Bierbacher K, Lichtenschopf A, et al. (2010) Sapovirus in adults in rehabilitation center, upper austria. Emerg Infect Dis 16: 1186–1187. 10.3201/eid1607.091789.

74. Lee LE, Cebelinski EA, Fuller C, Keene WE, Smith K, et al. (2012) Sapovirus outbreaks in long-term care facilities, Oregon and Minnesota, USA, 2002–2009. Emerg Infect Dis 18: 873–876. 10.3201/eid1805.111843.

75. Vandamme P, Pot B, Gillis M, de Vos P, Kersters K, et al. (1996) Polyphasic taxonomy, a consensus approach to bacterial systematics. Microbiol Rev 60: 407–438.

76. Collado L, Figueras MJ (2011) Taxonomy, epidemiology, and clinical relevance of the genus Arcobacter. Clin Microbiol Rev 24: 174–192. 10.1128/CMR.00034-10.

77. Lautenschlager K, Hwang C, Liu WT, Boon N, Koster O, et al. (2013) A microbiology-based multi-parametric approach towards assessing biological stability in drinking water distribution networks. Water Res 47: 3015–3025. 10.1016/j.watres.2013.03.002.

78. Pinto AJ, Xi C, Raskin L (2012) Bacterial community structure in the drinking water microbiome is governed by filtration processes. Environ Sci Technol 46: 8851–8859. 10.1021/es302042t.

79. Cox MJ, Cookson WO, Moffatt MF (2013) Sequencing the human microbiome in health and disease. Hum Mol Genet 22: R88–94. 10.1093/hmg/ddt398.

13

Shift in the Microbial Ecology of a Hospital Hot Water System following the Introduction of an On-Site Monochloramine Disinfection System

Julianne L. Baron[1,2]**, Amit Vikram**[3]**, Scott Duda**[2]**, Janet E. Stout**[2,3]**, Kyle Bibby**[3,4]*

1 Department of Infectious Diseases and Microbiology, University of Pittsburgh, Graduate School of Public Health, Pittsburgh, Pennsylvania, United States of America, **2** Special Pathogens Laboratory, Pittsburgh, Pennsylvania, United States of America, **3** Department of Civil and Environmental Engineering, University of Pittsburgh, Swanson School of Engineering, Pittsburgh, Pennsylvania, United States of America, **4** Department of Computational and Systems Biology, University of Pittsburgh Medical School, Pittsburgh, Pennsylvania, United States of America

Abstract

Drinking water distribution systems, including premise plumbing, contain a diverse microbiological community that may include opportunistic pathogens. On-site supplemental disinfection systems have been proposed as a control method for opportunistic pathogens in premise plumbing. The majority of on-site disinfection systems to date have been installed in hospitals due to the high concentration of opportunistic pathogen susceptible occupants. The installation of on-site supplemental disinfection systems in hospitals allows for evaluation of the impact of on-site disinfection systems on drinking water system microbial ecology prior to widespread application. This study evaluated the impact of supplemental monochloramine on the microbial ecology of a hospital's hot water system. Samples were taken three months and immediately prior to monochloramine treatment and monthly for the first six months of treatment, and all samples were subjected to high throughput Illumina 16S rRNA region sequencing. The microbial community composition of monochloramine treated samples was dramatically different than the baseline months. There was an immediate shift towards decreased relative abundance of Betaproteobacteria, and increased relative abundance of Firmicutes, Alphaproteobacteria, Gammaproteobacteria, Cyanobacteria and Actinobacteria. Following treatment, microbial populations grouped by sampling location rather than sampling time. Over the course of treatment the relative abundance of certain genera containing opportunistic pathogens and genera containing denitrifying bacteria increased. The results demonstrate the driving influence of supplemental disinfection on premise plumbing microbial ecology and suggest the value of further investigation into the overall effects of premise plumbing disinfection strategies on microbial ecology and not solely specific target microorganisms.

Editor: Stefan Bereswill, Charité-University Medicine Berlin, Germany

Funding: This project was funded by a grant from the Alfred P. Sloan Foundation (grant B2013-12). The funders had no role in study design, data collection and analysis, decision to publish, or preparation of the manuscript.

Competing Interests: The authors have declared that no competing interests exist.

* Email: BibbyKJ@pitt.edu

Introduction

Drinking water distribution systems, including premise plumbing, contain a diverse microbiological population [1]. Once new pipes have been added to an existing system, microbial colonization begins rapidly, with microbial communities being established in as little as one year [2]. For the purposes of this study, the 'microbial community' is defined as planktonic microbes within the hospital hot water system during the study period. The microbial ecology of drinking water distribution systems varies widely, depending upon system parameters such as disinfection scheme [3], hydraulic parameters [4], location in the system, age of the system [5], and pipe materials [6]. Microbes are capable of corroding pipes within distribution systems, possibly releasing harmful chemicals such as lead [7–9]. It is largely believed that within a drinking water distribution system, the disinfection scheme is one of the primary factors controlling the abundance and make-up of microbes [3,6,10]. Additionally, the effectiveness of disinfection in removing pathogens from drinking water is

mediated by the microbial ecology of the drinking water system [1]. However, the impact of on-site disinfection on premise plumbing microbial ecology is not well understood, motivating the current study.

The complex microbial ecology of premise plumbing systems can serve as a reservoir for opportunistic pathogens, such as *Legionella* spp., non-tuberculous Mycobacteria, *Pseudomonas* spp., *Acinetobacter* spp., *Stenotrophomonas* spp., *Brevundimonas* spp., *Sphingomonas* spp., and *Chryseobacterium* spp. [11–13]. Biofilms and amoeba within the water system can protect opportunistic pathogens from disinfection [1,14–16], and may even allow their regrowth and increase in pathogenicity [17–19]. As an example of the utility of microbial ecology-based approaches, a recent landmark microbial ecology-based study showed that biofilms in showerheads are actually enriched in opportunistic pathogens, creating the potential for an aerosol route of infection [20]. Additionally, antibiotic resistance genes have been detected in the biofilms of drinking water distribution systems [21,22]. Each of these points highlight the necessity for a greater understanding of premise plumbing microbial ecology.

Premise plumbing systems have an approximately ten-times greater microbial load than full-scale drinking water distribution systems, due to many factors including greater water stagnation and surface area to volume ratio [23,24]. Premise plumbing systems of hospitals are of particular concern, as hospitals may contain immunocompromised patients [25], who may not be protected by current drinking water monitoring standards [26], and who would be more susceptible to infections caused by opportunistic pathogens. To date, the majority of on-site disinfection systems have been installed in hospitals, creating a valuable testing ground to observe the impact of on-site disinfection systems on premise plumbing microbial ecology prior to more widespread application.

In addition to use in on-site systems, monochloramine as a secondary disinfectant has been advocated in the US as an effective method to reduce the production of disinfection-by-products [27,28] and control biofilm growth within water distribution systems [29]. While monochloramine is able to penetrate biofilms better than alternative disinfectants, this may not result in a reduction in biofilm growth [8]. Additionally, chloramine treatment requires the addition of an excess of ammonia, which may cause increased growth by ammonia-oxidizing bacteria [28], such as members of the genera *Nitrospira* spp. and *Nitrosomonas* spp. [30]. Bacterial nitrification is known to increase the degradation rate of monochloramine [31], thereby reducing the expected longevity and effectiveness of chloramine. Denitrifying bacteria have previously been identified in chlorami-nated drinking water systems [32]; however, this topic has not been fully explored in the literature.

The effectiveness of chloramination in removing opportunistic pathogens in premise plumbing remains unclear [27]. On-site monochloramine addition has been proposed as a disinfection strategy for the control of *Legionella* [33–36], but long-term studies have yet to been conducted [33,34]. Recently, a culture-based study of monochloramine on-site disinfection in a hospital's hot water system for the purpose of *Legionella* control demonstrated a significant reduction in *L. pneumophila* and no change in nitrate or nitrite levels [37]. Observed discrepancies in system performance are potentially due to differing microbial ecologies or water chemistries of the systems tested. A more holistic view of system microbial ecology, such as presented in this study, may allow more efficient application of supplemental disinfection.

Despite the obvious importance of the microbial ecology of drinking water systems in modulating disinfectant effectiveness and as a reservoir for opportunistic pathogens, there is a notable lack of studies detailing the shift in microbial diversity and composition in response to on-site disinfection. The objective of this study was to determine the effects of on-site monochloramine disinfection on the microbial ecology of a hospital hot water system. Both the microbial ecology of hot water systems and the response of premise plumbing microbial ecology to on-site disinfection are not currently well described in the literature. This study utilizes 216 samples taken from 27 sites and pooled into five composites for two time points prior to and six time points following the addition of on-site monochloramine addition. Samples were analyzed utilizing Illumina DNA sequencing of the microbial community 16S rRNA region and results demonstrate a dynamic shift of the microbial ecology of a hospital's hot water system in response to monochloramine addition.

Materials and Methods

Hospital setting

For these activities no specific permissions were required for these locations. This study took place in a 495-bed tertiary care hospital complex in Pittsburgh, PA. The building has 12 floors and receives chlorinated, municipal cold water. The hospital's hot water system was treated with the Sanikill monochloramine injection system (Sanipur, Lombardo, Flero, Italy). Monochlor-amine was dosed to a target concentration between 1.5 and 3.0 ppm as Cl_2. Details regarding monochloramine dosing and water chemistry are included in Text S1.

Sample collection and processing

Hot water was collected from 27 sites throughout the hospital at two time points before monochloramine injection (three months and immediately prior) and monthly for the first six months of monochloramine application. Water samples were collected from a variety of locations throughout the hospital (Table 1). Samples were taken from hot water tanks, the hot water return line, faucets in the intensive care units, rehabilitation suites including both automatic and standard faucets, and other patient rooms on the upper floors. The faucets in the intensive care units are located on the third, fourth, and fifth floors. The faucets in the rehabilitation suites are located on floors six and seven and represent both electronic sensor (automatic) faucets and standard faucets. The final grouping of sites was from short-term use patient rooms located on floors eight, nine, ten, eleven, and twelve. At each site, hot water was flushed for one minute prior to sample collection into sterile HDPE bottles with enough sodium thiosulfate to neutralize 20 ppm chlorine (Microtech Scientific, Orange, CA). For hot water tank sampling, the drain valve was opened, allowed to flush for one minute, then sampled into sterile HDPE bottles as described above. Following sampling, 100 mL of sample water was filtered through a 0.2 μm, 47 mm, polycarbonate filter membrane (Whatman, Florham Park, NJ), placed into 10 mL of the original water sample, and vortexed vigorously for 10 seconds as described in methods ISO Standards 11731:1998 and 11731:2004 for *Legionella* isolation. Five mL of each concentrated sample was frozen at −80°C until DNA extraction.

DNA extraction, PCR, and Sequencing

Frozen water samples were thawed and pooled as described in Table 1. The 27 samples were divided into five pools including the hot water tanks and hot water return line (HWT), floors 3–5 (the intensive care units, F3), floors 6 and 7 automatic faucets (the rehabilitation suites' automatic faucets, F6A), floors 6 and 7 standard faucets (the rehabilitation suites' standard faucets, F6S), and floors 8–12 (the short-term use patient rooms, F8). These samples were then filtered through 0.2 μm, 47 mm, Supor 200 Polyethersulfone membranes (Pall Corporation), housed in sterile Nalgene filter funnels (Thermo Scientific; Fisher). Filter mem-branes were subjected to DNA extraction using the RapidWater DNA Isolation Kit (MO-BIO Laboratories) as described by the manufacturer. PCR was performed in quadruplicate using 16S rRNA region primers 515F and 806R including sequencing and barcoding adapters as previously described [38]. These primers amplify an approximately 300 base pair region of the rRNA region spanning variable regions 3 and 4. The specificity of this primer set is considered to be well optimized and 'nearly universal' [39]; analysis of these primers against the 97% Greengenes 13.5 OTU database demonstrated a specificity of 99.9% and 98.3% for the 515f and 806r primers, respectively. Dreamtaq Mastermix (Thermo Scientific) was used and PCR product was checked on

Table 1. Sample pool description, abbreviation, and number of pooled sites.

Sample Description	Sample Abbreviation	Number of Pooled Sites
Outlets of Hot Water Tanks and Hot Water Return Line	HWT	3
Floors 3–5 Patient Room Faucets	F3	4
Floors 6 & 7 Patient Room Automatic Faucets	F6A	7
Floors 6 & 7 Patient Room Standard Faucets and Showers	F6S	7
Floors 8–12 Patient Room Faucets	F8	6
Technical Replicates of Floors 8–12 Patient Room Faucets	F8rep	6

Hot water was collected after a one-minute flush from the following locations throughout the hospital.

a 1% agarose gel. An independent negative control was run for each sample and primer set and all negative controls were negative for PCR amplification. PCR products were pooled and purified using the UltraClean PCR Clean-Up Kit (MO-BIO Laboratories). Each sample then underwent additional cleaning with the Agencourt AMPure XP PCR purification kit (Beckman Coulter) and quantified using the QuBit 2.0 Fluorometer (Invitrogen). Following quantification, 0.1 picomoles of each sample PCR product were pooled. The sample pool underwent two additional clean up steps with a 1.5:1 ratio of Agencourt AMPure XP beads followed by a 1.2:1 bead ratio (Beckman Coulter) to eliminate primer dimers. Samples were sequenced on an in-house Illumina MiSeq sequencing platform as previously described [38].

Data analysis

Data was analyzed within the MacQIIME (http://www.wernerlab.org/software/macqiime) implementation of QIIME 1.7.0 [40]. Sequences were parsed based upon sample-specific barcodes and trimmed to a minimum quality score of 20. Operational taxonomic units (OTUs) at 97% were then picked against the Greengenes 13.5 database using UCLUST [41] for taxonomic assignment. Following assignment, 7,000 successfully assigned sequences from each sample were chosen at random to allow for even downstream analyses and even cross-sample comparison. Observed OTUs were defined as observed species whereas unassigned sequences were removed from subsequent analyses (closed reference OTU picking). Alpha-diversity evenness was calculated using the 'equitability' metric within QIIME. Beta diversity analyses were conducted by UNIFRAC analysis [42]. OTUs were also open-reference picked, where unassigned sequences are placed in the taxa "other" and therefore not removed. Discussion and results from this open-reference OTU picking analysis is included in Text S1. Open-reference OTU picking did not result in a shift in any fundamental conclusions with the exception of the increase in the genus *Stenotrophomonas* spp. following monochloramine addition; closed-reference OTU picking is presented for higher-quality taxonomic assignment. Morisita-Horn indices were calculated as previously described [43,44]. Sequences are available under MG RAST accession numbers 4552832.3 to 4552878.3.

Results

Sequence Data

Sequencing reads were split by sample-specific barcodes, trimmed to a minimum quality score of 20, and placed into OTUs at 97% through comparison with the Greengenes 13.5 coreset. For each sample, 7,000 sequences with assigned taxonomy were selected to allow for even comparison across samples. Two

types of OTU picking were done for this study: closed reference (sequences were compared to a reference set of sequences for OTU clustering, sequences not matching one of these pre-defined sequences were discarded) and open reference (sequences were compared to each other for OTU picking, sequences not mapping to the reference database were grouped as 'other') in Text S1.

Alpha Diversity

Alpha diversity (number of observed OTUs) of samples treated with monochloramine was significantly higher than samples from the baseline months (Figure 1). Prior to treatment, the average number of observed OTUs at 97% similarity was 151.2 ± 39.7, whereas during treatment the average number of observed OTUs was 225.2 ± 61.2 ($p < 0.001$) (Figure 1). This shift was not associated with a statistically significant loss of sample evenness (Figure S1). The same statistical trends in alpha diversity were observed for open-reference picked OTUs (Figure S2).

Beta Diversity

Beta diversity (sample interrelatedness) was analyzed using weighted UNIFRAC [42]. The principal coordinate analysis (PCoA) plot from this analysis is shown in Figure 2. Samples from the first two months prior to treatment cluster together whereas those following disinfection tend to cluster by sample site more strongly than sample time (Figure 2). The same trend was observed for open-reference picked OTUs (Figure S3).

Taxonomic Comparison

Figure 3 shows the phyla-level taxonomy for each of the sample pools. Phyla <1.3% relative abundance are listed as 'minor phyla'. Prior to treatment, samples from all locations were similarly

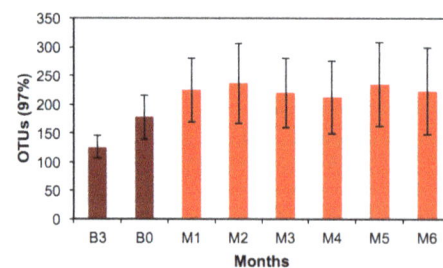

Figure 1. Comparison of the number of OTUs (97% similarity) for each month. Bars represent standard deviation. Each sample pool was normalized to 7,000 sequences. Samples from B3 and B0 represent those taken three months and immediately prior to monochloramine treatment, respectively. Samples from M1, M2, M3, M4, M5, and M6 were taken monthly during the first six months of treatment.

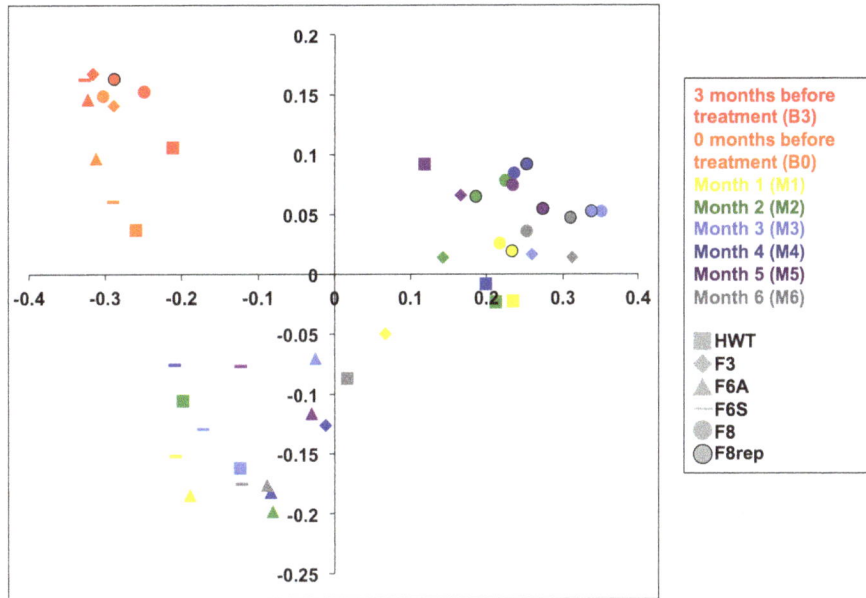

Figure 2. PCoA analysis of samples pools. Samples that cluster more closely together share a greater similarity in microbial community structure. Colors represent months sampled whereas shapes represent sample pool. Samples from B3 and B0 represent those taken three months and immediately prior to monochloramine treatment, respectively. Samples from M1, M2, M3, M4, M5, and M6 were taken monthly during the first six months of treatment.

structured, predominantly comprised of Betaproteobacteria, with lesser quantities of Firmicutes, Bacteroidetes, Alphaproteobacteria, and Gammaproteobacteria (Figure 3 Panels A–E). Following initiation of treatment (M1) there was a shift away from the predominance of Betaproteobacteria and towards a greater relative abundance of Firmicutes, Alphaproteobacteria, Gammaproteobacteria, and minor fractions of Cyanobacteria and Actinobacteria (Figure 3 Panels A–E). The same taxonomy trends were observed for open-reference picked data (Figure S4 Panels A–E).

The samples from the hot water tank (HWT) from pre-treatment months (B3 and B0) were approximately 60% Betaproteobacteria with approximately 35% Firmicutes, Bacteroidetes, Alphaproteobacteria, and Gammaproteobacteria in aggregate (Figure 3 Panel A). Following treatment the relative abundance of Betaproteobacteria was reduced to approximately 20% and Firmicutes, Alphaproteobacteria, and Gammaproteobacteria subsequently increased to comprise an average of 78% of the total relative abundance (Figure 3 Panel A).

The microbial community profile of samples from the lower floors of the hospital (intensive care units, F3) was slightly different than those of the hot water tank samples but a similar trend was observed (Figure 3 Panel B). Over 65% of pre-treatment samples were Betaproteobacteria with Firmicutes, Bacteroidetes, Alphaproteobacteria, and Gammaproteobacteria accounting for a combined 20% of community relative abundance (Figure 3 Panel B). Following treatment the amount of Betaproteobacteria and Bacteroidetes decreased to an average of 23% relative abundance, while the relative abundance of Firmicutes and Alphaproteobacteria increased sharply to approximately 68% (Figure 3 Panel B).

In spite of being from the same rooms, the taxonomic composition of samples from F6A and F6S differed after treatment (Figure 3 Panels C and D). Prior to treatment both the automatic (F6A) and standard faucets (F6S) in the rehabilitation suites contained 65–80% Betaproteobacteria, with Bacteroidetes, Alphaproteobacteria, Gammaproteobacteria, and Cyanobacteria

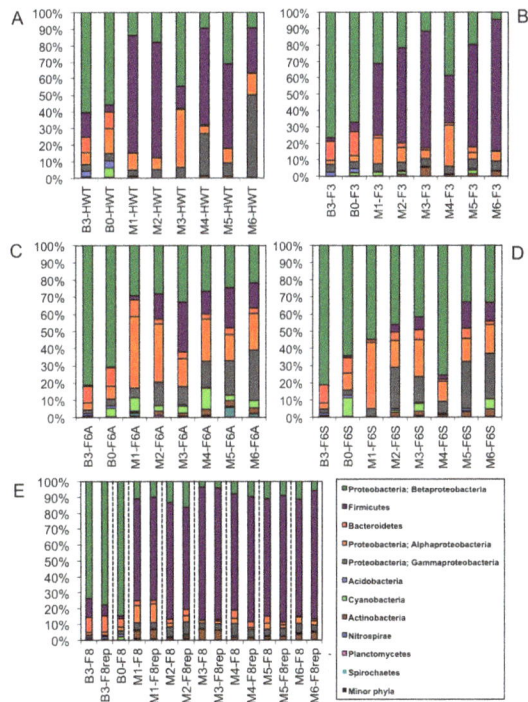

Figure 3. Taxonomic assignments of sequences from HWT (hot water tank samples) (Panel A), F3 (floors 3–5) (Panel B), F6A (floors 6 and 7 automatic faucets) (Panel C), F6S (floors 6 and 7 standard faucets) (Panel D), F8 (floors 8–12) and F8rep (replicate barcoded PCRs of samples from floors 8–12) (Panel E). Samples from B3 and B0 represent those taken three months and immediately prior to monochloramine treatment, respectively. Samples from M1, M2, M3, M4, M5, and M6 were taken monthly during the first six months of treatment. Black lines in Panel E separate pairs of replicates.

accounting for the other 20–35% of relative abundance (Figure 3 Panels C and D). However, after monochloramine application, the automatic faucets (F6A) underwent a 50% reduction in the total relative abundance of Betaproteobacteria and became enriched in Firmicutes, Alphaproteobacteria, Gammaproteobacteria, Actinobacteria, and Spirochaetes (Figure 3 Panel C). The standard faucets (F6S) lost only 26% of Betaproteobacteria, but also saw an increase in members of the Firmicutes, Alphaproteobacteria, Gammaprotobacteria, and Actinobacteria phyla from an average relative abundance of 10% before treatment to 46% after monochloramine addition (Figure 3 Panel D).

Prior to treatment, the microbial community in samples from the upper floors of the hospital (short-term use patient rooms, F8) resembled most of the other baseline samples with over 70% Betaproteobacteria and approximately 20% of Firmicutes, Bacteroidetes, Alphaproteobacteria, Gammaproteobacteria, Acidobacteria, and Cyanobacteria (Figure 3 Panel E). Following monochloramine treatment, the relative abundance of Betaproteobacteria was reduced from approximately 70% to 10% and replaced by Firmicutes, which increased from 7% of the relative abundance in the baseline months to 74% after treatment (Figure 3 Panel E). There was only a slight increase, from 2% to 9% relative abundance, in the amount of Gammaproteobacteria and Actinobacteria present (Figure 3 Panel E).

Sample Replicates

Separately amplified and barcoded technical replicates of sample pool F8 for 7 of the 8 sample pools were also sequenced to verify technical reproducibility. There is no replicate for month B0. UNIFRAC analysis demonstrated that the replicates from each month cluster very closely (Figure 2). All of the samples from F8 in samples M1–M6 and their replicates (circles and outlined circles) clustered together in the upper-right hand quadrant (Figure 2). Morisita-Horn analyses of replicates demonstrate high levels of community similarity, ranging from 0.990 (M2) to 0.9998 (M3). These results further validate the technical reproducibility of the methodology (Figure 3 Panel E) [43,44]. The open-reference picked UNIFRAC analysis and taxonomy also show replicates to have similar profiles to their original samples (Figure S3 and S4 Panel E). Morisita-Horn analyses of these samples showed similarly high levels of community similarity ranging from 0.991 (M2) to 0.9992 (M1).

Genera Containing Opportunistic Pathogens

Sequence data was further analyzed to observe the change in genera containing opportunistic pathogens of interest during treatment. Genera analyzed were: *Legionella* spp., *Pseudomonas* spp., *Acinetobacter* spp., and *Stenotrophomonas* spp. (Gammaproteobacteria group); *Brevundimonas* spp. and *Sphingomonas* spp. (Alphaproteobacteria group); *Chryseobacterium* spp. (Bacteroidetes group); and *Mycobacterium* spp. (Actinobacteria group). These genera are of special interest as some to all of the species contained within them are pathogens; however, the nature of short-read 16S rRNA region sequence analysis is such that species-level pathogens cannot be definitively identified. Trends demonstrated by this analysis could be used to direct future analyses targeting opportunistically pathogenic organisms more specifically. Analysis of the relative abundance of each of these organism groups over time shows a statistically significant increase in relative abundance for *Acinetobacter* ($p = 0.0054$), *Mycobacterium* ($p = 0.0017$), *Pseudomonas* ($p = 0.031$) and *Sphingomonas* ($p = 0.034$) as treatment progressed (Figure 4). *Brevundimonas*, *Chryseobacterium*, Legionellaceae, and *Stenotrophomonas* did not demonstrate a statistically significant increase in relative abundance following treatment (Figure 4).

The open-reference picked data demonstrated an increase in the same opportunistic pathogen containing genera as the closed-reference picked data, *Acinetobacter* ($p = 0.004$), *Mycobacterium* ($p = 0.002$), *Pseudomonas* ($p = 0.015$), and *Sphingomonas* ($p = 0.025$), but also showed a significant increase in the genera *Stenotrophomonas* ($p = 0.03$) (Figure S5).

Nitrification and Denitrification

Additionally, we investigated the shift in relative abundance of representative genera associated with nitrification and denitrification (Figure 5). There was no statistically significant difference in the relative abundance of the potential nitrifiers *Nitrospira* and Nitrosomonadaceae, before (mean = 0.0015 ± 0.0018) and after treatment (mean = 0.0005 ± 0.0011) ($p = 0.175$). Other nitrifier-containing genera such as *Nitrosococcus*, *Nitrobacter*, *Nitrospina*, or *Nitrococcus*, were not identified in any samples. The total relative abundance of genera containing denitrifiers (*Thiobacillus*, *Micrococcus*, and *Paracoccus*) underwent a statistically significant increase before (mean = 0.00005 ± 0.000074) and after treatment with monochloramine (mean = 0.0029 ± 0.0029) ($p = 0.026$). The denitrifier-containing genera *Rhizobiales* and *Rhodanobacter* were not identified in any samples. The same trends were observed in open-reference picked data (Figure S6).

Discussion

Our study objective was to examine the shift in the microbial ecology of a hospital hot water system associated with the introduction of on-site monochloramine addition. To evaluate the shift in microbial community structure we sampled 27 sites in a hospital and pooled samples into 5 groups for 8 sample time points. Sites were pooled based on their location and use in the hospital and faucet type (automatic versus standard). This study took place during the first U.S. trial of the Sanikill on-site monochloramine generation system (Sanipur, Brescia, Italy) [45–47]. These samples were subjected to DNA extraction, 16S rRNA region barcoded PCR, and Illumina sequencing to analyze the response of the microbial ecology to the addition of monochloramine.

The microbial population shift in response to monochloramine addition was immediate. The number of OTUs observed (alpha diversity) significantly increased following monochloramine treatment (Figure 1). It is possible that the overall loss of dominance of initially abundant microbial groups (e.g. Betaproteobacteria) allowed for a greater number of other bacterial species to grow, or for selected individuals to die off, thereby increasing the alpha diversity. Samples from different sites taken before monochloramine treatment were comprised of similar microbial populations and samples taken after treatment were distinct from samples taken in the baseline months (Figures 2 and 3, Figures S3 and S4). Interestingly, it appears that following monochloramine treatment the location of sampling matters more in sample similarity (beta diversity) than does the month they were taken (Figure 2, Figure S3). Microbial communities from the lower floors' intensive care units (F3) and the upper floors' short term patient rooms (F8) were more similar than to the floors 6 and 7's rehabilitation suites (F6A and F6S) automatic and standard faucet samples. These sites were located in single patient rooms in rehabilitation units and may experience as much use as some locations on the lower and upper floors, which include the trauma burn unit, the intensive care unit (ICU), the neonatal ICU, and the cardiovascular ICU. The HWT samples from earlier months of treatment closely resembled floors 6 and 7 (F6A and F6S) whereas the HWT microbial ecology from

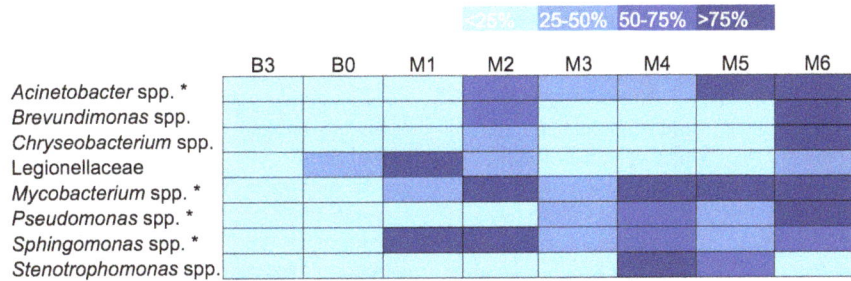

Figure 4. Relative abundance of different genera of opportunistic waterborne pathogens. Samples color coded into four groupings calculated by 25% of the maximum relative abundance for each organism. Months with the least relative abundance are lightest in color, whereas months with the highest relative abundance are darkest. *denotes a statistically significant increase in the relative abundance of this organism following treatment.

the later months was more related to the lower (F3) and upper floors (F8).

We investigated the possible differences in microbial ecology between automatic and standard faucets as it has been previously demonstrated that opportunistic pathogens, including *Legionella* [48] and *Pseudomonas aeruginosa* [49], are detected more frequently and in greater concentrations in automatic faucets. It has been suggested that the reason for the differences between automatic and standard faucets could be due to water flow, temperature, and structural issues. Automatic faucets may have diluted monochloramine concentrations due to low flow and poor flushing [48,49] and automatic faucets also contain mixing valves, which are made of materials such as rubber, polyvinylchloride, and plastic, which more easily support the growth of biofilms [48,49]. Potentially due to these biofilms, the increased colonization can persist even following disinfection with chlorine dioxide [48]. We observed a differential reduction in the relative abundance of Betaproteobacteria in standard and automatic faucets following treatment. The automatic faucets lost 50% of their relative abundance of Betaproteobacteria whereas the standard faucets only saw an average 26% reduction.

There was an overall shift towards less relative abundance of Betaproteobacteria, and more relative abundance of Firmicutes, Alphaproteobacteria, Gammaproteobacteria, Cyanobacteria and

Actinobacteria after monochloramine treatment. A previous microbial ecology study of a simulated drinking water distribution system treated with monochloramine demonstrated a different trend, with an increase in specific genera within the Actinobacteria, Betaproteobacteria, and Gammaproteobacteria phyla [3]. The dissimilarity of these studies may be due to the fact that the latter occurred in a cold water system whereas our study was in a hot water supply.

Several waterborne pathogen-containing genera were examined for changes in relative abundance due to monochloramine treatment. The relative abundance of a few of the waterborne pathogen-containing genera examined, including *Acinetobacter*, *Mycobacterium*, *Pseudomonas*, and *Sphingomonas*, showed an increase after monochloramine treatment. Other studies have described an increase in some of these organisms including *Legionella*, *Mycobacterium*, and *Pseudomonas* in chloraminated water [3,6] as well as biofilms treated with monochloramine [50]. Feazel et al. previously demonstrated that *Mycobacterium* spp. can be enriched in showerhead biofilms compared to the source water [20]. An increased relative abundance of *Mycobacterium* spp. due to monochloramine treatment is of concern, specifically if this increase in relative abundance is due to the presence of more viable mycobacterial cells. These microorganisms may pose a specific threat of aerosol exposure to immunocompromised

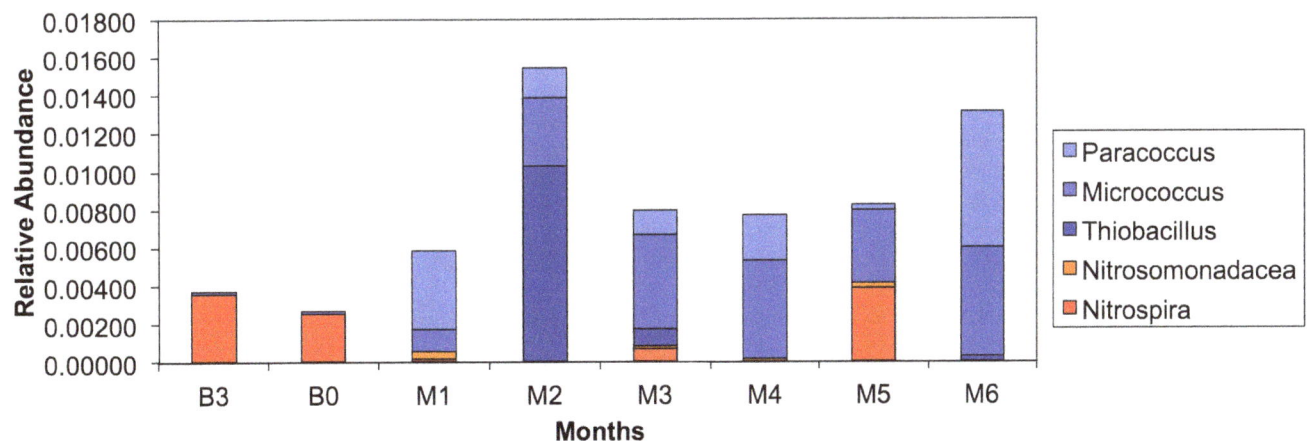

Figure 5. Relative abundance of genera containing nitrifying (*Nitrospira* and Nitrosomonadacea) and denitrifying bacteria (*Thiobacillus, Micrococcus,* and *Paracoccus*). No other genera associated with nitrification (*Nitrosococcus, Nitrobacter, Nitrospina,* or *Nitrococcus,*) or denitrification (*Rhizobiales* and *Rhodanobacter*) were found in any of our samples. The x-axis represents sampling months with months B3 and B0 being before monochloramine treatment and months M1–M6 representing the first six months of treatment. The y-axis represents the relative abundance.

patients who reside in buildings with an increased abundance of these organisms in hot water [20]. Interestingly, a recent study demonstrated that while the concentration of live bacteria is reduced after monochloramine treatment, only the viable microbial community structure is altered and genera containing opportunistic pathogens persist [51]. While we did not directly quantify microorganisms in the samples collected or verify that microorganisms detected were viable, our parallel culture-based study observed a statistically significant reduction in culturable total bacteria and *Legionella* species following monochloramine treatment (Table S1) [45–47,52].

Previous studies have found an increase in nitrification in chloraminated systems, which effectively decreased monochloramine concentration [6,31]. This chemical decay led to higher levels of *Legionella*, *Mycobacterium* spp., and *P. aeruginosa* at earlier water ages than in chlorinated simulated distribution systems [6]. A change in potentially nitrifying bacteria following monochloramine addition was not observed in the culture-based portion of this study [45–47], consistent with our molecular observations. Concentrations of nitrate and nitrite remained fairly stable throughout the study months, with the exception of a spike in nitrate levels in M6 (Table S1) [52]. We observed a statistically significant increase in the relative abundance of genera associated with denitrification in monochloramine treated samples. A previous study found a high absolute abundance, up to 200,000 cfu/mL, of potentially denitrifying bacteria in a chloraminated system even after regular flushing [32]. The highest relative abundance of bacterial genera associated with denitrification occurred during M6 when there was a spike in nitrate concentrations (Table S1) [52]. However, in months 1 and 2 there was also a large relative abundance of these bacteria present with fairly low nitrate concentrations, suggesting that some other factor might be important in their relative abundance. We do not believe that these trends were due to seasonality in our study as microbiological data were largely consistent across the study period. However, the possibility for seasonal effects cannot be excluded.

A notable increase in the relative abundance of the genus *Alicyclobacillus* spp. (Firmicutes phylum) was observed following monochloramine treatment, from an average of 4.1±4.5% of the microbial population prior to treatment to an average of 40.9±27.1% following treatment (p<0.001). This genera is comprised primarily of spore-formers that are of concern in food spoilage [53], and has previously been detected in drinking water [54]. The high relative abundance of *Alicyclobacillus* spp. suggests a potentially dominant role in chloraminated hot water system microbial ecology worthy of future investigation.

The incidence of reported Legionnaires' disease cases increased threefold from 2000 to 2009 [55]. This fact, coupled with an increasingly elderly and immunocompromised population [55], has lead to an increased concern about *Legionella* and other opportunistic waterborne pathogens. Additionally, the American Society of Heating, Refrigerating, and Air-Conditioning Engineers (ASHRAE) has recently proposed Standard 188P for the prevention of legionellosis associated with premise plumbing systems [56]. This standard serves to reduce the risk of *Legionella* infections through a risk management approach [56]. For these reasons, on-site disinfection has become progressively important to protect patients in hospitals and long-term care facilities from waterborne opportunistic pathogens. An increased understanding of the influence of on-site disinfection on premise plumbing microbial ecology is necessary to maximize effectiveness and to limit undesired side effects.

This study demonstrates that there exists the potential for unwanted consequences of supplemental disinfectant addition for the removal of *Legionella* such as the potential enrichment of other waterborne pathogens, including *Acinetobacter*, *Mycobacterium*, *Pseudomonas*, and *Sphingomonas*. Understanding the impact of supplemental disinfection on water system microbial ecology, through a holistic approach, is necessary to maximize disinfectant effectiveness and to ensure that supplemental disinfectant does not select for alternative opportunistic pathogens. A recent review emphasizes not only the role of disinfectants but also other system factors that may impact microbial ecology such as temperature, pipe material, organic carbon, presence of automatic faucets, and point-of-use filtration [24]. The authors suggest a probiotic approach to opportunistic pathogen control which would either add microbes that can outcompete these pathogens, remove key species, or using engineering controls to favor benign organisms that are antagonistic to opportunistic pathogens [24]. This systematic, probiotic, approach to premise plumbing opportunistic pathogen management is an inventive concept for dealing with the diverse microbial ecology of these systems, but requires a greater understanding of the drivers of premise plumbing microbial ecology, such as provided by this study.

In conclusion, we observed a shift in the microbial ecology of a hospital's hot water system treated with on-site chloramination. This shift occurred immediately following monochloramine treatment. Prior to treatment, the bacterial ecology of all samples was dominated by Betaproteobacteria; following treatment, members of Firmicutes and Alphaproteobacteria dominated. Differences in community composition were seen in different locations within the hospital as well as between automatic and standard faucets. This suggests that water from different locations and outlet types should be sampled to get a more thorough picture of the microbiota of a system. There was an increase in the relative abundance of several genera containing opportunistic waterborne pathogens following the onset of monochloramine treatment, including *Acinetobacter*, *Mycobacterium*, *Pseudomonas*, and *Sphingomonas* and genera associated with denitrification. The benefits and risks of each supplemental disinfection strategy should be evaluated before implementation in any building, especially in hospitals, long term care facilities, and other buildings housing immunocompromised patients. This work demonstrates the effects of a supplemental monochloramine disinfection system on the microbial ecology of premise plumbing biofilms. Given the importance of premise plumbing microbial ecology on opportunistic pathogen presence and persistence, understanding the driving influence of supplemental disinfectants on microbial ecology is a crucial component of any effort to rid premise plumbing systems of opportunistic pathogens. As additional facilities turn to on-site water disinfection strategies, more long-term studies on the effects of disinfectants on microbial ecology in premise plumbing are needed as well as those evaluating a probiotic approach to opportunistic pathogen eradication.

Supporting Information

Figure S1 Sample evenness for closed-reference OTU picking. No statistically significant different was observed for samples taken prior to or following monochloramine addition.

Figure S2 Alpha diversity for open-reference OTU picking. A statistically significant difference was observed for samples taken prior to or following monochloramine addition (p = 0.046).

Figure S3 Beta diversity for open-reference OTU picking. Samples from before monochloramine treatment clustered together whereas following treatment samples clustered by location more so than month of treatment.

Figure S4 Taxonomic assignment of sequences from HWT (hot water tank samples) (Panel A), F3 (floors 3–5) (Panel B), F6A (floors 6 and 7 automatic faucets) (Panel C), F6S (floors 6 and 7 standard faucets) (Panel D), F8 (floors 8–12) and F8rep (replicate barcoded PCRs of samples from floors 8–12) (Panel E) for open-reference OTU picking.

Figure S5 Relative abundance of waterborne pathogen containing genera for open-reference OTU picking. A statistically significant increase in *Acinetobacter* spp., *Mycobacterium* spp., *Pseudomonas* spp., *Sphingomonas* spp., and *Stenotrophomonas* spp. was observed following treatment.

Figure S6 Relative abundance genera containing nitrifying (*Nitrospira* and Nitrosomonadacea) and denitrifying bacteria (*Thiobacillus*, *Micrococcus*, and *Paracoccus*) for open-reference OTU picking. No other genera containing nitrifying bacteria (*Nitrosococcus*, *Nitrobacter*, *Nitrospina*, or *Nitrococcus*,) or denitrifying bacteria (*Rhizobiales* and *Rhodanobacter*) were found in our samples.

Table S1 Physicochemical data obtained during the study.

Text S1 Supplementary Information. Water chemistry and monochloramine dosing methods, description of minor phyla observed, and open-reference OTU picking results.

Acknowledgments

We would like to thank the staff of Special Pathogens Laboratory for their assistance in sample collection.

Author Contributions

Conceived and designed the experiments: JLB JES KB. Performed the experiments: JLB AV SD. Analyzed the data: JLB JES KB. Wrote the paper: JLB KB.

References

1. Berry D, Xi C, Raskin L (2006) Microbial ecology of drinking water distribution systems. Curr Opin Biotechnol 17: 297–302.
2. Martiny AC, Jorgensen TM, Albrechtsen HJ, Arvin E, Molin S (2003) Long-term succession of structure and diversity of a biofilm formed in a model drinking water distribution system. Appl Environ Microbiol 69: 6899–6907.
3. Gomez-Alvarez V, Revetta RP, Santo Domingo JW (2012) Metagenomic analyses of drinking water receiving different disinfection treatments. Appl Environ Microbiol 78: 6095–6102.
4. Douterelo I, Sharpe RL, Boxall JB (2013) Influence of hydraulic regimes on bacterial community structure and composition in an experimental drinking water distribution system. Water Res 47: 503–516.
5. Henne K, Kahlisch L, Brettar I, Hofle MG (2012) Analysis of structure and composition of bacterial core communities in mature drinking water biofilms and bulk water of a citywide network in Germany. Appl Environ Microbiol 78: 3530–3538.
6. Wang H, Masters S, Edwards MA, Falkinham JO, Pruden A (2014) Effect of Disinfectant, Water Age, and Pipe Materials on Bacterial and Eukaryotic Community Structure in Drinking Water Biofilm. Environ Sci Technol 48: 1426–1435.
7. White C, Tancos M, Lytle DA (2011) Microbial community profile of a lead service line removed from a drinking water distribution system. Appl Environ Microbiol 77: 5557–5561.
8. Zhang Y, Griffin A, Rahman M, Camper A, Baribeau H, et al. (2009) Lead contamination of potable water due to nitrification. Environ Sci Technol 43: 1890–1895.
9. Zhang Y, Triantafyllidou S, Edwards M (2008) Effect of nitrification and GAC filtration on copper and lead leaching in home plumbing systems. Journal of Environmental Engineering-Asce 134: 521–530.
10. Mathieu L, Bouteleux C, Fass S, Angel E, Block JC (2009) Reversible shift in the alpha-, beta- and gamma-proteobacteria populations of drinking water biofilms during discontinuous chlorination. Water Res 43: 3375–3386.
11. Squier C, Yu VL, Stout JE (2000) Waterborne Nosocomial Infections. Curr Infect Dis Rep 2: 490–496.
12. Perola O, Nousiainen T, Suomalainen S, Aukee S, Karkkainen UM, et al. (2002) Recurrent Sphingomonas paucimobilis-bacteraemia associated with a multibacterial water-borne epidemic among neutropenic patients. J Hosp Infect 50: 196–201.
13. Mondello P, Ferrari L, Carnevale G (2006) Nosocomial Brevundimonas vesicularis meningitis. Infez Med 14: 235–237.
14. Buse HY, Ashbolt NJ (2011) Differential growth of Legionella pneumophila strains within a range of amoebae at various temperatures associated with in-premise plumbing. Lett Appl Microbiol 53: 217–224.
15. Emtiazi F, Schwartz T, Marten SM, Krolla-Sidenstein P, Obst U (2004) Investigation of natural biofilms formed during the production of drinking water from surface water embankment filtration. Water Res 38: 1197–1206.
16. Berry D, Horn M, Xi C, Raskin L (2010) Mycobacterium avium Infections of Acanthamoeba Strains: Host Strain Variability, Grazing-Acquired Infections, and Altered Dynamics of Inactivation with Monochloramine. Appl Environ Microbiol 76: 6685–6688.
17. Swanson MS, Hammer BK (2000) Legionella pneumophila pathogesesis: a fateful journey from amoebae to macrophages. Annu Rev Microbiol 54: 567–613.
18. Lau HY, Ashbolt NJ (2009) The role of biofilms and protozoa in Legionella pathogenesis: implications for drinking water. J Appl Microbiol 107: 368–378.
19. van der Wielen PW, van der Kooij D (2013) Nontuberculous mycobacteria, fungi, and opportunistic pathogens in unchlorinated drinking water in The Netherlands. Appl Environ Microbiol 79: 825–834.
20. Feazel LM, Baumgartner LK, Peterson KL, Frank DN, Harris JK, et al. (2009) Opportunistic pathogens enriched in showerhead biofilms. Proc Natl Acad Sci U S A 106: 16393–16399.
21. Schwartz T, Kohnen W, Jansen B, Obst U (2003) Detection of antibiotic-resistant bacteria and their resistance genes in wastewater, surface water, and drinking water biofilms. FEMS Microbiol Ecol 43: 325–335.
22. Shi P, Jia S, Zhang XX, Zhang T, Cheng S, et al. (2013) Metagenomic insights into chlorination effects on microbial antibiotic resistance in drinking water. Water Res 47: 111–120.
23. NRC (2006) Drinking Water Distribution Systems: Assessing and Reducing Risks.
24. Wang H, Edwards MA, Falkinham JO, Pruden A (2013) Probiotic Approach to Pathogen Control in Premise Plumbing Systems? A Review. Environmental Science & Technology 47: 10117–10128.
25. Williams MM, Armbruster CR, Arduino MJ (2013) Plumbing of hospital premises is a reservoir for opportunistic pathogenic microorganisms: a review. Biofouling 29: 147–162.
26. Williams MM, Braun-Howland EB (2003) Growth of Escherichia coli in model distribution system biofilms exposed to hypochlorous acid or monochloramine. Appl Environ Microbiol 69: 5463–5471.
27. Wang H, Edwards M, Falkinham JO 3rd, Pruden A (2012) Molecular survey of the occurrence of Legionella spp., Mycobacterium spp., Pseudomonas aeruginosa, and amoeba hosts in two chloraminated drinking water distribution systems. Appl Environ Microbiol 78: 6285–6294.
28. Regan JM, Harrington GW, Noguera DR (2002) Ammonia- and nitrite-oxidizing bacterial communities in a pilot-scale chloraminated drinking water distribution system. Appl Environ Microbiol 68: 73–81.
29. LeChevallier MW, Cawthon CD, Lee RG (1988) Inactivation of biofilm bacteria. Appl Environ Microbiol 54: 2492–2499.
30. Hoefel D, Monis PT, Grooby WL, Andrews S, Saint CP (2005) Culture-independent techniques for rapid detection of bacteria associated with loss of chloramine residual in a drinking water system. Appl Environ Microbiol 71: 6479–6488.
31. Zhang Y, Edwards M (2009) Accelerated chloramine decay and microbial growth by nitrification in premise plumbing. Journal American Water Works Association 101: 51.
32. Nguyen C, Elfland C, Edwards M (2012) Impact of advanced water conservation features and new copper pipe on rapid chloramine decay and microbial regrowth. Water Res 46: 611–621.
33. Lin YE, Stout JE, Yu VL (2011) Controlling Legionella in hospital drinking water: an evidence-based review of disinfection methods. Infect Control Hosp Epidemiol 32: 166–173.

34. Stout JE, Goetz AM, Yu VL (2011) Hospital Epidemiology and Infection Control; Mayhall CG, editor: Lippincott Williams, & Wilkins.

35. Flannery B, Gelling LB, Vugia DJ, Weintraub JM, Salerno JJ, et al. (2006) Reducing Legionella colonization in water systems with monochloramine. Emerg Infect Dis 12: 588–596.

36. Pryor M, Springthorpe S, Riffard S, Brooks T, Huo Y, et al. (2004) Investigation of opportunistic pathogens in municipal drinking water under different supply and treatment regimes. Water Sci Technol 50: 83–90.

37. Marchesi I, Cencetti S, Marchegiano P, Frezza G, Borella P, et al. (2012) Control of Legionella contamination in a hospital water distribution system by monochloramine. American Journal of Infection Control 40: 279–281.

38. Caporaso JG, Lauber CL, Walters WA, Berg-Lyons D, Huntley J, et al. (2012) Ultra-high-throughput microbial community analysis on the Illumina HiSeq and MiSeq platforms. ISME J 6: 1621–1624.

39. Walters WA, Caporaso JG, Lauber CL, Berg-Lyons D, Fierer N, et al. (2011) PrimerProspector: de novo design and taxonomic analysis of barcoded polymerase chain reaction primers. Bioinformatics 27: 1159–1161.

40. Caporaso JG, Kuczynski J, Stombaugh J, Bittinger K, Bushman FD, et al. (2010) QIIME allows analysis of high-throughput community sequencing data. Nat Methods 7: 335–336.

41. Edgar RC (2010) Search and clustering orders of magnitude faster than BLAST. Bioinformatics.

42. Lozupone C, Knight R (2005) UniFrac: a new phylogenetic method for comparing microbial communities. Appl Environ Microbiol 71: 8228–8235.

43. Morisita M (1959) Measuring of the dispersion of individuals and analysis of the distributional patterns. Memoirs of the Faculty of Science, Kyushu University, Series E (Biology) 2.

44. Horn HS (1966) Measurement of "overlap" in comparative ecological studies. The American Naturalist 100: 419–424.

45. Stout JE, Duda S, Kandiah S, Hannigan J, Yassin M, et al. (2012) Evaluation of a new monochloramine generation system for controlling Legionella in building hot water systems. Association of Water Technologies Annual Convention and Exposition.

46. Kandiah S, Yassin MH, Hariri R, Ferrelli J, Fabrizio M, et al. (2012) Control of Legionella contamination with monochloramine disinfection in a large urban hospital hot water system. Association for Professionals in Infection Control and Epidemiology Annual Conference.

47. Duda S, Kandiah S, Stout JE, Baron JL, Yassin MH, et al. (2013) Monochloramine disinfection of a hospital water system for preventing hospital-acquired Legionnaires' disease: lessons learned from a 1.5 year study. The 8th International Conference on Legionella.

48. Sydnor ER, Bova G, Gimburg A, Cosgrove SE, Perl TM, et al. (2012) Electronic-eye faucets: Legionella species contamination in healthcare settings. Infect Control Hosp Epidemiol 33: 235–240.

49. Yapicioglu H, Gokmen TG, Yildizdas D, Koksal F, Ozlu F, et al. (2012) Pseudomonas aeruginosa infections due to electronic faucets in a neonatal intensive care unit. J Paediatr Child Health 48: 430–434.

50. Revetta RP, Gomez-Alvarez V, Gerke TL, Curioso C, Santo Domingo JW, et al. (2013) Establishment and early succession of bacterial communities in monochloramine-treated drinking water biofilms. FEMS Microbiol Ecol.

51. Chiao TH, Clancy TM, Pinto A, Xi C, Raskin L (2014) Differential resistance of drinking water bacterial populations to monochloramine disinfection. Environ Sci Technol 48: 4038–4047.

52. Duda S, Kandiah S, Stout JE, Baron JL, Yassin MH, et al. (2014) Evaluation of a new monochloramine generation system for controlling Legionella in building hot water systems. Submitted for publication.

53. Jensen N, Whitfield FB (2003) Role of Alicyclobacillus acidoterrestris in the development of a disinfectant taint in shelf-stable fruit juice. Letters in Applied Microbiology 36: 9–14.

54. Revetta RP, Pemberton A, Lamendella R, Iker B, Santo Domingo JW (2010) Identification of bacterial populations in drinking water using 16S rRNA-based sequence analyses. Water Research 44: 1353–1360.

55. Centers for Disease Control and Prevention (2011) Legionellosis–United States, 2000–2009. Morbidity and Mortality Weekly Report: 1083–1086.

56. BSR/ASHRAE (2011) Proposed New Standard 188P, Prevention of Legionellosis Associated with Building Water Systems. Atlanta, GA: American Society of Heating, Refrigerating, and Air-Conditioning Engineers, Inc.

14

Temporal Variations in the Abundance and Composition of Biofilm Communities Colonizing Drinking Water Distribution Pipes

John J. Kelly[1]*, Nicole Minalt[1], Alessandro Culotti[2], Marsha Pryor[3], Aaron Packman[2]

1 Department of Biology, Loyola University Chicago, Chicago, Illinois, United States of America, 2 Department of Civil and Environmental Engineering, Northwestern University, Evanston, Illinois, United States of America, 3 Pinellas County Utilities Laboratory, Largo, Florida, United States of America

Abstract

Pipes that transport drinking water through municipal drinking water distribution systems (DWDS) are challenging habitats for microorganisms. Distribution networks are dark, oligotrophic and contain disinfectants; yet microbes frequently form biofilms attached to interior surfaces of DWDS pipes. Relatively little is known about the species composition and ecology of these biofilms due to challenges associated with sample acquisition from actual DWDS. We report the analysis of biofilms from five pipe samples collected from the same region of a DWDS in Florida, USA, over an 18 month period between February 2011 and August 2012. The bacterial abundance and composition of biofilm communities within the pipes were analyzed by heterotrophic plate counts and tag pyrosequencing of 16S rRNA genes, respectively. Bacterial numbers varied significantly based on sampling date and were positively correlated with water temperature and the concentration of nitrate. However, there was no significant relationship between the concentration of disinfectant in the drinking water (monochloramine) and the abundance of bacteria within the biofilms. Pyrosequencing analysis identified a total of 677 operational taxonomic units (OTUs) (3% distance) within the biofilms but indicated that community diversity was low and varied between sampling dates. Biofilms were dominated by a few taxa, specifically *Methylomonas*, *Acinetobacter*, *Mycobacterium*, and Xanthomonadaceae, and the dominant taxa within the biofilms varied dramatically between sampling times. The drinking water characteristics most strongly correlated with bacterial community composition were concentrations of nitrate, ammonium, total chlorine and monochloramine, as well as alkalinity and hardness. Biofilms from the sampling date with the highest nitrate concentration were the most abundant and diverse and were dominated by *Acinetobacter*.

Editor: Ahmed Moustafa, American University in Cairo, Egypt

Funding: This work was supported by Water Research Foundation grant 4259 to Aaron Packman and John Kelly. The funders had no role in study design, data collection and analysis, decision to publish, or preparation of the manuscript.

Competing Interests: The authors have declared that no competing interests exist.

* E-mail: jkelly7@luc.edu

Introduction

The pipes that are used to transport drinking water through municipal drinking water distribution systems (DWDS) are challenging habitats for microorganisms. The transported water generally contains chemical disinfectants such as chlorine or chloramine, as well as very low concentrations of organic carbon and inorganic nutrients [1]. Despite these challenges, microbes frequently colonize the interior surfaces of DWDS pipes [1]. Indeed the pipe surfaces may represent the best possible microbial habitats within DWDS, as previous research has shown that surface attachment can enable bacteria to grow in oligotrophic habitats due to the accumulation of nutrients at the solid-liquid interface [2,3]. In addition, biofilm formation can provide bacteria with protection against chemical disinfectants [4–6].

Relatively little is known about the species composition and ecology of biofilms within DWDS. Obtaining samples from belowground pipes is difficult and expensive [7], and as a result most of the work that has been done on drinking water biofilms has been based on model systems run in the laboratory [6,8,9]. While these studies have provided valuable insight into biofilm formation

within drinking water, model systems often differ in significant ways from actual DWDS, including duration of biofilm growth, temporal variability, water flow conditions, diversity of pipe materials and the presence or absence of disinfectants. Additionally, most of the work on microbes within DWDS has focused on classical pathogens such as *Vibrio cholerae* and *Salmonella typhi*, emerging pathogens such as *Campylobacter jejuni* and *Legionella pneumophila*, or indicator organisms for fecal contamination, such as coliform bacteria [1]. Many of these studies have used culture-based techniques [1], which are able to assess only a small fraction of natural microbial diversity [10]. In contrast, recent studies using molecular approaches have demonstrated the predominance of nonpathogenic bacterial species within drinking water biofilms [7,11,12].

Information regarding the composition and ecology of biofilms within DWDS is valuable for several reasons. First, it improves our general understanding of microbial life in oligotrophic habitats, including built environments. Secondly, biofilms in DWDS are a concern for public health as they have been shown to harbor and protect pathogens from disinfectants and increase pathogen persistence in DWDS [13]. Thirdly, there is evidence that the activities of nitrifying microorganisms in DWDS can decrease monochloramine concentrations [14], which could lead to increased microbial growth and possibly increased persistence and transport of pathogens [15]. Finally, the presence of biofilms in DWDS can promote pipe corrosion [16] and cause taste and odor problems in the water [3]. Understanding the microbial composition and development of DWDS biofilms can suggest strategies for management of these problems.

We report here the analysis of biofilms found within pipe samples collected five times over an 18 month period from a DWDS in Pinellas County, FL, USA. Sections of below-ground pipe were cut and transported to the lab, and biofilm communities within the pipes were analyzed by heterotrophic plate counts and tag pyrosequencing of 16S rRNA genes.

Materials and Methods

Pipe samples were collected from the municipal drinking water distribution system in Pinellas County, FL, USA, which is operated by Pinellas County Utilities (PCU). The water supply for this system is a blend of groundwater and treated surface water, with desalinated seawater being used periodically as needed. Before entering the DWDS water is treated at the Tampa Bay Water Treatment Plant (TBW) by a four stage process: 1) clarification using the ACTIFLO system (Kruger Inc., Cary, NC), 2) ozone disinfection, 3) biologically active filtration, and 4) disinfection with chlorine. Disinfection within the PCU DWDS is based on maintenance of a chloramine residual, although the utility does switch to chlorine disinfection for a brief period in the summer each year to limit biofilm growth. The switch to chlorine occurred once during our sampling period, specifically between August 1 and September 11, 2011. The average flow through the system during our sampling period was 54.9 million gallons per month.

Sections of six-inch ductile-iron pipe from the main line in the Seminole, FL region within the Pinellas County DWDS were collected periodically over an 18 month period during planned replacement events. Specifically, pipe sections were collected on February 20, 2011, July 20, 2011, December 20, 2011, March 20, 2012, and August 8, 2012. All pipe samples were collected at approximately 9 am, before peak demand which occurs at approximately 10 am. The water main that we sampled was approximately 40–45 years old. The average water age for the main line during our sampling period was 3 to 4 days and the average velocity was 0.4 c.f.s. The sampling location is downstream of a large above ground storage (AGS) tank that was permanently shut down on June 19, 2012 due to concerns about nitrification occurring within the tank. A key outcome of the shut-down of this water tank was decreased water age at our sampling location for the August 2012 sampling date.

On each sampling date the road surface and soil above the water main were excavated using a backhoe. Soil was cleared from around the pipe by hand using a shovel and any soil adhering to the exterior of the pipe was removed using a brush or cloth. The exterior of the pipe was disinfected by pouring a 10% bleach solution over the pipe and simultaneously wiping the pipe exterior with a cloth saturated with 10% bleach solution. A section approximately 1 ft. long was cut from the 6-inch diameter pipe using a scoring-type pipe cutter. The pipe section was capped at one end using a flexible PVC cap with adjustable clamps. The capped pipe section was filled with dechlorinated water from the same water main, which was collected in a plastic carboy and dechlorinated on-site using sodium thiosulfite (~2 g/g Cl_2). The pipe section was filled until overflowing, capped at the other end,

placed in a cooler with ice packs, and shipped overnight to Northwestern University, Evanston, IL. Water was also collected from the main for chemical analysis and transported to the Pinellas County Utilities lab in a cooler.

Sample Processing

In the laboratory one cap was removed from the pipe section and the water was carefully poured out. The interior of the pipe section was rinsed gently with filter-sterilized tap water to remove unattached or settled solids. Biofilm samples were collected from three separate, evenly-spaced sections of the pipe interior. For each section, biofilm material was collected by scraping a 5.5 cm wide band around the entire interior circumference of the pipe with a sterile spatula, resulting in a sampling area of 263.3 cm^2. The collected biofilm material was transferred to a sterile 10 ml vial and 3 ml ultrapure water was added. The suspension was homogenized by vortexing and large particles such as corrosion byproducts were allowed to settle out of suspension. From this suspension 100 μl was used for the plate count assay (see method below) and 2 ml was used for molecular analysis of the biofilm communities (see methods below). The 2 ml for molecular analysis was transferred to a 2 ml microcentrifuge tube and centrifuged at 10,000×g for 10 minutes. The supernatant was then discarded and the remaining biofilm pellet was stored at -20°C.

Water Chemistry

All water chemistry analyses were done by Pinellas County Utilities. Analyses were performed based on either EPA Methods or Standard Methods for the Examination of Water and Wastewater [17]. The specific methods used for each assay and results of the water chemistry assays are listed in Table 1. Total haloacetic acid (HAA) concentrations in the source water were determined periodically by EPA method 552.2.

Plate Count Assay

R2A agar was purchased as a dried powder (Fisher Scientific, Pittsburgh, PA) and prepared according to the manufacturer's instructions. Biofilm suspensions were serially diluted from 10^2 to 10^4 in ultrapure water and 100 μl of all dilutions were spread on R2A agar plates. Plates were incubated at 37°C for 36 hours and counted. Counts were normalized based on the surface area of the pipe from which the biofilm had been collected.

Molecular Analysis of Biofilm Communities

DNA was extracted from the frozen biofilm pellets using the Power Biofilm DNA Kit (MoBio Laboratories, Carlsbad, CA) according to the manufacturer's instructions and stored at −20°C. For tag pyrosequencing of bacterial 16S rRNA genes the extracted DNA was sent to Research and Testing Laboratory (Lubbock, TX). Polymerase chain reaction (PCR) amplification was performed using primers 530F and 1100R [18]. The 530F primer was chosen in order to obtain sequences for the V4 hypervariable region, which has been shown to provide species richness estimates comparable to those obtained with the nearly full-length 16S rRNA gene [19]. Sequencing reactions utilized a Roche 454 FLX instrument (Roche, Indianapolis, IN) with Titanium reagents. Sequences were processed using MOTHUR software [20]. Briefly, any sequences containing ambiguities or homopolymers longer than 8 bases were removed. Remaining sequences were individually trimmed to retain only high quality sequence reads and sequences were aligned based on comparison to the SILVA-compatible bacterial alignment database available within MOTHUR. Aligned sequences were trimmed to a uniform length

Table 1. Water chemistry.

| Analyte | Method | Sampling Date | | | | |
		February 2011	July 2011	December 2011	March 2012	August 2012
Temperature (°C)	SM 2550 B	20.6	29.1	20.2	23.8	28.5
pH	SM 4500 H-B	7.6	7.58	7.65	7.52	7.69
Total Organic Carbon (mg L^{-1})	SM 5310-C	1.8	1.9	2.2	2	2.3
Total Phosphorous as P (mg L^{-1})	EPA 365.4	0.35	0.24	0.41	0.34	0.31
Nitrate as N (mg L^{-1})	EPA 300.0	0.08	0.11	0.09	0.08	0.43
Nitrite as N (mg L^{-1})[1]	EPA 300.0	BD	BD	BD	BD	BD
Free Ammonia as N (mg L^{-1})	SM 4500 NH3-F	0.19	0.42	0.21	0.16	0.14
Total chlorine (mg L^{-1})	SM 4500 CL-G	2.9	1.6	2.5	3	3
Monochloramine (mg L^{-1})	SM 4500 CL-G	3.2	1.6	2.2	2.6	2.7
Alkalinity as CaCO$_3$ (mg L^{-1})	SM 2320 B	170	190	170	180	150
Calcium Hardness (mg L^{-1})	SM 2340 B	210	202	202	205	192
Specific Conductance (umhos cm^{-1})	SM 2510 B	549	521	429	447	477
Aluminum (mg L^{-1})[2]	EPA 200.7-DW	BD	BD	BD	BD	BD
Calcium (mg L^{-1})	EPA 200.7-DW	84.3	80.7	81	81.9	65.1
Iron (mg L^{-1})	EPA 200.7-DW	0.055	0.079	0.381	0.046	0.014
Magnesium (mg L^{-1})	EPA 200.7-DW	7.35	6.51	6.39	6.53	7.14

BD = below detection limit.
[1] detection limit 0.02 mg L-1.
[2] detection limit 0.015 mg L-1.

of 127 bases and chimeric sequences were removed using UCHIME [21] run within MOTHUR. Sequences were grouped into phylotypes by comparison to the SILVA-compatible bacterial alignment database and algal chloroplast and mitochondrial sequences were removed from the data set. To avoid any biases associated with different numbers of sequences in each of the samples we randomly subsampled a total of 6,808 sequences from each sample, and used these subsampled sequences for all downstream analyses. Sequences were clustered into operational taxonomic units (OTUs) based on 97% sequence identity using the average neighbor algorithm. Rarefaction curves were produced using MOTHUR. The total OTU richness in each sample was calculated based on the Chao1 richness estimator [22]. The diversity of each sample was assessed based on the Shannon index [23] calculated using the Primer software package (Primer V.5, Primer-E Ltd., Plymouth, UK). The community composition of the individual samples was compared by using MOTHUR to calculate distances between sites based on the theta index [24]. The significance of differences in theta index scores between sites was assessed by analysis of molecular variance (AMOVA) run within MOTHUR. PC-ORD v. 6.08 (MjM Software, Gleneden Beach, Oregon, USA) was used to ordinate the theta index distance matrix via non-metric multidimensional scaling (nMDS) and to determine correlations between the water chemistry data and the axes in the nMDS ordination.

Statistical Analyses

Plate count data and diversity scores were analyzed by one-way analysis of variance (ANOVA) based on sampling date and pairwise comparisons were made by Tukey's post hoc test. Correlations were assessed by determining Pearson product-moment correlation coefficients and Bonferroni-corrected probabilities. Correlation between the relative abundance of *Methylomo-*

nas sequences and the concentration of total HAA in the source water was based on average abundance data for each biofilm sampling date and the HAA concentration from the closest source water sampling date. All statistical analyses were run using Systat 13 (Systat Software, Inc., San Jose, CA) and p values less than 0.05 were considered to be significant.

Data Sharing

All of the sequence data analyzed in this paper can be downloaded from the National Center for Biotechnology Information (NCBI) Sequence Read Archive (SRA) with accession number SRP038002.

Results

Water Source

The source water for the PCU DWDS is a blend of groundwater, surface water, and desalinated seawater. An approximately equal mix of groundwater and surface water was the most common during the study period, although there were some significant variations in the relative proportions of source waters (Fig. 1). For the four weeks preceding our February 2011 sampling date, the source water averaged 38% Groundwater, 48% surface water and 14% seawater. Between June 20, 2011 and March 20, 2012, a period which included three of our sampling dates, the source water averaged 50% Groundwater, 50% surface water and 0% seawater. Prior to our August 2012 sampling date there were some dramatic shifts in source waters. Between April 14, 2012 and June 9, 2012 the water was predominantly groundwater, with an average of 73% of the water coming from groundwater over that period. Finally, between July 6, 2012 and August 8, 2012 the water was predominantly surface water, with

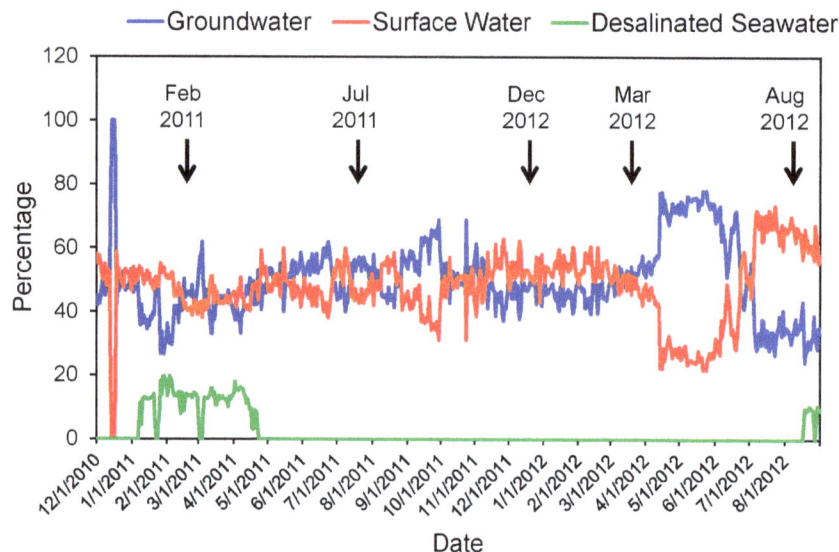

Figure 1. Relative percentages of source waters within PCU drinking water distribution system. Biofilm sampling dates are indicated by black arrows.

surface water representing an average of 67% of the source water over that period.

Water Chemistry

Water temperature varied seasonally in the water main from which the biofilm samples were obtained, being higher in summer months (July and August) than winter months (February, March and December) (Table 1). Some water chemistry parameters varied with sampling date but did not show seasonal trends. For example, there were large fluctuations in concentrations of nitrate (from 0.08 to 0.43 mg L^{-1}), free ammonia (from 0.14 to 0.42 mg L^{-1}) and iron (from 0.014 to 0.381 mg L^{-1}). The August 2012 sampling date had the highest concentration of nitrate, approximately four to five times higher than all other sampling dates, and the lowest concentration of free ammonia (Table 1). Analysis of the source water from TBW confirmed a high concentration of nitrate in the source water immediately prior to the August 2012 sampling date (data not shown). Analysis of the source water from TBW also showed fluctuations in concentrations of total haloacetic acids over the course of the study (Fig 2).

Plate Count Assay

Sampling date significantly affected the numbers of bacteria within the pipe biofilms as measured by heterotrophic plate count assay, with bacterial counts varying over 3 orders of magnitude between sampling dates (p<0.001; Table 2). Samples from the summer months (July and August) had significantly higher counts than the winter months (February, March and December), and there was a significant positive correlation between plate counts and water temperature ($R^2 = 0.605$; p<0.001). The August 2012 sampling date, which had the highest nitrate concentration, had bacterial counts that were significantly higher than all other sampling dates (p<0.001; Table 2), and there was a significant positive correlation between plate counts and nitrate concentration ($R^2 = 0.799$; p<0.001). There was no significant correlation between plate counts and phosphorous concentration ($R^2 = 0.249$; p = 0.058). There were also no significant correlations between plate counts and total chlorine ($R^2 = 0.005$, p = 0.802) or

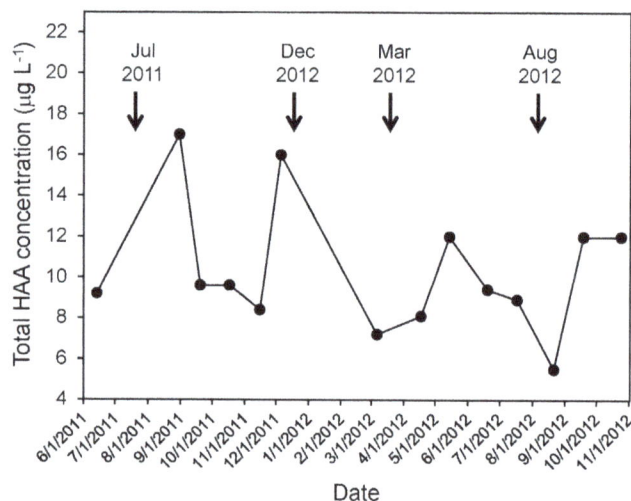

Figure 2. Concentrations of total haloacteic acids (HAA) in the source water for the PCU drinking water distribution system. Biofilm sampling dates are indicated by black arrows.

between plate counts and monochloramine ($R^2 = 0.285$, p = 0.548).

Bacterial Community Analysis

Tag pyrosequencing of 16S rRNA genes was used to profile the bacteria within the biofilms lining the drinking water pipes. Twelve samples representing three replicate biofilm samples from each of four sampling dates (July 2011, December 2011, March 2012 and August 2012) were sequenced successfully. Despite repeated attempts, DNA from the February 2011 pipe samples could not be amplified with the 530F and 1100R primers, so no sequence data were obtained for this sampling date. After processing, the data set included a total of 159,604 high-quality sequence reads. There was significant variation in the number of sequences obtained for each of the samples, from a low of 8,785 to

Table 2. Numbers of heterotrophic bacteria in pipe biofilms based on plate count assay.

Sampling Date	Number of Bacteria (cfu cm^{-2})[1]	
February 2011	215	a
July 2011	10,887	b
December 2011	2,013	a
March 2012	36	a
August 2012	23,167	c

[1]Data points represent mean values (n = 3) and data points followed by different letters are significantly different based on ANOVA and Tukey's HSD posthoc test (p< 0.05).

a high of 35,854. To avoid biases associated with unequal numbers of sequences, 6,808 sequences were randomly selected from each of the twelve samples using the subsample command in MOTHUR, producing a total of 81,696 high quality sequences that were used for all analyses of community composition. With this subsampled data set, rarefaction curves for all samples had reached plateaus (Fig. 3), suggesting that the sequencing depth obtained in this study was adequate to capture most of the diversity within these communities. Similarly, a comparison of the total number of OTUs observed in each sample and the estimated total number of OTUs present in each sample demonstrated that for all samples more than 50% of the estimated total number of OTUs in each sample were detected (Table 3).

nMDS ordination (Fig. 4) and AMOVA analysis (Table 4) indicated that there were significant differences between the biofilm bacterial communities from the different sampling dates. Biofilm bacterial communities from August 2012 were the most distinct and were significantly different from the communities from all other sampling dates (Fig. 4 and Table 4). The nMDS ordination also revealed relationships between the community composition and water chemistry parameters (Fig 4). The composition of the biofilm bacterial community from August 2012 was positively correlated with nitrate concentration and negatively correlated with calcium concentration and hardness. The August 2012 communities were also positively correlated with

pH and total organic carbon concentration and negatively correlated with alkalinity. In addition, the separation of the biofilm communities from July 2011 and March 2012 on the nMDS ordination was correlated with the concentrations of free ammonia, total chlorine and monochloramine (Fig. 4). Finally, the nMDS ordination demonstrated that there was variation in bacterial community composition between replicates from both the July and March sampling dates, whereas the December and August samples showed a high degree of similarity between replicates (Fig. 4).

Bacterial Community Diversity

Pyrosequencing analysis identified a total of 677 OTUs (3% distance) within these biofilms, and the number of OTUs per sample ranged from 21 to 199 (Table 3). Despite this large number of OTUs, the diversity of these communities was low, with three of the four sampling dates showing Shannon index scores below 1.2 (Fig 5A). The Shannon diversity index scores for the biofilm bacterial communities varied significantly between sampling dates (p<0.01) with the biofilms from August 2012 being the most diverse (Fig. 5A). There was also a significant positive correlation (p<0.01) between bacterial abundance in the biofilms and the diversity of the bacterial communities (Fig 5B). As indicated by the low diversity index scores, all of the biofilm communities were dominated by a small number of OTUs, with the ten most

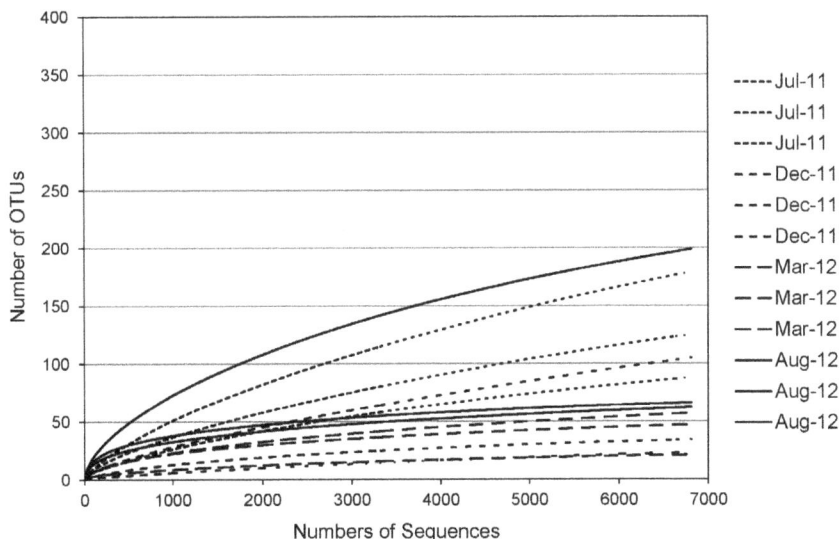

Figure 3. Rarefaction curves for biofilm bacterial communities based on tag pyrosequencing of 16S rRNA genes. OTUs were defined based on 3% distance.

Table 3. Comparison of the number of observed and estimated bacterial OTUs in pipe biofilm communities based on 16S tag pyrosequencing data.

Sampling Date	Observed OTUs[1]	Estimated OTUs[2]	Percent Coverage[3]
July 2011	88	158	55.7%
July 2011	179	332	53.9%
July 2011	125	239	52.3%
December 2011	34	42	81.0%
December 2011	105	194	54.1%
December 2011	23	38	60.5%
March 2012	47	67	70.1%
March 2012	57	87	65.5%
March 2012	21	26	80.8%
August 2012	199	297	67.0%
August 2012	66	74	89.2%
August 2012	62	75	82.7%

[1]OTUs were defined based on 3% distance.
[2]Total OTUs per sample were estimated based on Chao1 richness estimator.
[3]Percent coverage was calculated by dividing the number of observed OTUs by the number of estimated OTUs.

abundant OTUs accounting for 93% of the total sequences in the data set.

Bacterial Community Taxonomic Composition

Analysis of pyrosequencing data indicated that sequences corresponding to the genus *Methylomonas* were the most abundant within the biofilm communities, accounting for 41% of the sequences in the total data set (Table 5). Other abundant sequences corresponded to the genera *Acinetobacter*, *Mycobacterium*, *Pseudomonas*, and *Methylobacterium*, as well as an unclassified genus from the family Xanthomonadaceae and an unclassified genus from the class Betaproteobacteria. The relative abundance data

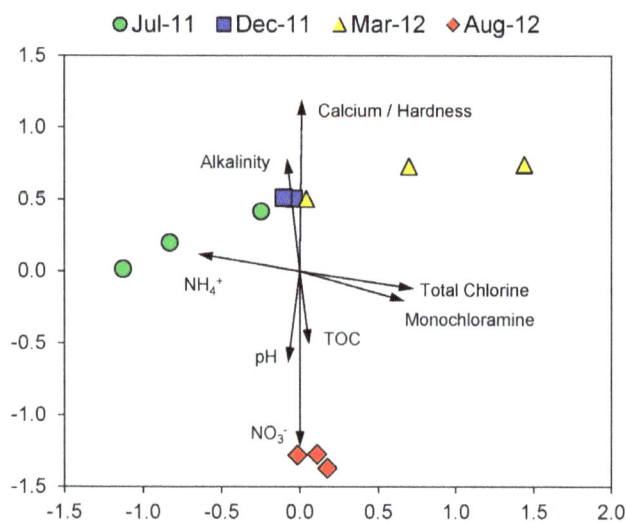

Figure 4. Non-metric multidimensional scaling ordination of biofilm bacterial communities based on tag pyrosequencing of 16S rRNA genes. Stress value of ordination is 0.129. Vector lines represent correlations between physical and chemical variables and the ordination axes. Variables with correlation values less than 0.5 for both axes are not shown.

indicated some commonalities between the biofilm bacterial communities from the four sampling dates (Table 5). For example, *Methylomonas* accounted for more than 30% of the sequences for three of the four sampling dates. However, the data also illustrate that there were dramatic differences in the composition of biofilm communities from the different sampling dates, as *Methylomonas* accounted for more than 95% of the sequences in December 2011, but less than 2% in August 2012. A previous study suggested a link between *Methylomonas* bacteria in drinking water and HAA [25], and there was a significant positive correlation between the relative abundance of *Methylomonas* sequences in our biofilms and the concentration of total HAA in the source water ($R^2 = 0.962$; $p = 0.027$). *Acinetobacter* abundance also varied significantly between sampling dates, accounting for 74% of the sequences in August 2012 biofilms, but representing less than 0.1% of the sequences from the other sampling dates.

Sequences corresponding to several genera containing pathogenic species were detected in the biofilms. In a total of 81,696 sequences from all four sampling dates, there were 18 *Escherichia* sequences, 5 *Clostridium* sequences and 3 *Streptococcus* sequences detected, so these genera were extremely rare. *Mycobacterium* represented 59% of the sequences in July 2011, while representing less than 1% of sequences from the other sampling dates. The genus *Mycobacterium* includes two well-known human pathogens, *M. tuberculosis* and *M. leprae*, although these species are generally not found in the environment [26]. The genus *Mycobacterium* also includes a large number of non-pathogenic or occasionally pathogenic species [26]. *Acinetobacter* accounted for 74% of the sequences in August 2012 but less than 1% of sequences from the other sampling dates. Bacteria from the genus *Acinetobacter* are a common cause of nosocomial infections among immunocompromised patients, with the most common example being respiratory infections of ventilated patients [27]. Due to the short length of the sequences obtained in our study, we were unable to discriminate any of the sequences down to the species level, so it is unclear whether the sequences from any of these genera represented pathogenic or non-pathogenic organisms.

The pipes analyzed in this study were ductile iron and did not show significant corrosion or the presence of tubercles. The

Table 4. AMOVA analysis of 16S tag pyrosequencing data from pipe samples.

Comparison	p value
August-July	<0.001
August-December	<0.001
August-March	0.0497
July-December	<0.001
July-March	0.0500
December-March	0.0487

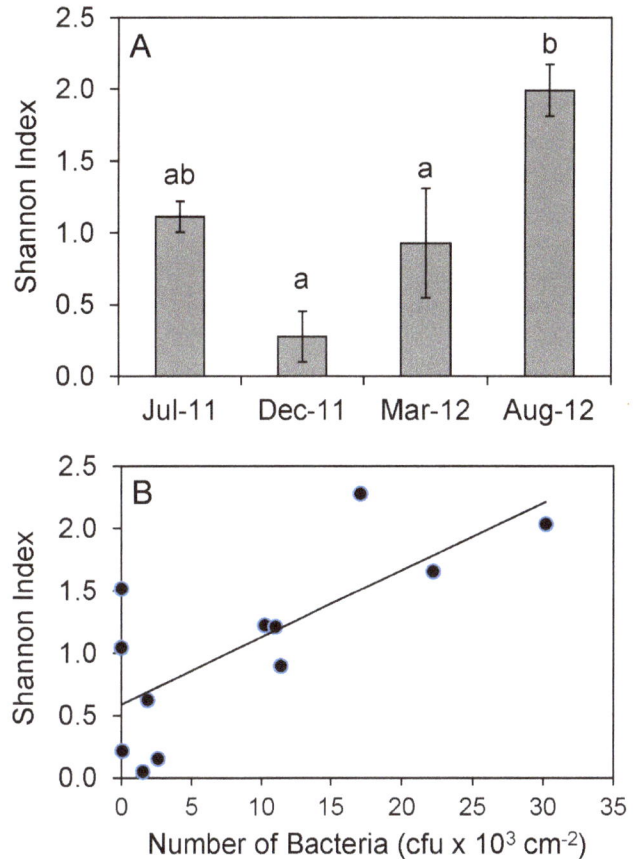

Figure 5. Diversity of bacterial biofilm communities and relationship between bacterial abundance and diversity. (A) Shannon index scores calculated using 16S rRNA gene tag pyrosequencing data. Each data point represents mean value (n = 3) with standard error bars. ANOVA indicated a significant effect of sampling date (p<0.01). Data points with different letters are significantly different based on Tukey's posthoc test (p<0.05). (B) Relationship between biofilm bacterial abundance and diversity. Linear regression indicated $R2 = 0.541$. Pearson correlation analysis indicated a significant correlation between numbers of bacteria and diversity concentration and resistance (p<0.01).

pyrosequencing data identified only a handful of sequences corresponding to genera known to contain iron-oxidizing species: *Acidovorax* (1 sequence), *Aquabacterium* (1 sequence) and *Thiobacillus* (4 sequences) [28,29]. In addition, no sequences corresponding to any known ammonia oxidizing bacterial genera were detected. However, a few sequences (19 total) from a known nitrite oxidizing genus, *Nitrospira*, were detected, and all of these sequences were found in the July 2011 samples, which also showed the highest concentration of free ammonia (Table 1).

Discussion

Pipe samples were collected from the same region of a drinking water distribution system in Pinellas County, FL on five dates over an 18-month period between February 2011 and August 2012. Water from these pipes showed seasonal variations in temperature and some large fluctuations in concentrations of nitrate, free ammonia and iron. The August 2012 sample had a much higher concentration of nitrate and a lower concentration of free ammonia relative to other sampling dates. The shut-down of the upstream above ground storage tank prior to the August 2012 sampling date probably contributed to the observed differences in water chemistry between the sampling dates. The above ground storage tank was shut down because of nitrification occurring in the tank, which could have contributed to the high nitrate and low free ammonia concentrations in the August 2012 sample. The August 2012 samples also had a different source-water mixture than other dates: for one month prior to our sampling date the source water was composed of a higher percentage of surface water and a lower percentage of groundwater than typical for this system. Analysis of the TBW source water at its point of entry to the DWDS confirmed the high nitrate and low alkalinity of the source water at the time of our August 2012 sampling, indicating that source water was a key driver of the unique aspects of the August 2012 water chemistry.

The abundance of bacteria in the biofilms varied greatly across the sampling dates, with higher bacterial numbers in the summer months and lower bacterial numbers in the winter months. These differences in abundance may have been driven by the strong seasonal differences in temperature in the system, as the water in the summer months was on average 7.3°C warmer than in the winter months, and a temperature change of this magnitude can significantly increase the growth rates of mesophilic bacteria [30]. The connection between bacterial abundance in the biofilms and water temperature was further supported by a statistically significant correlation between these two parameters. The differences in bacterial abundance may also have been related to the availability of inorganic nutrients in the drinking water, specifically nitrogen. We found a significant correlation between

bacterial cell numbers and nitrate concentrations. Other studies have indicated that inorganic nutrients can be a limiting factor for bacterial growth in DWDS [1]. In our samples, nitrate concentrations were higher in the summer months than in the winter months and were highest in the August 2012 sample. Finally, our results indicated no significant relationship between the abundance of bacteria within the biofilms and the concentrations of total chlorine or monochloramine. These results suggest that low concentrations of chlorine disinfectants may not be effective at limiting biofilm growth within DWDS, owing to the protection provided by the biofilm matrix, as has been demonstrated by previous studies [4,5,6]. It should be noted that we assessed bacterial abundance within the pipe biofilms using heterotrophic plate counts. While the limitations of plate counts as estimates of bacterial abundance are well known [31], this method is commonly used to assess bacterial loads in DWDS, so we chose to use this method to make our results comparable to existing data.

Tag pyrosequencing analysis revealed a total of 677 OTUs in the biofilm bacterial communities, with the estimated total numbers of bacterial OTUs per sample ranging from 26 to 332.

Table 5. Relative abundance of most numerically dominant bacterial genera[1].

Bacterial genus	All Samples	July 2011	December 2011	March 2012	August 2012
Methylomonas	40.8	33.6±15.6	95.5±3.5	32.4±21.3	1.3±0.6
Acinetobacter	18.5	0.0±0.0	0.0±0.0	0.0±0.0	74.2±2.4
Mycobacterium	14.7	59.1±14.3	0.2±0.2	0.1±0.1	0.0±0.0
Unclass. Xanthomonadaceae	13.5	0.0±0.0	2.6±2.4	50.4±24.4	0.1±0.1
Pseudomonas	4.7	0.6±0.3	0.3±0.2	15.5±8.6	2.1±0.2
Unclass. Betaproteobacteria	2.3	0.3±0.1	0.1±0.1	0.0±0.0	9.0±1.6
Methylobacterium	1.2	3.7±2.3	0.2±0.2	0.1±0.0	0.8±0.1
Unclass. Bacteria	1.0	0.5±0.2	0.3±0.1	0.0±0.0	3.2±2.8
Massilia	0.9	0.0±0.0	0.0±0.0	0.0±0.0	3.4±0.7
Unclass. Gammaproteobacteria	0.4	0.7±0.4	0.2±0.2	0.7±0.5	0.0±0.0

[1]Values for each sampling period represent mean values (n = 3) ± standard error.
doi:10.1371/journal.pone.0098542.t005

A previous pyrosequencing survey of drinking water meters revealed a similarly high number of total bacterial OTUs [7]. Despite the large numbers of OTUs observed in our pipe biofilms, these communities were not very diverse, with all samples having Shannon index scores below 2. These low diversity scores reflect the fact that these communities were dominated by a small number of taxa, specifically *Methylomonas*, *Acinetobacter* and *Mycobacterium*. The diversity of the bacterial biofilm communities was significantly correlated with bacterial abundance, suggesting that when conditions within the DWDS were favorable for bacterial growth (e.g. higher temperature and higher nitrate concentrations) a wider range of bacterial taxa were able to proliferate within the biofilms, whereas when conditions were not favorable (e.g. lower temperature and lower nitrate concentrations), a more limited range of bacteria were able to persist. Within our data set the August 2012 samples seemed to represent the most favorable conditions for biofilm growth, as this sampling date had the highest nitrate concentration, one of the highest water temperatures, and supported biofilms with the highest bacterial numbers and the highest bacterial diversity.

Methylomonas was the most numerically dominant bacterial genus within the biofilms, with its sequences accounting for more than 40% of all of the sequences recovered. *Methylomonas* is a genus of type I methanotrophic bacteria, which obtain their carbon and energy from the oxidation of methane. *Methylomonas* sequences have been detected previously in drinking water [32], and a recent study of water meter biofilms also detected sequences corresponding to the family *Methylococcaceae*, the bacterial family that includes the genus *Methylomonas*, [7]. *Methylobacterium*, a genus of methylotrophic bacteria that oxidize methyl compounds such as methanol but cannot metabolize methane, was also one of the most commonly detected taxa in our biofilms, although it represented just over 1% of the total sequences. Methylotrophic bacteria, specifically *Methylophilus*, were also detected in biofilms within drinking water meters [7]. The factors favoring high abundance of methanotrophic and methylotrophic taxa within drinking water distribution systems are unclear. We would not expect high concentrations of methane or methanol in drinking water, however these compounds could be produced in anoxic sites within DWDS via anaerobic processes such as methanogenesis or fermentation. Another process that might have supported the growth of methylotrophic bacteria is utilization of haloacetic acid, which is a common by-product of chlorination of drinking water. A recent study isolated a *Methylobacterium* strain from a DWDS

biofilm that was capable of growth with haloacetic acid as the sole carbon source [25]. The results of our study, which showed a significant correlation between the relative abundance of *Methylomonas* sequences and HAA concentration in the source water, lends further support to this hypothesis.

Our results demonstrated that there was significant variation in the taxonomic composition of the biofilm bacterial communities within the pipe sections across our sampling dates. nMDS and AMOVA analyses indicated that the August 2012 bacterial communities were the most distinct in terms of their composition, and further examination of the composition of these biofilms revealed that the August 2012 communities were dominated by *Acinetobacter* sequences (74% of total sequences), while *Acinetobacter* sequences represented less than 0.1% of sequences from the July, December and March samples. *Acinetobacter* is a genus of Gram-negative, heterotrophic bacteria [33] that is commonly found in soils and groundwater [34–36]. *Acinetobacter* is also one of the most common groups of bacteria isolated from drinking water [1], and a number of *Acinetobacter* species have been shown to produce biofilms [37–39]. Therefore the presence of *Acinetobacter* in the pipe biofilms was not surprising. However, the dramatic variation in *Acinetobacter* abundance that we observed between August 2012 and the other sampling dates was remarkable. There were several unique features of the August 2012 sampling date that may have contributed to its distinct biofilm composition. First, the source of the drinking water within the distribution system changed prior to August 2012 to predominantly groundwater for a period of three weeks, with this period of groundwater dominance occurring approximately two months prior to the August 2012 sampling. Since *Acinetobacter* are regularly detected in soil and groundwater, this switch to a groundwater dominated system prior to August 2012 might have provided an additional inoculum of *Acinetobacter* that were able to become established within the pipe biofilms. Another related feature of the August 2012 samples was that the drinking water at that sampling time had a much higher nitrate concentration (at least four times higher) than all of the other sampling dates. The nMDS analysis indicated that nitrate concentration was one of the main drivers of the composition of the bacterial communities within the August 2012 biofilms. The August 2012 biofilms also showed bacterial counts that were more than two times higher than any of the other sampling dates, suggesting more biofilm mass which could have generated more anoxic microsites. Although bacteria from the genus *Acinetobacter* are generally aerobes, there are some species within the genus that can utilize nitrate as an electron acceptor when oxygen is not present [40]. In contrast, *Methylomonas*, which was one of the predominant taxa in July 2011, December 2011 and March 2012 but was less than 2% of total sequences in August 2012, is strictly aerobic [41]. Therefore, the higher drinking water nitrate concentration and possibly more anoxic microsites caused by higher bacterial biofilm growth on the August 2012 sampling date may have provided *Acinetobacter* with a competitive advantage over *Methylomonas*.

Mycobacterium abundance also varied considerably over time, as this genus represented 59% of the sequences from the July 2011 sampling date but less than 1% for all other sampling dates. *Mycobacteria* are frequently detected in DWDS and are considered a significant public health issue [42]. The genus *Mycobacterium* consists of approximately 100 species, including a large number of species that are either non-pathogenic or pathogenic under certain situations [26]. For example, nontuberculosis *Mycobacterium* are a major cause of opportunistic infections in immunocompromised hosts [42]. Mycobacteria have been detected previously in this DWDS via culturing, and *M. gordonae* and *M. intracellulare* were the most frequently detected *Mycobacterium* species [43]. *M. gordonae* is among the most frequently reported mycobacteria in drinking water and in DWDS [26] and is generally considered non-pathogenic [44]. *M. intracellulare*, which is part of the *Mycobacterium avium* complex (MAC), has also been detected in DWDS [26]. MAC is the group of non-tuberculosis *Mycobacterium* most commonly associated with human disease, causing primarily pulmonary infections in individuals who are immunocompromised [45]. In this study, we were unable to discriminate the *Mycobacterium* sequences down to the species level, so it is unclear whether the sequences we detected represented potentially pathogenic species.

Several characteristics of *Mycobacteria* enhance their survival in DWDS, including their ability to grow under oligotrophic conditions [46], form biofilms and resist chlorine disinfection [26]. Several recent studies have detected species related to *Mycobacterium* in chlorinated drinking water [47–49]. Previous work at the DWDS considered here indicated that the frequency of detection of *Mycobacteria* increased when the disinfectant was switched from chlorine to chloramine in 2002 [43], suggesting that *Mycobacteria* might be less sensitive to chloramine than chlorine. However, a recent study using a model DWDS observed that the relative sensitivity of *Mycobacterium avium* biofilms to chlorine and monochloramine depended on the pipe material [50]. Specifically, *M. avium* was more sensitive to chlorine than chloramine when biofilms were grown on copper pipe, but the reverse was true for *M. avium* biofilms on iron [50], possibly due to corrosion products interfering with free chlorine [51]. Here we found that *Mycobacterium* sequences were most abundant in biofilms from the July 2011 sampling date, which also had the highest level of free ammonia and the lowest levels of total chlorine and monochloramine. These constituents are related, as monochloramine is reductively dehalogenated to ammonia, and monochloramine is the major component of total chlorine in this system. The results of our nMDS analysis indicate that the monochloramine concentration strongly influenced bacterial community composition for the July 2011 sampling date. Therefore, the fact that *Mycobacterium* was predominant only on the sampling date with the lowest level of monochloramine suggests that within the ductile iron pipes in this DWDS monochloramine significantly reduced the abundance of *Mycobacterium* within the biofilms.

An unclassified genus from the family Xanthomonadaceae also varied considerably in abundance across the sampling dates, representing 50% of the sequences from the March 2012 samples but only a small fraction of the sequences from the other sampling dates. The Xanthomonadaceae are obligate aerobic chemoorganotrophs, and this family includes some well-known plant pathogens [33]. Organisms from the family Xanthomonadaceae family were not detected in a previous pyrosequencing survey of drinking water biofilms [7], but these organisms have been isolated from drinking water and from drinking water pipe biofilms using culture based techniques [52]. In this study, we were unable to discriminate the Xanthomonadaceae sequences down to the genus or species level, and the reason for their high abundance in the March 2012 samples is unclear.

The biofilm communities from July 2011 and March 2012 showed much higher variation in composition among replicates than did biofilm communities from December 2011 and August 2012. The December and August samples were each dominated by a single bacterial genus (*Methylomonas* in December and *Acinetobacter* in August) with very little variation between replicates. These data suggest that the environmental conditions in December and August each favored one specific bacterial genus that dominated all of the biofilms on that sampling date. In contrast, biofilms from July and March had several dominant bacterial genera that showed high variations between the replicates, suggesting that conditions during those months produced greater variability by enabling several genera to compete for dominance within the biofilms.

Nitrification is a significant concern in drinking water distribution systems that use chloramine as the secondary residual, as nitrification can lead to a decrease in the chloramine residual, an increase in the growth of heterotrophic bacteria and an increase in concentrations of nitrate and nitrite, which pose risks to human health [53]. Multiple studies have identified nitrifying bacteria in DWDS [53–55]. No sequences corresponding to any known ammonia oxidizing bacterial genera were detected in the biofilms analyzed in this study. However, a few sequences from a known nitrite oxidizing genus, *Nitrospira*, were detected in the July 2011 samples, but not in samples from any of the other sampling dates. July 2011 also showed the highest concentration of free ammonia (at least two times higher than all other sampling dates) and unpublished data from PCU confirm that there was a peak in nitrification activity in June and July of 2011. The nMDS analysis indicated that ammonia concentration was a significant driver of the composition of the biofilm bacterial communities from July 2011, and previous studies have indicated that the presence of free ammonia is the principal cause of nitrification in DWDS [56]. Therefore, these data suggest that the high free ammonia concentration combined with the high temperature in July 2011 enabled nitrification to occur within the pipes; however, the lack of detection of ammonia oxidizing bacteria suggests that ammonia oxidation within the biofilms may have been driven by ammonia oxidizing archaea (AOA). Previous studies have detected AOA in drinking water distribution systems [54], but archaea would not have been detected by the bacterial primers used in the current study.

In summary, the results of our study demonstrate that the biofilms within the DWDS pipes were dominated by a few bacterial taxa, specifically *Methylomonas*, *Acinetobacter* and *Mycobacterium*, and that the dominant taxa within the biofilms varied dramatically between sampling times. It is likely that these differences in dominant taxa were driven by differences in environmental conditions, and our analysis suggests that nitrate, ammonium, total chlorine, and monochloramine concentrations were key drivers of biofilm bacterial community composition.

Another possibility is that these differences in dominant taxa could have been the result of the founder effect, which stipulates that the founding member of a biofilm will have an advantage over subsequent colonizers and will remain dominant. The founder effect has been suggested as a possible driver of biofilm community composition in a variety of habitats [57–60] and could have been a contributing factor to the differences in dominant community members in our biofilms. Further experimental work, which is ongoing in our lab, will be needed to explore the relative contributions of environmental factors and founder effects on the composition of biofilms within drinking water distribution systems.

Acknowledgments

The authors thank Sharon Waller for coordinating the pipe sampling effort and Kesha Baxi for assistance with data analysis at the start of the project. The authors thank Tim LaPara for his helpful comments on an earlier version of this manuscript.

Author Contributions

Conceived and designed the experiments: JK MP AP. Performed the experiments: NM AC. Analyzed the data: JK NM AC. Contributed reagents/materials/analysis tools: JK MP AP. Wrote the paper: JK NM AC MP AP.

References

1. Szewzyk U, Szewzyk R, Manz W, Schleifer K (2000) Microbiological safety of drinking water. Annual Reviews in Microbiology 54: 81–127.
2. Marshall KC (1988) Adhesion and growth of bacteria at surfaces in oligotrophic habitats. Can J Microbiol 34: 503–506.
3. Zacheus OM, Lehtola MJ, Korhonen LK, Martikainen PJ (2001) Soft deposits, the key site for microbial growth in drinking water distribution networks. Water Res 35: 1757–1765.
4. LeChevallier MW, Cawthon CD, Lee RG (1988) Factors promoting survival of bacteria in chlorinated water supplies. Appl Environ Microbiol 54: 649–654.
5. Ridgway H, Olson B (1982) Chlorine resistance patterns of bacteria from two drinking water distribution systems. Appl Environ Microbiol 44: 972–987.
6. Liu Y, Zhang W, Sileika T, Warta R, Cianciotto NP, et al. (2011) Disinfection of bacterial biofilms in pilot-scale cooling tower systems. Biofouling 27: 393–402.
7. Hong P, Hwang C, Ling F, Andersen GL, LeChevallier MW, et al. (2010) Pyrosequencing analysis of bacterial biofilm communities in water meters of a drinking water distribution system. Appl Environ Microbiol 76: 5631–5635.
8. Bois FY, Fahmy T, Block J, Gatel D (1997) Dynamic modeling of bacteria in a pilot drinking-water distribution system. Water Res 31: 3146–3156.
9. Martiny AC, Jørgensen TM, Albrechtsen H, Arvin E, Molin S (2003) Long-term succession of structure and diversity of a biofilm formed in a model drinking water distribution system. Appl Environ Microbiol 69: 6899–6907.
10. Amann RI, Ludwig W, Schleifer K (1995) Phylogenetic identification and in situ detection of individual microbial cells without cultivation. Microbiol Rev 59: 143–169.
11. Kalmbach S, Manz W, Szewzyk U (1997) Isolation of new bacterial species from drinking water biofilms and proof of their in situ dominance with highly specific 16S rRNA probes. Appl Environ Microbiol 63: 4164–4170.
12. Schmeisser C, Stöckigt C, Raasch C, Wingender J, Timmis K, et al. (2003) Metagenome survey of biofilms in drinking-water networks. Appl Environ Microbiol 69: 7298–7309.
13. Parsek MR, Singh PK (2003) Bacterial biofilms: An emerging link to disease pathogenesis. Annual Reviews in Microbiology 57: 677–701.
14. Hoefel D, Monis PT, Grooby WL, Andrews S, Saint CP (2005) Culture-independent techniques for rapid detection of bacteria associated with loss of chloramine residual in a drinking water system. Appl Environ Microbiol 71: 6479–6488.
15. Eichler S, Christen R, Höltje C, Westphal P, Bötel J, et al. (2006) Composition and dynamics of bacterial communities of a drinking water supply system as assessed by RNA-and DNA-based 16S rRNA gene fingerprinting. Appl Environ Microbiol 72: 1858–1872.
16. Marshall K, Blainey BL (1991) Role of bacterial adhesion in biofilm formation and biocorrosion. In: Anonymous Biofouling and Biocorrosion in Industrial Water Systems: Springer. pp. 29–46.
17. Rice EW, Baird RB, Eaton AD, Clesceri LS, editors (2012) Standard methods for the examination of water and wastewater, 22nd ed. Denver, CO: American Water Works Association.
18. Lane DJ (1991) 16S/23S rRNA sequencing. In: Stackebrandt E, Goodfellow M, editors. Nucleic acid techniques in bacterial systematics. Chichester, UK: John Wiley.
19. Youssef N, Sheik CS, Krumholz LR, Najar FZ, Roe BA, et al. (2009) Comparison of species richness estimates obtained using nearly complete fragments and simulated pyrosequencing-generated fragments in 16S rRNA gene-based environmental surveys. Appl Environ Microbiol 75: 5227–5236.
20. Schloss PD, Westcott SL, Ryabin T, Hall JR, Hartmann M, et al. (2009) Introducing mothur: Open-source, platform-independent, community-supported software for describing and comparing microbial communities. Appl Environ Microbiol 75: 7537–7541.
21. Edgar RC, Haas BJ, Clemente JC, Quince C, Knight R (2011) UCHIME improves sensitivity and speed of chimera detection. Bioinformatics 27: 2194–2200.
22. Chao A (1984) Nonparametric estimation of the number of classes in a population. Scandinavian Journal of Statistics 11: 265–270.
23. Shannon CE (2001) A mathematical theory of communication. ACM SIGMOBILE Mobile Computing and Communications Review 5: 3–55.
24. Yue JC, Clayton MK (2005) A similarity measure based on species proportions. Communications in Statistics-Theory and Methods 34: 2123–2131.

25. Zhang P, LaPara TM, Goslan EH, Xie Y, Parsons SA, et al. (2009) Biodegradation of haloacetic acids by bacterial isolates and enrichment cultures from drinking water systems. Environ Sci Technol 43: 3169–3175.
26. Vaerewijck MJ, Huys G, Palomino JC, Swings J, Portaels F (2005) Mycobacteria in drinking water distribution systems: Ecology and significance for human health. FEMS Microbiol Rev 29: 911–934.
27. Forster D, Daschner F (1998) Acinetobacter species as nosocomial pathogens. European Journal of Clinical Microbiology and Infectious Diseases 17: 73–77.
28. Hedrich S, Schlomann M, Johnson DB (2011) The iron-oxidizing proteobacteria. Microbiology 157: 1551–1564.
29. Emerson D, Fleming EJ, McBeth JM (2010) Iron-oxidizing bacteria: An environmental and genomic perspective. Annu Rev Microbiol 64: 561–583.
30. Madigan MT, Martinko JM, Dunlap PV, Clark DP (2009) Brock biology of microorganisms. San Francisco, CA: Pearson/Benjamin Cummings.
31. Staley JT, Konopka A (1985) Measurement of in situ activities of nonphotosynthetic microorganisms in aquatic and terrestrial habitats. Annual Reviews in Microbiology 39: 321–346.
32. Revetta RP, Pemberton A, Lamendella R, Iker B, Santo Domingo JW (2010) Identification of bacterial populations in drinking water using 16S rRNA-based sequence analyses. Water Res 44: 1353–1360.
33. Brenner D, Krieg N, Staley J, Garrity G, Boone D, et al. (2005) The proteobacteria, part B, the gammaproteobacteria. Bergey's manual of systematic bacteriology 2.
34. McKeon DM, Calabrese JP, Bissonnette GK (1995) Antibiotic resistant gram-negative bacteria in rural groundwater supplies. Water Res 29: 1902–1908.
35. Bifulco JM, Shirey J, Bissonnette G (1989) Detection of acinetobacter spp. in rural drinking water supplies. Appl Environ Microbiol 55: 2214–2219.
36. Shirey JJ, Bissonnette GK (1991) Detection and identification of groundwater bacteria capable of escaping entrapment on 0.45-micron-pore-size membrane filters. Appl Environ Microbiol 57: 2251–2254.
37. Hansen SK, Rainey PB, Haagensen JA, Molin S (2007) Evolution of species interactions in a biofilm community. Nature 445: 533–536.
38. Marin M, Pedregosa A, Laborda F (1996) Emulsifier production and microscopical study of emulsions and biofilms formed by the hydrocarbon-utilizing bacteria acinetobacter calcoaceticus MM5. Appl Microbiol Biotechnol 44: 660–667.
39. Tomaras AP, Dorsey CW, Edelmann RE, Actis LA (2003) Attachment to and biofilm formation on abiotic surfaces by acinetobacter baumannii: Involvement of a novel chaperone-usher pili assembly system. Microbiology 149: 3473–3484.
40. Wentzel M, Lötter L, Loewenthal R, Marais G (1986) Metabolic behaviour of acinetobacter spp. in enhanced biological phosphorus removal- a biochemical model. Water S A 12: 209–224.
41. Hanson RS, Hanson TE (1996) Methanotrophic bacteria. Microbiol Rev 60: 439–471.
42. Covert TC, Rodgers MR, Reyes AL, Stelma GN (1999) Occurrence of nontuberculous mycobacteria in environmental samples. Appl Environ Microbiol 65: 2492–2496.
43. Pryor M, Springthorpe S, Riffard S, Brooks T, Huo Y, et al. (2004) Investigation of opportunistic pathogens in municipal drinking water under different supply and treatment regimes. Water Science & Technology 50: 83–90.
44. Weinberger M, Berg SL, Feuerstein IM, Pizzo PA, Witebsky FG (1992) Disseminated infection with mycobacterium gordonae: Report of a case and critical review of the literature. Clinical infectious diseases 14: 1229–1239.
45. Desforges JF, Horsburgh CR Jr (1991) Mycobacterium avium complex infection in the acquired immunodeficiency syndrome. N Engl J Med 324: 1332–1338.
46. Carson LA, Petersen NJ, Favero MS, Aguero SM (1978) Growth characteristics of atypical mycobacteria in water and their comparative resistance to disinfectants. Appl Environ Microbiol 36: 839–846.
47. Beumer A, King D, Donohue M, Mistry J, Covert T, et al. (2010) Detection of mycobacterium avium subsp. paratuberculosis in drinking water and biofilms by quantitative PCR. Appl Environ Microbiol 76: 7367–7370.
48. Falkinham JO, Norton CD, LeChevallier MW (2001) Factors influencing numbers of mycobacterium avium, mycobacterium intracellulare, and other mycobacteria in drinking water distribution systems. Appl Environ Microbiol 67: 1225–1231.

49. Gomez-Alvarez V, Revetta RP, Santo Domingo JW (2012) Metagenomic analyses of drinking water receiving different disinfection treatments. Appl Environ Microbiol 78: 6095–6102.
50. Norton CD, LeChevallier MW, Falkinham JO III (2004) Survival of< i> mycobacterium avium</i> in a model distribution system. Water Res 38: 1457–1466.
51. LeChevallier MW, Lowry CD, Lee RG, Gibbon DL (1993) Examining the relationship between iron corrosion and the disinfection of biofilm bacteria. Journal-American Water Works Association 85: 111–123.
52. Critchley M, Fallowfield H (2001) The effect of distribution system bacterial biofilms on copper concentrations in drinking water. Water Sci Technol Water Supply 1: 247–252.
53. Lipponen MT, Suutari MH, Martikainen PJ (2002) Occurrence of nitrifying bacteria and nitrification in finnish drinking water distribution systems. Water Res 36: 4319–4329.
54. Cunliffe DA (1991) Bacterial nitrification in chloraminated water supplies. Appl Environ Microbiol 57: 3399–3402.
55. Regan JM, Harrington GW, Noguera DR (2002) Ammonia- and nitrite-oxidizing bacterial communities in a pilot-scale chloraminated drinking water distribution system. Appl Environ Microbiol 68: 73–81.
56. Wilczak A, Jacangelo JG, Marcinko JP, Odell LH, Kirmeyer GJ, et al. (1996) Occurrence of nitrification in chloraminated distribution systems. Journal-American Water Works Association 88: 74–85.
57. McKew BA, Taylor JD, McGenity TJ, Underwood GJ (2010) Resistance and resilience of benthic biofilm communities from a temperate saltmarsh to desiccation and rewetting. The ISME journal 5: 30–41.
58. Harrison F (2007) Microbial ecology of the cystic fibrosis lung. Microbiology 153: 917–923.
59. Boomer SM, Noll KL, Geesey GG, Dutton BE (2009) Formation of multilayered photosynthetic biofilms in an alkaline thermal spring in yellowstone national park, wyoming. Appl Environ Microbiol 75: 2464–2475.
60. Ledder R, Gilbert P, Pluen A, Sreenivasan P, De Vizio W, et al. (2006) Individual microflora beget unique oral microcosms. J Appl Microbiol 100: 1123–1131.

15

Carbon, Nitrogen and Phosphorus Accumulation and Partitioning, and C:N:P Stoichiometry in Late-Season Rice under Different Water and Nitrogen Managements

Yushi Ye[1], Xinqiang Liang[1]*, Yingxu Chen[2], Liang Li[1], Yuanjing Ji[1], Chunyan Zhu[2]

1 Institute of Environmental Science and Technology, College of Environmental and Resource Sciences, Zhejiang University, Hangzhou, China, **2** Zhejiang Province Key Laboratory for Water Pollution Control and Environmental Safety, Hangzhou, China

Abstract

Water and nitrogen availability plays an important role in the biogeochemical cycles of essential elements, such as carbon (C), nitrogen (N) and phosphorus (P), in agricultural ecosystems. In this study, we investigated the seasonal changes of C, N and P concentrations, accumulation, partitioning, and C:N:P stoichiometric ratios in different plant tissues (root, stem-leaf, and panicle) of late-season rice under two irrigation regimes (continuous flooding, CF; alternate wetting and drying, AWD) and four N managements (control, N0; conventional urea at 240 kg N ha^{-1}, UREA; controlled-release bulk blending fertilizer at 240 kg N ha^{-1}, BBF; polymer-coated urea at 240 kg N ha^{-1}, PCU). We found that water and N treatments had remarkable effects on the measured parameters in different plant tissues after transplanting, but the water and N interactions had insignificant effects. Tissue C:N, N:P and C:P ratios ranged from 14.6 to 52.1, 3.1 to 7.8, and 76.9 to 254.3 over the rice growing seasons, respectively. The root and stem-leaf C:N:P and panicle C:N ratios showed overall uptrends with a peak at harvest whereas the panicle N:P and C:P ratios decreased from filling to harvest. The AWD treatment did not affect the concentrations and accumulation of tissue C and N, but greatly decreased those of P, resulting in enhanced N:P and C:P ratios. N fertilization significantly increased tissue N concentration, slightly enhanced tissue P concentration, but did not affect tissue C concentration, leading to a significant increase in tissue N:P ratio but a decrease in C:N and C:P ratios. Our results suggested that the growth of rice in the Taihu Lake region was co-limited by N and P. These findings broadened our understanding of the responses of plant C:N:P stoichiometry to simultaneous water and N managements in subtropical high-yielding rice systems.

Editor: Xiujun Wang, University of Maryland, United States of America

Funding: This work was funded by the National Natural Science Foundation of China (No. 41271314, 21077088) and the National Key Science and Technology Project: Water Pollution Control and Treatment (2014ZX07101-012). The funders had no role in study design, data collection and analysis, decision to publish, or preparation of the manuscript.

Competing Interests: The authors have declared that no competing interests exist.

* Email: liang410@zju.edu.cn

Introduction

Nitrogen (N) is one of the most important mineral nutrients. It promotes large leaf area index [1], long duration of photosynthesis [2], high nutrient uptake [3], and ultimately high crop productivity [4,5]. Along with N, phosphorous (P) is another vital mineral nutrient influencing plant photosynthesis assimilation and biomass production [2,6,7]. To meet the challenge of food security, a large amount of chemical fertilizers has been applied to the rice cropping systems, particularly in China [8]. Long-term use of high rates of fertilization with improper water and nutrient managements has resulted in low water and nutrient use efficiencies, leading to detrimental effects on ecology, environment and human health [4,8]. The seasonal absorption, accumulation and allocation of N and P in rice may deserve special attention for implementing sound water and nutrient management practices in sustainable rice production systems.

Carbon (C), which provides the structural basis and constitutes a fairly stable 50% of a plant's dry mass, can also act as a limiting element for plant [9]. Rice crop may play an important role in terrestrial C cycle by both C sequestrations through photosynthesis and C releases through residues decomposition and/or root respiration [10,11]. Management practices such as irrigation and fertilization can influence crop physio-ecological activities [1], hence affect the sequestration and emission of C in paddy fields, which presumably have an effect on the mitigation of global warming and the stability of food security [12]. Most previous studies have focused on the accumulation and partitioning of dry matter of rice in response to elevated [CO_2] (free-air CO_2 enrichment) [5,7,13,14]. However, little has been done to evaluate the effects of water and N managements on the assimilation, accumulation and distribution of C in rice plant.

Since C, N and P are strongly coupled in their biochemical functioning [9,15] and their balance generally affects crop production and food-web dynamics [16], the C:N:P stoichiometry is the most investigated factor in ecological interactions. To date, C:N:P stoichiometry has been widely applied in diverse ecological processes, and successfully incorporated into explain many phenomena at all levels of biology, from genes and molecules to whole organisms and even to ecosystems and the biosphere [16–19]. Some measurements have already been made on the spatiotemporal variations, biological regulation mechanisms and the ecological implications of C:N:P stoichiometric ratios in soil, plant and litter at different trophic levels on a regional, national and even global scale [9,17–20]. Elucidating changes in C:N:P ratios during plant growth could be useful in calibrating plant mechanistic models and developing terrestrial biogeochemical models [21,22]. However, the seasonal changes of C:N:P ratios in rice plant, particularly with their responses to different water and N managements have not been well characterized.

In an earlier study, we investigated the effects of two controlled-release N fertilizers (CRNFs) (controlled-release bulk blending fertilizer and polymer-coated urea: BBF and PCU) under two irrigation regimes (continuous flooding and alternate wetting and drying: CF and AWD) in comparison with urea on the dry matter accumulation and partitioning, grain yield, and water and N use efficiencies in late-season rice in the Taihu Lake region of China, and found the agronomic performances played individually or jointly by the irrigation and N managements [4]. The objectives of this research were to: (1) investigate the seasonal changes of C, N and P concentrations, accumulation, allocation, and stoichiometric ratios in different plant tissues under different water and N managements, (2) get the relationships between tissue C:N:P ratios and rice grain yield, and (3) evaluate the limiting patterns of nutrients via C:N:P stoichiometry in rice production systems.

Materials and Methods

(This work was unrelated to ethics issues, and no specific permit was required for the described field study, and we confirmed that the field study did not involve endangered or protected species).

Site description

This study was conducted at Yuhang town of Zhejiang province, Taihu Lake region of China ($30°21'50''$ N, $119°53'17''$ E) in 2010 and 2011. The study site has a subtropical monsoon climate with an average temperature of $16.2°C$ and an annual precipitation of 1290 mm. The soil type of the experimental field is hydromorphic paddy soil (Ferric-accumulic Stagnic Anthrosols). Initial soil properties of the plow layer (0–20 cm) were: pH 5.8 (1:5, soil/water), soil organic C 21.75 g kg^{-1}, total N 3.46 g kg^{-1}, mineral N 24.04 mg kg^{-1}, and total P 0.32 g kg^{-1}. Single cropping of late-season rice has been widely adopted in the region. The average routine rate of fertilization is 240 kg N ha^{-1} (as urea, 46% N), 120 kg P_2O_5 ha^{-1} (as superphosphate, 12% P_2O_5), and 120 kg K_2O ha^{-1} (as potassium chloride, 60% K_2O) for one rice season.

Experimental design

The field experiment was arranged in a split-plot design with three replications in both years. Main plots consisted of two irrigation regimes: CF and AWD. All plots were flooded during the first 10–14 days after transplanting (DAT), allowing seedlings to recover from the shock of transplanting prior to the imposition of AWD treatments. Further details on the application of CF and AWD were described by Ye et al. [4]. The daily temperature, rainfall, irrigation, and field water depth from transplanting to harvest under CF and AWD irrigation in 2010 and 2011 were also reported in detail in our previous publication [4]. Subplots consisted of four N treatments: the control (N0), a urea application of 240 kg N ha^{-1} (UREA), and two basal CRNF treatments both at the rate of 240 kg N ha^{-1} (BBF and PCU). The PCU (42% N) is one of the most widely used coated granular CRNF fertilizers, and the BBF (24% N-12% P_2O_5-12% K_2O) is a compound CRNF product in which N source is made up of 70% controlled-release N and 30% ordinary quick-acting N. Both BBF and PCU (Kingenta

Ecological Engineering Co., Ltd., Shandong, China) are 90 days of N release period and need only one-off fertilization.

Each plot was 6 m×3 m in size. All bunds were mulched with plastic film to minimize lateral seepage between adjacent plots. Individual inlet and outlet in the boundary side of the bunds were established in each plot for irrigation and drainage. A local late-season rice cultivar named Xiushui 134 (*Oryza sativa* L.) with high-yielding potential and pest resistance was used. Urea was applied in three splits for the late-season rice, 40% was basally applied (0 DAT), 40% was topdressed at the tillering stage (32 DAT), and the remaining 20% was topdressed at the panicle initiation stage (63 DAT). Full doses of superphosphate and potassium chloride in the N0, UERA and PCU treatments were applied basally at rates of 120 kg P_2O_5 ha^{-1} and 120 kg K_2O ha^{-1}, respectively. The 3-week-old seedlings were transplanted at a spacing of 25 cm×16 cm with three seedlings per hill on 23 June 2010 and 1 July 2011. Plots were regularly hand-weeded until canopy leaves were extremely crowded to prevent weed damage. The pests and diseases managements followed the local tradition. Final harvests were done on 9 November 2010 and 17 November 2011, and the growth duration was 140 days in all treatments for either year.

Plant sampling and measurements

Five hills of plants (except border plants) in each plot were dug out by a shovel at seedling, tillering, booting, filling, and maturity stages (harvest), 1, 35, 56, 91, 140 DAT in both years. All visible root tissues were collected from the soil. Plants were washed free of soil and separated into two parts: root, stem-leaf before heading (75–80 DAT) and three parts: root, stem-leaf, and panicle after heading. All plant sub-samples were oven-dried to a constant weight at $70°C$, weighted, and ground to sift through a 0.15 mm sieve for chemical measurements. Tissue C and N concentrations were determined with combustion technique on a Vario MAX CNS elemental analyzer (Elementar Analysensysteme GmbH, Hanau, Germany) [10]. Tissue P concentration was analyzed colorimetrically by the molybdenum blue method following digestion in concentrated H_2SO_4 and H_2O_2 [23].

Calculations and statistical analysis

Accumulation of C, N and P in root, stem-leaf, and panicle at various growth stages was calculated from the element concentration multiplied by the dry matter. The element accumulation rate (kg ha^{-1} day^{-1}) at each growth stage was obtained from the element accumulation divided by the number of days of the growth stage. Stoichiometric ratios of C:N, N:P and C:P in plant tissues were calculated on mass basis.

All statistical analyses were performed using PASW Statistics 18.0 (SPSS Inc. Chicago, USA). Combined analysis of variance using data from two years indicated that the interactions between years and irrigation regimes, and between years and N managements on the measured parameters (C, N and P concentrations, accumulation, partitioning, and stoichiometric ratios) were not significant in both seasons. Therefore, the data were arithmetic averaged across two years (a total of six replications) for further analyses [1]. Irrigation methods and N managements were considered as fixed factors. Two-way ANOVA were used to assess the effects of both water regimes (CF vs. AWD) and N managements (N0 vs. UREA vs. BBF vs. PCU) on the analysis of variance (*F*-value) of C, N and P concentrations, accumulation, partitioning, and C:N:P stoichiometric ratios in root, stem-leaf, and panicle at various growth stages, and also to test the interactions of water regimes × N managements. The least significant difference (LSD) test at the 0.05 probability level was used to compare significant differences among treatment means.

Spearman rank correlation was performed to reveal the interrelations between C:N:P ratios of plant tissues and the grain yield at harvest.

Results

Water and N managements did alter the patterns of C, N and P concentrations, accumulation and partitioning in different plant tissues after transplanting, but no significant water by N interaction effect was found at any stage of sampling ($P>0.05$). Besides, the effects of different N treatments on tissue C, N and P concentrations, accumulation and partitioning were always more significant than those of different irrigation regimes (Tables S1 and S2 in File S1). Because irrigation and N treatments had remarkable effects on the assimilation, accumulation and allocation of C, N and P in different organs, the plant C:N:P stoichiometry was greatly affected by the water and N managements, particularly during the late-cultivation period. However, the water × N interactions on tissue C:N:P ratios were generally not significant (except root N:P ($F = 3.08$, $P<0.05$) and stem-leaf N:P ($F = 3.02$, $P<0.05$) at tillering stage) through the rice growing seasons (Table S3 in File S1).

C, N and P concentrations

C concentration ranged from 295.3 to 343.2 g kg^{-1} in root, 374.1 to 403.7 g kg^{-1} in stem-leaf, and 401.9 to 414.2 g kg^{-1} in panicle over the rice growing seasons (Fig. 1). Root C concentration showed a remarkable increase from seedling (299.0 g kg^{-1}) to filling (325.5 g kg^{-1}), and then decreased to maturity (312.2 g kg^{-1}). Stem-leaf C concentration increased only up to tillering (398.3 g kg^{-1}), and then decreased to harvest (378.0 g kg^{-1}). Panicle C concentration showed a slight decrease from filling (411.5 g kg^{-1}) to maturity (403.9 g kg^{-1}). Tissue C concentration was not obviously affected by the two irrigation regimes or by the three N-fertilized treatments, and UREA, BBF and PCU did not give consistently higher C concentrations than N0 from transplanting to harvest.

N concentration ranged from 6.6 to 16.3 g kg^{-1} in root, 7.1 to 27.4 g kg^{-1} in stem-leaf, and 11.7 to 19.5 g kg^{-1} in panicle over the planting seasons (Fig. 2). Both root and stem-leaf N concentrations in N0 treatment showed clearly decreasing trends from seedling (12.9 and 18.6 g kg^{-1}) to harvest (6.7 and 7.2 g kg^{-1}), while those in N-fertilized treatments first increased from seedling (12.9 and 18.8 g kg^{-1}) to tillering (15.6 and 25.5 g kg^{-1}) and then decreased to harvest (10.8 and 11.7 g kg^{-1}). Panicle N concentration dropped from filling (17.0 g kg^{-1}) to maturity (13.4 g kg^{-1}) irrespective of N addition. There was no significant difference in tissue N concentration between CF and AWD at any stage of observation. As expected, N fertilization dramatically increased tissue N concentration after seedling, particularly during the late growth period. At harvest, N concentration was significantly increased by 64.7%, 61.2% and 59.0% in root, 62.6%, 63.9% and 65.7% in stem-leaf, and 20.4%, 22.2% and 11.9% in panicle in UREA, BBF and PCU compared with those of N0, respectively. However, no consistent significant differences among the three N-fertilized treatments were observed through the growing seasons.

P concentration ranged from 1.3 to 4.2 g kg^{-1} in root, 1.5 to 4.3 g kg^{-1} in stem-leaf, and 2.9 to 3.9 g kg^{-1} in panicle over the planting seasons (Fig. 3). Both root and stem-leaf P concentrations displayed slight increases from seedling (3.4 and 3.6 g kg^{-1}) to tillering (3.8 and 4.1 g kg^{-1}), then gradual decreases until harvest (1.5 and 1.8 g kg^{-1}). Interestingly, panicle P concentration increased from filling (3.2 g kg^{-1}) to maturity (3.6 g kg^{-1}),

exhibiting an opposite trend to those of panicle C and N concentrations. Water regimes did alter tissue P concentration, particularly during the late growth period. Compared with CF, AWD significantly decreased P concentration by 9.1%, 5.8%, and 5.6% at filling and 9.6%, 12.5%, and 7.8% at maturity in root, stem-leaf, and panicle, respectively. Tissue P concentration was not obviously affected by the N fertilization before heading, hereafter enhanced by 5.7–38.6% at filling and 2.6–30.4% at maturity in those N-fertilized treatments compared with N0, though the differences among the four N treatments were not always significant at the $P<0.05$ level.

C, N and P accumulation

For the N0 control, both root and stem-leaf C accumulation peaked at booting stage (506 and 1824 kg ha^{-1}). However, root C accumulation peaked at filling stage (656–695 kg ha^{-1}) and stem-leaf C accumulation peaked at harvest (2904–3052 kg ha^{-1}) for the N-fertilized treatments (Fig. 4A). Panicle C accumulation peaked at harvest irrespective of N-fertilizer application (1996, 3120, 3132, and 3548 kg ha^{-1} in N0, UREA, BBF, and PCU, respectively). The maximum rates of average C accumulation across all treatments were observed at booting in root and stem-leaf (13.8 and 78.1 kg ha^{-1} day^{-1}) and at filling in panicle (46.8 kg ha^{-1} day^{-1}). Tissue C accumulation did not differ remarkably between CF and AWD or among UREA, BBF and PCU at any stage of sampling, but increased significantly in the N-fertilized treatments compared with N0 after tillering. At harvest, C accumulation was enhanced by 59.1%, 71.3% and 62.5% in root, 69.5%, 62.1% and 70.4% in stem-leaf, and 56.3%, 56.9% and 77.8% in panicle in UREA, BBF and PCU compared with those of N0, respectively.

The maximum N accumulation in root and stem-leaf was observed at booting in N0 control (16.1 and 60.2 kg ha^{-1}), but at filling in treatments with N addition (25.5, 24.0 and 26.0 kg ha^{-1} for root and 109.6, 109.5 and 118.9 kg ha^{-1} for stem-leaf in UREA, BBF and PCU, respectively) (Fig. 4B). Panicle N accumulation showed a clear uptrend from filling (68.4 kg ha^{-1}) to maturity (98.9 kg ha^{-1}) regardless of N addition. The maximum rates of average N accumulation across all treatments were observed at tillering in root (0.39 kg ha^{-1} day^{-1}), booting in stem-leaf (3.03 kg ha^{-1} day^{-1}), and filling in panicle (1.96 kg ha^{-1} day^{-1}). There was no significant difference in tissue N accumulation between CF and AWD at all observation stages. Obviously, N addition greatly improved plant N uptake after seedling, and the difference between fertilized and unfertilized plots expanded continuously with crop growth. At harvest, N accumulation was enhanced by 155.8%, 157.5% and 167.4% in root, 170.4%, 162.8% and 179.1% in stem-leaf, and 109.4%, 111.5% and 116.0% in panicle in UREA, BBF and PCU compared with those of N0, respectively. Besides, PCU obtained higher plant N accumulation than BBF and UREA after tillering, suggesting that PCU tended to be more effective in promoting the N transformation from soil to plant during rice growth period.

P accumulation increased remarkably up to booting in both root and stem-leaf (6.1 and 20.0 kg ha^{-1}), and up to harvest in panicle (26.5 kg ha^{-1}) (Fig. 4C). The maximum rates of average P accumulation across all treatments were observed at booting in root and stem-leaf (0.12 and 0.70 kg ha^{-1} day^{-1}) and at filling in panicle (0.36 kg ha^{-1} day^{-1}). AWD decreased tissue P accumulation after seedling, particularly at booting and filling stages ($P< 0.05$). N fertilization had a positive effect on plant P uptake which increased with the increase of plant maturity. At harvest, P accumulation was significantly enhanced by 67.9%, 108.3% and 75.6% in root, 89.2%, 91.7% and 90.5% in stem-leaf, and 64.9%,

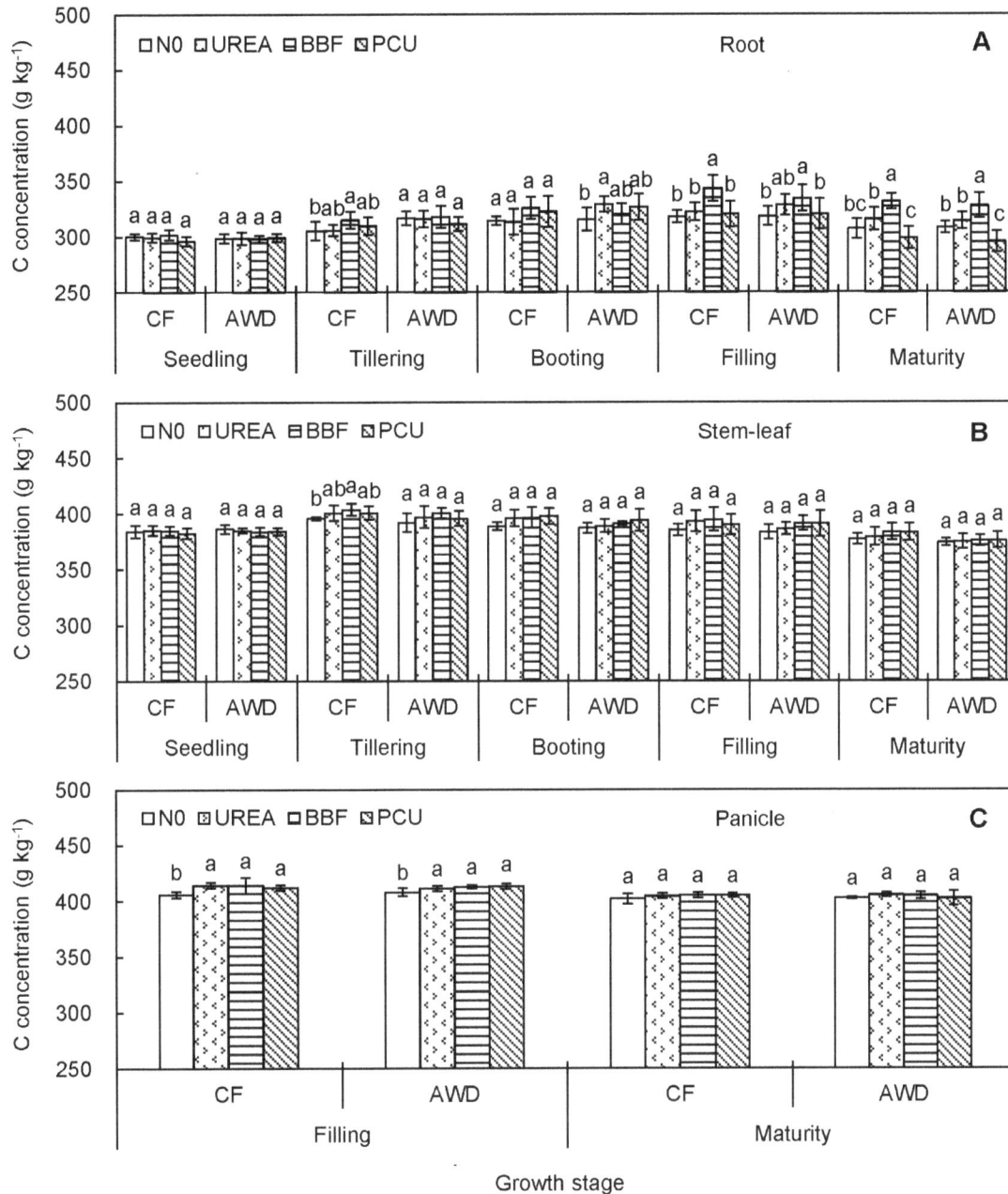

Figure 1. Seasonal changes of carbon concentration in root (A), stem-leaf (B), and panicle (C) of late-season rice under different water and N managements (2-year average). Vertical bars represent ± standard deviation of the mean (n = 6). The different letters listed above bars represent significant differences at $P<0.05$.

68.3% and 81.6% in panicle in UREA, BBF and PCU compared with those of N0, respectively. However, no consistent significant differences among the three N-fertilized treatments were observed through the whole rice seasons.

C, N and P partitioning

Crop element composition depends on the process of accumulation, translocation and allocation of the elements. The seasonal variations of tissue C, N and P partitioning were shown in Table 1. C partitioning displayed a sharp decrease from transplanting (37.6%) to harvest (8.3%) in root but a remarkable increase from filling (34.8%) to maturity (47.9%) in panicle. Stem-leaf C

partitioning initially increased up to booting (78.5%) and then decreased with the increase of plant maturity (43.8% at harvest). Tissue C partitioning was comparable between the two irrigation treatments at seedling, tillering and maturity stages, but increased significantly in root (7.7% and 7.9%) vs. decreased in stem-leaf (2.0% and 2.1%) in AWD compared with CF at booting and filling stages. No significant differences in tissue C partitioning among the four N treatments were found before ripening (110–130 DAT), while the highest C partitioning of root, stem-leaf, and panicle was observed in BBF (8.9%), UREA (45.3%), and PCU (49.5%) at harvest, respectively.

Tissue N partitioning exhibited a similar seasonal pattern to C partitioning (Table 1). N partitioning was also significantly higher

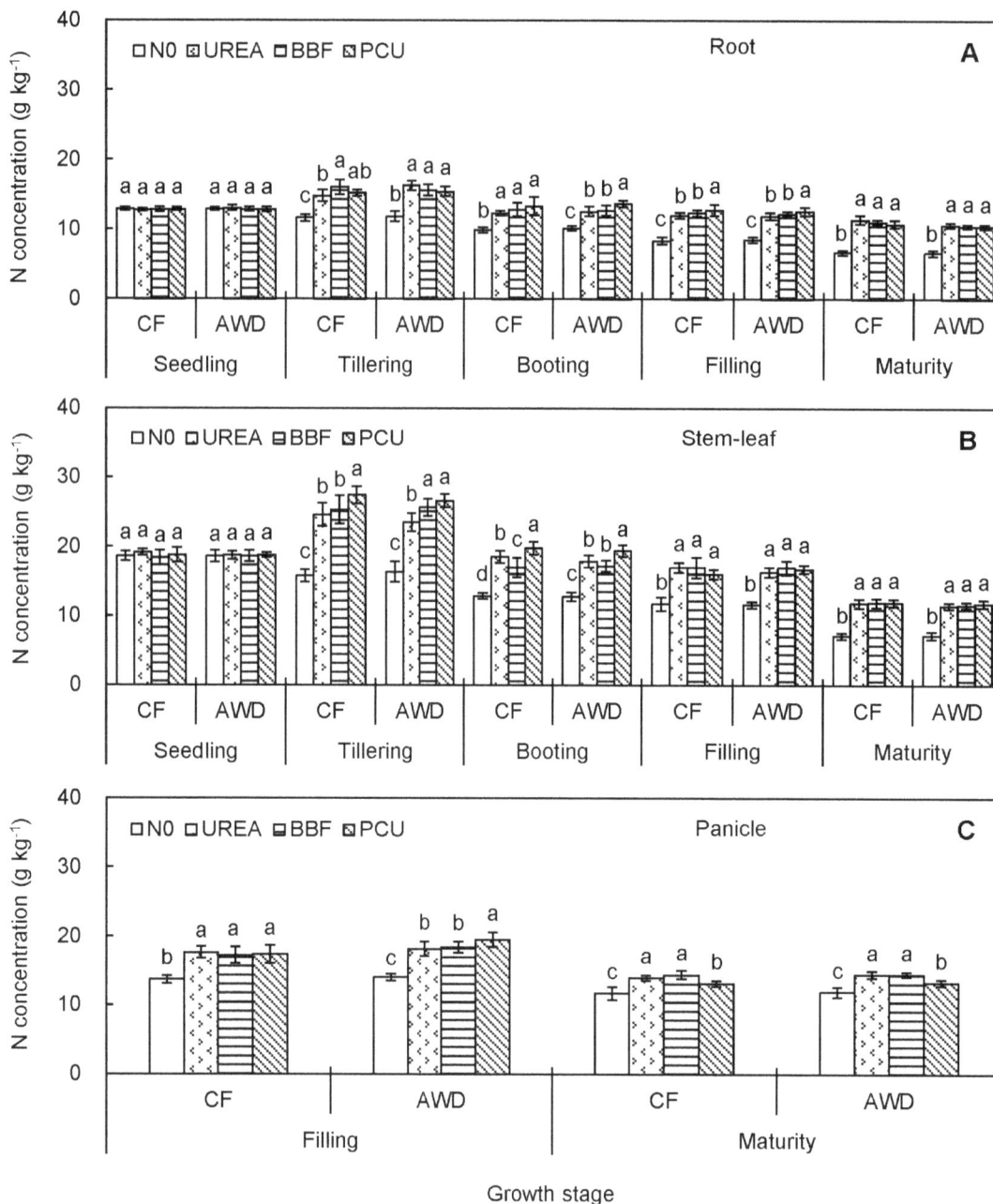

Figure 2. Seasonal changes of nitrogen concentration in root (A), stem-leaf (B), and panicle (C) of late-season rice under different water and N managements (2-year average). Vertical bars represent ± standard deviation of the mean (n = 6). The different letters listed above bars represent significant differences at $P<0.05$.

in root (8.1% and 5.3%) but lower in stem-leaf (2.0% and 3.9%) in AWD than CF at booting and filling stages. There were no clear differences in tissue N partitioning among the four N treatments before ripening. However, N0 resulted in the lowest N partitioning in root (7.5%) and stem-leaf (34.1%) but the highest one in panicle (58.4%) at harvest.

The seasonal changes of P partitioning in root, stem-leaf, and panicle were similar to those of C and N partitioning (Table 1). Tissue P partitioning was unaffected by the two water regimes at seedling, tillering and filling stages, but increased significantly by 6.9% in root at booting and 2.2% in panicle in AWD compared with CF at harvest. Tissue P partitioning was comparable among

the four N treatments at all observation stages, indicating that N fertilization had no significant influence on P allocation.

C:N:P stoichiometric ratios

C:N ratio ranged from 19.4 to 46.6 in root, 14.6 to 52.1 in stem-leaf, and 21.3 to 34.6 in panicle over the planting seasons (Fig. 5). Tissue C:N ratio increased gradually with crop growth after seedling, and peaked at harvest. Notably, C:N ratio was comparable between root and stem-leaf during the rice seasons, though C and N concentrations were much higher in stem-leaf than root (Figs. 1 and 2). Tissue C:N ratio did not differ significantly between CF and AWD or among UREA, BBF and

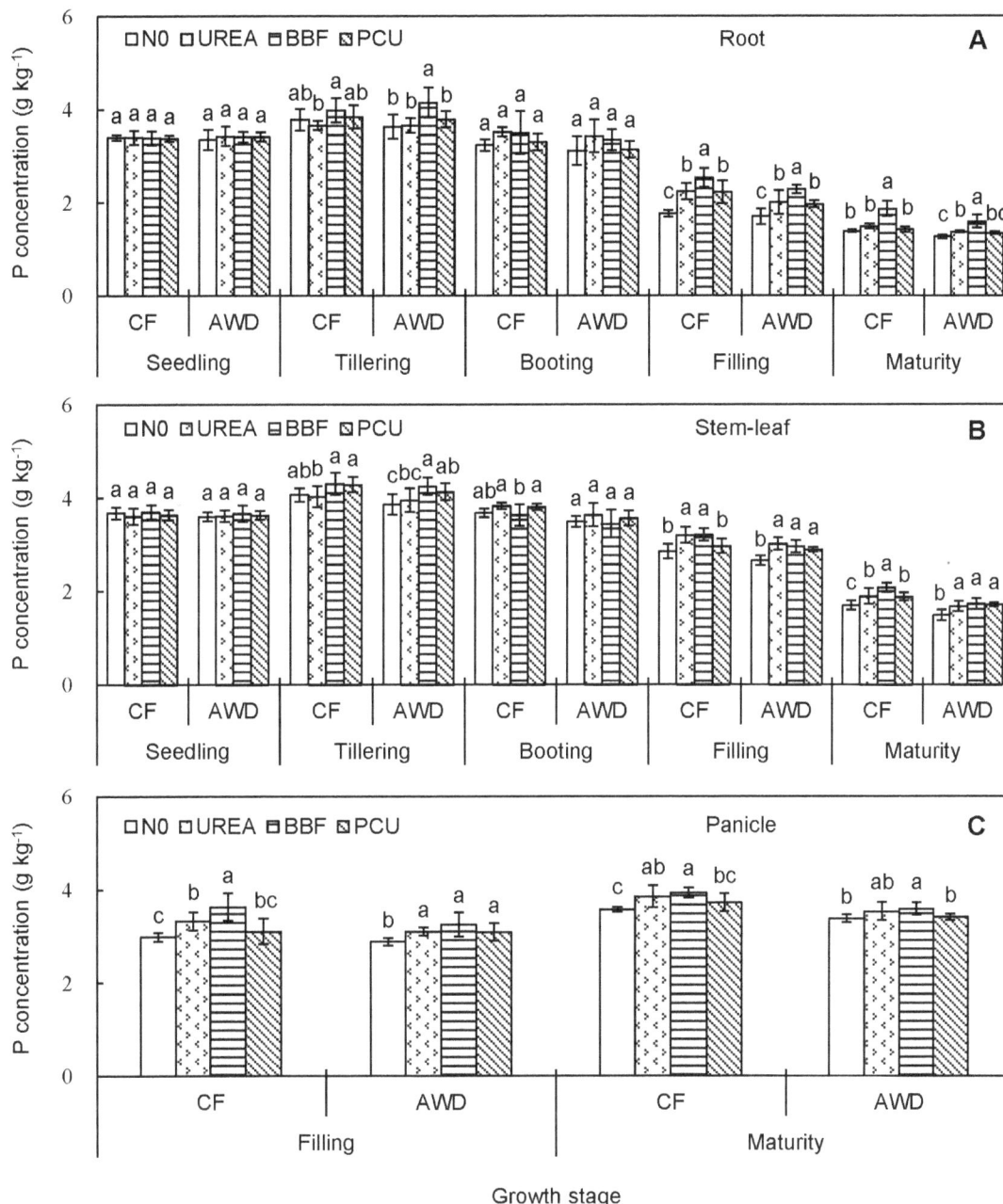

Figure 3. Seasonal changes of phosphorus concentration in root (A), stem-leaf (B), and panicle (C) of late-season rice under different water and N managements (2-year average). Vertical bars represent ± standard deviation of the mean (n = 6). The different letters listed above bars represent significant differences at $P<0.05$.

PCU at any stage of observation, but decreased significantly by 23.4–40.1%, 17.8–33.3%, 21.0–32.9%, and 10.6–39.3% in the N-fertilized treatments compared with those of N0 at tillering, booting, filling, and maturity stages, respectively.

N:P ratio ranged from 3.1 to 7.8 in root, 3.5 to 6.9 in stem-leaf, and 3.3 to 6.3 in panicle over the planting seasons (Fig. 6). Root N:P ratio showed no obvious difference from seedling (3.8) to tillering (3.7), then increased remarkably up to maturity (6.7). Stem-leaf N:P ratio exhibited a slight increase from booting (4.7) to maturity (6.0). Panicle N:P ratio displayed a 31.0% average reduction from filling (5.4) to maturity (3.7). The N:P ratio of panicle was significantly lower than those of root and stem-leaf at harvest (81.0% and 62.1%), though N and P concentrations of

panicle were the highest among the three plant parts (Figs. 2 and 3). Tissue N:P ratio was comparable between the two irrigation regimes before heading, but enhanced significantly by 9.5% and 4.9% in root, 5.8% and 12.1% in stem-leaf, and 12.1% and 10.2% in panicle in AWD compared with CF at filling and maturity stages, respectively. N fertilization significantly increased tissue N:P ratio after seedling. The average N:P ratio was 52.2%, 24.6% and 52.5% higher in root, 44.6%, 37.3% and 46.0% higher in stem-leaf, and 13.7%, 13.2% and 9.2% higher in panicle in UREA, BBF and PCU than those of N0 at harvest, respectively. No consistent significant differences in N:P ratios of stem-leaf and panicle among the three N-fertilized treatments were observed through the rice growing seasons, except that UREA and PCU

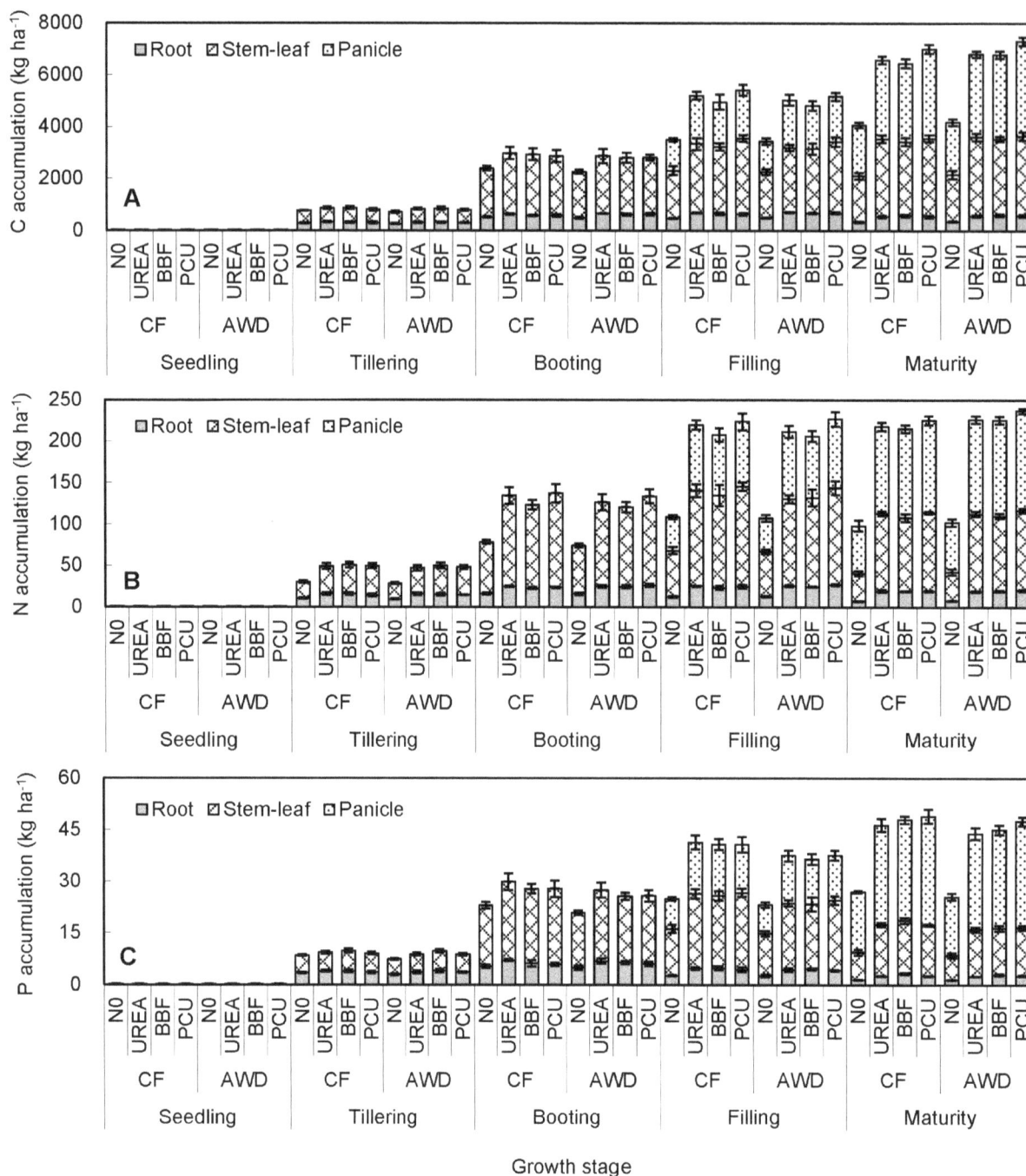

Figure 4. Seasonal changes of carbon (A), nitrogen (B) and phosphorus (C) accumulation in root, stem-leaf, and panicle of late-season rice under different water and N managements (2-year average). Vertical bars represent ± standard deviation of the mean (n = 6).

obtained significantly greater root N:P ratios (22.2% and 22.4%) than BBF at harvest.

C:P ratio ranged from 76.9 to 244.1 in root, 94.0 to 254.3 in stem-leaf, and 103.1 to 142.0 in panicle over the growing seasons (Fig. 7). Both root and stem-leaf C:P ratios fluctuated at around 100 before booting and then increased to 215.4 and 216.6 at harvest. As with panicle N:P ratio, panicle C:P ratio also decreased from filling (130.7) to maturity (111.9). The C:P ratio of panicle was significantly lower than those of root and stem-leaf at harvest (92.5% and 93.6%), though panicle possessed the highest C and P concentrations (Figs. 1 and 3). Tissue C:P ratio did not change

significantly in response to irrigation nor to N fertilization before heading, but decreased by 5.3–9.4% at filling and 8.4–12.6% at maturity in CF compared with AWD, and reduced by 3.8–28.6% at filling and 2.0–17.1% at maturity in the N-fertilized treatments compared with N0 (although not in all cases significantly).

Correlation studies

Tissue C:N ratio displayed significantly positive correlation with tissue C:P ratio but negative correlation with tissue N:P ratio (Table 2). Meanwhile, C:N ratios of root, stem-leaf, and panicle were significantly and positively correlated with each other.

Table 1. Seasonal changes of carbon, nitrogen and phosphorus partitioning in root, stem-leaf, and panicle of late-season rice under different water and N managements (2-year average).

Growth stage	Water regime	Nitrogen treatment	C partitioning (%)			N partitioning (%)			P partitioning (%)		
			Root	Stem-leaf	Panicle	Root	Stem-leaf	Panicle	Root	Stem-leaf	Panicle
Seedling	CF	N0	38.0 a	62.0 a	-	35.2 a	64.8 a	-	42.2 a	57.8 a	-
		UREA	37.1 a	62.9 a	-	33.6 a	66.4 a	-	42.0 a	58.0 a	-
		BBF	36.3 a	63.7 a	-	33.7 a	66.3 a	-	40.1 a	59.9 a	-
		PCU	38.5 a	61.5 a	-	35.7 a	64.3 a	-	43.0 a	57.0 a	-
		Avg.	37.5 A	62.5 A	-	34.6 A	65.4 A	-	41.7 A	58.3 A	-
	AWD	N0	37.3 a	62.7 a	-	34.8 a	65.2 a	-	41.8 a	58.2 a	-
		UREA	37.2 a	62.8 a	-	34.7 a	65.3 a	-	42.0 a	58.0 a	-
		BBF	37.9 a	62.1 a	-	35.1 a	64.9 a	-	42.2 a	57.8 a	-
		PCU	38.7 a	61.3 a	-	35.6 a	64.4 a	-	43.3 a	56.7 a	-
		Avg.	37.8 A	62.2 A	-	35.1 A	64.9 A	-	42.3 A	57.7 A	-
Tillering	CF	N0	36.4 a	63.6 a	-	35.5 a	64.5 c	-	40.9 a	59.1 a	-
		UREA	37.9 a	62.1 a	-	32.5 b	67.5 b	-	42.0 a	58.0 a	-
		BBF	35.9 a	64.1 a	-	31.3 bc	68.7 ab	-	39.9 a	60.1 a	-
		PCU	36.4 a	63.6 a	-	29.1 c	70.9 a	-	39.8 a	60.2 a	-
		Avg.	36.6 A	63.4 A	-	32.1 A	67.9 A	-	40.7 A	59.3 A	-
	AWD	N0	36.3 a	63.7 a	-	33.8 a	66.2 b	-	39.9 a	60.1 a	-
		UREA	37.5 a	62.5 a	-	34.4 a	65.6 b	-	41.1 a	58.9 a	-
		BBF	36.4 a	63.6 a	-	30.5 b	69.5 a	-	41.3 a	58.7 a	-
		PCU	37.4 a	62.6 a	-	30.5 b	69.5 a	-	41.1 a	58.9 a	-
		Avg.	36.9 A	63.1 A	-	32.3 A	67.7 A	-	40.8 A	59.2 A	-
Booting	CF	N0	21.6 a	78.4 a	-	20.8 a	79.2 b	-	23.1 a	76.9 a	-
		UREA	21.4 a	78.6 a	-	18.6 b	81.4 a	-	24.0 a	76.0 a	-
		BBF	19.6 a	80.4 a	-	18.3 b	81.7 a	-	22.3 a	77.7 a	-
		PCU	20.3 a	79.7 a	-	17.5 b	82.5 a	-	21.5 a	78.5 a	-
		Avg.	20.7 B	79.3 A	-	18.8 B	81.2 A	-	22.7 B	77.3 A	-
	AWD	N0	21.8 a	78.2 a	-	21.4 a	78.6 a	-	23.4 a	76.6 a	-
		UREA	22.9 a	77.1 a	-	19.8 a	80.2 a	-	24.9 a	75.1 a	-
		BBF	22.1 a	77.9 a	-	20.4 a	79.6 a	-	25.2 a	74.8 a	-
		PCU	22.4 a	77.6 a	-	19.7 a	80.3 a	-	23.5 a	76.5 a	-
		Avg.	22.3 A	77.7 B	-	20.3 A	79.7 B	-	24.3 A	75.7 B	-
Filling	CF	N0	13.6 a	52.3 a	34.1 a	11.6 a	51.3 a	37.1 a	10.6 b	54.3 a	35.1 a
		UREA	13.1 a	51.1 a	35.8 a	11.6 a	52.8 a	35.6 a	11.5 a	52.3 ab	36.2 a
		BBF	13.2 a	52.3 a	34.7 a	11.2 a	53.7 a	35.1 a	11.7 a	51.6 b	36.7 a

Table 1. Cont.

Growth stage	Water regime	Nitrogen treatment	C partitioning (%)			N partitioning (%)			P partitioning (%)		
			Root	Stem-leaf	Panicle	Root	Stem-leaf	Panicle	Root	Stem-leaf	Panicle
	AWD	PCU	11.6 b	54.1 a	34.3 a	11.2 a	53.9 a	34.9 a	10.8 b	54.9 a	34.3 a
		Avg.	12.8 B	52.5 A	34.7 A	11.4 B	52.9 A	35.7 A	11.1 A	53.2 A	35.7 A
		N0	14.2 a	51.4 ab	34.4 a	12.1 a	50.3 a	37.6 a	11.2 b	52.8 a	36.0 a
		UREA	14.1 a	49.4 b	36.5 a	12.2 a	49.6 a	38.2 a	11.6 b	51.5 a	36.9 a
		BBF	14.0 a	51.5 ab	34.5 a	11.9 a	52.1 a	36.0 a	12.6 a	51.5 a	35.9 a
		PCU	13.2 a	52.9 a	33.9 a	11.8 a	51.7 a	36.5 a	11.2 b	53.9 a	34.9 a
		Avg.	13.9 A	51.3 B	34.8 A	12.0 A	50.9 B	37.1 A	11.6 A	52.5 A	35.9 A
Maturity	CF	N0	8.2 b	43.4 b	48.4 a	7.4 b	34.2 c	58.4 a	5.6 b	29.4 b	65.0 a
		UREA	8.1 b	45.7 a	46.2 b	8.9 a	42.9 a	48.2 b	5.4 b	32.2 a	62.4 b
		BBF	9.0 a	44.2 b	46.8 b	8.9 a	41.1 b	50.0 b	6.7 a	32.2 a	61.1 b
		PCU	7.7 b	43.3 b	49.0 a	8.7 a	41.9 ab	49.4 b	5.3 b	30.3 ab	64.4 a
		Avg.	8.3 A	44.1 A	47.6 A	8.5 A	40.0 A	51.5 A	5.8 A	31.0 A	63.2 B
	AWD	N0	8.5 b	43.2 b	48.3 b	7.5 b	34.0 b	58.5 a	5.7 b	27.9 b	66.4 a
		UREA	8.3 b	44.8 a	46.9 c	8.4 a	41.3 a	50.3 b	5.6 b	31.0 a	63.4 b
		BBF	8.9 a	43.4 b	47.7 bc	8.6 a	40.2 a	51.2 b	6.5 a	29.9 a	63.6 b
		PCU	7.9 c	42.1 c	50.0 a	8.6 a	40.5 a	50.9 b	5.5 b	29.3 ab	65.2 ab
		Avg.	8.4 A	43.4 A	48.2 A	8.3 A	39.0 A	52.7 A	5.8 A	29.6 B	64.6 A

Within a column for each growth stage, means followed by the same letter are not significantly different at $P < 0.05$ by LSD test. Lowercase and uppercase letters indicate comparisons among four N treatments and between two irrigation regimes, respectively.

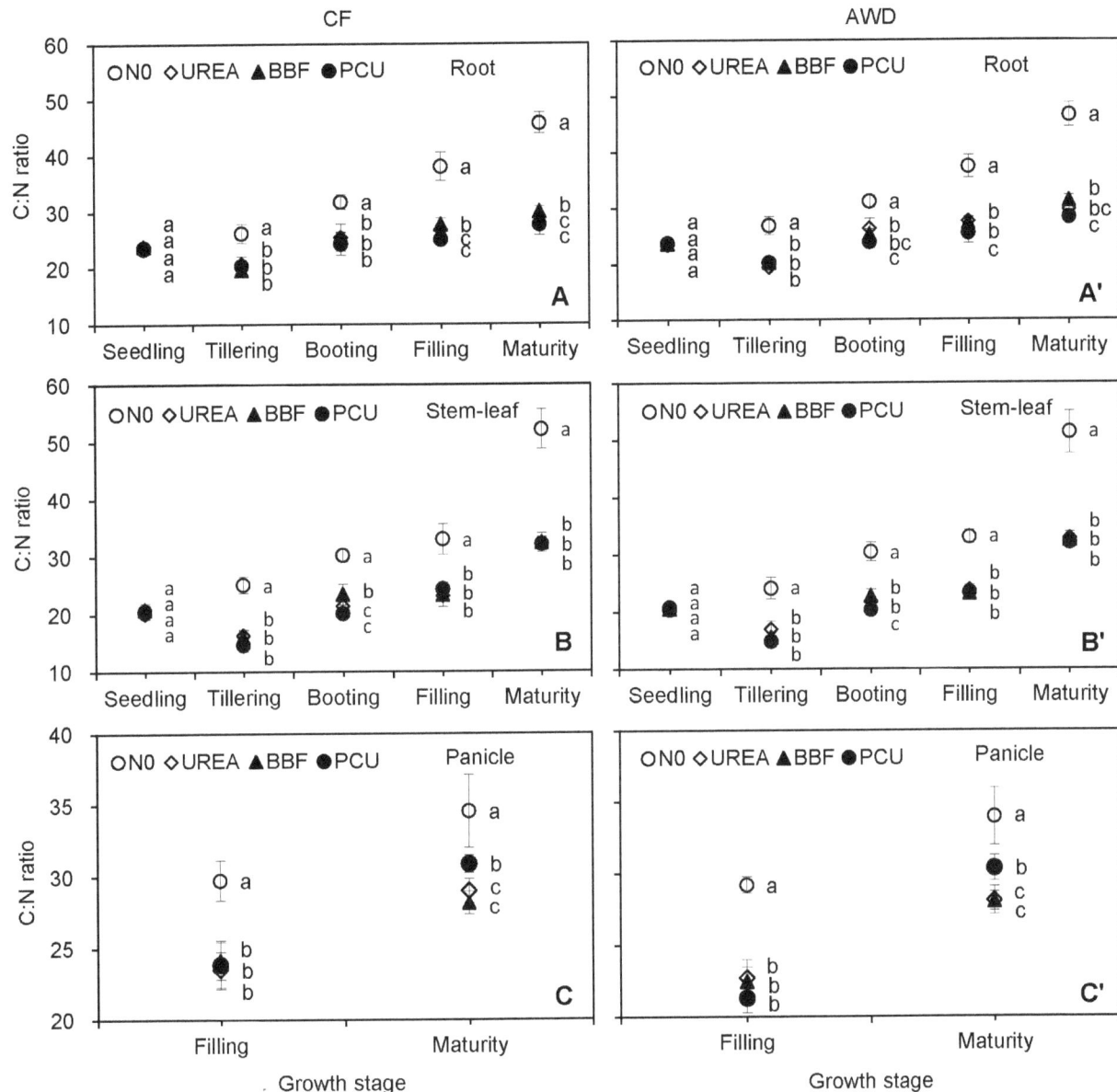

Figure 5. Seasonal changes of C:N ratio in root (A, A'), stem-leaf (B, B'), and panicle (C, C') of late-season rice under different water and N managements (2-year average). Vertical bars represent ± standard deviation of the mean (n = 6). The different letters listed around bars represent significant differences at $P<0.05$.

Similar patterns of correlation were also found for N:P and C:P ratios. These results indicated that C:N:P stoichiometric ratios were highly correlated both within and across plant tissues.

Tissue N:P ratio was significantly and positively correlated with grain yield, while tissue C:N ratio got the opposite correlations. No obvious correlations were found between tissue C:P ratio and grain yield. Herein, tissue C:N and N:P ratios could have much more important implications than C:P ratio for expressing ecological stoichiometric relations in rice crop.

Discussion

Seasonal changes of C, N and P concentrations under different water and N managements

The C, N and P concentrations differed remarkably with plant tissues and crop growth stages. The overall increases of N and P concentrations in root and stem-leaf from seedling to tillering likely reflected the rapid soil exploration and high nutrient uptake by the crop roots. Besides, N and P concentrations in root increased only up to tillering while C concentration increased up to filling (Fig. 1), indicating that root C concentration was less affected by crop senescence during the late growth period. For stem-leaf, there exhibited systematic reductions in C, N and P concentrations from tillering to harvest (Figs. 1, 2 and 3). Decreases of nutrient concentration in vegetative parts with ontogenetic development of individual plants were documented not only in rice [5,7], but also in corn, wheat, barley, and soybean [3,24], resulting from an analogous dilution effect caused by increased plant size and biomass [13,25,26,27]. Notably, P concentration in panicle increased from filling to maturity (Figs. 1, 2 and 3), showing an opposite trend to those of C and N concentrations. This could be explained by the fact that rapidly growing organ needs relatively

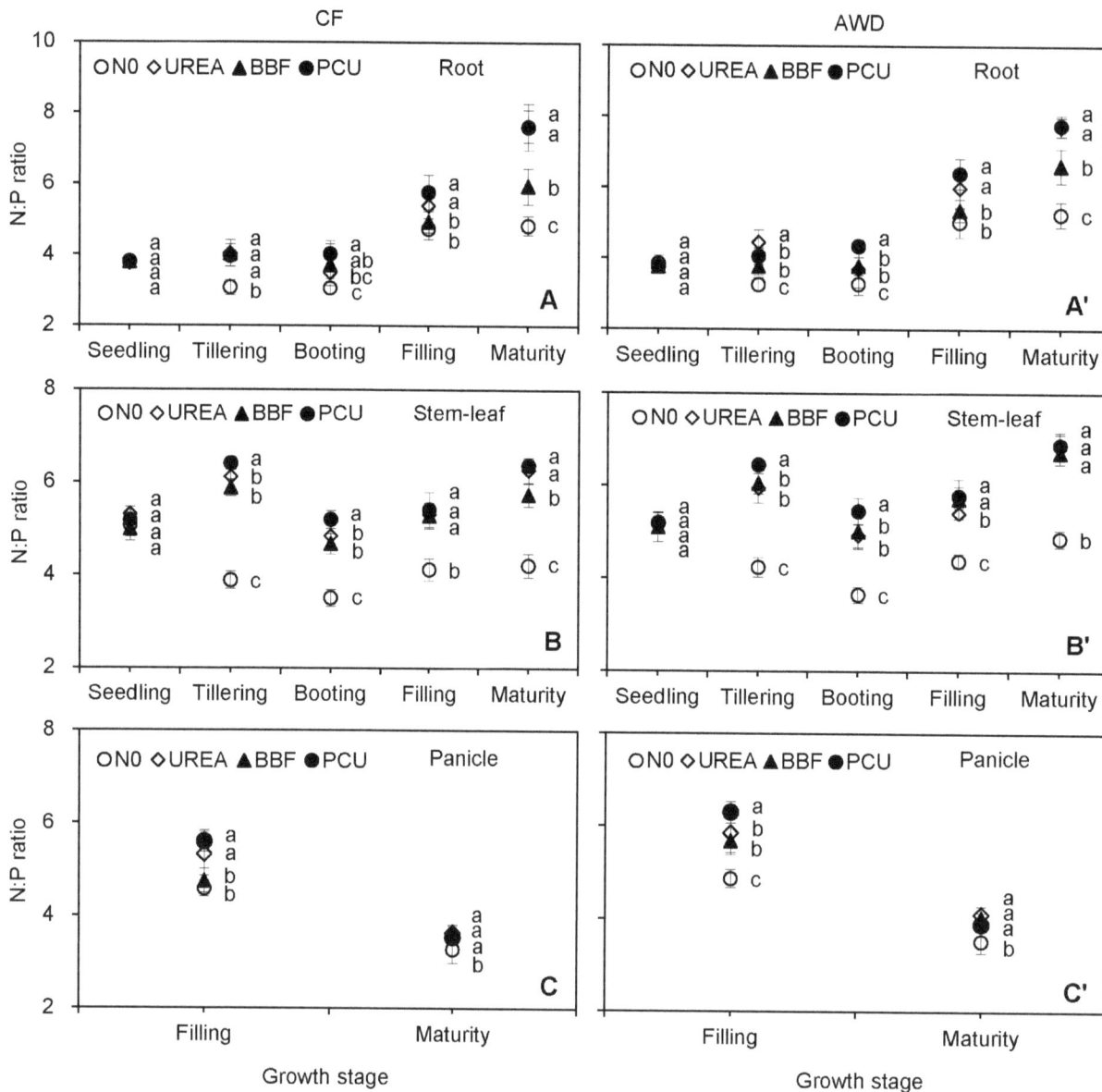

Figure 6. Seasonal changes of N:P ratio in root (A, A'), stem-leaf (B, B'), and panicle (C, C') of late-season rice under different water and N managements (2-year average). Vertical bars represent ± standard deviation of the mean (n = 6). The different letters listed around bars represent significant differences at $P<0.05$.

more P-rich ribosomal RNA (approx. 9% by mass) to maintain rapid rate of protein synthesis [28,29].

We noted different responses of C, N and P concentrations to CF and AWD in different plant tissues and at different growth stages, confirming that water availability played a key role in mediating element status in rice plant. Tissue C and N concentrations were comparable between CF and AWD (Figs. 1 and 2), implying that the water-saving irrigation had only a marginal effect on crop C and N assimilation. However, tissue P concentration was significantly reduced by AWD after heading (Fig. 3), probably because soil drying increased soil P sorption and led to less P availability for plant [30]. These results implied that the process governing tissue P concentration might be independent of those determining C and N concentrations. Indeed, in terms of plant, P is mainly derived from soil whereas C and N have diversified sources (e.g. CO_2 enrichment, SOC mineralization, N

deposition and N fixation). Therefore, soil water conditions could have more influence on P compared with C and N.

Tissue C, N and P concentrations responded differently to N fertilization. C concentration was almost unaffected by N fertilization (Fig. 1), in line with previous researches [18,31], mainly due to the stable plant carbon composition and structural basis [9]. N and P concentrations were positively affected by the N application for most stages after tillering (Figs. 2 and 3), consistent with the reports of Bélanger et al. [25] who stated that increases in N concentration from increased N fertilization resulted in increased P concentration. This phenomenon was partly attributed to the increased capacity of roots to absorb more nutrients [24], and partly because of the stimulated P mobilization by the enhanced extracellular phosphatase activity [6,32] and/or mycorrhizal colonization [27,33].

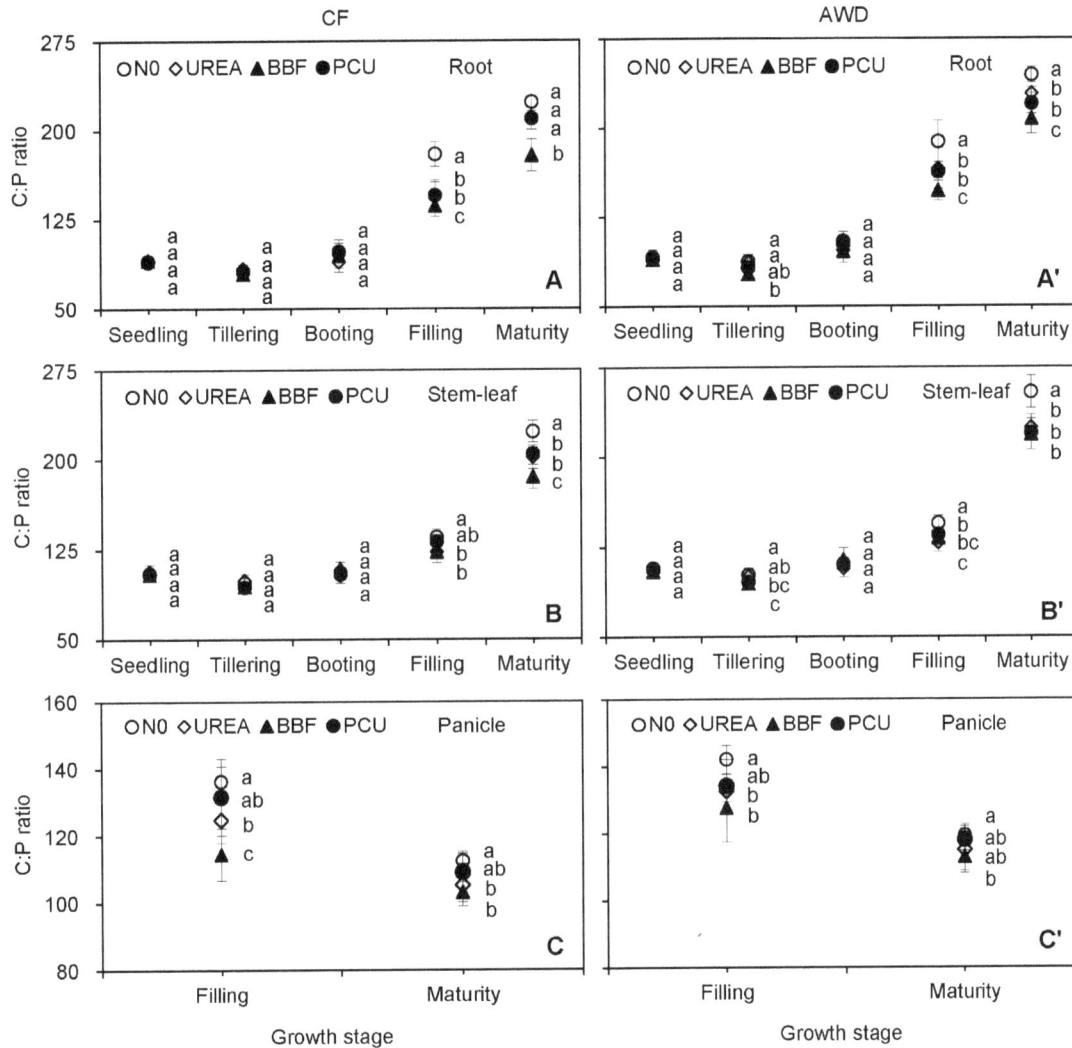

Figure 7. Seasonal changes of C:P ratio in root (A, A'), stem-leaf (B, B'), and panicle (C, C') of late-season rice under different water and N managements (2-year average). Vertical bars represent ± standard deviation of the mean (n = 6). The different letters listed around bars represent significant differences at P<0.05.

Seasonal changes of C, N and P accumulation and partitioning under different water and N managements

An unsynchronized tissue C, N and P accumulation was observed during rice growth. The highest accumulation of C, N and P in root, stem-leaf, and panicle emerged at different growth stages. N and P uptakes in root and stem-leaf were much lower at maturity than those at filling or even booting stage, while C, N and P accumulation in panicle all peaked at final harvest (Fig. 4). These results illustrated that carbohydrate provided by photosynthesis and nutrient derived by N and P remobilization were translocated from the senesced vegetative parts to the newly generated reproductive parts during the reproductive period [3]. The different behaviors among plant C, N and P accumulation were possibly due to the high flexibility of element composition and migration as a result of the trade-off between elements uptake and storage efficiency by the plant [34].

The complete knowledge of element allocation among plant organs is critical to evolutionary explanations of plant functional diversity [35]. As for the C, N and P partitioning of late-season rice in this study (Table 1), seasonal decreasing trends in root vs.

increasing trends in panicle were exhibited. Stem-leaf was more like a transmission part between root and panicle, as its element partitioning initially increased during the early vegetative period and then decreased during the late reproductive period. Such seasonal allocation patterns of C, N and P were closely linked to (1) the obvious remobilization and retranslocation of C, N and P from root to aboveground parts before heading, and from stem-leaf to panicle hereafter, (2) the corresponding seasonal changes of biomass partitioning in different organs [4], and (3) the combined changes in plant nutrient absorbability, leaf phenology, and soil nutrient availability [32,33]. Besides, C, N and P proportions in panicle increased substantially from 34.8%, 36.5% and 35.7% at filling to 47.9%, 52.1% and 63.9% at maturity (Table 1). The rapid increases in panicle N and P fractions were in line with the data reported by Yang et al. [5,7] who found that the average N and P proportions were 10% and 7% at heading and 61% and 56% at maturity in panicle to the total aboveground rice plant. A similar pattern of element allocation was also found in maize [2] and wheat [3], demonstrating that the grain rather than stem-leaf was the major and most active sink for assimilated C, N and P in field crops. Furthermore, the reproductive allocation of N (48.2–

Table 2. Spearman rank correlations among tissue C:N:P stoichiometric ratios and grain yield of late-season rice at harvest.

Trait[a]	$C{:}N_R$	$C{:}N_{SL}$	$C{:}N_P$	$N{:}P_R$	$N{:}P_{SL}$	$N{:}P_P$	$C{:}P_R$	$C{:}P_{SL}$	$C{:}P_P$	GY
$C{:}N_R$	1									
$C{:}N_{SL}$	0.648**	1								
$C{:}N_P$	0.336*	0.509**	1							
$N{:}P_R$	-0.870**	-0.618**	-0.390**	1						
$N{:}P_{SL}$	-0517**	-0.559**	-0.532**	0.667**	1					
$N{:}P_P$	-0.315*	-0.332*	-0.661**	0.385**	0.677**	1				
$C{:}P_R$	0.327*	0.365*	0.366*	0.031ns	-0.073ns	-0.010ns	1			
$C{:}P_{SL}$	0.494**	0.605**	0.374**	-0.281ns	0.076ns	0.072ns	0.637**	1		
$C{:}P_P$	0.306*	0.302*	0.307*	-0.114ns	0.163ns	0.437**	0.504**	0.651**	1	
GY	-0.699**	-0.633**	-0.302*	0.852**	0.832**	0.458**	0.200ns	-0.158ns	0.094ns	1

Data from both CF and AWD treatments in both years were included (n = 48).

[a]$C{:}N$, $N{:}P$, and $C{:}P$: C:N, N:P and C:P ratios in different plant tissues, i refers to root (R), stem-leaf (SL), and panicle (P), respectively. GY: grain yield (data was drawn from Ye et al. [4]).

*Significant at $P<0.05$.

**Significant at $P<0.01$.

ns No significant.

58.5%) and P (61.1–66.4%) to the panicle exceeded that of C (46.2–48.4%) at harvest (Table 1), implying that the grain reproduction required higher fractions of N and P than C in rice.

Since intermittent irrigation has certain influence on the microclimatic conditions of paddy fields [1,36], AWD would affect crop C assimilation and nutrient uptake. In fact, plant C and N accumulation was not remarkably affected by irrigation regimes through the rice growing seasons (Figs. 4A and 4B), whereas plant P accumulation was decreased by AWD compared with CF after seedling (Fig. 4C), resulting from the decreased available P in soil under the intermittent irrigation. This phenomenon implied that the AWD irrigation influenced P accumulation in a different way from its effect on the C and N accumulation. Water regimes affected tissue C, N and P partitioning after tillering (Table 1). In general, AWD obtained lower proportions of C, N and P in stem-leaf but higher ones in root and panicle (although not in all cases significantly) especially where N supply was abundant, mainly due to the functional shifts of plant organs and the distributional differences of biomass and nutrients [33] in response to the water-saving irrigation.

N fertilization greatly enhanced tissue C, N and P accumulation after transplanting (Fig. 4), resulting from both increased tissue element concentration (Figs. 1, 2 and 3) and biomass [4]. Meanwhile, N fertilizer (especially the PCU) application delayed the emerging of peak values of C and N accumulation in root and stem-leaf, owing to the extended soil N availability. In addition, the maximum rates of average C, N and P accumulation always appeared at the booting stage, mostly attributing to the fact that fast-growing plant tissues need relatively more elements to support rapid rate of cell proliferation. In order to take full advantage of the high uptake rate of N during the middle growth period and facilitate N harvesting during the late growth period, the delay of N release was necessary, which could be realized by the CRNFs [5]. In fact, the PCU enhanced not only N uptake but also C and P accumulation in panicle, shoot and whole plant compared with BBF and UREA at an equivalent N rate (Fig. 4). As a compound CRNF product, BBF failed to achieve a comparative effect on the increment of element accumulation as the PCU, owing to the lower components of controlled-release N source (70%) and the less effectiveness in controlling N release than the PCU [4]. N fertilization also altered N partitioning in different plant tissues. Zero N addition resulted in the lowest N partitioning in root (7.5%) and stem-leaf (34.1%) but the highest one in panicle (58.4%) at harvest (Table 1), which was in agreement with Yang et al. [5], suggesting that rice has evolved some internal regulation and conveyance strategies [37] to promote the absorbed N preferentially transferred from vegetative part to the reproductive organ when confronted with soil N deficiency.

Seasonal variations of C:N:P ratios in different tissues under different water and N managements

C:N:P stoichiometry is an important and sensitive index reflecting diverse ecological processes. Knowledge on the C:N ratio of crop residues is of great importance for modelling C and N dynamics in agricultural systems [38]. We found that the stem-leaf C:N ratio showed an overall uptrend after seedling (Fig. 5), agreeing with the results of Ruan et al. [39] who observed that leaf C:N ratio increased from 20.8 at heading to 30.6 at filling and then to 32.0 at maturity in hybrid rice varieties. The stem-leaf C:P ratio exhibited a similar increasing trend as stem-leaf C:N ratio, owing to the reduced allocation of N and P to the senesced leaves. Panicle, which had higher C, N and P concentrations than root and stem-leaf, presented the lowest C:N, N:P and C:P ratios at harvest (Figs. 5, 6 and 7). Gan et al. [38] found that C:N ratio of seed (6–17) was significantly lower than those of straw (14–55) and root (17–75) at maturity in oilseed and pulse crops. Bélanger et al. [25] reported that grain N:P ratio (2.6–7.4) was lower and less variable than those of whole shoot (3.6–12.9) and the uppermost collared leaf (6.8–16.6) in maize. These results revealed that the storage-related tissues (panicle/seed) which optimized N and P use relative to C assimilation for grain production possessed lower C:N:P ratios than the growth-related tissues (stem/leaf).

Water availability that influences leaf phenology and photosynthesis rate could affect plant growth, nutritional status, and ultimately the stoichiometric ratios [32]. In this study, AWD irrigation did not significantly affect the tissue C:N ratio (Fig. 5), because neither tissue C nor N concentrations varied significantly under water-saving irrigation (Figs. 1 and 2). Besides, rice was unlikely to be water-limited under current experimental conditions because the AWD irrigation experienced in this study was within the 'Safe AWD' threshold (water level at 150 mm below the soil surface), and the soil in AWD plots was kept relatively wet and saturated during the non-submerged periods [4]. Plant N:P and C:P ratios were significantly enhanced by AWD irrigation at filling and maturity stages (Figs. 6 and 7), resulting from the reduced tissue P concentration under the alternate submergence-nonsubmergence (Fig. 3). These results indicated that the C:N ratio was much less sensitive to water conditions than the other two stoichiometric ratios, and confirmed that the AWD irrigation had important implications for plant-mediated C, N and P biogeochemical cycling in rice production systems. Lü et al. [32] noted that when water availability was enhanced, higher tissue P and unaltered N concentrations resulted in lower foliar N:P ratio in grassland plant species. Sardans & Peñuelas [11] pointed out that plants under drought tended to have high C:P ratio in leaves and litters.

Terrestrial plants change their C:N and N:P ratios in response to changes in N availability [20,32]. N dynamics have been proved to drive stoichiometric shifts in both plant tissues and mineral soils [31]. It has, in fact, become a widely stated view that increasing soil N availability through N fertilization or atmospheric N deposition would increase N:P but decrease C:N ratios in plants [20,25,32,39]. For instance, a meta-analysis demonstrated that N addition significantly increased the N:P but reduced the C:N ratios of the photosynthetic tissues of woody and herbaceous plants [20]. Our results were in accordance with these studies because N fertilization obtained higher increases in tissue N concentration than P concentration but had no visible effect on tissue C concentration (Figs. 1, 2 and 3). Although the responses of plant C:N and N:P ratios to N addition have been widely investigated, the response of C:P ratio to N addition has received less attention. We found that N application decreased tissue C:P ratio in the late growth period (Fig. 7), reflecting that N availability had significant impacts on P uptake (Fig. 3) but little effects on C assimilation (Fig. 1).

Implication of the C:N:P stoichiometry for evaluating nutrient limitation in rice production systems

Plant C:N:P stoichiometry determines plant community composition and structure (resource allocation), trophic dynamics, and nutrient limitation [19,20,33]. Among those stoichiometric ratios, N:P ratio is the most widely investigated because it reflects important biochemical constraints on relative investments in N-

rich proteins (approx. 16% by mass) and the P-rich ribosomal RNA used to generate them [19]. Though fluctuating across a broad range (approx. 1–100) in individual measurements [15], N:P ratio, with high diagnostic value for nutrient limitation [22], offers a simple and cost-effective tool for evaluating the limiting patterns of nutrient from individual plant species to terrestrial ecosystems [6,19,40,41]. Based on theoretical considerations and laboratory results, the 'Liebig's law of the minimum' for N and P suggests that there is a 'critical (also called ideal or optimal) N:P ratio' below which plant growth is limited only by N and above which growth is limited only by P, and within which growth is co-limited by both elements [15,22,40,42]. Tremendous researches were carried out on this critical N:P ratio in natural vegetations. Koerselman & Meuleman [40] suggested an optimal aboveground biomass N:P ratio of 14–16 in wetland plant species by a review of 40 fertilization studies. Güsewell [15] proposed a broader range of 10–20 as the ideal N:P ratio in terrestrial plants from short-term fertilization experiments. Knecht & Göransson [42] sorted out an average optimal N:P ratio of about 10.0 in field-grown terrestrial plants from published data. Sadras et al. [22] reported that the critical N:P ratio was 4.5 for oilseed crops (n = 81), 5.6 for cereals (n = 134) and 8.7 for legumes (n = 52), and stated that over 40% of cereal and oilseed crops attaining maximum yield had N:P ratios in a relatively narrow range between 4 and 6. Aulakh & Malhi [43] summarized that the main cereal crops, such as rice, wheat, and corn, typically had optimal N:P ratios in both grain and straw in a fairly narrow range of 4.2–6.7. As for rice, Witt et al. [44] simulated balanced nutrient uptakes of 14.7 kg N and 2.6 kg P per ton of grain with a corresponding N:P ratio of 5.7. In this study, the rice aboveground plant N:P ratio (excluding N0 treatment) ranged from 4.4 to 6.4 through the growing seasons, which was somewhat lower than those critical N:P ratios in other plant species mentioned above (10–20) but still remained within the normal range (4.2–6.7), suggesting that the growth of rice in this region was co-limited by N and P. Collectively, these findings provided valuable insights into the mechanisms underpinning plant essential elements cycling in response to simultaneous changes in water and N availability, and broadened the knowledge of the C:N:P stoichiometry in subtropical high-yield rice systems.

Supporting Information

File S1 Table S1, Combined analysis of variance (F values) for C, N and P concentrations in root, stem-leaf, and panicle of late-season rice at various growth stages under different water and N managements in 2010–2011. **Table S2,** Combined analysis of variance (F values) for C, N and P accumulation and partitioning in root, stem-leaf, and panicle of late-season rice at various growth stages under different water and N managements in 2010–2011. **Table S3,** Combined analysis of variance (F values) for C:N:P stoichiometric ratios in root, stem-leaf, and panicle of late-season rice at various growth stages under different water and N managements in 2010–2011.

Author Contributions

Conceived and designed the experiments: XL. Performed the experiments: YY. Analyzed the data: YY. Contributed reagents/materials/analysis tools: YC YJ LL CZ. Wrote the paper: YY.

References

1. Mahajan G, Chauhan BS, Timsina J, Singh PP, Singh K (2012) Crop performance and water- and nitrogen-use efficiencies in dry-seeded rice in response to irrigation and fertilizer amounts in northwest India. Field Crops Research 134: 59–70.

2. Ning P, Li S, Yu P, Zhang Y, Li C (2013) Post-silking accumulation and partitioning of dry matter, nitrogen, phosphorus and potassium in maize varieties differing in leaf longevity. Field Crops Research 144: 19–27.

3. Dordas C (2009) Dry matter, nitrogen and phosphorus accumulation, partitioning and remobilization as affected by N and P fertilization and source-sink relations. European Journal of Agronomy 30: 129–139.

4. Ye YS, Liang XQ, Chen YX, Liu J, Gu JT, et al. (2013) Alternate wetting and drying irrigation and controlled-release nitrogen fertilizer in late-season rice. Effects on dry matter accumulation, yield, water and nitrogen use. Field Crops Research 144: 212–224.

5. Yang LX, Huang HY, Yang HJ, Dong GC, Liu HJ, et al. (2007a) Seasonal changes in the effects of free-air CO_2 enrichment (FACE) on nitrogen (N) uptake and utilization of rice at three levels of N fertilization. Field Crops Research 100: 189–199.

6. Ågren GI, Martin Wetterstedt JÅ, Billberger MFK (2012) Nutrient limitation on terrestrial plant growth - modeling the interaction between nitrogen and phosphorus. New Phytologist 194: 953–960.

7. Yang LX, Wang YL, Huang JY, Zhu JG, Yang HJ, et al. (2007b) Seasonal changes in the effects of free-air CO_2 enrichment (FACE) on phosphorus uptake and utilization of rice at three levels of nitrogen fertilization. Field Crops Research 102: 141–150.

8. Ju XT, Xing GX, Chen XP, Zhang SL, Zhang LJ, et al. (2009) Reducing environmental risk by improving N management in intensive Chinese agricultural systems. Proceedings of the National Academy of Sciences of the United States of America 106: 3041–3046.

9. Ågren GI (2008) Stoichiometry and nutrition of plant growth in natural communities. Annual Review of Ecology, Evolution, and Systematics 39: 153–170.

10. Pampolino MF, Laureles EV, Gines HC, Buresh RJ (2008) Soil carbon and nitrogen changes in long-term continuous lowland rice cropping. Soil Science Society of America Journal 72: 798–807.

11. Sardans J, Peñuelas J (2012) The role of plants in the effects of global change on nutrient availability and stoichiometry in the plant-soil system. Plant Physiology 160: 1741–1761.

12. Lal R (2004) Soil carbon sequestration impacts on global climate change and food security. Science 304: 1623–1627.

13. Kim HY, Lim SS, Kwak JH, Lee DS, Lee SM, et al. (2011) Dry matter and nitrogen accumulation and partitioning in rice (*Oryza sativa* L.) exposed to experimental warming with elevated CO_2. Plant and Soil 342: 59–71.

14. Seneweera S (2011) Effects of elevated CO_2 on plant growth and nutrient partitioning of rice (*Oryza sativa* L.) at rapid tillering and physiological maturity. Journal of Plant Interactions 6: 35–42.

15. Güsewell S (2004) N:P ratios in terrestrial plants: variation and functional significance. New Phytologist 164: 243–266.

16. Elser JJ, Fagan WF, Denno RF, Dobberfuhl DR, Folarin A, et al. (2000) Nutritional constraints in terrestrial and freshwater food webs. Nature 408: 578–580.

17. Ågren GI, Weih M (2012) Plant stoichiometry at different scales: element concentration patterns reflect environment more than genotype. New Phytologist 194: 944–952.

18. Elser JJ, Fagan WF, Kerkhoff AJ, Swenson NG, Enquist BJ (2010) Biological stoichiometry of plant production: metabolism, scaling and ecological response to global change. New Phytologist 186: 593–608.

19. Sterner RW, Elser JJ (2002) Ecological Stoichiometry: the Biology of Elements from Molecules to the Biosphere. Princeton University Press, Princeton.

20. Sardans J, Rivas-Ubach A, Peñuelas J (2012) The C:N:P stoichiometry of organisms and ecosystems in a changing world: A review and perspectives. Perspectives in Plant Ecology, Evolution and Systematics 14: 33–47.

21. Greenwood DJ, Karpinets TV, Zhang K, Bosh-Serra A, Boldrini A, et al. (2008) A unifying concept for the dependence of whole-crop N:P ratio on biomass: theory and experiment. Annals of Botany 102: 967–977.

22. Sadras VO (2006) The N:P stoichiometry of cereal, grain legume and oilseed crops. Field Crops Research 95: 13–29.

23. Bao SD (2000) Soil and Agricultural Chemistry Analysis, 3rd ed. China Agriculture Press, Beijing, 268–270. (in Chinese).

24. Ziadi N, Bélanger G, Cambouris AN, Tremblay N, Nolin MC, et al. (2007) Relationship between P and N concentrations in corn. Agronomy Journal 99: 833–841.

25. Bélanger G, Claessens A, Ziadi N (2012) Grain N and P relationships in maize. Field Crops Research 126: 1–7.

26. Zhang HY, Wu HH, Yu Q, Wang ZW, Wei CZ, et al. (2013) Sampling date, leaf age and root size: implications for the study of plant C:N:P stoichiometry. Plos One 8: e60360.

27. Gifford RM, Barrett DJ, Lutze JL (2000) The effects of elevated [CO_2] on the C:N and C:P mass ratios of plant tissues. Plant and Soil 224: 1–14.

28. Yu Q, Wu HH, He NP, Lü XT, Wang ZP, et al. (2012) Testing the growth rate hypothesis in vascular plants with above- and below-ground biomass. Plos One 7: e32162.

29. Matzek V, Vitousek PM (2009) N:P stoichiometry and protein:RNA ratios in vascular plants: an evaluation of the growth-rate hypothesis. Ecology Letters 12: 765–771.

30. Haynes RJ, Swift RS (1989) The effects of pH and drying on adsorption of phosphate by aluminium-organic matter associations. Journal of Soil Science 40: 773–781.

31. Yang YH, Luo YQ, Lu M, Schadel C, Han WX (2011) Terrestrial C:N stoichiometry in response to elevated CO_2 and N addition: a synthesis of two meta-analyses. Plant and Soil 343: 393–400.

32. Lü XT, Kong DL, Pan QM, Simmons ME, Han XG (2012) Nitrogen and water availability interact to affect leaf stoichiometry in a semi-arid grassland. Oecologia 168: 301–310.

33. Zheng SX, Ren HY, Li WH, Lan ZC (2012) Scale-dependent effects of grazing on plant C:N:P stoichiometry and linkages to ecosystem functioning in the Inner Mongolia grassland. Plos One 7: e51750.

34. Abbas M, Ebeling A, Oelmann Y, Ptacnik R, Roscher C, et al. (2013) Biodiversity effects on plant stoichiometry. Plos One 8: e58179.

35. Kerkhoff AJ, Fagan WF, Elser JJ, Enquist BJ (2006) Phylogenetic and growth form variation in the scaling of nitrogen and phosphorus in the seed plants. The American Naturalist 168: E103–E122.

36. Mao Z (2001) Water efficient irrigation and environmentally sustainable irrigated rice production in China. ICID website. Available: http://www.icid.org/wat_mao.pdf. Accessed 2014 Jun 12.

37. Rentsch D, Schmidt S, Tegeder M (2007) Transporters for uptake and allocation of organic nitrogen compounds in plants. FEBS Letters 581: 2281–2289.

38. Gan YT, Liang BC, Liu LP, Wang XY, McDonald CL (2011) C:N ratios and carbon distribution profile across rooting zones in oilseed and pulse crops. Crop & Pasture Science 62: 496–503.

39. Ruan XM, Shi FZ, Luo ZX (2011) Effects of nitrogen application on the leaf C:N and nitrogen uptake and utilization at later developmental stages in different high yield hybrid rice varieties. Soil and Fertilizer Sciences in China (2): 35–38. (in Chinese).

40. Koerselman W, Meuleman AFM (1996) The vegetation N:P ratio: a new tool to detect the nature of nutrient limitation. Journal of Applied Ecology 33: 1441–1450.

41. Tessier JT, Raynal DJ (2003) Use of nitrogen to phosphorus ratios in plant tissue as an indicator of nutrient limitation and nitrogen saturation. Journal of Applied Ecology 40: 523–534.

42. Knecht MR, Göransson A (2004) Terrestrial plants require nutrients in similar proportions. Tree Physiology 24: 447–460.

43. Aulakh MS, Malhi SS (2005) Interactions of nitrogen with other nutrients and water: effect on crop yield and quality, nutrient use efficiency, carbon sequestration, and environmental pollution. Advances in Agronomy 86: 341–409.

44. Witt C, Dobermann A, Abdulrachman S, Gines HC, Guanghuo W, et al. (1999) Internal nutrient efficiencies of irrigated lowland rice in tropical and subtropical Asia. Field Crops Research 63: 113–138.

16

Waterborne Outbreak of Gastroenteritis: Effects on Sick Leaves and Cost of Lost Workdays

Jaana I. Halonen[1]*, Mika Kivimäki[1,2], Tuula Oksanen[1], Pekka Virtanen[3], Mikko J. Virtanen[4], Jaana Pentti[1], Jussi Vahtera[1,5]

1 Finnish Institute of Occupational Health, Helsinki, Finland, **2** Department of Epidemiology and Public Health, University College of London, London, United Kingdom, **3** Tampere School of Health Sciences, University of Tampere, Tampere, Finland, **4** Epidemiologic Surveillance and Response Unit, National Institute for Health and Welfare, Helsinki, Finland, **5** Department of Public Health, University of Turku, and Turku University Hospital, Turku, Finland

Abstract

Background: In 2007, part of a drinking water distribution system was accidentally contaminated with waste water effluent causing a gastroenteritis outbreak in a Finnish town. We examined the acute and cumulative effects of this incidence on sick leaves among public sector employees residing in the clean and contaminated areas, and the additional costs of lost workdays due to the incidence.

Methods: Daily information on sick leaves of 1789 Finnish Public Sector Study participants was obtained from employers' registers. Global Positioning System-coordinates were used for linking participants to the clean and contaminated areas. Prevalence ratios (PR) for weekly sickness absences were calculated using binomial regression analysis. Calculations for the costs were based on prior studies.

Results: Among those living in the contaminated areas, the prevalence of participants on sick leave was 3.54 (95% confidence interval (CI) 2.97–4.22) times higher on the week following the incidence compared to the reference period. Those living and working in the clean area were basically not affected, the corresponding PR for sick leaves was 1.12, 95% CI 0.73–1.73. No cumulative effects on sick leaves were observed among the exposed. The estimated additional costs of lost workdays due to the incidence were 1.8–2.1 million euros.

Conclusions: The prevalence of sickness absences among public sector employees residing in affected areas increased shortly after drinking water distribution system was contaminated, but no long-term effects were observed. The estimated costs of lost workdays were remarkable, thus, the cost-benefits of better monitoring systems for the water distribution systems should be evaluated.

Editor: Qamaruddin Nizami, Aga Khan University, Pakistan

Funding: This study was funded by the Academy of Finland (projects 124271, 124322, 129262, and 132944). Mika Kivimäki is supported by the BUPA Foundation, UK; and the New OSH ERA research programme. The funders had no role in study design, data collection and analysis, decision to publish, or preparation of the manuscript.

Competing Interests: The authors have declared that no competing interests exist.

* E-mail: jaana.halonen@ttl.fi

Introduction

Regardless of the developed systems for drinking water purification and distribution, and the active legislation towards safe water [1], microbial contamination of drinking water and waterborne infections keep occurring also in developed societies [2,3,4,5,6,7,8]. Large populations may become affected if public water systems are contaminated. These incidences may be due to system deficiencies, improper water treatment or contamination of one part of, or the whole, water supply [3,4,7]. In most developed countries with public water distribution systems, however, rapid actions for managing the incidences can be taken to limit the harms; clean water can often be provided by temporary arrangements and the distribution systems can be disinfected. Severities of the waterborne outbreaks are often evaluated by calculating excess cases or risk rates for morbidity among the exposed [2,4,5]. However, the effects of the waterborne infections

on acute and cumulative sickness absences (*Campylobacter*, *Salmonella* and *Giardia* infections may also have long-term health effects [9,10,11,12]) and the related costs of lost workdays have rarely been determined [13].

In late November 2007 part of the municipal drinking water system in a Finnish town Nokia, a municipality participating in the Finnish Public Sector Study, was contaminated with wastewater effluent (including *Campylobacter* sp., Norovirus, *Giardia*, and *Salmonella* sp.) causing an outbreak of gastroenteritis [6]. In this study, we determined how this epidemic affected the sick leaves immediately and in the following year among employees of the Public Sector Study cohort who resided in Nokia. We hypothesized that employees living in the contaminated areas would be more affected than those living in the non-contaminated "clean" areas. We also estimated the additional costs of lost workdays for all branches of industry during the acute phase of the epidemic.

Methods

Ethics statement

All the register data obtained from the national registers were based on the legal permissions granted by institutes maintaining these registers, and data were analyzed anonymously. According to Finnish Personal Data Act (523/1999, Chapter 4, Section 14) [14], written consent was not needed for research that uses register data. The Coordinating Ethics Committee of Helsinki and Uusimaa Hospital District has approved the study.

The Finnish Public Sector Study cohort consists of employees working for 10 municipalities (including Nokia and neighbouring town Tampere) and 21 hospitals (including Tampere University Hospital). All men and women employed by these organizations for more than six months in any year between 1991 and 2005 were eligible for the cohort (n = 151 618). Of this cohort, 1789 members were employed and resided in Nokia during the outbreak (weeks 48/2007–8/2008 referred to as week of outbreak = 0, and weeks after +1, +2, etc. up to week +12) and formed the analytic sample of this study. Because in Finland female-dominated occupations are common in the public sector (nurses, kindergarten teachers, cleaners etc.) a majority of the study population were women. All information about cohort members' job contracts, age, sex and sick leaves was obtained from the employers' registers. For each day we obtained data on all sick leaves (due to own illness), and sick-child leaves (due to illness of a child <10 years), however, diagnoses for the illnesses were not available as they are not recorded by the employers. Sick days (up to 60 days) and sick-child days (allowed absence up to four days at one time) are paid by the employer and, thus, carefully recorded. Residential addresses of the cohort members and Global Positioning System (GPS)-coordinates of these addresses were obtained from Population Register Center. Coordinates of the buildings alongside the contaminated water line were in the National Institute for Health and Welfare. Using these coordinates the cohort members' residences we linked to the buildings to which contaminated drinking water was distributed. The locations of cohort participants' residences in relation to areas where contaminated and clean drinking water were supplied are shown in Figure 1.

The incidence has been described in detail previously [6]. In short, during maintenance work at the Nokia wastewater plant on November 28th 2007 (week 0) a valve connecting wastewater effluent line and drinking water line was accidentally opened, and in some areas inhabitants started to receive contaminated municipal drinking water (referred to as contaminated areas: inner pattern in Figure 1). This area was approximately 18 km^2 in size (whole town 347 km^2), and defined based on the addresses and GPS-coordinates of the buildings along the branch of the contaminated distribution system. Two days later, an increase in the number of patients with symptoms of gastroenteritis was observed at the local health care center. An estimated 450 m^3 of wastewater effluent had been mixed to the drinking water and distributed to the customers, until the water contamination was realized. Recommendation for boiling tap water was issued the same day, clean water distribution organized by municipal officials, the military, and volunteers was started, as well as increased chlorination of the distribution system. After the incident schools and day care centres were closed due to Christmas holiday on weeks +4 and +5. The last restrictions in water use were canceled on February 18th 2008 (week +12).

Statistical analyses

We calculated the weekly prevalence of study population taking at least one sick day or sick-child day during each week for seven weeks before, during, and 12 weeks after the incidence, separately for those living in the contaminated and clean areas. For contaminated and clean areas cumulative sick days per person year were also calculated for the 52 weeks following the incidence. This was because there is evidence of long-term health effects of *Campylobacter* infections (rheumatological complaints, symptoms of arthritis, inflammatory bowel disease) [9,10,15], *Giardia* infections (functional gastrointestinal diseases) [12], and *Salmonella* infections (inflammatory bowel disease) [15].

We calculated the prevalence ratios (PR) and 95% confidence intervals (CI) for weekly sickness absences using log-binomial regression analyses with GENMOD procedure of SAS [16], adjusting for age and sex. The reference period used in the analyses was the mean weekly prevalence of participants taking at least one sick day during the seven weeks prior to the incidence (weeks −7 to −1). The same method was used for the subgroup analyses of the cohort members consisting of those employed by schools and day care centers (n = 361). For them, the data included the location of their residence and workplace, and information of whether these were in the contaminated or clean area. Thus, we could estimate more explicitly the role of clean and contaminated areas in the sickness absences.

The additional cost of lost workdays on weeks 0 to +3 was estimated for the total number of employees residing in Nokia based on the number of additional (compared to reference period) personal and sick-child days per person among the cohort participants. The costs were estimated at 280 and 330 euros per day; the mean estimated costs of sickness absence for the service sector (200–250 €/day), industry (300–350 €/day) [17] and government employees (340–380 €/day) [18] in Finland. Of the total workforce in Nokia in 2007 (n = 10,719), 30% was employed by the government and 32% by industry. These estimations considered the contamination status of the residential area, and were calculated by sex (n of employed women in town = 5020, n of employed men in town = 5699) because the sex distribution of the analytic sample was skewed and women generally have higher rates of sick leaves than men [19].

Results

The total population in Nokia was 30 016 of which 9538 (32%) lived in the area of contaminated water distribution system. Correspondingly, of the 1789 cohort members 586 (33%) lived in the contaminated area. Of the study population 1419 (79.5%) were women, and the mean age was 45.9 years (standard deviation 9.4). The number of cohort members who worked at schools and day care centers in Nokia was 361, of which 174 lived and worked in the clean area, and 187 lived or worked, or both, in the contaminated area.

Sick leaves increased sharply after the contamination among those who lived in the contaminated area, but not among those living in the clean area for whom the prevalence returned back to the pre-incident level in three weeks (Figure 2). The prevalence of cohort members taking at least one sick day during the week immediately after the contamination (week +1) was 35%, whereas in the week before the contamination the corresponding figure was 10% (week −1). The prevalence of participants at sick-child leave also peaked in the contaminated area on week +1, being 4.3%, whereas the week before (week −1) the level was 1.3% (data not shown). The cumulative number of sick days grew at a faster rate during the first post-incident weeks among those who lived in the contaminated compared to clean area (Figure 3). However, later the curves for cumulative sick days developed similarly for the

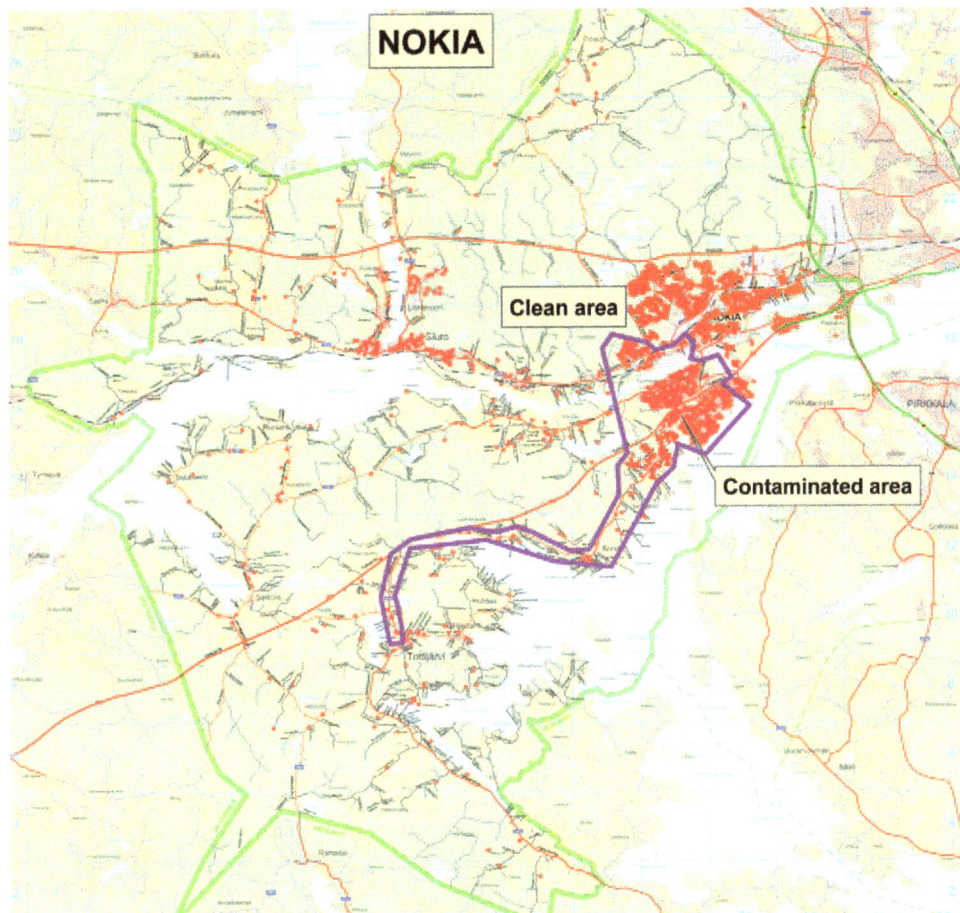

Figure 1. Distribution of cohort participants within areas receiving contaminated and clean drinking water. Dots in the inner pattern represent cohort participants' residences receiving contaminated water.

exposed and non-exposed and by the end of the year the difference had disappeared.

Compared to the reference period, the proportion of participants at sick leave in the contaminated area was the highest on the week following the incidence after which the prevalence ratio decreased gradually (Table 1). A slight increase in the sick leaves

was observed also among those living in the clean area with the highest prevalence ratio for week +1. In the subgroup of personnel of schools and day care centers, we found that for those living and working in clean areas, the prevalence ratios for sick leaves remained at the same level after, compared to prior, the contamination (Table 2).

For the cost estimations we calculated weekly additional sick days (compared to the reference period) in this study population. For the clean and contaminated areas, the sum of additional sick

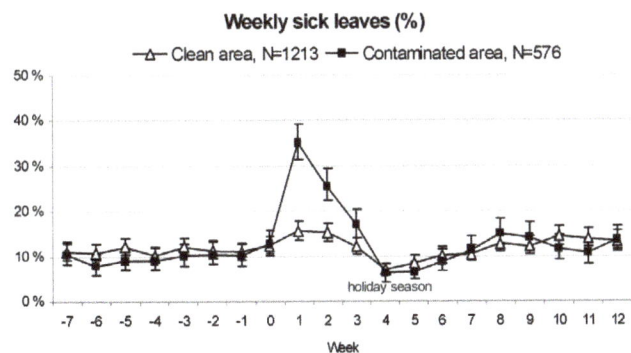

Figure 2. Weekly percentages of participants at sick leave before and after the contamination of water distribution system. Percentages are given by the contamination status of the participants' residence.

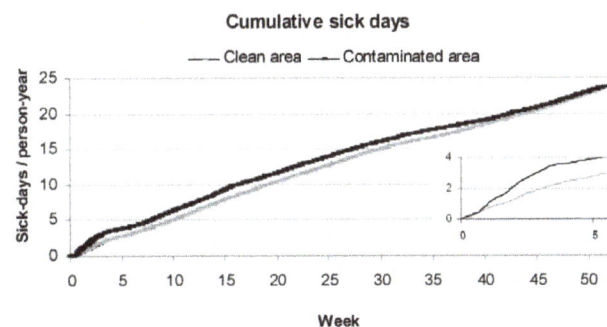

Figure 3. Cumulative sick days for the 52 weeks following the contamination by the contamination status of residence.

Table 1. Prevalence ratios (PR, 95% confidence intervals) for weekly sick leaves among public sector employees after the contamination of drinking water distribution system by contamination status of residence.

Week	Clean area (n = 1203[a])			Contaminated area (n = 586[b])		
	PR[c]	95% CI		PR[c]	95% CI	
−7 to −1 (ref)	1			1		
0	1.16	1.02	1.31	1.23	0.99	1.52
+1	1.41	1.23	1.61	3.54	2.97	4.22
+2	1.41	1.23	1.62	2.63	2.18	3.16
+3	1.17	1.01	1.36	1.88	1.52	2.31
+4–12	1.12	1.00	1.26	1.18	0.99	1.40

[a]Number of participants living in clean area;
[b]Number of participants living in contaminated area,
[c]Models adjusted for age and sex.

Table 2. Prevalence ratios (PR, 95% confidence intervals) of weekly sick leaves among the personnel of schools and day care centres after the contamination of drinking water distribution system by contamination status of residence and workplace.

Week	Clean area[a] (n = 174)			Contaminated area[b] (n = 187)		
	PR[c]	95% CI		PR[c]	95% CI	
−7 to −1 (ref)	1			1		
0	0.90	0.56	1.42	1.21	0.80	1.85
+1	1.12	0.73	1.73	3.43	2.58	4.55
+2	1.10	0.71	1.73	2.89	2.14	3.89
+3	1.03	0.67	1.58	1.36	0.93	2.00
+4–12	1.16	0.88	1.55	1.23	0.96	1.61

[a]Residence and workplace in the clean area;
[b]Residence or workplace, or both, in the contaminated area;
[c]Models adjusted for age and sex.

days per person on weeks 0 to +3 was 0.71 for women and 0.42 for men. Thus, the estimated auxiliary cost of lost workdays due to sick leaves among all employees residing in Nokia were 1.00–1.18 and 0.67–0.80 million euros for women and men, respectively. Similarly, the sum of additional sick-child days per person were 0.07 and 0.02, and the estimated cost of lost workdays 0.11–0.13 and 0.03–0.04 million euros for women and men, respectively. These resulted in a total of 1.8–2.1 million euros from lost workdays due to own illness and illness of one's child(ren).

Discussion

We studied the effect of microbial contamination of drinking water distribution system on sick leaves among public sector employees. Sick- and sick-child leaves were found to increase sharply after the contamination, and this increase was restricted to people either residing or working in the contaminated areas. However, 1-year accumulation of sick days was similar for exposed and unexposed suggesting the effects of the incidence were short-term. The estimated additional costs of lost workdays due to the incidence were 1.8–2.1 million euros.

Contaminations of water distribution systems in the scale of this study are rare, although some have been reported [2,4,20]. As far as we are aware, the effects of these incidences on sick leaves have scarcely been studied. We found that those residing in the contaminated area were heavily affected, while those residing and working in the clean areas were unaffected. The peak in the proportion of participants at sick leave was observed during the week after the contamination, which is in agreement with the peak in the detected cases of gastroenteritis related to this incidence [6]. This delay is also in line with other studies on waterborne outbreaks [21,22].

For the following year, we found no difference in the sick leave accruals between those living in the contaminated and clean areas during the incidence. This suggests the acute epidemic had no residual effects at a population level. Supporting evidence has been reported in regard of occurrence of reactive arthritis in this study area half a year after the incidence [23]. However, prior studies have shown that the effects of gastroenteritis from drinking water contaminated with *Campylobacter* may materialize later -with a

delay of up to fifteen years- as increase in arthritis symptoms or inflammatory bowel disease [10,15,24], or in case of *Giardia* infection as functional gastrointestinal diseases [12]. Thus, we cannot exclude the possibility that the exposed may still be at increased risk of developing chronic conditions.

We estimated that the excess costs due to sick leaves related to this incidence was 1.8–2.1 million euros, a remarkable part of the total expenditure of the whole incidence that reaches possibly five to six million euros. The other expenses include e.g. the distribution of clean water for the inhabitants, cleaning of the water distribution system, and overtime work of the town and water works employees. As in occupational health in general [25], cost-benefit analyses are needed to estimate whether actions such as online monitoring systems of the water distribution networks would be beneficial in preventing and limiting the magnitude of waterborne outbreaks and all harms related to them. We found that only one study, almost two decades ago, has estimated costs of a waterborne gastroenteritis outbreak. The authors concluded the outbreak affecting 2000–3000 people caused costs of 1.6 million Danish krones (~215 000 euros in current exchange rate) [13]. However, the algorithm of this estimate was not reported, and because of the long delay between studies the comparison of the costs is difficult. We also found that sick leaves and sick-child leaves were more common among women than men, a finding in line with prior research [19]. Indeed, only a fourth of the costs of sick-child leaves were observed among men, which suggests that female-dominated sectors suffer considerably more from the illnesses of children than male-dominated sectors.

Major strengths of this study are the reliable and comprehensive records on the cohort members' sickness absences and information whether contaminated water was distributed to their residences. Additional information on the locations of workplaces for a sub-population of the cohort enabled us a more detailed examination of the differences in sick leaves between the exposed and non-exposed. One limitation of this study is that we could not determine the portion of the sick leaves that was actually related to the increase in infections of the digestive system because data on diagnoses were not available; however, it is unlikely that the observed short-term increase in the sick leaves is due to illnesses not related to the incidence. Because e.g. salaries and insurance costs vary by industry, we also emphasize that the calculations for the costs of sick leaves depend on the country and employer in question. Finally, due to differences in infrastructure and public services, the generalizability of these findings in less developed regions may be limited.

In summary, we have shown that the prevalence of sickness absences among public sector employees residing in affected areas increased considerably after drinking water distribution system was contaminated. The estimated costs of lost workdays were remarkable, thus, the cost-benefits of better monitoring systems for the water distribution systems should be evaluated.

Acknowledgments

The authors thank town of Nokia for providing the map of contaminated and clean areas to be used in this article.

Author Contributions

Conceived and designed the experiments: JIH MK TO JP JV. Analyzed the data: JIH JP MJV. Contributed reagents/materials/analysis tools: MK JP MJV JV PV. Wrote the paper: JIH TO JV. Critically reviewed the article: MV MJV JP PV.

References

1. Zacheus O, Miettinen IT (2011) Increased information on waterborne outbreaks through efficient notification system enforces actions towards safe drinking water. J Water Health 9: 763–772.
2. Yang Z, Wu X, Li T, Li M, Zhong Y, et al. (2011) Epidemiological survey and analysis on an outbreak of gastroenteritis due to water contamination. Biomed Environ Sci 24: 275–283.
3. Craun GF, Brunkard JM, Yoder JS, Roberts VA, Carpenter J, et al. (2010) Causes of outbreaks associated with drinking water in the United States from 1971 to 2006. Clin Microbiol Rev 23: 507–528.
4. Riera-Montes M, Brus Sjolander K, Allestam G, Hallin E, Hedlund KO, et al. (2011) Waterborne norovirus outbreak in a municipal drinking-water supply in Sweden. Epidemiol Infect 20: 1–8.
5. Daly ER, Roy SJ, Blaney DD, Manning JS, Hill VR, et al. (2009) Outbreak of giardiasis associated with a community drinking-water source. Epidemiol Infect 138: 491–500.
6. Laine J, Huovinen E, Virtanen MJ, Snellman M, Lumio J, et al. (2010) An extensive gastroenteritis outbreak after drinking-water contamination by sewage effluent, Finland. Epidemiol Infect 15: 1–9.
7. Brunkard JM, Ailes E, Roberts VA, Hill V, Hilborn ED, et al. (2011) Surveillance for waterborne disease outbreaks associated with drinking water— United States, 2007–2008. MMWR Surveill Summ 60: 38–68.
8. Breitenmoser A, Fretz R, Schmid J, Besl A, Etter R (2011) Outbreak of acute gastroenteritis due to a washwater-contaminated water supply, Switzerland, 2008. J Water Health 9: 569–576.
9. Locht H, Krogfelt KA (2002) Comparison of rheumatological and gastrointestinal symptoms after infection with Campylobacter jejuni/coli and enterotoxigenic Escherichia coli. Ann Rheum Dis 61: 448–452.
10. Garg AX, Pope JE, Thiessen-Philbrook H, Clark WF, Ouimet J (2008) Arthritis risk after acute bacterial gastroenteritis. Rheumatology (Oxford) 47: 200–204.
11. Mann EA, Saeed SA (2012) Gastrointestinal infection as a trigger for inflammatory bowel disease. Curr Opin Gastroenterol 28: 24–29.
12. Hanevik K, Dizdar V, Langeland N, Hausken T (2009) Development of functional gastrointestinal disorders after Giardia lamblia infection. BMC Gastroenterol 9: 27.
13. Laursen E, Mygind O, Rasmussen B, Ronne T (1994) Gastroenteritis: a waterborne outbreak affecting 1600 people in a small Danish town. J Epidemiol Community Health 48: 453–458.
14. Finnish Ministry of Justice (1999) Personal Data Act (523/1999). Available: http://www.finlex.fi/en/laki/kaannokset/1999/en19990523.pdf. Accessed 2012 15 Feb.
15. Gradel KO, Nielsen HL, Schonheyder HC, Ejlertsen T, Kristensen B, et al. (2009) Increased short- and long-term risk of inflammatory bowel disease after salmonella or campylobacter gastroenteritis. Gastroenterology 137: 495–501.
16. Spiegelman D, Hertzmark E (2005) Easy SAS calculations for risk or prevalence ratios and differences. Am J Epidemiol 162: 199–200.
17. Pohjola Pankki Oyj, (Company offering bank and insurance services) Sairauspoissaolokustannus [Sick leave costs]. Available: https://www.pohjola.fi/pohjola/yritys-ja-yhteisoasiakkaat/vakuutukset/vakuutustuotteet/henkilovakuutukset/sairauspoissaolokustannukset?id = 321811&srcpl = 4. Accessed 2012 16 Jan.
18. Lehtonen V Miten hallita sairauspoissaoloja? [in Finnish]. Available: http://www.vm.fi/vm/fi/04_julkaisut_ja_asiakirjat/01_julkaisut/06_valtion_tyomarkkinalaitos/miten_hallita_sairauspoissaoloja.pdf. Accessed 2011 15 Nov. Helsinki: Valtiovarainministeriö [Ministry of finance].
19. Bekker MH, Rutte CG, van Rijswijk K (2009) Sickness absence: A gender-focused review. Psychol Health Med 14: 405–418.
20. Koroglu M, Yakupogullari Y, Otlu B, Ozturk S, Ozden M, et al. (2011) A waterborne outbreak of epidemic diarrhea due to group A rotavirus in Malatya, Turkey. New Microbiol 34: 17–24.
21. Auld H, MacIver D, Klaassen J (2004) Heavy rainfall and waterborne disease outbreaks: the Walkerton example. J Toxicol Environ Health A 67: 1879–1887.
22. Public Health Agency of Canada (2000) Waterborne outbreak of gastroenteritis associated with a contaminated municipal water supply, Walkerton, Ontario, May–June 2000. Can Commun Dis Rep 26: 170–173.
23. Uotila T, Antonen J, Laine J, Kujansuu E, Haapala AM, et al. (2011) Reactive arthritis in a population exposed to an extensive waterborne gastroenteritis outbreak after sewage contamination in Pirkanmaa, Finland. Scand J Rheumatol 40: 358–362.
24. Bremell T, Bjelle A, Svedhem A (1991) Rheumatic symptoms following an outbreak of campylobacter enteritis: a five year follow up. Ann Rheum Dis 50: 934–938.
25. Burdorf A (2011) Do costs matter in occupational health? Occup Environ Med 68: 707–708.

17

Artificial Regulation of Water Level and Its Effect on Aquatic Macrophyte Distribution in Taihu Lake

Dehua Zhao*, Hao Jiang, Ying Cai, Shuqing An

Department of Biological Science and Technology, Nanjing University, Nanjing, China

Abstract

Management of water levels for flood control, water quality, and water safety purposes has become a priority for many lakes worldwide. However, the effects of water level management on the distribution and composition of aquatic vegetation has received little attention. Relevant studies have used either limited short-term or discrete long-term data and thus are either narrowly applicable or easily confounded by the effects of other environmental factors. We developed classification tree models using ground surveys combined with 52 remotely sensed images (15–30 m resolution) to map the distributions of two groups of aquatic vegetation in Taihu Lake, China from 1989–2010. Type 1 vegetation included emergent, floating, and floating-leaf plants, whereas Type 2 consisted of submerged vegetation. We sought to identify both inter- and intra-annual dynamics of water level and corresponding dynamics in the aquatic vegetation. Water levels in the ten-year period from 2000–2010 were 0.06–0.21 m lower from July to September (wet season) and 0.22–0.27 m higher from December to March (dry season) than in the 1989–1999 period. Average intra-annual variation (CV_a) decreased from 10.21% in 1989–1999 to 5.41% in 2000–2010. The areas of both Type 1 and Type 2 vegetation increased substantially in 2000–2010 relative to 1989–1999. Neither annual average water level nor CV_a influenced aquatic vegetation area, but water level from January to March had significant positive and negative correlations, respectively, with areas of Type 1 and Type 2 vegetation. Our findings revealed problems with the current management of water levels in Taihu Lake. To restore Taihu Lake to its original state of submerged vegetation dominance, water levels in the dry season should be lowered to better approximate natural conditions and reinstate the high variability (i.e., greater extremes) that was present historically.

Editor: Christopher Fulton, The Australian National University, Australia

Funding: This work was financially supported by the National Natural Science Foundation of China (No. 31000226)(http://www.nsfc.gov.cn) and the State Key Development Program for Basic Research of China (2008CB418004). The funders had no role in study design, data collection and analysis, decision to publish, or preparation of the manuscript.

Competing Interests: The authors have declared that no competing interests exist.

* E-mail: dhzhao@nju.edu.cn

Introduction

Because of the important ecological and socioeconomic functions of aquatic macrophytes, such as stabilization of sediments, regulation of the nutrient cycle, slowing of water currents and fishery maintenance, numerous studies over the past three decades have focused on the dynamics of aquatic macrophytes in freshwater ecosystems and identification of the forces driving their abundances and distributions [1–4]. Water quality degradation of the world's freshwater ecosystems over the past decades has led to extensive decreases in the area occupied by aquatic macrophytes as well as species losses [5,6]. Promoting the recovery of aquatic macrophytes has become a critical step in the restoration and rehabilitation of these degraded aquatic ecosystems [7–9].

Water levels, which are controlled by both natural conditions (e.g., meteorological and catchment characteristics) and local human activities (e.g., flood-control projects and artificial water transfer) [10], have been thought to be responsible for the variability in biomass and species composition of aquatic macrophytes in many freshwater ecosystems of the world [10–16]. Although artificial management and manipulation of water levels have been practiced widely, the effect of managed water levels on aquatic macrophytes has not been fully understood in most cases because of the complex relationship between macrophytes and water level [10,14,17,18].

Taihu Lake is the third-largest freshwater lake in China, occupying a surface area of 2,425 km^2 [19–21]. Due to rapid industrialization and urbanization, nutrient concentrations have increased continually during the past decades, and eutrophication has become a dominant water quality problem [22]. In an effort to recover the degraded aquatic ecosystem of Taihu Lake, numerous costly water conservation projects have been implemented in recent years. Planting and restoration of aquatic macrophytes for the purpose of removing excess nutrients are key facets of most of these projects [8,23,24].

Meanwhile, large amounts of water have been flushed into the lake from the Yangtze River since 2001 under the premise of "conquering the unmoving with the moving, diluting the polluted with the clean, supplementing low flow with ample flow" to improve water quality and control algal blooms [25,26]. Following the notorious blue-green algal bloom that occurred in the summer of 2007 and which resulted in serious drinking water shortages in Wuxi City [27,28], one of the most economically developed cities in Jiangsu Province, even more water was pumped into the lake [19]. Concurrently, more than 28,000 km of sea walls, river banks, embankments and polder dikes were built to control flooding [19].

As a result, water levels and their dynamics, especially intra-annual dynamics, have changed substantially in Taihu Lake.

Despite the considerable changes in the water levels in Taihu Lake, little attention has been focused on the effects on aquatic macrophytes, even though inter- and intra-annual water levels have been identified as one of the most important forces driving variability in aquatic macrophyte distribution [10,14]. Because aquatic macrophytes are distributed over such a large area (i.e. hundreds of square kilometers) [22], slight variations in water levels are likely to have precipitated a decrease or increase in the distribution of the macrophytes on the scale of tens of kilometers, greatly affecting their combined ability to remove nutrients and act as a source of food for fisheries [29,30]. Many water conservation projects focusing on planting aquatic macrophytes have been conducted [23,24,31], but it is likely more economical to protect and restore the existing communities of aquatic macrophytes. Protection and restoration, however, requires that increased attention be focused on understanding the effects of inter- and intra-annual water levels on aquatic macrophytes in the lake.

Although some authors have found correlations between the variation in aquatic vegetation and water levels in regard to aquatic systems at large temporal scales, most of those studies were based on either limited short-term or discrete long-term data [12,13,32], and thus the results are either narrowly applicable or easily confounded by the cumulative effect of other environmental factors with gradual temporal variation, such as trophic status. In this project, we mapped aquatic vegetation distribution between 1989 and 2010 based on remote sensing images with spatial resolutions from 15 to 30 m. Our objective was to determine the effect of managed water level on the distribution and composition of aquatic macrophytes in Taihu Lake.

Materials and Methods

2.1 Ethics Statement

No specific permits were required for the described field studies. The location studied is not privately-owned or protected in any way, and the field studies did not involve endangered or protected species.

2.2 Study Area

The Taihu Lake catchment plays an important role in China's political economy, containing 3.7% of the country's population, creating 11.6% of Chinese gross domestic product (GDP) and contributing 19% of total revenue while comprising only 0.4% of the land area of China [19]. Since the 1950s and especially since the 1980s, human activities have placed increased pressure on the lake's ecological components [21]. Our study area was limited to the areas identified in the remote sensing images as being covered by water in winter, the season when water levels were lowest and the topsoil of most patches of emergent vegetation was dry. As a result, most emergent vegetation was excluded from this work. We chose to exclude most emergent vegetation in order to reduce the effects of human activities on the relationship between aquatic vegetation and water levels since human activities have drastically altered the emergent vegetation of Taihu Lake through large-scale construction of embankments and buildings, as well as vegetation restoration or destruction [31].

Pen-fishing has also had a large influence on the distribution of aquatic vegetation due to farmers' activities such as planting and harvest. The area subject to pen-fishing has varied dramatically over the past two decades, i.e. a gradual increase between 1990 and 2005 followed by a sudden decrease after 2007, when an extensive blue-green algal bloom occurred and resulted in serious drinking water shortages in Wuxi City [27,28]. However, pen-fishing activities have been limited primarily to the East Bay of Taihu Lake [31,33]. Therefore, the East Bay was also excluded from this study to minimize the confounding influence of pen-fishing on aquatic macrophyte distribution.

From 1960 to 2000, human activities resulted in a worsening of the water quality of Taihu Lake at an approximate rate of one grade every 10–15 years [23]. To improve water quality, much effort has been expended on lake restoration, especially after 2000. Artificial management of water levels through pumping of water as well as construction of embankments and dams was a common strategy for improving water quality and controlling blue-green algal bloom. In particular, increased amounts of water were pumped into Taihu Lake from the Yangtze River after 2000 [19,25]. Therefore, our study period ranged from 1989 to 2010 to encompass the ten years before and after 2000, i.e., Period 1 (1989–1999) and Period 2 (2000–2010). However, high-quality remote sensing images were available for only sixteen of the years between 1989 and 2010.

2.3 Field Surveys

The aquatic vegetation of Taihu Lake was grouped into four types: emergent, floating-leaf, floating and submerged [22,31]. Dominant species in the lake included *Phragmites communis*, *Zizania latifolia*, *Nymphoides peltatum*, *Trapa natans*, *Potamogeton malaianus* and *Vallisneria spiralis*, as identified by field observations as well as previously published studies [30,31]. We divided the aquatic vegetation into two types according to their spectral characteristics. Type 1 represented the typical green vegetation identified in remote sensing images by lower red band reflectance paired with higher near-infrared band reflectance than other ground cover types and included emergent, floating-leaf and floating vegetation having some green leaves above the water surface. As previously noted, most emergent vegetation was excluded from Type 1. Type 2 consisted of the submerged vegetation, which had all green leaves submerged beneath the water surface, thus distinguishing it from typical green vegetation in remote sensing images. Because Type 1 vegetation had a higher signal intensity than Type 2, areas containing both Type 1 and Type 2 vegetation were classified as emergent vegetation.

To obtain data for developing and validating models to identify aquatic vegetation, we conducted field surveys on 14–15 September 2009 and 27 September 2010. A total of 783 samples were collected in open water or aquatic vegetation (mostly floating-leaf or submerged) of Taihu Lake, including the East Bay. An additional 182 samples of reed (emergent) vegetation or terrestrial areas (e.g., shoreline roads and buildings such as docks, businesses and factories) were obtained from a 1:50,000 land use and land cover map due to logistical difficulties in maneuvering a boat in the dense reed vegetation. A total of 426 and 539 ground truth samples were collected in 2009 and 2010, respectively (Fig. 1). At each field sampling plot, photographs were taken using a digital camera (IXUS 950, Canon) held at about 1.2 m above the water surface, with the camera axis angled about 30 degrees down from the horizon. The position of each photograph was geo-located using a portable GPS receiver with an accuracy of 3 m. In the laboratory, all the photographs were interpreted visually and classified as Type 1, Type 2 or open water sediment.

2.4 Image Processing

Multispectral TM, ETM+, SPOT-4, HJ, and CBERS remote sensing images were used in this study, with spatial resolutions ranging from 15 to 30 meters. Following the recommended standard for aquatic remote sensing [34], we selected images

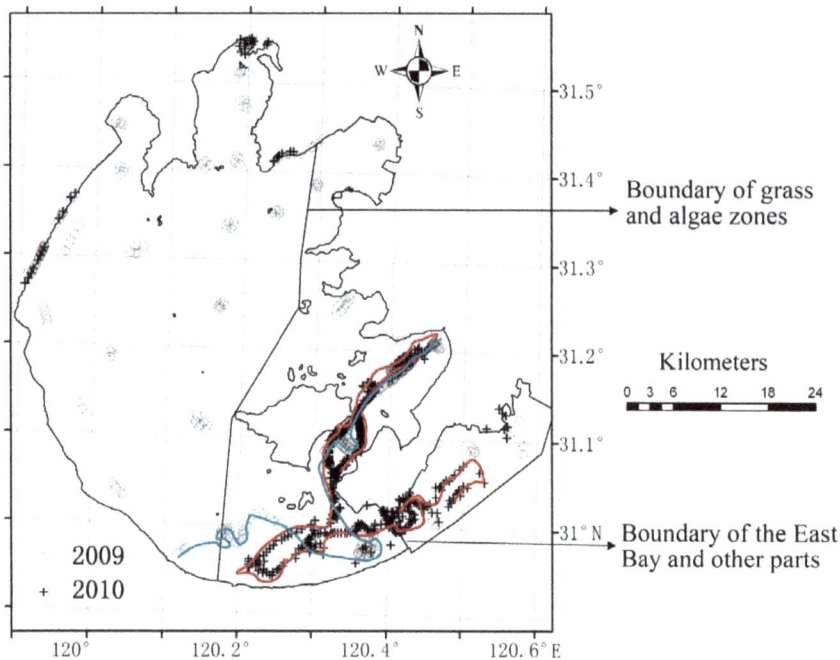

Figure 1. The study area showing the distribution of 965 ground truth samples (426 in 2009 and 539 in 2010) in Taihu Lake, China.

containing no more than 10% cloud cover at the study area. The cosine approximation model (COST; Chavez, 1996), which has been implemented successfully in other aquatic remote sensing studies [35,36], was used to apply atmospheric corrections to all the images used. Prior to atmospheric correction, cloud-contaminated pixels were removed from all images using interactive interpretation. Geometric correction was applied using second-order polynomials with accuracy higher than 0.5 pixel.

Table 1. Remote sensing images used in this study with associated dates.

Years	Image-1	Image-2	Image-3	Image-4
1989	TM, 1/14	TM, 7/17	TM, 10/21	
1991	TM, 1/12	TM, 7/23	TM, 8/24	
1992	TM, 2/16	TM, 7/25	TM, 8/10	
1995	TM, 2/24	TM, 8/3	TM, 8/19	
1996	TM, 1/10	TM, 7/20	TM, 9/6	
1998	TM, 1/31	TM, 8/11	TM, 7/10	
2000	ETM+, 3/18	ETM+, 8/8	CBERS, 9/16	ETM+, 10/12
2001	ETM+, 1/15	ETM+, 7/26	ETM+, 9/28	
2002	ETM+, 2/3	ETM+, 7/13	TM, 8/22	
2003	ETM+, 2/6	ETM+, 8/1	SPOT, 8/23	
2004	ETM+, 2/9	ETM+, 8/3	CBERS, 8/8	ETM+, 8/19
2005	ETM+, 3/31	ETM+, 6/19	ETM+, 9/7	
2006	ETM+, 3/2	CBERS, 8/6	TM, 9/18	ETM+, 9/28
2008	ETM+, 2/20	CBERS, 7/24	ETM+, 8/14	HJ, 9/23
2009	ETM+, 1/13	ETM+, 8/25	HJ, 9/10	
2010	ETM+, 3/13	ETM+, 8/20	ETM+, 9/21	

Because they contain information on seasonal dynamics of both aquatic vegetation and related environmental factors, multiple intra-annual remote sensing images can provide higher accuracy for the identification of aquatic vegetation than a single image [37,38]. Therefore, we used combinations of winter (between January and March when the biomass of aquatic vegetation was lowest) and summer (between June and October when the biomass of aquatic vegetation was highest) images from each study year between 1989 and 2010. For each study year, at least three clear Landsat images were selected (one from winter and the others from summer) and formed into at least two pairs in which the winter image was paired with each summer image. A total of 36 pairs were used in this study (Table 1).

Aquatic vegetation was mapped using each image combination, so at least two vegetation maps were obtained for each year. Maps for the same year were superimposed and combined according to the following rules: (1) In the grass type zone of Taihu Lake (i.e. the eastern portion in Fig. 1), if a pixel was classified as aquatic vegetation in either map within a single study year, it was classified as aquatic vegetation in the final map. This rule was established primarily because human activities such as harvesting of aquatic vegetation might decrease the distribution of aquatic vegetation, and because particulate matter that is suspended very high in the water column might obscure the submerged vegetation, resulting in underestimation of submerged vegetation [22] at some time during the growing season. (2) In the algae type zone (i.e. the remaining portions of the lake not in the grass type zone in Fig. 1), a pixel was regarded as aquatic vegetation only when it was classified as aquatic vegetation in all maps for a single study year. This rule aimed to reduce classification interference from algal blooms, which occur frequently between May and October [27,28,39] and is based on the probability being much lower that algal blooms will appear twice in the same location than that aquatic macrophytes will. The ground truth samples from 2009 and 2010 were used to evaluate the accuracy of the final aquatic vegetation classifications derived from the 2009 and 2010 image pairs.

2.5 Analytical Methods

2.5.1 Identification of aquatic vegetation in remote sensing images.
Classification tree (CT) analysis, which uses recursive dichotomous partitioning of the data according to calculated thresholds, has been used successfully for the identification of aquatic vegetation because of its flexibility with regard to the inclusion of data from multiple sources and of multiple types, such as spectral signals, environmental variables and other variables related to aquatic vegetation growth [38,40–44]. Zhao et al. (2012) developed an improved CT modeling algorithm for identifying emergent, floating-leaf and submerged vegetation from remote sensing images both from different times [45] and from different sensor (Zhao et al. A method for application of classification tree models to map aquatic vegetation using remotely sensed images from different sensors and dates. Sensors, Re-submitted after revision). Because we divided the aquatic vegetation into only two types in this study (i.e. Type 1 and Type 2), minor modifications were made to the CT model structure (Fig. 2).

We obtained the quantitative thresholds for the CT base model structure and thus the final CT models by applying CT analysis to the 2009 image pairs (Fig. 2), attaining an overall classification accuracy of 94.0%, with classification accuracies of 95.6% and 88.8% for Type 1 and Type 2 vegetation, respectively. When the CT models were applied to the image pairs of 2010, overall accuracy was 93.3%, with classification accuracies of 94.2% and 87.9% for Type 1 and Type 2 vegetation, respectively (Table 2). These results suggested that our CT model could be used to effectively identify the aquatic vegetation in Taihu Lake. Therefore, we used the models to map the distribution of aquatic vegetation in Taihu Lake from 1989 to 2010.

2.5.2 Evaluation of water level effects on aquatic vegetation.
The annual Coefficient of Variation (CV_a) was calculated to describe the inter-annual fluctuation of water levels:

$$CV_a = \frac{\sqrt{\frac{\sum(x-\bar{x})^2}{n-1}}}{\bar{x}} \times 100\% \qquad (1)$$

where x and $\bar{x}$ are average monthly water level and average annual water level, respectively.

We used regression analysis to investigate the effects of water level variation and CV_a fluctuation on the distribution of aquatic vegetation through time. However, using un-transformed values for water level and area of aquatic vegetation is unlikely to reflect the true relationship because of the inevitable temporal autocorrelation of aquatic vegetation as well as the confounding influence of gradual changes in environmental factors such as water nutrient content, chemical oxygen demand (COD) and water clarity. Therefore, we transformed the water level and aquatic vegetation area variables using the variability from one year to the next before performing the regression analysis. This was accomplished by subtracting the previous year values from the focus year values using the same time period. Thus, water level variability was calculated as:

$$V_{wl} = WL_i - WL_{i-1} \qquad (2)$$

Where WL_i and WL_{i-1} are the average water levels or CV_a in the focus year (i) and the year previous to the focus year (i−1), respectively. However, if data for the previous year were missing, data from two years prior to the focus year (i−2), or three years prior (i−3) if data were also missing for i−2, were used instead. Variability in aquatic vegetation area was calculated as:

$$VA_{ava} = \frac{(AVA_i - AVA_{i-1})}{AVA_{i-1}} \qquad (3)$$

Where AVA_i and AVA_{i-1} are aquatic vegetation areas in the focus year (i) and year previous to the focus year (i−1), respectively. Similar to Equation (2), if aquatic vegetation data were not available for the previous year, it was replaced by the data for the closest year for which data were available.

Because most of the images used to map aquatic vegetation distribution were dated prior to October with only two exceptions (Table 1), water levels between October and December in a certain year did not influence aquatic vegetation area of the same year. However, October-December water levels probably influenced the aquatic vegetation of the following year. Therefore, October through December water levels of the year previous to the focus year were used to analyze relationships between monthly water levels and aquatic vegetation.

Results

3.1 Temporal Dynamics of Water Level
Between 1989 and 2010, annual average water levels fluctuated between 2.86 m and 3.33 m, with no significant inter-annual trend (Fig. 3A). The average water level in Period 2

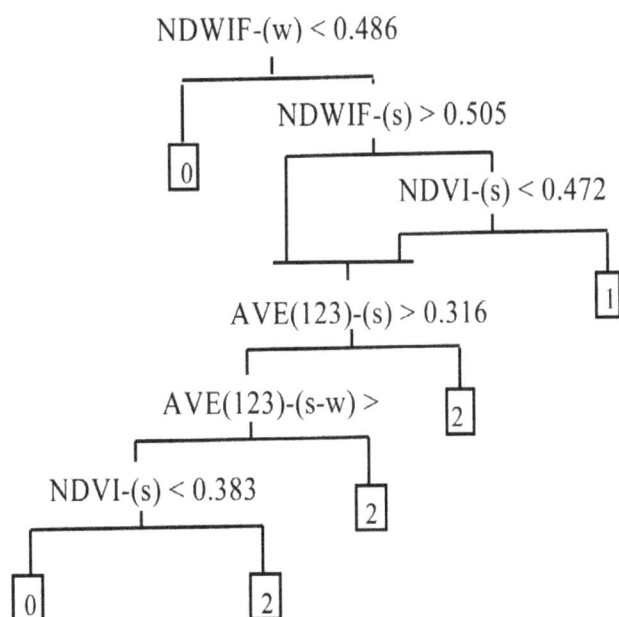

Figure 2. Classification tree models established for Type 1 and Type 2 aquatic vegetation. The numbers 1 and 2 in the end nodes of the classification trees represent Type 1 and Type 2 vegetation, respectively, whereas 0 represents other types. Variables used are the Modified Normalized Difference Water Index (MNDWI), the Normalized Difference Vegetation Index (NDVI) and the average reflectance of the blue, green and red image bands (AVE123). Variables were calculated by season (s = summer, w = winter) or differences among seasonal values (e.g., s-w).

Table 2. Confusion matrix of the CT models developed in this paper as applied to 2009 and 2010 data, respectively (in number of field samples).

			Prediction				
			Type 1	Type 2	Other types	Classification accuracy (%)	Overall accuracy (%)
2009	Truth	**Type 1**	130	4	2	95.6	94.0
		Type 2	5	103	8	88.8	
		Other types	1	4	145	96.7	
2010	Truth	**Type 1**	175	7	4	94.1	93.3
		Type 2	6	102	8	87.9	
		Other types	0	8	186	95.9	

(2000–2010) was 3.18 m, slightly higher than that in Period 1 (i.e. 1989–1999, average = 3.10 m). However, substantially different intra-annual dynamics were observed between the two temporal periods (Fig. 3B), with more stable water levels in Period 2 than in Period 1. In Period 1, monthly water levels ranged from 2.57 m to 4.61 m with the annual Coefficient of Variation (CV_a) ranging from 3.06% (1994) to 18.41% (1999) and averaging 10.21%. However, in Period 2, monthly water levels ranged from 2.76 m (2006) to 3.98 m (2009), with CV_a ranging from 2.65% to 7.94% and averaging 5.41% (Fig. 3C). CV_a in Period 1 was significantly higher than that in Period 2 ($p = 0.01$). For July, August and September, monthly water levels in Period 2 were 0.064 to 0.21 m lower than those in Period 1, whereas monthly water levels in Period 2 for the remaining months were 0.042 to 0.27 m higher than those in Period 1. Thus, our results indicate differences in intra-annual dynamics of water level between Period 1 and Period 2.

Monthly precipitation in Period 1 was slightly higher (1.49 to 30.48 mm) than that in Period 2 for January, March, June, July, August and September; for the other six months, monthly precipitation in Period 1 was slightly lower (0.03 to 9.80 mm) than that in Period 2 (Fig. 3D). Annual precipitation in Period 1 (1175.4 mm) was slightly higher than that in Period 2 (1132.1 mm). Thus, precipitation and water level showed different intra-annual variation patterns between the two periods. These results suggested that climatic conditions were not responsible for the greater stability of intra-annual water level in Period 2 than in Period 1.

3.2 Temporal Dynamics of Aquatic Vegetation Distribution Area

From 1989 to 2010, aquatic vegetation was distributed primarily in the eastern part of the lake (Fig. 4). The spatial pattern of distribution experienced some changes, with aquatic vegetation shifting gradually from the northeast to the southeast.

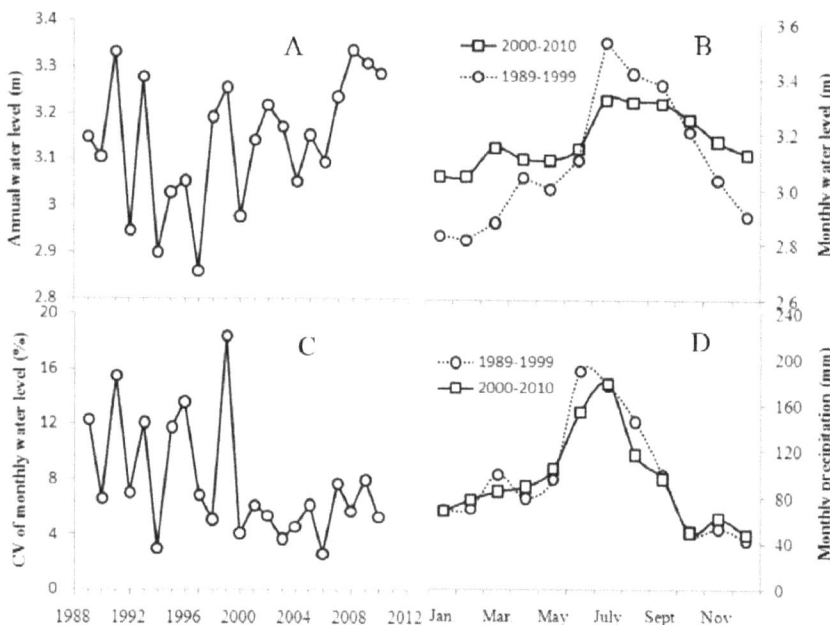

Figure 3. Inter- and intra- annual dynamics of water level and precipitation in Taihu Lake. (A) annual water level dynamics between 1989 and 2010; (B) average monthly water levels for the two 10-year time periods examined in this study; (C) intra-annual fluctuation of water levels (CV_a) from 1989–2010; and (D) average monthly precipitation for the two 10-year time periods examined in this study.

Substantial changes were observed in both distribution area and composition of aquatic vegetation during the study period (Fig. 5). The area covered by Type 1 vegetation ranged from 5.80 km^2 (1996) to 142.5 km^2 (2009), with an average of 57.3 km^2, whereas Type 2 vegetation covered an area ranging from 68.3 km^2 (1991) to 190.8 km^2 (2001), with an average of 137.4 km^2. The total aquatic vegetation area (i.e., the sum of Type 1 and Type 2 vegetation) ranged from 77.9 km^2 (1991) to 282.0 km^2 (2005), with an average of 194.7 km^2, and the ratio of Type 2 to Type 1 vegetation ranged from 0.94 (2010) to 23.4 (1996), with an average of 6.07. Significant temporal dynamics were observed for each variable from 1989 to 2010 (p = 0.01). Both total aquatic vegetation area and area of Type 1 vegetation increased significantly over the study period (p = 0.01). Area of Type 2 vegetation increased before 2001 and then decreased (p = 0.01), and the ratio of Type 2 to Type 1 vegetation decreased steadily over the study period (p = 0.01).

The area covered by aquatic vegetation also differed between the two inter-annual temporal periods examined. Average Type 2 vegetation area increased from 99.2 km^2 in Period 1 to 160.4 km^2 in Period 2, an increase of 61.6%, and average Type 1 vegetation area increased 7.14-fold, from 10.5 km^2 in Period 1 to 85.4 km^2 in Period 2. Total vegetation area increased 124.0%, from 109.7 km^2 in Period 1 to 245.8 km^2 in Period 2. Finally, the ratio of Type 2 to Type 1 vegetation decreased from 11.8 in Period 1 to 2.62 in Period 2.

3.3 Relationship between Water Level and Aquatic Vegetation Area

Firstly, we investigated whether the annual average water levels and intra-annual fluctuation of water levels (CV_a) influenced inter-annual aquatic vegetation area and its derivations. No significant correlations were found between V_{wl} and VA_{ava} using either annual averages or CV_a for either vegetation type, total vegetation area, or the ratio of Type 2 to Type 1 vegetation (Table 3).

Secondly, we tested the correlations between monthly average water levels and aquatic vegetation (Fig. 6). Significant positive correlations were found between Type 1 vegetation area and monthly water levels from December to March (p = 0.05), whereas significant negative correlations were found between Type 2 vegetation area and monthly water levels from January to April (p = 0.05). Total vegetation area was negatively correlated with monthly water levels from March to April (p = 0.05), and the ratio of Type 2 to Type 1 vegetation was negatively correlated with monthly water levels from November to March (p = 0.05). These results suggested that water levels in late winter and early spring (traditional dry season) significantly influenced aquatic vegetation area.

Discussion

4.1 Temporal Changes

We found significant relationships between water level from December to March and the area of Type 1 vegetation, with

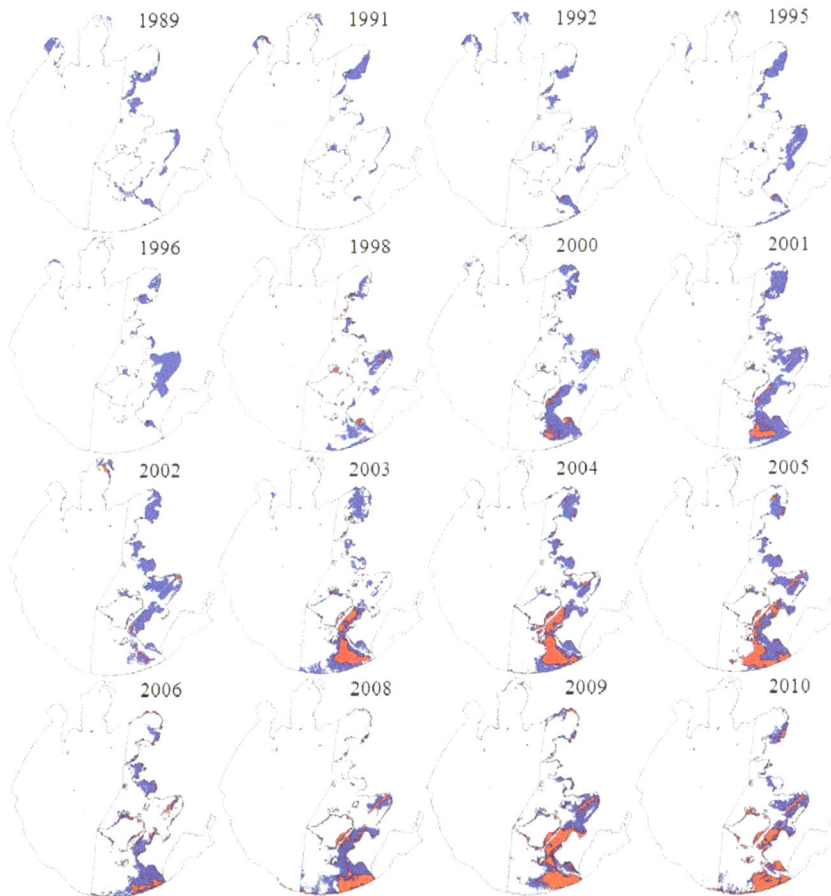

Figure 4. Distribution of aquatic vegetation of Type 1 (red) and Type 2 (blue) between 1989 and 2010. Type 1 vegetation consisted of emergent, floating-leaf and floating vegetation, whereas Type 2 consisted of submerged vegetation.

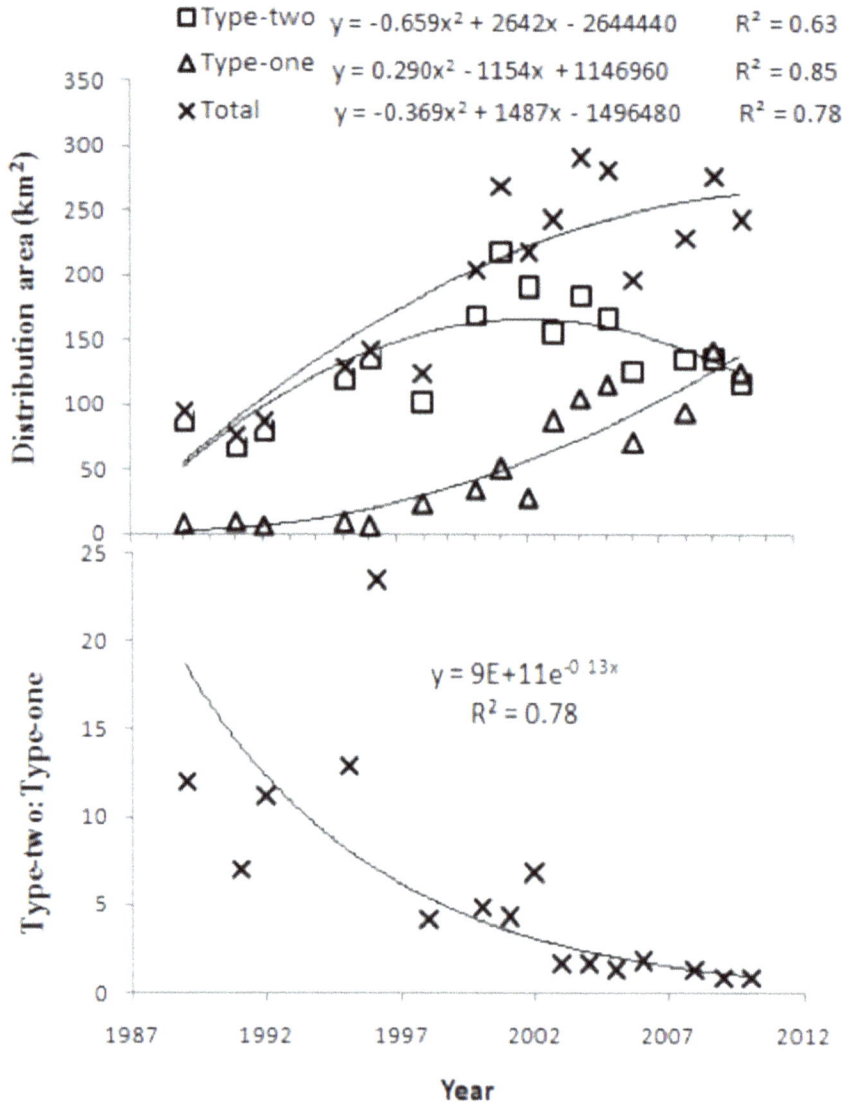

Figure 5. Temporal trends (1989–2010) of the distribution area of aquatic vegetation components and derivatives. Type 1 vegetation consisted of emergent, floating-leaf and floating vegetation, whereas Type 2 consisted of submerged vegetation.

increases in Type 1 vegetation occurring steadily from 1989 to 2010 (and consequently for the 2000–2010 period relative to the 1989–1999 period) and coinciding with increased water levels during the dry season. The aquatic macrophyte communities in

Table 3. Linear correlation coefficients between variation of aquatic vegetation (VA$_{ava}$) and variation of water level (V$_{wl}$) (n = 15).

Aquatic Vegetation Parameter	Annual average	CV$_a$
Type 2	−0.33	0.29
Type 1	0.21	−0.34
Total	−0.25	0.33
Type 2:Type 1	−0.40	0.02

Type 1 vegetation consisted of emergent, floating-leaf and floating vegetation, whereas Type 2 consisted of submerged vegetation.

very shallow lakes such as Taihu Lake [19] may be especially sensitive to variation in water levels. In addition, according to the stable state theory proposed by Scheffer et al. (2003) [46], nutrient enrichment should shift the vegetation in freshwater systems toward dominance by floating plants; in fact, nutrient concentrations, as well as area of Type 1 vegetation, have shown a gradual increase in Taihu Lake over the past 20 years [23,47]. Therefore, nutrient enrichment and higher dry-season water levels may represent two of the most important factors responsible for the temporal shifts in Type 1 vegetation distribution observed in this study.

We found significant negative relationships between Type 2 vegetation and dry-season monthly water level as well as slightly higher dry-season water levels in 2000–2010 than in 1989–1999. Therefore, water level couldn't explain the 61.6% increase of Type 2 in 2000–2010 over that in 1989–1999. According to an alternative equilibrium theory [48,49], submerged vegetation responds in a non-linear way to eutrophication, first increasing then decreasing. Our findings support this theory, with nutrient enrichment increasing over the past 20 years in Taihu Lake and

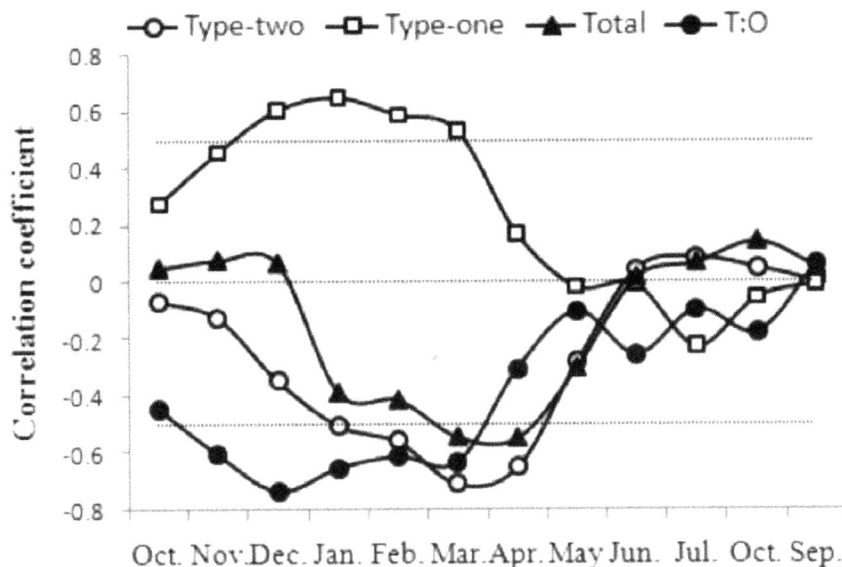

Figure 6. Linear correlation coefficients between monthly water level (in V_{wl}) and the distribution area of aquatic vegetation components (in VA_{ava}) during the time period 1989–2010. Type 1 vegetation consisted of emergent, floating-leaf and floating vegetation, whereas Type 2 consisted of submerged vegetation. T:O is the ratio of Type 2 to Type 1 vegetation area.

Type 2 vegetation increasing in area until about 2002 and decreasing thereafter. Nutrient enrichment is thought to be responsible for the expansion of aquatic vegetation in numerous lakes throughout the world and probably acted as one of most important driving factors of vegetation dynamics in our study [50,51]. We speculate that water level probably acted as the dominant factor determining Type 2 vegetation area on a one- to two-year temporal scale, whereas gradual changes in eutrophication probably acted as the dominant driving factor on a longer (~10 years) temporal scale.

The methodology used for identifying the effect of water level on aquatic vegetation can have a large influence on the results obtained. Despite a few successful reports [10,11], using non-transformed values of water level and aquatic vegetation area in regression models is unlikely to reveal the actual temporal relationship between the two variables in periods when confounding factors such as nutrient concentration vary considerably. Using transformed parameters such as the V_{wl} and VA_{ava} variables we used in this study instead of non-transformed water level and aquatic vegetation area values has two advantages: (1) it can alleviate the problem of temporal autocorrelation in aquatic vegetation area and its derivatives; and (2) it can help distinguish influences of water level from gradually and continuously increasing or decreasing factors such as water eutrophication [23].

4.2 Mechanisms for the Influence of Water Level on Aquatic Vegetation

Underwater light availability, which is affected strongly by water level, is an important mechanism influencing aquatic vegetation. Decreases in available light have been found to be of the most important factors resulting in species disappearance and biodiversity loss both for aquatic and terrestrial vegetation [52,53]. Many studies have found underwater light to be closely correlated with water level [4,10,13,16], especially for turbid waters such as Taihu Lake [25,54]. Generally, increases in water level will reduce underwater light availability, especially at the bottom. If under-

water light decreases below the threshold of a species' minimum light requirements, the species will disappear from the community.

Changes in light availability resulting from water level variability can easily explain the negative correlation found between area of Type 2 vegetation, which is submerged, and water level in traditional annual dry seasons. However, the positive correlation between Type 1 vegetation area and water level during the dry season cannot be explained as a direct result of light availability because of the positive effect of light availability on aquatic vegetation growth [55]. A more plausible explanation of the positive correlation would be due to competitive interactions between Type 1 and Type 2 vegetation [56]. According to observations from our field campaign, Type 1 species such as *Nymphoides peltatum* and Type 2 species such as *Potamogeton malaianus* grew widely in mixed communities in Taihu Lake [30], promoting strong competition for space and light. Because Type 2 vegetation is more sensitive to underwater light restrictions, which can be exacerbated by increases in water level, than Type 1 vegetation [55,57,58], water level increases can strengthen competitive ability of Type 1 vegetation by inhibiting the growth of Type 2 vegetation, ultimately resulting in the correlation patterns observed.

Our results indicated that water level influenced aquatic vegetation in dry seasons more so than in rainy seasons. This may be a consequence of the phenology of aquatic plants, which are most sensitive to light conditions in the germination and initial growth stages [59]. Most species in Taihu Lake survive winter with tuber-like buds in silt and germinate in the early spring [30,31]. Water level can directly influence the germination of buds by altering light availability at the lake bottom. Upon entering the rapid growth period, *Nymphoides peltatum* and *Potamogeton malaianus*, two of the most widely distributed species, become less sensitive to water level variability because of their strong morphological plasticity [60]. Our results were consistent with Blindow et al. (1993) and Paillisson and Marion (2006) [11,49], who found that spring water level influenced the growth of aquatic vegetation.

4.3 Management Implications

Our results suggest that regulation of the distribution of aquatic vegetation is feasible through management of water levels in Taihu Lake. Since 2000, and especially since 2007 when a severe blue-green algal bloom resulted in serious drinking water shortages in Wuxi City [27,28], multiple costly water conservation projects have been conducted in Taihu Lake. Restoration of aquatic vegetation through artificial planting or other means has been one of most common approaches [31] because of the purification function of aquatic vegetation. The significantly better water quality in the eastern coastal area of Taihu Lake relative to other parts of the lake is a solid example of this purification function [60]. Our results indicate that decreasing water levels in the dry season could increase the area occupied by aquatic vegetation in tens of square kilometers, which is a difficult goal to achieve using other restoration strategies such as direct planting.

In addition to regulation of the area occupied by aquatic vegetation generally, our results suggest that artificial control of distinct aquatic vegetation components is possible by regulating water levels. This finding is potentially very useful for lake management because of the different ecological and socioeconomic functions performed by the different aquatic vegetation types such as submerged vs. floating-leaf vegetation [29,30]. For example, submerged vegetation is one of the most important food resources for breeding crabs, which represent an annual harvest value from Taihu Lake of more than two hundred million dollars [29]. Farmers usually must plant submerged vegetation in the lake to act as a food source for the crabs, but water level regulation may represent a more economical and effective alternative. Additionally, decreasing water levels in the dry season will bring intra-annual water level fluctuations closer to natural conditions (i.e. large intra-annual fluctuations [14]) while restoring Taihu Lake to its original state of dominance by submerged vegetation.

Finally, our results suggest that regulation of water levels could be used to better control algal blooms in the lake. One of the main objectives of the current water level regulation strategies, such as the flushing of water into the lake from the Yangtze River, is to control algal blooms [19,25,26]. Because algal blooms usually occur between May and October in Taihu Lake, decreases in water level between late winter and early spring will not reduce the effectiveness of the current strategy for controlling algal blooms. On the contrary, careful reductions in water level between late winter and early spring are beneficial to the control of algal blooms in the lake because of their ability to increase the distribution of aquatic vegetation, which will in turn reduce nutrient levels.

Author Contributions

Conceived and designed the experiments: DZ SA. Performed the experiments: DZ HJ. Analyzed the data: DZ YC. Wrote the paper: DZ HJ.

References

1. Orth RJ, Moore KA (1983) Chesapeake Bay: an unprecedented decline in submerged aquatic vegetation. Science 222: 51–53.
2. Jackson JBC, Kirby MX, Berger WH, Bjorndal KA, Botsford LW, et al. (2001) Historical overfishing and the recent collapse of coastal ecosystems. Science 293: 629.
3. Franklin P, Dunbar M, Whitehead P (2008) Flow controls on lowland river macrophytes: A review. Sci Tot Environ 400: 369–378.
4. van der Heide T, van Nes EH, van Katwijk MM, Olff H, Smolders AJP (2011) Positive feedbacks in seagrass ecosystems–evidence from large-scale empirical data. PloS one 6: e16504.
5. Gullström M, Lundén B, Bodin M, Kangwe J, Ohman MC, et al. (2006) Assessment of changes in the seagrass-dominated submerged vegetation of tropical Chwaka Bay (Zanzibar) using satellite remote sensing. Estuar Coast Shelf Sci 67: 399–408.
6. Chambers P, Lacoul P, Murphy K, Thomaz S (2008) Global diversity of aquatic macrophytes in freshwater. Hydrobiologia 595: 9–26.
7. Qin B (2009) Lake eutrophication: Control countermeasures and recycling exploitation. Ecol Eng 35: 1569–1573.
8. Li EH, Li W, Wang XL, Xue HP, Xiao F (2010) Experiment of emergent macrophytes growing in contaminated sludge: Implication for sediment purification and lake restoration. Ecol Eng 36: 427–434.
9. Lorenz AW, Korte T, Sundermann A, Januschke K, Haase P (2012) Macrophytes respond to reach-scale river restorations. J Appl Ecol 49: 202–212.
10. Paillisson JM, Marion L (2011) Water level fluctuations for managing excessive plant biomass in shallow lakes. Ecol Eng 37: 241–247.
11. Paillisson JM, Marion L (2006) Can small water level fluctuations affect the biomass of Nymphaea alba in large lakes? Aquat Bot 84: 259–266.
12. Liira J, Feldmann T, Mäemets H, Peterson U (2010) Two decades of macrophyte expansion on the shores of a large shallow northern temperate lake-a retrospective series of satellite images. Aquat Bot 93: 207–215.
13. O'Farrell I, Izaguirre I, Chaparro G, Unrein F, Sinistro R, et al. (2011) Water level as the main driver of the alternation between a free-floating plant and a phytoplankton dominated state: a long-term study in a floodplain lake. Aquat Sci 73: 275–287.
14. Van Geest G, Coops H, Roijackers R, Buijse A, Scheffer M (2005) Succession of aquatic vegetation driven by reduced water-level fluctuations in floodplain lakes. J Appl Ecol 42: 251–260.
15. Geest GJV, Wolters H, Roozen F, Coops H, Roijackers R, et al. (2005) Water-level fluctuations affect macrophyte richness in floodplain lakes. Hydrobiologia 539: 239–248.
16. Bain MB, Singkran N, Mills KE (2008) Integrated ecosystem assessment: Lake Ontario water management. PloS one 3: e3806.
17. Coops H, Vulink JT, Van Nes EH (2004) Managed water levels and the expansion of emergent vegetation along a lakeshore. Limnologica 34: 57–64.
18. Wantzen KM, Rothhaupt KO, Mörtl M, Cantonati M, Tóth LG, et al. (2008) Ecological effects of water-level fluctuations in lakes: an urgent issue. Hydrobiologia 595: 1–4.
19. An S, Wang RR (2009) The human-induced driver on the development of Lake Taihu: In Lee, Xuhui ed. Lectures on China's Environment, Yale School of Forestry and Environmental Studies.
20. Zhang YL, Dijk MA, Liu ML, Zhu GW, Qin BQ (2009) The contribution of phytoplankton degradation to chromophoric dissolved organic matter (CDOM) in eutrophic shallow lakes: Field and experimental evidence. Water Res 43: 4685–4697.
21. Zhao DH, Cai Y, Jiang H, Xu DL, Zhang WG, et al. (2011) Estimation of water clarity in Taihu Lake and surrounding rivers using Landsat imagery. Adv Water Resour 34: 165–173.
22. Ma RH, Duan HT, Gu XH, Zhang SX (2008) Detecting aquatic vegetation changes in Taihu Lake, China using multi-temporal satellite imagery. Sensors 8: 3988–4005.
23. Qing BQ (2009) Progress and prospect on the eco–environmental research of Lake Taihu. J Lake Sci 21: 445–455.
24. Pan G, Yang B, Wang D, Chen H, Tian B, et al. (2011) In-lake algal bloom removal and submerged vegetation restoration using modified local soils. Ecol Eng: 302–308.
25. Hu W, Zhai S, Zhu Z, Han H (2008) Impacts of the Yangtze River water transfer on the restoration of Lake Taihu. Ecol Eng 34: 30–49.
26. Li Y, Acharya K, Yu Z (2011) Modeling impacts of Yangtze River water transfer on water ages in Lake Taihu, China. Ecol Eng 37: 325–334.
27. Guo L (2007) Doing battle with the green monster of Taihu Lake. Science 317: 1166.
28. Yang M, Yu J, Li Z, Guo Z, Burch M, et al. (2008) Taihu Lake not to blame for Wuxi's woes. Science 319: 158.
29. Gu X, Zhang S, Bai X, Hu W, Hu Y, et al. (2005) Evolution of community structure of aquatic macrophytes in East Taihu Lake and its wetlands. Acta Ecol Sinica 25: 1541–1548.
30. He J, Gu XH, Liu GF (2008) Aquatic macrophytes in East Lake Taihu and its interaction with water environment. J Lake Sci 20: 790–795.
31. Liu WL, Hu WP, Chen YG, Gu XH, Hu ZX, et al. (2007) Temporal and spatial variation of aquatic macrophytes in west Taihu Lake. Acta Ecol Sinica 27: 159–170.
32. Otahelová H, Otahel J, Pazúr R, Hrivnák R, Valachovic M (2011) Spatio-temporal changes in land cover and aquatic macrophytes of the Danube floodplain lake. Limnologica 41: 316–324.
33. Yang Y, Jiang N, Yin L, Hu B (2005) RS-based dynamic monitoring of lake area and enclosure culture in East Taihu Lake. J Lake Sci 17: 133–138.
34. Kloiber SM, Brezonik PL, Bauer ME (2002) Application of Landsat imagery to regional-scale assessments of lake clarity. Water Res 36: 4330–4340.
35. Chavez PS (1996) Image-based atmospheric corrections-revisited and improved. Photogramm Eng Rem Sens 62: 1025–1035.
36. Wu G, de Leeuw J, Skidmore AK, Prins HHT, Liu Y (2007) Concurrent monitoring of vessels and water turbidity enhances the strength of evidence in remotely sensed dredging impact assessment. Water Res 41: 3271–3280.
37. Ozesmi SL, Bauer ME (2002) Satellite remote sensing of wetlands. Wetlands Ecol Manage 10: 381–402.

38. Davranche A, Lefebvre G, Poulin B (2010) Wetland monitoring using classification trees and SPOT-5 seasonal time series. Remote Sens Environ 114: 552–562.

39. Lu N, Hu WP, Deng JC, Zhai SH, Chen XM, et al. (2009) Spatial distribution characteristics and ecological significance of alkaline phosphatase in water column of Tahihu Lake. Environ Sci 30: 2898–2903.

40. Brown EC, Story MH, Thompson C, Commisso K, Smith TG, et al. (2003) National Park vegetation mapping using multitemporal Landsat 7 data and a decision tree classifier. Remote Sens Environ 85: 316–327.

41. Baker C, Lawrence R, Montagne C, Patten D (2006) Mapping wetlands and riparian areas using Landsat ETM+ imagery and decision-tree-based models. Wetlands 26: 465–474.

42. Wright C, Gallant A (2007) Improved wetland remote sensing in Yellowstone National Park using classification trees to combine TM imagery and ancillary environmental data. Remote Sens Environ 107: 582–605.

43. Wei A, Chow-Fraser P (2007) Use of IKONOS imagery to map coastal wetlands of Georgian Bay. Fisheries 32: 167–173.

44. Midwood JD, Chow-Fraser P (2010) Mapping floating and emergent aquatic vegetation in coastal wetlands of Eastern Georgian Bay, Lake Huron, Canada. Wetlands 30: 1–12.

45. Zhao DH, Jiang H, Yang TW, Cai Y, Xu DL, et al. (2012) Remote sensing of aquatic vegetation distribution in Taihu Lake using an improved classification tree with modified thresholds. J Environ Manage 95: 98–107.

46. Scheffer M, Szabó S, Gragnani A, Van Nes EH, Rinaldi S, et al. (2003) Floating plant dominance as a stable state. Proceedings of the national academy of sciences 100: 4040–4045.

47. Zhu GW (2008) Eutrophic status and causing factors for a large, shallow and subtropical Lake Taihu, China. J Lake Sci 20: 21–26.

48. Scheffer M, Hosper S, Meijer M, Moss B, Jeppesen E (1993) Alternative equilibria in shallow lakes. Trends Ecol Evol 8: 275–279.

49. Blindow I, Andersson G, Hargeby A, Johansson S (1993) Long-term pattern of alternative stable states in two shallow eutrophic lakes. Freshwater Biol 30: 159–167.

50. Scheffer M, Carpenter S, Foley JA, Folke C, Walker B (2001) Catastrophic shifts in ecosystems. Nature 413: 591–596.

51. Schindler DW (2006) Recent advances in the understanding and management of eutrophication. Limnol Oceanogr 51: 356–363.

52. Hautier Y, Niklaus PA, Hector A (2009) Competition for light causes plant biodiversity loss after eutrophication. Science 324: 636–638.

53. Michelan TS, Thomaz S, Mormul RP, Carvalho P (2010) Effects of an exotic invasive macrophyte (tropical signalgrass) on native plant community composition, species richness and functional diversity. Freshwater Biol 55: 1315–1326.

54. Zhang S, Liu A, Ma J, Zhou Q, Xu D, et al. (2010) Changes in physicochemical and biological factors during regime shifts in a restoration demonstration of macrophytes in a small hypereutrophic Chinese lake. Ecol Eng 36: 1611–1619.

55. Squires M, Lesack L, Huebert D (2002) The influence of water transparency on the distribution and abundance of macrophytes among lakes of the Mackenzie Delta, Western Canadian Arctic. Freshwater Biol 47: 2123–2135.

56. Szabo S, Scheffer M, Roijackers R, Waluto B, Braun M, et al. (2010) Strong growth limitation of a floating plant (*Lemna gibba*) by the submerged macrophyte (*Elodea nuttallii*) under laboratory conditions. Freshwater Biol 55: 681–690.

57. Jin X, Yan C, Xu Q (2007) The community features of aquatic plants and its influence factors of lakeside zone in the north of Lake Taihu. J Lake Sci 19: 151–157.

58. Havens KE (2003) Submerged aquatic vegetation correlations with depth and light attenuating materials in a shallow subtropical lake. Hydrobiologia 493: 173–186.

59. Tuckett R, Merritt D, Hay F, Hopper S, Dixon K (2010) Dormancy, germination and seed bank storage: a study in support of ex situ conservation of macrophytes of southwest Australian temporary pools. Freshwater Biol 55: 1118–1129.

60. Liu WL, Hu WP, Chen Q (2007) The phenotypic plasticity of *Potamogeton malaianus* Miq. on the effect of sediment shift and Secchi depth variation in Taihu Lake. Ecol Environ 16: 363–368.

18

Drinking Water in Transition: A Multilevel Cross-sectional Analysis of Sachet Water Consumption in Accra

Justin Stoler[1,2]*, John R. Weeks[3], Richard Appiah Otoo[4]

1 Department of Geography and Regional Studies, University of Miami, Coral Gables, Florida, United States of America, **2** Department of Public Health Sciences, Miller School of Medicine, University of Miami, Miami, Florida, United States of America, **3** Department of Geography, San Diego State University, San Diego, California, United States of America, **4** Ghana Urban Water Limited, Head Office Operations, Accra, Ghana

Abstract

Rapid population growth in developing cities often outpaces improvements to drinking water supplies, and sub-Saharan Africa as a region has the highest percentage of urban population without piped water access, a figure that continues to grow. Accra, Ghana, implements a rationing system to distribute limited piped water resources within the city, and privately-vended sachet water–sealed single-use plastic sleeves–has filled an important gap in urban drinking water security. This study utilizes household survey data from 2,814 Ghanaian women to analyze the sociodemographic characteristics of those who resort to sachet water as their primary drinking water source. In multilevel analysis, sachet use is statistically significantly associated with lower overall self-reported health, younger age, and living in a lower-class enumeration area. Sachet use is marginally associated with more days of neighborhood water rationing, and significantly associated with the proportion of vegetated land cover. Cross-level interactions between rationing and proxies for poverty are not associated with sachet consumption after adjusting for individual-level sociodemographic, socioeconomic, health, and environmental factors. These findings are generally consistent with two other recent analyses of sachet water in Accra and may indicate a recent transition of sachet consumption from higher to lower socioeconomic classes. Overall, the allure of sachet water displays substantial heterogeneity in Accra and will be an important consideration in planning for future drinking water demand throughout West Africa.

Editor: Lawrence Kazembe, Chancellor College, University of Malawi, Malawi

Funding: This research was funded in part by grant number R01 HD054906 from the Eunice Kennedy Shriver National Institute of Child Health and Human Development ("Health, Poverty and Place in Accra, Ghana," John R. Weeks, Project Director/Principal Investigator). The content is solely the responsibility of the authors and does not necessarily represent the official views of the National Institute of Child Health and Human Development or the National Institutes of Health. Additional funding was provided by Hewlett/PRB ("Reproductive and Overall Health Outcomes and Their Economic Consequences for Households in Accra, Ghana," Allan G. Hill, Project Director/Principal Investigator). The 2003 Women's Health Study of Accra was funded by the World Health Organization, the United States Agency for International Development, and the Fulbright New Century Scholars Award (Allan G. Hill, Principal Investigator). The funders had no role in study design, data collection and analysis, decision to publish, or preparation of the manuscript.

Competing Interests: The authors have declared that no competing interests exist.

* E-mail: stoler@miami.edu

Introduction

Population growth in the developing world continues to put a strain on drinking water supplies, even amid declining fertility. As of 2010, approximately 884 million people–over a third of whom live in sub-Saharan Africa–still did not have access to an improved drinking water source [1]. Sub-Saharan Africa is the only region not on track to meet the Millennium Development Goal target of halving the proportion of the population without sustainable access to safe drinking water [1]; the percentage of individuals with access to piped water in the dwelling, yard, or plot stagnated from 1990–2010 and fell in urban areas as urban populations grew by over 30% [2,3]. Despite international efforts to extend access, morbidity and mortality attributable to inadequate water and sanitation remain high, particularly for children under five [4]. In Ghana, rapid urbanization continues to erode the government's ability to provide municipal water to its urban centers. The percentage of the urban population with access to an improved water source increased from 84% in 1990 to 90% in 2008, yet the percentage with access to piped water decreased steadily from 41% to 30% [1].

In Accra, Ghana's coastal capital, drinking water shortages are not driven by lack of surface or ground water *per se*, but are historically attributable to poor governance and improper water resource management [5]. Population growth and urban water mismanagement have resulted in water demand in Accra that far exceeds the water production capabilities of the two local water treatment plants. Ghana Urban Water Ltd. (GUWL; a subsidiary of the Ghana Water Company Ltd.) has instituted a rationing program for water distribution within city limits [6], and by one estimate 75% of Accra lacks 24-hour water access while another 10% has no access at all [7]. Municipal water rationing varies both geographically and socioeconomically by neighborhood in Accra [8], and these service gaps have been linked to the recent ubiquity of packaged "sachet water" sold in sealed 500 ml plastic sleeves. Over the last five years, sachet water has become an important primary drinking water source for the urban poor, and may even confer a health benefit when sachets replace the consumption of improperly stored water in the home [8]. As piped water access in Accra continues to decline, it remains unclear whether sachet water is strictly a phenomenon of the poor, or if sachets are being consumed by a more socioeconomically diverse population and with similar protective health effects.

The most recent Joint Monitoring Report from WHO/UNICEF [3] indicates that urban access to piped drinking water is on the decline throughout sub-Saharan Africa, a trend previously observed over three decades in East Africa [9], but the report attributes these service losses solely to population growth. International development agencies continue to ignore West Africa's transition to sachet water with no mention of it in several recent regional reports [1,3,10–12]. WHO's recent report on household water treatment and storage [13] discusses routine and emergency distribution of flocculent-disinfectant sachets for household use, but there is no discussion of pre-treated sachet water in the form addressed here. Because urban drinking water options have evolved over just the last few years, a fuller understanding of West Africa's urban drinking water patterns is crucial for catching up on missed Millennium Development Goals for sustainable drinking water and sanitation in the region.

This study utilizes survey data from the Women's Health Study of Accra (WHSA) to analyze the socioeconomic and geographic patterns associated with sachet water consumption and the effect of water rationing on sachet consumption. We present data on primary drinking water sources and socioeconomic characteristics for a sample of 2,814 women in Accra. Because GUWL water rationing is managed by local water districts, we expect that a household's drinking water options are influenced by the degree of neighborhood rationing after accounting for individual differences. We specifically test three hypotheses: (1) that residents of lower socioeconomic status are more likely to consume sachet water as their primary water source, (2) that residents enduring a greater number of days of water rationing in the neighborhood are more likely to consume sachets, and (3) that the interaction of higher rationing and lower socioeconomic status will produce the highest rates of sachet use. In light of sachet water's massive appeal throughout West Africa, we comment on the sustainability of sachets as an urban drinking water solution, as well as how the phenomenon might better inform water provision projects in the region.

Methods

Study Population

Accra is typical of many developing urban areas in West Africa both for its progress in health care delivery and its overall epidemiologic transition. The Accra metropolitan area, with its southern border on the Gulf of Guinea coast, extends about 11 km north just beyond the University of Ghana at Legon, and is roughly 20 km east to west. The metropolitan area contained 1.66 million people and 373,540 households according to the March 2000 Accra census, and is approaching 2.4 million people as of the 2010 census, while still growing at 3% annually [2]. Local governance of this population rests with the Accra Metropolitan Assembly, and logistical coordination with this single administrative body facilitated implementation of the study instrument.

Study Instrument

The WHSA is a community-based longitudinal population study conducted in Accra by the Harvard School of Public Health, the University of Ghana at Legon, and San Diego State University. The first round of the WHSA was conducted in 2003, with a follow-up survey in 2008–2009. The survey provides a comprehensive snapshot of health, disease, and related risk factors in each study year among a representative sample of women ≥18 years of age in the Accra metropolitan area [14,15]. Participants were originally selected through a two-stage cluster probability sample stratified by socioeconomic status (SES) quartiles, as determined by the 2000 census, and by four age groups. The sample was drawn from a population-weighted sample of 200 of Accra's 1,731 enumeration areas (EAs), and study women were selected according to the probability of being in one of 16 cells formed by the stratification of SES and age group [14]. To improve our understanding of neighborhood context, EAs are aggregated into 108 neighborhood units called *field modified vernacular neighborhoods* (FMVN) [16], which are field-validated configurations of previously-defined *vernacular neighborhoods* [17]. The study sample and spatial hierarchy are depicted in Figure 1.

The study interviewed 3,172 women in 2003, but due to significant challenges tracking women between surveys (primarily the lack of a formal address system to trace those who moved, and

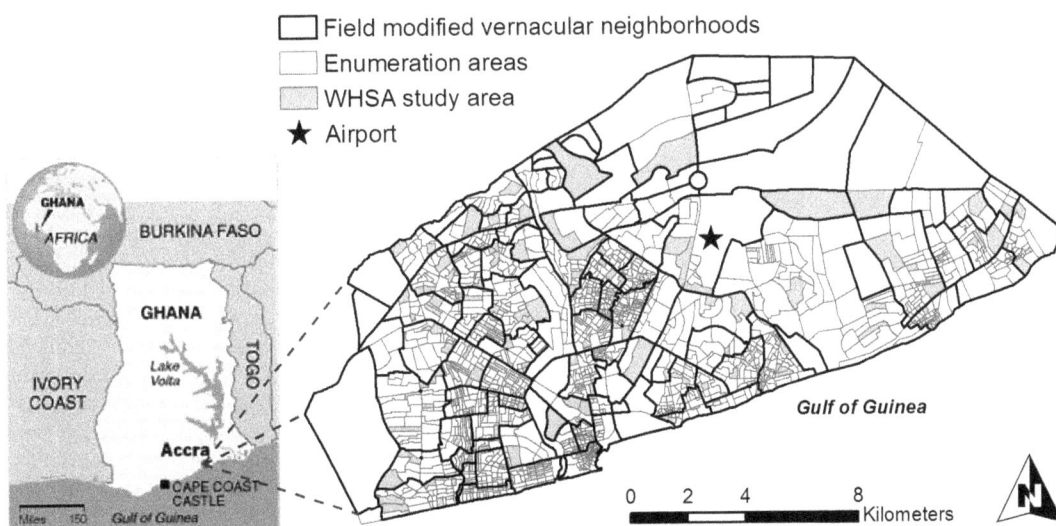

Figure 1. Digital layer of the study site in Accra, Ghana, depicting WHSA-II study sample with enumeration area (EA) and field modified vernacular neighborhood (FMVN) levels of analysis.

also Institutional Review Board prohibition of collecting original household location data via Global Positioning System receivers in 2003), 39% of women were lost to follow-up in 2009, in addition to 5% that were deceased. The 2008–2009 round of the WHSA (henceforth referred to as WHSA-II) interviewed 2,814 women, 995 of whom were replacements–matched by age and enumeration area–for those lost to follow-up. This paper explores the recent sachet water phenomenon, which was not an important factor in 2003 when 97.5% of WHSA respondents reported piped water access (either in or outside the home) as their primary drinking water source. In order to analyze the new dynamics of sachet water using the broadest sample possible, we restrict ourselves to a cross-sectional application using only the WHSA-II data set with an explicit acknowledgment of limitations on interpreting causality from the results.

All WHSA-II participants provided written informed consent; literate women provided their personal signature, while illiterate women provided an ink fingerprint in lieu of a signature. Institutional Review Board approval for all consent procedures, data collection, and analysis was granted by the respective boards of Harvard University, University of Ghana-Legon, and San Diego State University (SDSU).

Statistical Analyses

To test hypothesis 1, we use X^2 and F tests to compare socioeconomic and demographic characteristics between sachet drinkers and those using all other drinking water sources. To test hypothesis 2, we implement a series of iterative multilevel logistic regression models to separate compositional and contextual effects associated with sachet consumption as a primary drinking water source, with neighborhood factors treated as the exposures of interest. We use exploratory forward and backward stepwise regression models to advise the introduction of independent measures in the multilevel model-building process. The full model parsimoniously maximizes higher-level covariance parameters while minimizing the model fit statistic (−2 restricted log pseudo-likelihood). To test hypothesis 3, we introduce cross-level interactions with rationing into the full model, as well as individual-level interactions for other significant measures. We implement a random intercept model using the GLIMMIX procedure in SAS 9.2 with parameters fitted to the FMVN, EA, and/or woman as described below.

The theoretical underpinnings of measuring health and poverty through socioeconomic indicators have long been established in the social epidemiology and international development literature [18–21]. In this analysis, the individual effects modeled include respondent demographics such as age, ethnicity, and education; household characteristics such as dwelling type, and access to toilets and waste disposal services; and an individual-level measure of water pipe density within 500 m of the household (generated from a kernel density surface estimation of neighborhood water pipe penetration). We control for SES quartiles at the EA level, then introduce neighborhood, i.e. FMVN-level, factors that have been previously linked to adverse health outcomes: infrastructure conditions such as pipe density and days of water rationing [8], the proportion of vegetated land cover as a proxy for socioeconomic status [22], and a slum index and housing quality index constructed from 2000 census data to infer socioeconomic status at a finer level [16,22]. The proportion of vegetated land cover was summarized for each FMVN and EA from a classification of a high spatial resolution (2.4 m) QuickBird image captured in 2010. GUWL rationing metrics were recorded in July 2009 and are drawn from schedules of days per week of water service as implemented by local water districts in 2009. Rationing data are

managed by GUWL as a GIS point layer and mapped as a surface of Theissen polygons; mean rationing values are summarized at the FMVN scale by an area-weighted algorithm. All FMVN-level measures were also calculated (or disaggregated) at the EA-level for comparison purposes, and the multilevel models were computed as 3-level (FMVN, EA, individual) and 2-level (EA, individual) models.

Results

Table 1 summarizes the individual characteristics of women from the WHSA-II interviews. Among 2,814 women in the study, 6.8% named sachet water as their primary drinking water source. The demographic profile of these women, compared with non-sachet users, was about 3 years younger on average, disproportionately more likely to live in an EA that ranks in the lowest SES quartile, and more likely to be any ethnicity other than Ga, particularly ethnic groups such as Mole-Dagbani that are grouped into the *other* category. There were no statistically significant differences observed between sachet drinkers and their counterparts in education, though sachet drinkers were significantly more likely to report lower levels of overall health. A profile of sachet consumers as younger, lower-income ethnic minorities begins to emerge, but additional socioeconomic details in Table 1 offer conflicting results.

Women relying on sachet water were less likely to be connected to a sewage system for liquid waste disposal (11.6% vs. 16.7%), an observation that is consistent with the "low income" profile, but was not statistically significant. Conversely, sachet drinkers were significantly more likely to report using a collection service for solid waste disposal (43.7%) than those drinking from other water sources (34.1%), yet access to a waste collection service is generally associated with higher-income populations in Accra. Solid waste disposal was the sole household socioeconomic variable that yielded statistically significant differences between sachet drinkers and the rest of the study population ($p = 0.009$). There were other non-significant hints of better amenities among sachet drinkers in Table 1: sachet drinkers were more likely than their counterparts to use their own bathroom for bathing (39.5% vs. 33.3%), and rated higher on a wealth score of durable goods ownership (mean 0.11 vs. -.01), a difference that translates to the 62nd vs. 59th percentile. There were no significant differences in dwelling type, rooms per dwelling, type of bathing facility, and toilet access. All of these variables except toilet access have previously differentiated sachet drinkers as being of lower means than other women in Accra [8]. The density of GUWL pipe infrastructure was calculated within varying buffers from each household, and sachet users had lower pipe density values within 500 m than non-sachet users (819 vs. 920 mm/km^2, $p = 0.03$).

Table 2 summarizes the overall mean FMVN-level characteristics and Pearson's correlations for 71 neighborhoods that encompass the WHSA-II study population, and selected characteristics are mapped in Figure 2. The neighborhood mean GUWL pipe density was 1,074 mm/km^2, a figure higher than the household means because major pipelines tend to co-locate with major roads that are not necessarily residential centers; the kernel density surface of water pipe density is depicted in Figure 2A. Each FMVN averaged about 5 days per week of running water service due to the GUWL rationing regime as shown in Figure 2B. Neighborhood pipe density was weakly correlated with days of running water, implying that water service was truly more dependent on a local water district's rationing policy than on local infrastructure capacity. The raw neighborhood average for sachets as the primary drinking water source was 7.5%; to account

Table 1. Women's individual characteristics from the WHSA-II.

Characteristic	Sachet Water		Other Water Source	
	Freq.	% or mean (95% CI)	Freq.	% or mean (95% CI)
Individual characteristics (%, n = 2,814)	190	6.8	2,624	93.2
Age (years) *		43.5 (41.1–46.0)		46.5 (45.8–47.2)
Major Ethnic Group (%) ***				
Akan	65	34.2	850	32.4
Ewe	33	17.4	358	13.6
Ga	53	27.9	1,085	41.3
Other	39	20.5	331	12.6
Education (%) ~				
None, other, religious	43	22.8	568	21.8
Primary	29	15.3	306	11.7
Middle	57	30.2	1,043	40.0
Secondary	36	19.0	435	16.7
Higher	24	12.7	255	9.8
Self-reported overall health (%) ***				
Excellent	5	2.6	319	12.3
Very good	33	17.5	650	25.0
Good	108	57.1	1,238	47.6
Fair or poor	43	22.8	396	15.2
Socioeconomic status quartile of EA ***				
Lower class	78	41.1	707	26.9
Lower middle class	36	18.9	605	23.1
Upper middle class	36	18.9	681	26.0
Higher class	40	21.1	631	24.0
Type of dwelling (%)				
House, semi-detached, flat	63	33.2	820	31.3
Compound house	125	65.8	1772	67.6
Hut, tent, kiosk, business, other	2	1.1	28	1.1
Number of rooms in dwelling		2.5 (2.3–2.8)		2.5 (2.4–2.6)
Solid waste disposal (%) **				
Collection service	83	43.7	894	34.1
Public dump	88	46.3	1,516	57.8
Burnt, buried, dumped elsewhere, other	19	10.0	214	8.2
Liquid waste disposal (%) ~				
Sewage system	22	11.6	438	16.7
Thrown in street, gutter, compound, other	168	88.4	2,186	83.3
Type of toilet access (%)				
WC or another house	73	38.4	978	37.3
KVIP or public toilet	88	46.3	1268	48.3
Pit latrine, bucket/pan, other, none	29	15.3	378	14.4
Type of bathing facility (%)				
Own bathroom	75	39.5	873	33.3
Shared with other households	109	57.4	1,635	62.3
Cubicle, open space, other	6	3.2	116	4.4
Wealth score ~		0.11 (−0.02–0.24)		−0.01 (−0.05–0.03)
Pipe density (mm/km^2) within 500 m *		819 (732–906)		920 (892–947)

~$p<0.10$;
*$p<0.05$;
**$p<0.01$;
***$p<0.001$.
Note: p-values for categorical measures are from X^2 test; p-values for continuous measure are from Welch F test of equality of means to account for variance heterogeneity.

Table 2. Characteristics and Pearson's correlations of 71 Field Modified Vernacular Neighborhoods (FMVN) comprising the WHSA-II study population.

Characteristic	Mean	SE	Range	Pearson's Correlation						
				WPD	H2O	SAC1	SAC2	VEG	SI	HQI
GUWL water pipe density (mm/km^2) (WPD)	1074	73	207–3018	1.00						
Days per week of running water (no rationing) (H2O)	4.92	0.20	1–7	0.16	1.00					
Sachets as primary water source (raw %) (SAC1)	7.53	1.81	0–100	0.16	0.09	1.00				
Sachets as primary water source (smoothed %) (SAC2)	6.59	0.75	0.62–35.57	0.09	−0.04	0.73*	1.00			
Vegetated land cover in 2010 (%) (VEG)	19.86	2.00	0–83.38	−0.14	0.05	0.12	0.12	1.00		
Slum index (SI)	2.00	0.04	1.13–2.57	−0.16	−0.07	−0.04	0.00	−0.65*	1.00	
Housing quality index (HQI)	2.44	0.07	1.58–3.51	0.06	0.01	0.04	0.02	0.79*	−0.92*	1.00

*$p < 0.01$ (2-tailed).

for small sample denominators in several neighborhoods, we applied an empirical Bayes smoothing algorithm and mapped the smoothed sachet rates in Figure 2C (smoothed mean = 6.6%). As shown in Table 2, neither neighborhood-level sachet metric was correlated with any of the other neighborhood measures. Three additional FMVN-level measures of socioeconomic variation were considered: the proportion of vegetated land cover (mean 19.86%, SE 2.00), a slum index (2.00, 0.04), and a housing quality index (2.44, 0.07). While these socioeconomic proxies were statistically significantly correlated with each other, none yielded significant correlations with the water-related neighborhood-level measures in Table 2.

Table 3 contains the empty and full 2-level multilevel models. The empty model is the simplest form of the multilevel model (no covariates), and we estimated this form as a baseline for assessing the change in higher-level variance as we added independent measures. We began by fitting 3-level models with the FMVN atop the spatial hierarchy, but the introduction of interaction terms resulted in unstable models with an unexpected inflation of variance. This lack of fit may indicate that the FMVN scale is not optimal for assessing the role of rationing, despite a significant relationship between rationing and sachet use in these models (not shown). In lieu of "overfitting the data," the model was simplified to two levels with neighborhood parameters fitted at the EA level.

The total variance at the EA level was 2.040 in the empty model (Model 1). Model building began with the EA-level rationing and SES measures, and we proceeded to iteratively test combinations of covariates and first-order interaction terms until arriving at the most parsimonious model shown in Table 3. After adjusting for covariates, the full model without interactions (Model 2) minimized EA variance at 1.691, a decrease of 17.1%. Substantial variance remains, and this result suggests that living in a particular neighborhood strongly influenced the likelihood that a woman relied on sachet drinking water.

Table 3 gives the restricted pseudo-likelihood estimates for determinants of sachet use. We observed no support for our second hypothesis that neighborhood rationing was driving sachet consumption; the rationing variable was not statistically significant, though it did contribute to minimizing higher-level variance in the model. There is, as seen in Table 1, mixed support for our first hypothesis. At the individual level, lower self-reported overall health was strongly associated with sachet consumption: we observed a trend of increasing odds of sachet use for each category of overall health, with women reporting fair or poor health being significantly more likely to have used sachets than women

reporting excellent health after adjusting for covariates ($p < 0.001$). Age was also statistically significantly associated with sachet use, and the negative coefficient suggests that women were slightly less likely to use sachet water as their primary water source for each additional year of age, after controlling for covariates ($p < 0.01$). The EA-level SES measure was also statistically significant; women in the lower class SES quartile were more likely ($p < 0.01$) to use sachets than women in the upper class quartile, and we observed a similar trend of increasing odds of sachet use for each progressively lower SES quartile as seen with the *overall health* measure. Women without access to a sewer connection were also more likely to rely on sachets, but this difference was not statistically significant. These results affirm the profile of the "younger, less-well-off" sachet consumer, except that the wealth score was again positively associated with sachet use: women were more likely to use sachets ($p < 0.01$) for each unit-increase in wealth score, which translated to a 20-percentile change when the score is centered on zero. This implies a massive relative increase in wealth score to produce a modest increase in the odds of sachet consumption, but it does indicate that even after controlling for covariates–particularly SES quartile–relative wealth still did impact the use of sachet water. Evidence of a possible interaction between SES and wealth appears in Table 4 where we stratified the mean wealth score for sachet users and non-users by SES quartile. Only the lower class quartile yielded a statistically significant difference in wealth score, as sachet users scored higher than expected (−0.15 vs. −0.58)–a difference that translated to the 54th vs. 35th percentile–and higher than sachet users in the lower-middle quartile (−0.34).

Models 3–5 in Table 3 summarize the beta coefficients and standard errors for three additional models that introduce interaction terms to the full model. Model 3 contains cross-level interactions between rationing and proportion of vegetated land cover, and between rationing and SES, but these relationships were not statistically significant. We added cross-level interactions in Model 4 to test if rationing and SES interacted with overall health and wealth score. Only the interaction between lowest-quartile SES and wealth score–the relationship depicted in Table 4–was statistically significantly associated with higher sachet use ($p < 0.01$). While most of the interaction terms in Model 4 were not statistically significant, their presence in the model rendered the beta for *days of rationing* marginally significant ($p < 0.10$) and *proportion of vegetated land cover* statistically significant ($p < 0.01$), thus lending initial, though weak, support to hypothesis 2 that increased rationing was associated with increased sachet use. At the

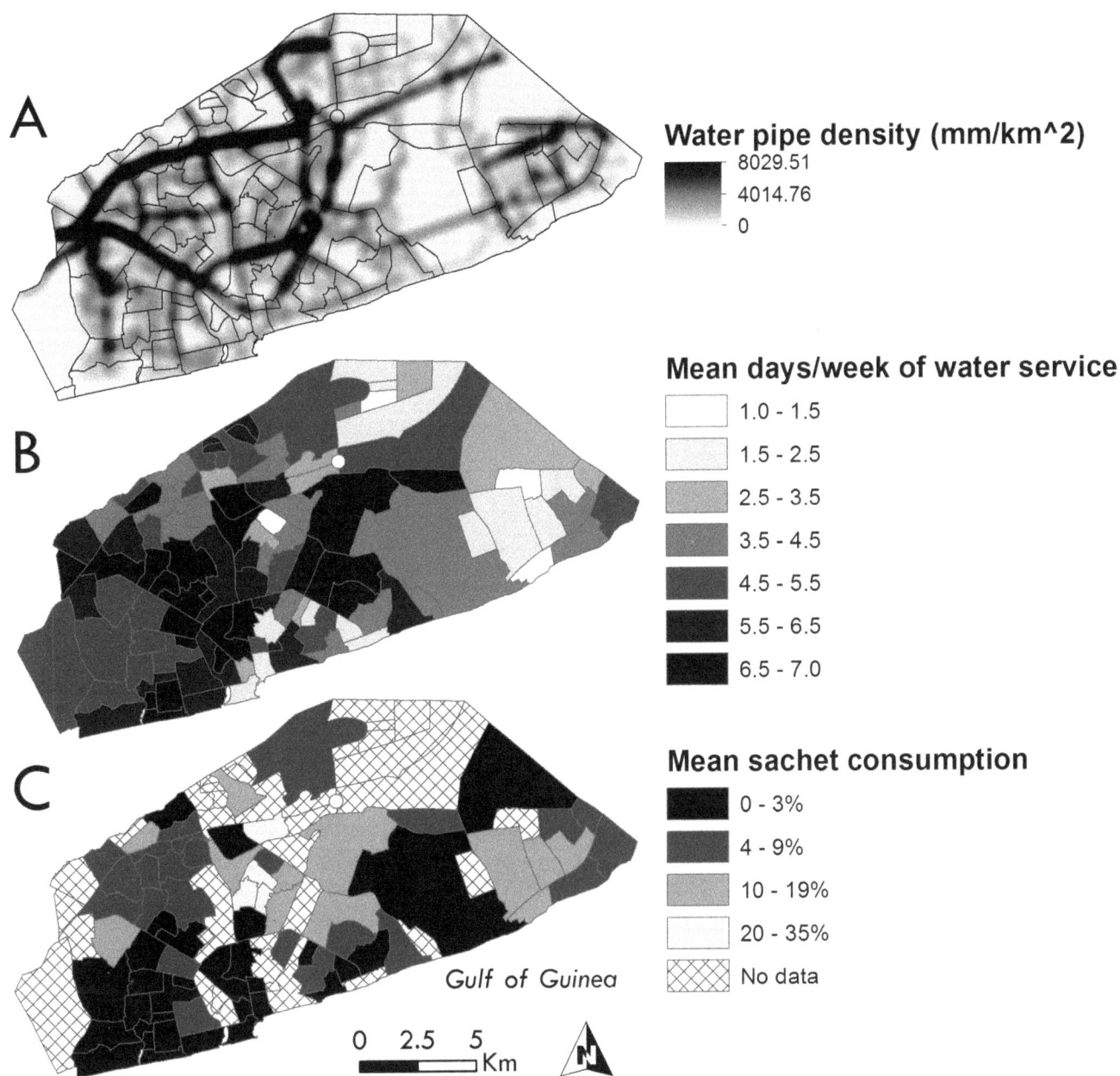

Figure 2. Accra metropolitan area neighborhoods (FMVN) depicting (A) mean density of 2009 GUWL water pipe infrastructure (mm/km2) using a 500 m kernel, (B) mean days per week of running water service according to the 2009 GUWL rationing regime, and (C) smoothed mean percentage of women reporting sachet water as the primary drinking water source for 71 FMVN sampled in the 2008–2009 WHSA-II.

individual level, living in the lower class SES quartile, worse overall health, and younger age were all associated with higher sachet use just as in Model 2. There was no strong evidence in Model 4 for hypothesis 3, which posits that the interaction between higher rationing and lower SES quartile–or any proxy for lower SES such as worse overall health or lower wealth–was associated with higher sachet use.

Model 5 contains additional individual-level interactions between overall health and wealth score, and between overall health and age. The interaction between overall health and age was statistically significant: for each additional year of age, women with fair or poor health ($p<0.01$), or very good health ($p<0.10$), were less likely to use sachets than women with excellent health,

and the beta coefficients revealed a stronger effect for age than in Models 2–4. Just as in Models 3 and 4, worse overall health, living in a lower class EA, higher rationing, and a higher proportion of vegetated land cover were predictive of sachet use.

Models 3–5 present a more nuanced picture of the predictors of sachet use than Model 2, though the general conclusion is similar: younger, poorer women (or women living in poorer EAs) in neighborhoods experiencing higher rationing were most likely to rely on sachet water. The sole factor with a relationship contrary to hypothesis 1 was the significance of vegetated land cover at the EA level: a higher percentage of vegetation–which was statistically significantly associated with higher-SES quartiles in Table 5–was also predictive of sachet use. Previous research has shown a higher

Table 3. Beta coefficients (SE) for random intercept models assessing the association between rationing and sachet use among Accra women.

Characteristic	Model 1 Empty (no covariates)		Model 2 Full model of EA- and individual-level factors		Model 3 Full model with level-2 interactions		Model 4 Full model with level-2 and cross-level interactions		Model 5 Full model with level-1 interactions	
Intercept	-3.077***	(0.149)	-5.133***	(0.825)	-5.830***	(0.942)	-6.061***	(1.138)	-6.678***	(1.807)
Interactions: Level-2 (EA)										
Rationing×Vegetation					-1.420	(0.875)	-1.648~	(0.889)		
Rationing×Upper middle class SES					-0.404~	(0.241)	-0.389	(0.243)		
Rationing×Lower middle class SES					-0.072	(0.230)	-0.047	(0.238)		
Rationing×Lower class SES					-0.176	(0.230)	-0.109	(0.236)		
Interactions: Cross-level										
Rationing×OH Very good							-0.106	(0.216)		
Rationing×OH Good							-0.150	(0.205)		
Rationing×OH Fair or poor							-0.151	(0.217)		
Rationing×wealth score							0.084~	(0.050)		
Upper middle class SES×wealth score							0.156	(0.218)		
Lower middle class SES×wealth score							-0.405	(0.309)		
Lower class SES×wealth score							0.627**	(0.231)		
Interactions: Level-1 (Individual)										
Wealth score×OH Very good									0.213	(0.634)
Wealth score×OH Good									0.612	(0.606)
Wealth score×OH Fair or poor									0.896	(0.628)
Age×OH Very good									-0.035~	(0.019)
Age×OH Good									-0.007	(0.007)
Age×OH Fair or poor									-0.029**	(0.011)
Level-1 (Individual) variables										
Self-reported overall health										
Excellent†										
Very good			1.264*	(0.502)	1.259*	(0.502)	1.593~	(0.822)	2.955~	(1.745)
Good			1.853***	(0.483)	1.855***	(0.484)	2.312**	(0.796)	2.610	(1.646)
Fair or poor			2.151***	(0.516)	2.139***	(0.517)	2.633**	(0.833)	3.988*	(1.711)
Solid waste disposal										
Burnt, buried, dumped elsewhere, other†										
Public dump			-0.355	(0.303)	-0.326	(0.304)	-0.391	(0.305)	-0.428	(0.307)

Table 3. Cont.

Characteristic	Model 1 Empty (no covariates)	Model 2 Full model of EA- and individual-level factors	Model 3 Full model with level-2 interactions	Model 4 Full model with level-2 and cross-level interactions	Model 5 Full model with level-1 interactions
Collection service		0.104 (0.323)	0.109 (0.325)	0.078 (0.327)	0.038 (0.327)
Liquid waste disposal					
Sewage system†					
Thrown in street, gutter, compound, other		0.592~ (0.333)	0.578~ (0.334)	0.497 (0.344)	0.484 (0.354)
Major ethnic group					
Other		0.024 (0.300)	0.015 (0.310)	0.004 (0.304)	−0.019 (0.309)
Ewe		−0.255 (0.290)	−0.275 (0.291)	−0.312 (0.293)	−0.292 (0.296)
Ga		0.012 (0.269)	−0.007 (0.270)	−0.011 (0.273)	0.004 (0.275)
Akan		−0.016** (0.005)	−0.016** (0.005)	−0.015** (0.006)	– (–)
Age (years)		0.380** (0.125)	0.392** (0.126)	– (–)	– (–)
Wealth score					
Level-2 (EA) variables					
SES quartile					
Upper class†					
Upper middle class		−0.057 (0.449)	0.827 (0.645)	0.718 (0.659)	0.720 (0.664)
Lower middle class		0.523 (0.482)	0.674 (0.695)	0.431 (0.712)	0.381 (0.717)
Lower class		1.354** (0.489)	1.776* (0.712)	1.703* (0.724)	1.704* (0.728)
% Vegetated land cover		2.069 (1.290)	4.881* (1.964)	5.452** (1.988)	5.464** (2.010)
Days of rationing		0.087 (0.069)	0.395~ (0.233)	0.514~ (0.311)	0.531~ (0.312)
Random Effects					
Level-2 variance (EA)	2.040 (0.368)	1.691 (0.349)	1.636 (0.354)	1.668 (0.361)	1.711 (0.368)
Δ (%) in level-2 variance	–	−17.1%	−19.8%	−18.2%	−16.1%
−2 res log pseudo-likelihood	15,945.99	15,791.81	15,905.96	16,083.04	16,285.13
Generalized chi-square	1,427.79	1,429.90	1,483.29	1,504.83	1,527.98

†Reference Category;
~ $p<0.10$;
*$p<0.05$;
**$p<0.01$;
***$p<0.001$.

Table 4. Differences in wealth score between sachet users and non-users stratified by SES quartile.

Characteristic	Sachet Water		Other Water Source	
	Freq.	mean (95% CI)	Freq.	mean (95% CI)
Individual characteristics (%, n = 2,814)	190	6.8%	2,624	93.2%
Wealth score ~		0.11 (−0.02–0.24)		−0.01 (−0.05–0.03)
By SES quartile				
Lower class *	78	−0.15 (−0.31–0.02)	707	−0.58 (−0.63– −0.53)
Lower middle class	36	−0.34 (−0.56– −0.11)	605	−0.27 (−0.34– −.021)
Upper middle class	36	0.35 (0.04–0.66)	681	0.16 (0.08–0.23)
Higher class	40	0.79 (0.47–1.11)	631	0.71 (0.62–0.79)

~ p<0.10;
*p<0.001.
Note: p-values are from Welch F test of equality of means to account for variance heterogeneity.

proportion of vegetated land cover to be associated with higher-status areas, not poverty [22], so this relationship may be an artifact of the transition in sachet consumption from higher to lower socioeconomic classes. The modeling of interaction terms in Table 3 lent weak support for hypothesis 2, as days of rationing persists as a marginally significant neighborhood-level influence on sachet use. Models 3 and 4 explain a slightly greater proportion of EA-level variance than Model 2, but the use of individual interaction terms in Model 5 does not improve overall results. There is no strong support for hypothesis 3; there are no significant interactions between rationing and any SES indicator.

Discussion and Conclusion

Over the last decade sachet water has risen from relative obscurity to become an important source of drinking water throughout West Africa. Previous research has explored sachet water in select neighborhoods of Accra [8,23] but this is the first study to analyze sachet water consumption in a West African metropolis using a socioeconomic and geographic cross-sectional approach. This study uses data collected in 2008–2009 to examine intra-urban variability of drinking water selection in Accra, and factors that influence these choices. We observe evidence of our first hypothesis, as sachet consumers tend to be younger, of poorer overall health, and of lower socioeconomic means, and yet those living in the economically poorest areas may be slightly better-off than their immediate neighbors. We also observe limited support for our second hypothesis that water rationing at the neighbor-

hood scale positively influences sachet use, though rationing may be less important than individual-level factors. A smaller neighborhood unit was more appropriate for modeling sachet use, though there is still a considerable unexplained neighborhood effect. These neighborhood-level findings are consistent with recent reports from Accra [8]. The analysis does not support our third hypothesis regarding an interaction between rationing and lower socioeconomic status.

Two other recent population-based surveys, Measure DHS's Ghana Demographic and Health Survey (DHS) and the Harvard/SDSU Housing and Welfare Study of Accra (HAWS), each report higher rates of sachet consumption than those reported in the WHSA-II [8,23]. The last two Ghana DHS indicate that between 2003 and 2008 the percentage of urban households using sachet water as the primary water source increased from 6% to 37% nationally [24]. In Greater Accra, 87% of sachet-using households were in the top wealth quintile in the 2003 survey; by 2008 this rate had fallen to 71% and sachet use started trickling down into the middle and lower quintiles, but with scant evidence that sachets were a phenomenon of poverty [23,24]. The WHSA-II data were collected between October 2008 and March 2009 and yield only 10% (78/785) of lowest-SES quartile participants relying on sachets; this survey seems to have been administered amidst a widespread transition to sachet water by lower-income populations. The HAWS data, collected between September 2009 and March 2010 exclusively in Accra slum neighborhoods, report 50% of households depending on sachet water with lowest-income residents as the most likely consumers [8].

Table 5. Enumeration area (EA) differences in rationing, proportion of vegetated land cover, and sachet use stratified by SES quartile.

SES Quartile of EA	N	Days of Rationing mean (SE)	Vegetation (%)* mean (SE)	Sachet Use (%) mean (SE)
Lower class	48	2.18 (0.26)	5.03 (0.91)	12.06 (3.00)
Lower middle class	50	2.12 (0.25)	8.03 (1.27)	6.69 (1.95)
Upper middle class	53	1.75 (0.20)	15.17 (1.85)	5.62 (2.13)
Higher class	44	1.84 (0.23)	25.06 (2.42)	8.53 (2.73)
Total	195	1.97 (0.12)	13.24 (1.01)	8.14 (1.23)

*p<0.001 from Welch F test of equality of means to account for variance heterogeneity.

Both 2008 and 2010 were especially bad years for water service delivery as power outages and construction projects at both of Accra's water treatment plants led to multiple-week service interruptions in many neighborhoods. Sachet water was already a fairly ubiquitous product in Accra in 2008. It is plausible that sachet water's higher unit price prevented it from becoming a *primary* drinking water source for many low-income communities until severe piped water shortages finally made sachets a necessity, even if temporarily. This study's finding that sachet users in the lowest SES quartile score higher on a wealth index than sachet users in the lower-middle quartile underscores the reality that, in lieu of abject poverty, sachets may still be a discretionary, but increasingly attractive, choice to younger, poorer urban residents.

There are also several geographical discrepancies between the sachet patterns observed in the WHSA-II and those seen in the 2008 DHS and HAWS. In particular, the WHSA-II reports no sachet use among 270 women interviewed in the coastal Ga neighborhoods of Gbegbeyise, Chorkor, Korle Gonno, and Jamestown in western Accra. The DHS and HAWS data sets report significant sachet reliance in these and other neighborhoods where WHSA-II reports no sachet use. These coastal Ga neighborhoods in particular are known to have built much of their own water infrastructure (in a mix of legal and pirated networks), and the lack of maintenance by GUWL may increase the frequency of service disruptions. Given the geographic and temporal heterogeneity of water service in Accra, it is possible that some neighborhoods were interviewed in atypically water-abundant or water-scare periods, thus potentially biasing responses during any of these three surveys. It is also possible that there may have been some inadvertent bias in how the WHSA-II question about primary drinking source was framed to participants, despite all three surveys using the same DHS format. This mismatch between surveys may account for the lack of significant effects among neighborhood-level measures in this study, particularly days of rationing as observed in 2009–2010 [8]. More broadly speaking, it also highlights the difficulty of accuracy assessment for primary data collected in a developing urban setting. These survey discrepancies may soon be reconciled by the forthcoming release of 2010 Ghana Census results, as the 2010 questionnaire split the traditional household water question into separate items for drinking water sources and all other household water sources. The 2010 census may provide the most detailed picture yet of Ghana's evolving drinking water trends.

This analysis is limited by the cross-sectional nature of the WHSA-II data. Despite compiling a broad array of women's health outcomes, the WHSA-II did not measure diarrhea prevalence or other outcomes that might specifically be attribut-

able to drinking water quality, thus it is difficult to assess the strong association between sachet use and overall health. Higher self-reported overall health has previously been positively associated with higher socioeconomic amenities in Accra [16,25,26], and ill-health has long been linked to chronic poverty [27], so the overall health measure can reasonably be interpreted as a SES proxy. Future research should directly target the relationship between sachet consumption and health outcomes suggested in the HAWS project [8].

Urban drinking water patterns in the developing world remain understudied, as urban areas are generally presumed to have sufficient piped water coverage. Accra is emblematic of many developing cities where this assumption is no longer valid [3,9,28–30]. Piped water access is lagging in urban sub-Saharan Africa [3], and residents are increasingly turning to privately vended water such as sachets. Problems with the unsustainable amount of plastic waste generated by sachet water consumption have been reported elsewhere [8,23,31], and have resulted in increased local media coverage of municipal water delivery issues faced by Ghana's Ministry of Water Resources, Works and Housing. Sachet water continues to spread throughout West Africa, and many cities are likely experiencing similar transitions in drinking water patterns without a complete understanding of potential health effects or an appreciation for the environmental sanitation consequences of all the plastic waste. Global safe water and sanitation goals face many challenges [32], but in Ghana improvements are primarily constrained by financial resources; the capital investments required for an adequate water supply and sanitation infrastructure are estimated at $1.3 billion for the rehabilitation and expansion of urban water infrastructure alone [33]. Local governments and non-governmental organizations interested in alternative sustainable water provision would be well-served to understand the demographic appeal of sachet water, as well as the geographic and economic realities that turn citizens into sachet customers.

Acknowledgments

The authors thank David Nunoo and Michael Nyoagbe at Ghana Urban Water Ltd., Head Office Operations, Accra, for providing important GIS data management and assistance.

Author Contributions

Conceived and designed the experiments: JS JRW. Performed the experiments: JS. Analyzed the data: JS. Contributed reagents/materials/analysis tools: JS JRW RAO. Wrote the paper: JS.

References

1. WHO/UNICEF (2010) Progress on sanitation and drinking water: 2010 update. Geneva: World Health Organization.

2. United Nations (2010) World Urbanization Prospects: The 2009 Revision. New York, NY: United Nations Population Division, Department of Economic and Social Affairs.

3. WHO/UNICEF (2011) Drinking water: Equity, safety and sustainability. Geneva: World Health Organization.

4. Fink G, Günther I, Hill K (2011) The effect of water and sanitation on child health: evidence from the demographic and health surveys 1986–2007. International Journal of Epidemiology 2011: 1–9.

5. Nsiah-Gyabaah K (2001) The looming national dilemma of water crisis in peri-urban areas in Ghana. Second Workshop on Peri-Urban Natural Resources Management Project at the Watershed Level. Kumasi, Ghana: DFID.

6. Van-Rooijen DJ, Spalthoff D, Raschid-Sally L (2008) Domestic water supply in Accra: How physical and social constraints to planning have greater consequences for the poor. Accra, Ghana. 5p.

7. WaterAid (2005) National Water Sector Assessment: Ghana. London, UK. 8 p.

8. Stoler J, Fink G, Weeks JR, Appiah Otoo R, Ampofo JA, et al. (2012) When urban taps run dry: Sachet water consumption and health effects in low income neighborhoods of Accra, Ghana. Health & Place 18: 250–262.

9. Thompson J, Porras IT, Wood E, Tumwine JK, Mujwahuzi MR, et al. (2000) Waiting at the tap: Changes in urban water use in East Africa over three decades. Environment and Urbanization 12: 37–52.

10. International Development Committee (2007) Sanitation and Water: Sixth Report of Session 2006–07, House of Commons. London, UK: The Stationery Office Limited.

11. United Nations (2008) The Millennium Development Goals Report 2008. New York, NY: United Nations.

12. United Nations Development Programme (2006) Human Development Report 2006: Beyond scarcity: Power, poverty and the global water crisis. New York, NY: United Nations.

13. WHO, Clasen TF (2009) Scaling up household water treatment among urban populations. Geneva: WHO. 84p.

14. Hill AG, Darko R, Seffah J, Adanu RMK, Anarfi JK, et al. (2007) Health of urban Ghanaian women as identified by the Women's Health Study of Accra. International Journal of Gynecology & Obstetrics 99: 150–156.

15. WHSA-II Writing Team (2011) Final report on the Women's Health Study of Accra, Wave II. Accra, Ghana: Institute for Statistical, Social and Economic Research, University of Ghana. 144 p.

16. Weeks JR, Getis A, Stow D, Hill AG, Rain D, et al. (2012) Connecting the dots between health, poverty and place in Accra, Ghana. Annals of the Association of American Geographers 102: 932–941.

17. Weeks JR, Getis A, Hill AG, Agyei-Mensah S, Rain D (2010) Neighborhoods and fertility in Accra, Ghana: An AMOEBA-based approach. Annals of the Association of American Geographers 100: 558–578.

18. Krieger N (1992) Overcoming the absence of socioeconomic data in medical records: validation and application of a census-based methodology. American Journal of Public Health 82: 703–710.

19. Krieger N, Williams D, Moss N (1997) Measuring social class in US public health research: concepts, methodologies and guidelines. Annual Review of Public Health 18: 341–378.

20. Lynch J, Kaplan G (2000) Socioeconomic Position. In: Berkman LF, Kawachi I, editors. Social Epidemiology. Oxford: Oxford University Press. pp. 13–35.

21. UNDP-UNEP Poverty and Environment Initiative (2008) Poverty and Environment Indicators. Cambridge: Capability and Sustainability Centre, St. Edmund's College.

22. Weeks JR, Hill AG, Stow DA, Getis A, Fugate D (2007) Can we spot a neighborhood from the ground? Defining neighborhood structure in Accra, Ghana. GeoJournal 69: 9–22.

23. Stoler J, Weeks JR, Fink G (2012) Sachet drinking water in Ghana's Accra-Tema metropolitan area: Past, present, and future. Journal of Water, Sanitation and Hygiene for Development 2: 223–240.

24. Macro International Inc (2011) MEASURE DHS STATcompiler.

25. Arku G, Luginaah I, Mkandawire P, Baiden P, Asiedu AB (2011) Housing and health in three contrasting neighborhoods in Accra, Ghana. Social Science & Medicine 72: 1864–1872.

26. Stoler J, Weeks JR, Getis A, Hill AG (2009) Distance threshold for the effect of urban agriculture on elevated self-reported malaria prevalence in Accra, Ghana. American Journal of Tropical Medicine and Hygiene 80: 547–554.

27. Sverdlik A (2011) Ill-health and poverty: A literature review on health in informal settlements. Environment and Urbanization 23: 123–155.

28. Collignon B, Vézina M (2000) Independent water and sanitation providers in African cities. Full report of a ten-country study. Washington, DC: Water and Sanitation Program.

29. Dominguez Torres C (2012) The future of water in African cities: Why waste water? Urban access to water supply and sanitation in sub-Saharan Africa, background report. Washington DC: World Bank. 35p.

30. Kessides C (2005) The urban transition in sub-Saharan Africa: Implications for economic growth and poverty reduction. Africa Region Working Paper Series No 97. Washington DC: World Bank. 84p.

31. Stoler J (2012) Improved but unsustainable: Accounting for sachet water in post-2015 goals for global safe water. Tropical Medicine & International Health 17: 1506–1508.

32. Moe CL, Rheingans RD (2006) Global challenges in water, sanitation and health. Journal of Water and Health 4: 41–57.

33. AfDB/OECD (2007) African Economic Outlook. Paris: AfDB/OECD. 618 p.

19

Performance and Reliability Analysis of Water Distribution Systems under Cascading Failures and the Identification of Crucial Pipes

Qing Shuang, Mingyuan Zhang*, Yongbo Yuan

Department of Construction Management, Dalian University of Technology, Dalian, Liaoning, China

Abstract

As a mean of supplying water, Water distribution system (WDS) is one of the most important complex infrastructures. The stability and reliability are critical for urban activities. WDSs can be characterized by networks of multiple nodes (e.g. reservoirs and junctions) and interconnected by physical links (e.g. pipes). Instead of analyzing highest failure rate or highest betweenness, reliability of WDS is evaluated by introducing hydraulic analysis and cascading failures (conductive failure pattern) from complex network. The crucial pipes are identified eventually. The proposed methodology is illustrated by an example. The results show that the demand multiplier has a great influence on the peak of reliability and the persistent time of the cascading failures in its propagation in WDS. The time period when the system has the highest reliability is when the demand multiplier is less than 1. There is a threshold of tolerance parameter exists. When the tolerance parameter is less than the threshold, the time period with the highest system reliability does not meet minimum value of demand multiplier. The results indicate that the system reliability should be evaluated with the properties of WDS and the characteristics of cascading failures, so as to improve its ability of resisting disasters.

Editor: Vanesa Magar, Plymouth University, United Kingdom

Funding: This work is supported by the National Natural Science Foundation of China under Grant No. 51208081. The funders had no role in study design, data collection and analysis, decision to publish, or preparation of the manuscript.

Competing Interests: The authors have declared that no competing interests exist.

* E-mail: myzhang@dlut.edu.cn

Introduction

The stability and reliability of Water distribution systems (WDSs) is one of the important factors in ensuring public safety and the continuous operation of urban functions. Such functions include water supply, infrastructure construction and industrial development, etc. It is also the key field for infrastructure construction. The WDS is a large scale network system with complex topological structure [1]. Its functions are designed to convey volumes of water to customers under adequate pressure. Nowadays, along with the increased population and population density, WDS is developing into wide-range supply which carries fluid under high or less pressure. A WDS can be represented as a spatially networks of multiple interconnected components. Pipes can be represented as links. Junctions, reservoirs and consumers can be represented as a collection of nodes. With the link-node representation of physical components in WDS, complex network analysis can be applied to evaluate the system reliability.

Complex networks are an essential part in the understanding of many natural systems [2]. A complex network is a network with non-trivial topological features, which often occur in real life. Complex networks analysis provides a way to understand the meaning and functions of the system [3]. It focuses on predict the networked system behavior on the basis of measured structure. Albert et al. [4] have found that the scale-free networks have strong robustness under random disturbance, but it is very vulnerable under intentional attacks. These important discoveries

have made the network security under abnormal conditions become a hot issue in this field. Cascading failures is a conductive failure process in the field of network security [5]. When the network encounters natural or man-made disasters, i.e. network attacks and random failures, the minor anomalous event of a point may spread to the whole system through cascade reaction, leading to large-scale consequences and secondary failures. Many models have been provided to investigate the cascading failures. The present studies mainly focus on: (1) the network reliability and topology structure after remove some nodes or links [6–8]; (2) the formation conditions and reasons of cascading phenomena and the network dynamics in networks or weighted networks [9–10]; and (3) the metrics of network robustness and the sequent network optimization and design [11].

For the real-world networks, the cascading failures of infrastructure systems have been proposed as well. The power grids of North America and the Western United States are two key studies in this field [12–15]. Besides, the Internet network [16], the power grids of European [17–18] and Italian [19], and other kinds of power systems [20] and traffic networks [21] are also the focus of studies. In these studies, most cascading failures use the virtual network simulation method which measures the network load by the topological property, such as the betweenness and degree. The betweenness is defined as the total number of the shortest paths that pass through the vertex [22]. The degree is defined as the number of edges connected to the vertex [23]. Therefore betweenness and degree basically measure the topological structure of a network. This method fits for the disaster simulation under uncertainties and the rapid assessment on disasters. However, it ignores the properties of city lifeline system as an entity network. Using the betweenness or degree to represent the actual network flow cannot guarantee the accuracy of the calculation results. Therefore, the results cannot be directly applied to the decision-making.

System reliability is defined as the ability of the system to complete the scheduled functions in a certain period under the given working state [24–25]. There are two failure mechanisms of a WDS [26]: mechanical and hydraulic failures. Mechanical reliability is defined as the probability that the WDS and its components are operational. Mechanical reliability focuses on a topological perspective. Hydraulic reliability is defined as the probability that the WDS meet flow and pressure requirements. Hydraulic reliability considers failures in meeting consumer demand. The reliability in this paper combined these two types of reliability. With the topological dynamics changing, the scheduled functions refer to ensuring the water demand and hydraulic pressure required in the daily life of customs in normal conditions; under failure conditions, the water supply and hydraulic pressure would not be lower or higher than the specified limit.

The purpose of this paper is to study the propagation characteristics of cascading failures in the WDS and put forward the methods to identify the crucial pipes. The methods from complex networks and engineering are adopted. The simulation of the cascading failures in WDS is required to meet the equilibrium of water supply and demand. The main task is adopting the numerical simulation technology to depict the damage process of WDS in cascading failures. The network dynamics are used with failure propagation. The uncertain factors, i.e., the nodal head bounds, daily demand multipliers and the water demand have been taken into consideration. The crucial pipe is identified by its vulnerability and sensitivity to cascading failures. By identifying the most crucial pipes, one can effectively protect the network to avoid cascading failures and build attack-robust networks.

Methods

1. Reliability Assessment

The reliability is defined as the probability that the WDS meet flow and pressure requirements under the possible mechanical failure scenarios (e.g., pipe breaks). The definition of system reliability given by Zhuang et al. [27] is adopted. Mathematically, the reliability can be expressed as the ratio of the available flow to the require demand. The condition includes normal condition and failure condition. The reliability of node and system is expressed as:

$$R_j = \frac{Q_j^{avl}}{Q_j^{req}} \tag{5}$$

$$R_{sys} = \frac{\sum_{j=1}^{N_{Node}} Q_j^{avl}}{\sum_{j=1}^{N_{Node}} Q_j^{req}} \tag{6}$$

where R_j is the jth node reliability; R_{sys} is the reliability of WDS; N_{Node} is the number of nodes; Q_j^{avl} is the available flow at jth node

(L/s); Q_j^{req} is the require demand when jth node is under normal condition.

2. WDS Topological Structure Expression

Before analogue simulation and calculation, the first step is to input the graphic information of the system into the computer and set up the model. A WDS can be analyzed by the methods of graph theory based on its topology structure. Graph theory is the study of graphs. A graph consists of a set of nodes and links, representing the interactions among them. A graph is customarily depicted the nature of the links between nodes. A directed graph is one in which links have orientation [28]. A WDN is a directed graph due to the operational flow and pressure requirements. The reservoirs, junctions and customers are described as nodes; while the pipes, pumps and valves are represented as links. The adjacent nodes in the graph are connected by the links in most of the cases.

The matrix is used as an effective tool to depict the properties of the graph in network modeling. The adjacency matrix and incidence matrix are the most common ones. The adjacency matrix A is used to represent the relationship between the nodes in the network. The values for the element a_{ij} are: 0 and 1. When the node i and node j is connected by a pipe, the value is 1; when the node i and node j is unconnected, the value is 0.

$$a_{ij} = \begin{cases} 1, & Node\ i\ and\ node\ j\ connected \\ 0, & Node\ i\ and\ node\ j\ unconnected \end{cases} \tag{1}$$

Incidence matrix N is for describing the relationship between nodes and pipes. The row represents the nodes and the column represents the pipes, respectively. In the network graph, each node and pipe is numbered by consecutive number from 1. The information of node i is recorded in the ith row and that of pipe j is recorded in the jth column. The element n_{ij} is expressed as:

$$n_{ij} = \begin{cases} 1, & Node\ i\ is\ the\ initial\ point\ of\ pipe\ j \\ -1, & Node\ i\ is\ the\ terminal\ point\ of\ pipe\ j \\ 0, & Node\ i\ is\ unconnected\ with\ pipe\ j \end{cases} \tag{2}$$

3. Modeling of WDS Based on Cascading Failures

3.1 The Load of WDS Nodes. In view of the actual physical meaning of WDS, we adopt the nodal pressure head as the load. The load of nodes is a relevant quantity, which can be material, information and energy [5], and can be concrete or abstract. With the passage of time, the load of the nodes exchanges along the connected edge between each of the node pair. The WDS is a kind of material loading network. It distributes water from the reservoirs to the customers. In order to meet the customers' daily needs, each component of the WDS must be able to provide required water demand and pressure head under both normal and failure conditions. Therefore, we define the initial load of the node as the service head H_s. The service head ensures that all the imposed demands can be satisfied.

3.2 The Relationship between Nodal Capacity and Initial Load. The nodal capacity is the maximum load the node can bear. In the place of residence or business, the nodal pressure should not be too high or too low. This because the low pressure leads to flow reductions or blanking; high pressure causes the pipe leakage and even the burst of ageing pipes so that losing its service functions. In general, for each node, there are three kinds of pressure head:

(1) H_{min}, the acceptable minimum level of pressure. If the nodal pressure head is lower than H_{min}, the node loses its service function. Therefore, the minimum capacity is defined as the nodal minimum head.

(2) H_s, the service level of pressure to meet all the imposed demands. Only when each nodal pressure head is higher than service head, the WDS can be performing normally;

(3) H_{max}, the acceptable maximum level of pressure that a node can bear. Then, define the maximum capacity of the node. In a man-made system, the nodal capacity is severely limited by cost. For WDS, the pressure head, H, at each demand node is always within a specified range of a minimum head H_{min} and a maximum H_{max} [29] when it's in normal operating conditions. Besides, there is a direct relationship between pipe leakage and service pressure. The water leakage increases with pressure [30]. In order to avoid the leakage of ageing pipes caused by over-high pressure, the maximum capacity of the node should be defined. Suppose α is the tolerance parameter. It is possible to assume that the maximum capacity is proportional to the initial load H_s. The maximum capacity can be expressed as:

$$H_j^{\max} = (1 + \alpha)H_j^s \qquad (3)$$

where H_j^{max} is the jth nodal maximum head capacity; H_j^s is the jth nodal service head. α allows a systematic evaluation of the aggregated performance of water distribution network element during cascading propagation [31]. The bigger α is, the higher the capacity constraint will be, which means less likely the node would fail.

To represent actual flows supplied to customers under abnormal condition, the available nodal demand is expressed as a function of nodal pressure head. Considering the formulation proposed by Wagner et al. [32], and take the maximum head capacity into account, the function can be expressed as follows:

$$Q_j^{avl} = \begin{cases} 0 & H_j < H_j^{\min} \\ Q_j^{req}\sqrt{\dfrac{H_j - H_j^{\min}}{H_j^s - H_j^{\min}}} & H_j^{\min} \le H_j \le H_j^s \\ Q_j^{req} & H_j^s < H_j \le H_j^{\max} \\ 0 & H_j > H_j^{\max} \end{cases} \qquad (4)$$

where Q_j^{avl} is the flow delivered to the jth node (L/s); Q_j^{req} is the require demand when jth node is under normal condition. H_j, H_j^{min}, H_j^{max} and H_j^s represent the calculated head, the minimum head, the maximum head and the service head at jth node, respectively.

For the above three kinds of pressure heads, there are four states of water supply: (1) nodes are completely shut off when the pressure head is lower than the minimum head; (2) the customers' demands is supplied at a reduction level when the pressure head is higher than the minimum head and lower than the service head; (3) nodes meet the customers' demand when the pressure is higher than the service head; and (4) nodes are closed when the pressure head is higher than the maximum head.

3.3 The Cascading Dynamics. The load on the network is in dynamic changes, especially when the network structure transforms. For example, the load is redistributed due to some nodes or pipes shut off. In general, the network node and edge have a limited bearing capacity. If the maximum load (capacity) is exceeded, the network equilibrium will be broken and the load will be redistributed.

When a pipe of the WDS is closed for failure condition, it is equivalent to removing an edge of the network. Then it triggers the network flow to be redistributed among all nodes. The artificial time step t ($t = 1, 2, \ldots$) is introduced to monitor the process of cascading failures. At time t, if the nodal pressure head H_j exceeds its capacity (above the maximum head or below the minimum head), this node fails to provide required water. The failure triggers the reduction of its downstream pipes. The WDS pressure is a spatial vector and the pressure on each node is interdependent. The pressure of one node changes leads to other node pressure changes to varying degrees. In this situation, a new round of load redistribution occurs and leads to cascading failures. The iterative process continues until there are no failure nodes or pipes produced, which implies the cascading can be considered stopped. The iterative process is described in Figure 1.

In the operation of WDS, two or more components come to failure together rarely happens [33]. Therefore, this paper only considers the situation of single-pipe failure. Note that this model can be easily extended to multiple-pipe failure.

4. Hydraulic Simulation

The EPANET [34] simulation engine has been used for the WDS hydraulic simulation. EPANET is a water distribution system modeling software developed by the United States Environmental Protection Agency's (EPA). EPANET is available as an open-source toolkit. Its Programmer's Toolkit is a dynamic link library (DLL) of functions that allow developers to customize EPANET to their own needs.

In actual operation, the network distributes water along the pipelines. The water demands of the customers are time-varying and uncertain. Therefore, the water demand multipliers need to be considered as an uncertain and dynamic variable. To evaluate the system reliability with the time-varying demands, a water demand pattern is established by the extended period hydraulic simulations module of EPANET. The time step is set as one hour. Use MATLAB 2010a to implement modeling of different stages and the EPANET Toolkits' functions.

5. Assumptions and Algorithm Flowchart

Assumptions:

1. The pipes of the WDS have just two states: operation or failure. Demand nodes have three states: operation, failure or intermediate state. The intermediate state means the water demand at the node is partly met. The amount of water flow is available but less than the require demand.

2. A pipe is operational if the water can flow smoothly from its initial node to its terminal node.

3a. A node is operational if its service function is effective, i.e., the pressure head of the node is no less than the minimum head and no higher than the maximum capacity.

3b. When a node is in failure, its downstream connected edge is in failure also.

3c. If the upstream pipes of a node are all in failure, the node is failed as an unintended isolated node due to it has no source of water supply.

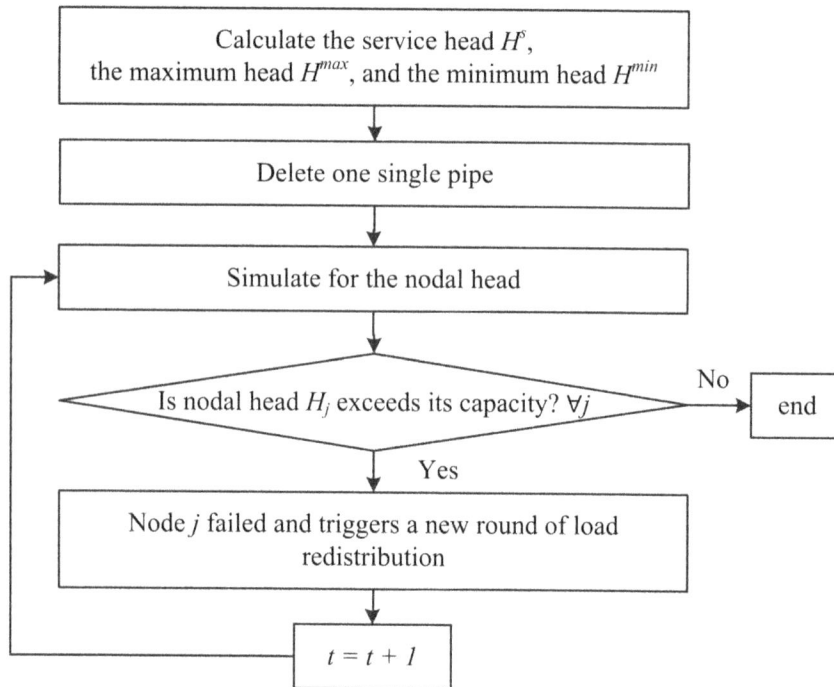

Figure 1. Iterative process of the cascading dynamics of WDS.

For a WDS, three kinds of failure states are existed based on the assumptions. (1) The pressure head at a node exceeds its maximum capacity; (2) The pressure head at a node is lower than its minimum capacity; and (3) A node is unintended isolated as a result of all its upstream pipes are in failure state.

According to the simulation flow of cascading failures, the algorithm flowchart is described as follows (Fig. 2):

1. Input the basic information of a WDS, including diameters, lengths and roughness coefficients of pipes, nodal elevations, base demands, and demand multipliers. Load the network topologic and construct the adjacency matrices A and the incidence matrices N.

2. Calculate the initial load (service pressure head) and its maximum and minimum capacity.

3. Adjust the require water demand of customers according to the demand multipliers. Set the initial time step as $t = 1$. $t = 1$ indicates that each node of a WDS meet its pressure head. $t = 2$ describes the hydraulic state after a certain pipe failed. $t \geq 3$ represents each stage of the cascading failures where new nodes or pipes fail as a result of flow redistribution

4. Assume each of the pipes is in failure successively. Define FailureNodeList and FailureLinkList as the set of failed nodes and failed pipes, respectively. Set FailureNodeList = {} and FailureLinkList = {}. If there is a subsequent failure, the node or pipe index is appended to the FailureNodeList and FailureLinkList, respectively. These two sets are used to update the system topological structure.

5. Run hydraulic analysis in failure conditions. Simulate the running state of the water distribution network by EPANET and obtain the pressure head on each node. A node is recognized as a subsequent failure node if its load exceeds the range of capacity. Update the available flow according to Eq. (4).

6. Update the topological structure under failure conditions. Close the downstream pipes of failure nodes. Judge whether there is unintended isolated node; if so, this node is in failure. Store the calculated failure nodes and pipes in FailureNodeList and FailureLinkList respectively and update the topological structure of the WDS.

7. Calculate the system and nodal reliability. Repeat steps (3)–(6) until the failure of all the pipes have been simulated. Use Eq. (5) and (6) to calculate the reliability of each node and the whole network, respectively.

Results and Discussion

The proposed methodology is applied to a WDS from Islam et al. [35–36]. The network includes two reservoirs, twenty-five water demand nodes and forty pipes. The topological structure, nodal elevation, base demand, pipe diameter and length as well as other basic information of the network has been shown in Figure 3. The total length of all pipes is 19.5 km. The water is supplied by gravity from the elevated reservoirs (reservoir 26 and reservoir 27) with the total heads of 90 m and 85 m, respectively. The pipe lengths range from 100 m to 680 m and the diameters vary from 200 mm to 700 mm. The basic demand is in the range of 33.33 l/s to 133.33 l/s. The demand multipliers (DMs) are considered as ranging from 0.38 at 2am to 1.53 at 7am. The Hazen-Williams formula is used to calculate the head loss.

As the original literature has not given the minimum head, we use the EPANET to solve hydraulic calculation on the network under normal condition (no pipe failure occurred). The pressure head at each node is shown in Table 1. We can find from the results that the pressure heads of node 24 and 25 are lower than 65 m for their long distance away from the reservoirs. Considering the minimum head have a certain tolerance, here assume $H_{min} = 50$ m. The maximum head can be calculated by Eq. (1)

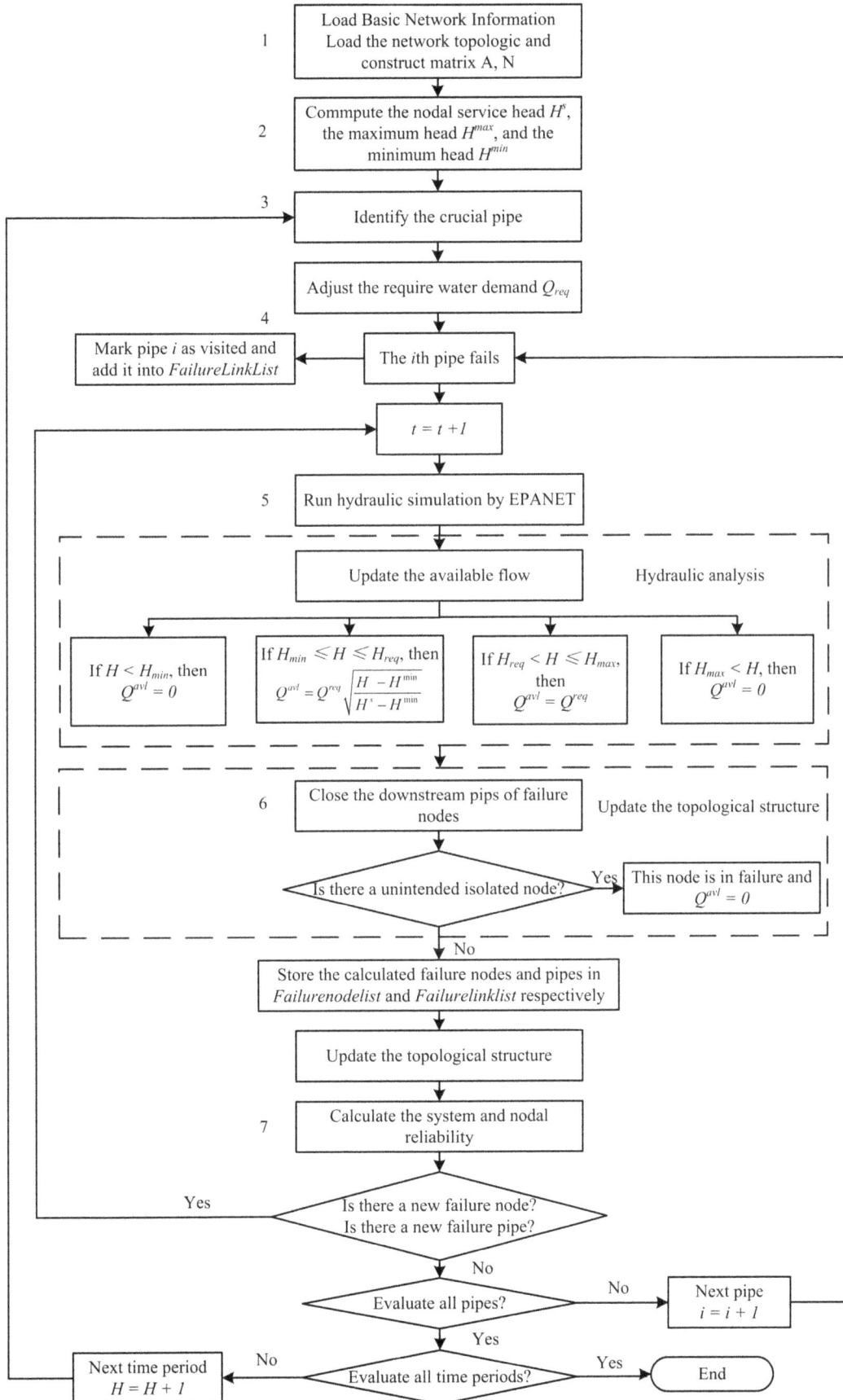

```
┌─────────────────────────────────┐
│  Load Basic Network Information  │
1 │  Load the network topologic and │
│      construct matrix A, N       │
└─────────────────────────────────┘
             ↓
┌─────────────────────────────────┐
│ Commpute the nodal service head Hˢ, │
2 │   the maximum head Hᵐᵃˣ, and the   │
│       minimum head Hᵐⁱⁿ            │
└─────────────────────────────────┘
```

1. Load Basic Network Information Load the network topologic and construct matrix A, N

2. Commpute the nodal service head H^s, the maximum head H^{max}, and the minimum head H^{min}

3. Identify the crucial pipe

Adjust the require water demand Q_{req}

4. The ith pipe fails

Mark pipe i as visited and add it into *FailureLinkList*

$t = t + 1$

5. Run hydraulic simulation by EPANET

Hydraulic analysis

Update the available flow

If $H < H_{min}$, then $Q^{avl} = 0$

If $H_{min} \leqslant H \leqslant H_{req}$, then $Q^{avl} = Q^{req} \sqrt{\dfrac{H - H^{min}}{H^s - H^{min}}}$

If $H_{req} < H \leqslant H_{max}$, then $Q^{avl} = Q^{req}$

If $H_{max} < H$, then $Q^{avl} = 0$

Update the topological structure

6. Close the downstream pips of failure nodes

Is there a unintended isolated node? — Yes → This node is in failure and $Q^{avl} = 0$

No

Store the calculated failure nodes and pipes in *Failurenodelist* and *Failurelinklist* respectively

Update the topological structure

7. Calculate the system and nodal reliability

Is there a new failure node? Is there a new failure pipe? — Yes

No

Evaluate all pipes? — No → Next pipe $i = i + 1$

Yes

Next time period $H = H + 1$ ← No — Evaluate all time periods? — Yes → End

Figure 2. Simulation flowchart of pipe failure under WDS cascading failures.

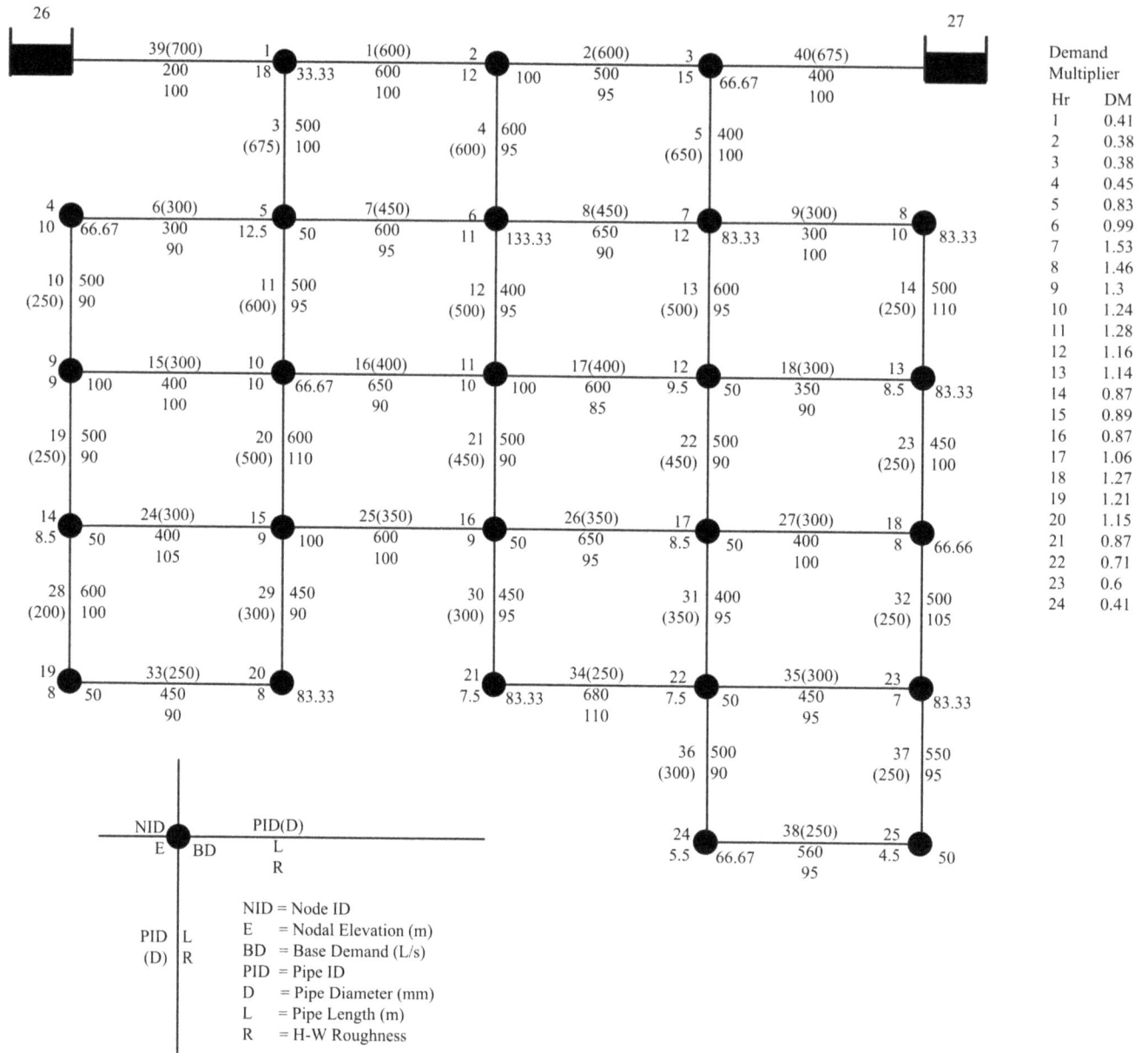

Figure 3. The layout of the example WDS.

Table 1. Pressure heads of each node under the normal condition.

Node ID	Pressure Head (m)	Node ID	Pressure Head (m)	Node ID	Pressure Head (m)
1	86.91	10	79.30	19	69.65
2	84.30	11	78.88	20	70.27
3	84.35	12	77.89	21	69.91
4	78.66	13	74.54	22	68.82
5	82.92	14	74.91	23	66.85
6	82.14	15	76.20	24	64.70
7	82.16	16	75.88	25	64.36
8	77.09	17	74.24		
9	76.14	18	71.53		

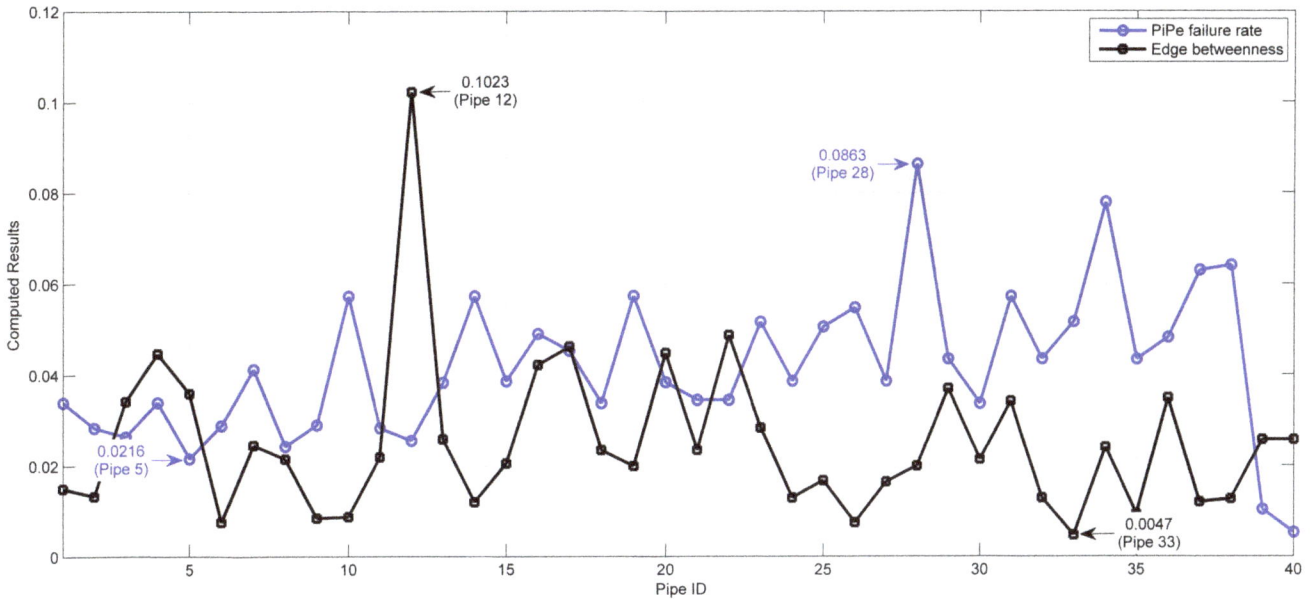

Figure 4. Failure rate and edge betweenness of WDS.

1. Failure rate and edge betweenness

Figure 4 shows the failure rate of each pipe and the edge betweenness of pipe network. The failure rate per year per unit length of pipe is calculated by the function proposed by Su and Mays [37].

Calculate the edge betweenness of the network by Matlab BGL toolbox, and then normalize the results. It is obvious that the pipe 39 and 40 which are directly connected with the reservoirs are the most crucial pipes. The failures of these two pipes directly lead to the failure of the downstream water demand nodes. Therefore, in the study of pipes, it is necessary to research the reliability of pipes except pipe 39 and 40. Figure 4 shows pipe failure rate and the edge betweenness. It is visible that except pipe 39 and 40, the minimum pipe failure rate is 0.0216 (pipe 5) and the maximum one is 0.0863 (pipe 28). So, it is difficult to distinguish the crucial pipe. From the edge betweenness, we can see that the top three maximums are 0.1023 (pipe 12), 0.0486 (pipe 22) and 0.0461 (pipe17), and the minimum is 0.0047 (pipe 33).

2. Peak and Period of Cascading Failures in in WDS

The value of α measures the limit of capacity caused by cost factors in the initial stage of WDS construction, or that caused by ageing and corrosion in the operation stage of WDS. Despite the simple meaning of α, it provides the method to evaluate the overall performance of the system in the cascading failures. Figure 5 shows the reliability of the system when DM value is in the range of 0.3~1.5 and α is ranging from 0 to 0.3. Figure 5 (a) presents the 3D view, (b) is an overhead view. Suppose a certain pipe fails, the simulation is carried out. The system reliability is calculated for the whole system if no subsequent failures are found. Simulate the 38 pipes successively. The system reliability in Figure 5(a) is the average reliability of these 38 pipes. The simulation result shows that, (1) the larger α is, the better the invulnerability of WDS against cascading failures will be. This is because under the given load redistribution strategies, the higher capacity of network design means a stronger potential ability of the system in assimilating and accommodating local failure and a better ability of the network to deal with of cascading failure. (2) as the

computational condition of H_{max}, α enlarges the node pressure from maximum 86.91 m to 112.98 m, the relative amplification is 30%. With the increase of α, the system reliability increases in the whole with a remarkable change from 0 to 1.0 and the relative growth is 100%, 70% higher than the maximum pressure head. In the limited capacity of the node, the small disturbance of WDS has a significant effect on water supply.

Analysis from the perspective of DM shows that DM has a great effect on the propagation of cascading failures in WDS. The simulation results shows, (1) when $DM>1$, system reliability is low after coming to the stable state and the stable value becomes smaller with the increase of DM. For example, when $DM=1.3$, the stable value of system reliability is 0.7179; $DM=1.5$, the stable value of system reliability is 0.5557; when $DM<1$, the stable value of system reliability is much higher. For example, when $DM=0.5$, the stable value of system reliability is 0.9993. (2) when α is given, there exists a threshold DM_c of DM making the system reliability always reach a peak over a period of time. When DM is at its threshold DM_c, the system reliability levels off to DM_c and rises rapidly. After reaching the peak, it decreases rapidly. Moreover, DM_c decreases with the increase of α. For example, when $\alpha=0$ and $DM=1.08$, the peak is $Rsys=0.7161$; when $\alpha=0.3$ and $DM=0.33$, $Rsys=1.0$. (3) When $DM<1$, as DM decreases, the system reliability needs more time to be stable and this time gets longer with the decrease of DM. For example, when $DM=0.7$, the system reliability gets stable at $\alpha=0.18$; when $DM=0.3$, the system reliability has not yet reached a stable value at $\alpha=0.3$. Therefore, in the simulation of the pipe network, the DM should be considered to make the evaluation result more accurate.

3. Identification of Crucial Pipe

The WDS simulation is carried out based on the flowchart in Figure 2. α is set in six types from 0.05 to 0.3. Steps 3–7 in Figure 2 are carried out after setting α and DM to a certain value. The subsequent failures are considered. If no new failures are found, the cascading failure stops and the system remain stable. The system reliability can be obtained when the system come to stable state again. The 38 pipes are simulated successively. The lowest

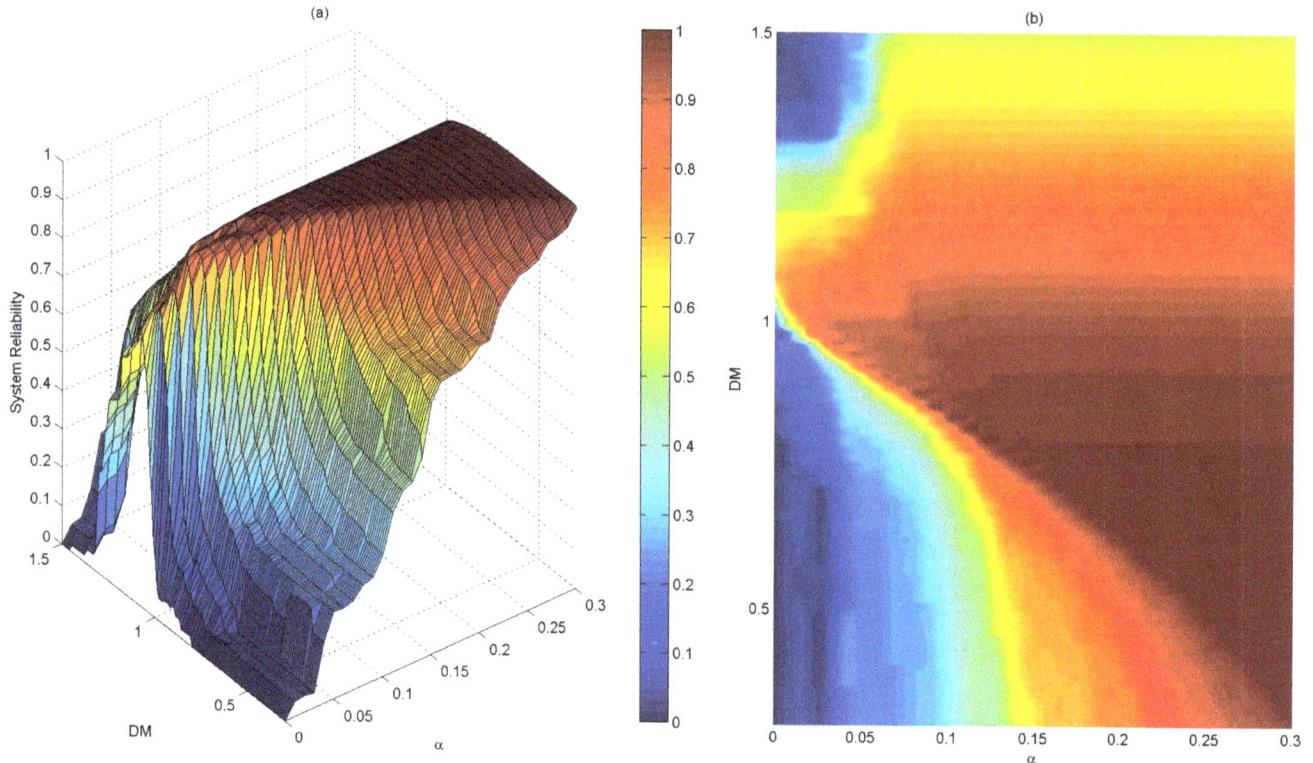

Figure 5. Variation diagram of WDS reliability with *DM* and *α*.

system reliability of this period and its related pipe ID are recorded in Table 2.

The simulation results are shown in Table 2. The first column is the *DM* per hour; the second column is hours; the column 3, 5, 7, 9, 11, and 13 list the ID of crucial pipes under different tolerance parameters *α* and their corresponding system reliability R_{sys}, respectively.

It can be found in Table 2 that the vulnerable pipes ID are 1, 2, 3, 5, 8, 9, 11, 22, 30, 32, and 36. Among them, in order to find which one is the most crucial pipe, we calculate the frequency of each pipe ID which appears in Table 2. It is done by calculating the sum of each pipe ID over sum of all pipes ID. The comparison of the frequency of crucial pipes, the failure rate and the edge betweenness is shown in Figure 6. We can see that pipe 3 is the most crucial pipes because its frequency is significantly higher than that of other pipes. The distribution period of pipe 3 covers 5–21, the corresponding DM≥0.83 and the maximum value reaches 1.53. In addition to pipe 3, the other crucial pipes are pipe 32, 5, 2, 1, (9, 30, 36) (figures in brackets denote the equally important). Besides, from the geographical distribution of pipes, pipe 1, 2, 3, 5 are relatively closer to the reservoirs while pipe 30, 32, 36 are relatively far away from the reservoirs.

Compare the pipe failure rate and edge betweenness shown in Figure 4, we can see that the crucial pipes 3, 32, 5, 2, and 1 got from the simulation of cascading failure are not the maximum values in the failure rate or the edge betweenness. On the contrary, pipe 2, 3, 5 have a relative small failure rate which lead to serious network avalanche. On the other hand, betweenness used as a way to measure network characteristics has been widely seen as the index of the importance. The existing studies on the robustness of the lifeline system have taken the edge or node betweenness as network initial load [12,14,38], based on what the further analysis on network characteristics and emergency strategies have been

proposed. The simulation results show that the failure of pipe 12 with the largest betweenness does not cause a wide range of network avalanche. Using betweenness or degree instead of lifeline network entity flow is a method that analyses different kinds of lifeline networks (WDS, transportation, communication, power grid, etc.) in the same way. It ignores the characteristics of the network flow, service function and constraints. The results cannot be directly applied to adjusting the lifeline network relationship or adjusting the network flow.

4. Discussion of the Most Reliable Time Period

Except the multiple identical minimum values occurred at *α* = 0.05, the minimum value under other conditions are within the time period of *H* = 7 period (R_{sys} = 0.1961). The *DM = 1.53* at 7am is the largest water demand during the day. It can be verified that the peak of water usage is the most vulnerable period of a WDS. System reliability decreases with the increase of *DM*.

Table 3 presents the maximum value (relative maximum) of the minimum values of system reliability in one day. The value is selected by the maximum value of system reliability from each column in Table 2. The time period 2 and 3 has the minimum *DM* (*DM = 0.38*). Except for the situation of *α* = 0.3, the relative maximum values of system reliability do not meet the minimum *DM*, but 0.41, 0.6, 0.71, 0.87, 0.99 and 1.21, respectively.

In order to make further analysis on the most reliable period in a day, Figure 7 censuses the frequency of R_{sys} = 1 in each period. In order to avoid the deviation of results, the frequency of R_{sys} = 1 when *α*>0.3 is calculated and analyzed. The statistical results show that the frequency of R_{sys} = 1 does not change with *α* increased when *α*≥0.3. Hence, in this example, there is a threshold of tolerance parameter $α_c$ = 0.3. When *α*>$α_c$, the period with the highest system reliability meets the time period with

Table 2. Crucial pipes ID (CPID) under six types of α and their corresponding system reliability R_{sys}.

| | | $\alpha = 0.3$ | | $\alpha = 0.25$ | | $\alpha = 0.2$ | | $\alpha = 0.15$ | | $\alpha = 0.1$ | | $\alpha = 0.05$ | |
DM	Hour	CPID	R_{sys}	CPID	R_{sys}	CPID	R_{sys}	CPID	R_{sys}	CPID	R_{sys}	CPID	R_{sys}
0.41	1	32	0.9956	32	0.8861	1	0.7407	2	0.6482	2	0.2593	5	0.1111
0.38	2	32	0.9993	32	0.8885	1	0.7407	5	0.6019	2	0.2593	5	0.1111
0.38	3	32	0.9993	32	0.8885	1	0.7407	5	0.6019	2	0.2593	5	0.1111
0.45	4	32	0.9881	32	0.9233	1	0.7407	5	0.6759	2	0.2593	5	0.1111
0.83	5	3	0.9151	3	0.9151	3	0.9151	3	0.9151	32	0.463	2	0.1574
0.99	6	3	0.7651	3	0.7651	3	0.7651	3	0.7651	30	0.75	30	0.2315
1.53	7	3	0.1961	3	0.1961	3	0.1961	3	0.1961	3	0.1961	3	0.1961
1.46	8	3	0.2285	3	0.2285	3	0.2285	3	0.2285	3	0.2285	3	0.2285
1.3	9	3	0.3181	3	0.3181	3	0.3181	3	0.3181	3	0.3181	5	0.2315
1.24	10	3	0.3793	3	0.3793	3	0.3793	3	0.3793	3	0.3793	9	0.2315
1.28	11	3	0.3336	3	0.3336	3	0.3336	3	0.3336	3	0.3336	5	0.2315
1.16	12	3	0.4921	3	0.4921	3	0.4921	3	0.4921	3	0.4921	9	0.2315
1.14	13	3	0.5214	3	0.5214	3	0.5214	3	0.5214	3	0.5214	11	0.2315
0.87	14	3	0.8833	3	0.8833	3	0.8833	3	0.8833	36	0.463	2	0.1574
0.89	15	3	0.8662	3	0.8662	3	0.8662	3	0.8662	30	0.75	8	0.1574
0.87	16	3	0.8833	3	0.8833	3	0.8833	3	0.8833	36	0.463	2	0.1574
1.06	17	3	0.6649	3	0.6649	3	0.6649	3	0.6649	3	0.6649	22	0.2315
1.27	18	3	0.3465	3	0.3465	3	0.3465	3	0.3465	3	0.3465	5	0.2315
1.21	19	3	0.4299	3	0.4299	3	0.4299	3	0.4299	3	0.4299	5	0.2315
1.15	20	3	0.508	3	0.508	3	0.508	3	0.508	3	0.508	9	0.2315
0.87	21	3	0.8833	3	0.8833	3	0.8833	3	0.8833	36	0.463	2	0.1574
0.71	22	32	0.9259	32	0.9259	32	0.9259	2	0.7037	5	0.3148	5	0.1111
0.6	23	32	0.9305	32	0.9305	32	0.787	1	0.7037	5	0.2593	5	0.1111
0.41	24	32	0.9956	32	0.8861	1	0.7407	2	0.6482	2	0.2593	5	0.1111

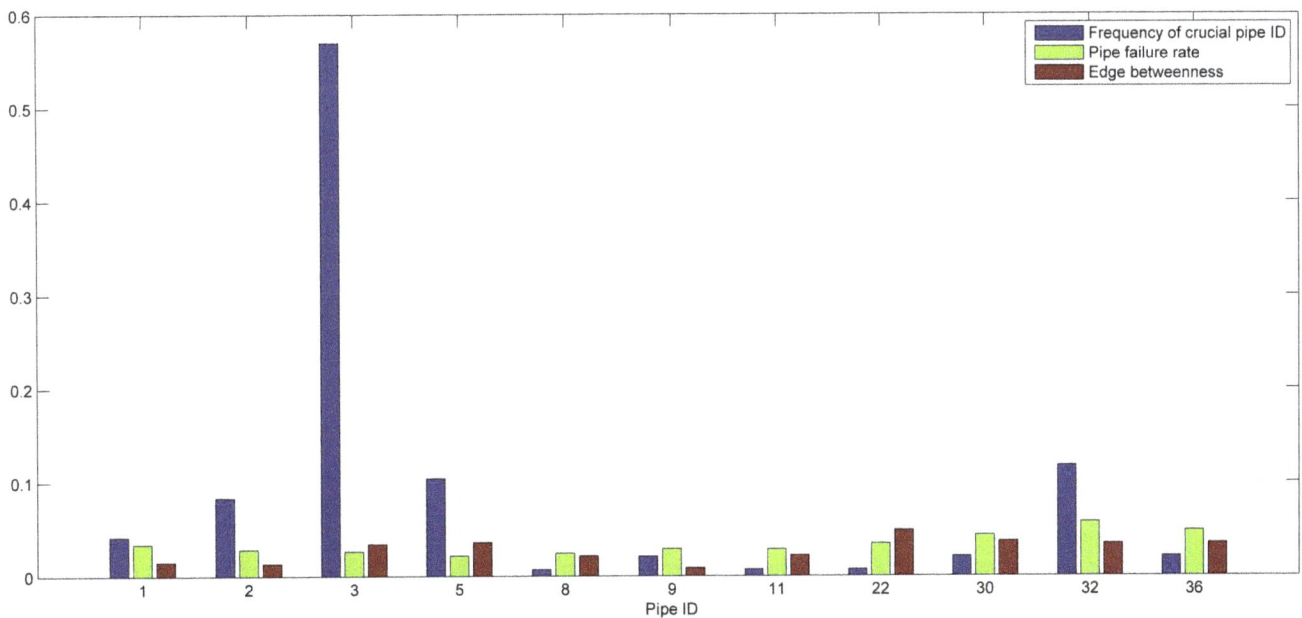

Figure 6. Frequency diagram of crucial pipes.

Table 3. The relative maximum of the minimum values of system reliability under six states of α.

$\alpha=0.3$		$\alpha=0.25$		$\alpha=0.2$	
Hour (DM)	R_{sys}	Hour (DM)	R_{sys}	Hour (DM)	R_{sys}
2 (0.38), 3 (0.38)	0.9993	23 (0.6)	0.9305	22 (0.71)	0.9259
$\alpha=0.15$		$\alpha=0.1$		$\alpha=0.05$	
Hour (DM)	R_{sys}	Hour (DM)	R_{sys}	Hour (DM)	R_{sys}
5 (0.83)	0.9151	6 (0.99)	0.75	6 (0.99)	0.2315

$DM<1$; when $\alpha<\alpha_c$, the period with the highest system reliability does not meet the minimum DM. Therefore, when select the repair period of a WDS, one cannot simply choose the period with the minimum DM, but choose the period with the highest system reliability based on the operation status. What's more, the crucial pipes apt to causing large-scale cascading failure should be avoided also.

Conclusions

The cascading dynamics of WDS in failure condition and the identification of crucial pipes have been discussed in this paper. The propagation of cascading failures in WDS is measured by the damage of certain pipe. The identify factor of crucial pipes is the system reliability after the network restores to stable state. The cascading failure simulation of WDS has taken the nodal pressure head, available water flow, daily demand multipliers and the topological structure in to account. Based on this method, using MATLAB to call EPANET source program is realized. The case

study has demonstrated the applicability of this method. The results verified that this method is suitable for WDS and can effectively identify the crucial pipes.

In the network cascading dynamics modeling, it is generally assumed that once the load of a node or edge in the network exceeds its maximum capacity, the corresponding node or edge is avalanched and out of function, triggering the redistribution of network load and cascading failures. However, in the real-world network, there is always some kind of emergency mechanism and emergency response. When a failure occurs, the external emergency power can be involved in to exert its ability of emergency processing so as to repair the network and ensure its normal operation. Therefore, starting from the protection of critical infrastructure network, further research should focus on how to improve the utilization of limited emergency resources and resist the propagation of cascading failure in the whole network.

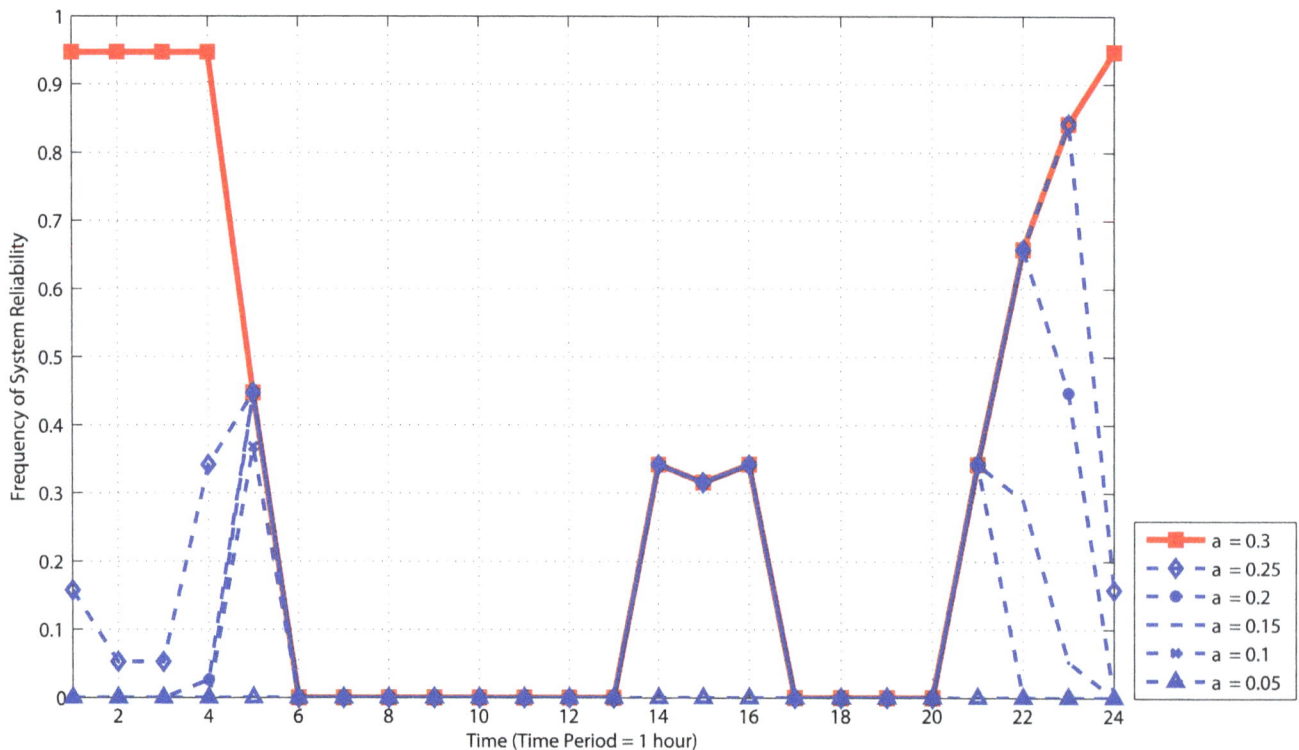

Figure 7. Frequency diagram of the maximum value ($R_{sys}=1$) of system reliability within 24 hours.

Acknowledgments

The authors wish to thank the anonymous reviewer whose constructive comments were very helpful for strengthening the presentation of this paper.

References

1. Yazdani A, Jeffrey P (2011) Complex network analysis of water distribution systems. Chaos: Interdisciplinary J Nonlinear Sci 21: 016111.
2. Pastor-Satorras R, Vazquez A, Vespignani A (2001) Dynamical and correlation properties of the Internet. Phys Rev E 87: 258701.
3. Newman MEJ (2003) The structure and function of complex networks. SIAM Rev 45: 167–256.
4. Albert R, Jeong H, Barabasi AL (2000) Error and attack tolerance of complex networks. Nature 409: 378–382.
5. Motter AE, Lai YC (2002) Cascade-based attacks on complex networks. Phys Rev E 66: 065102.
6. Holme P, Kim BJ, Yoon CN, Han SK (2002) Attack vulnerability of complex networks. Phys Rev E 65: 056109.
7. Buldyrev SV, Shere N, Cwilich GA (2011) Interdependent networks with correlated degrees of mutually dependent nodes. Phys Rev E 83: 016112
8. Buldyrev SV, Parshani R, Paul G, Stanley HE, Havlin S, et al. (2010) Catastrophic cascade of failures in interdependent networks. Nature 464: 1025–1028.
9. Simonsen I, Buzna L, Peters K, Bornholdt S, Helbing D (2008) Transient dynamics increasing network vulnerability to cascading failures. Phys Rev Lett 100: 218701.
10. Wang WX, Chen G (2008) Universal robustness characteristic of weighted networks against cascading failure. Phys Rev E 77: 026101.
11. Mishkovski I, Biey M, Kocarev L (2011) Vulnerability of complex networks. Commun Nonlinear Sci Numer Simul 16: 341–349.
12. Wang JW, Rong LL (2009) Cascade-based attack vulnerability on the US power grid. Saf Sci 47: 1332–1336.
13. Bompard E, Masera M, Napoli R, Xue F (2009) Assessment of structural vulnerability for power grids by network performance based on complex networks. Crit Inform Infrastruc Security 5508: 144–154.
14. Wang JW, Rong LL (2011) Robustness of the western United States power grid under edge attack strategies due to cascading failures. Saf Sci 49: 807–812.
15. Kinney R, Crucitti P, Albert R (2005) Modeling cascading failures in the North American power grid. Eur Phys J B 46: 101–107.
16. Xia YX, Hill DJ (2008) Attack vulnerability of complex communication networks. IEEE Trans Circuits Syst II Express Briefs 55: 65–69.
17. Rosas-Casals M, Valverde S, Sole RV (2007) Topological vulnerability of the European power grid under errors and attacks. Int J Bifurc Chaos 17: 2465–2475.
18. Sole RV, Rosas-Casals M, Corominas-Murtra B, Valverde S (2008) Robustness of the European power grids under intentional attack. Phys Rev E 77: 026102.
19. Cracitti P, Latora V, Marchiori M (2004) A topological analysis of the Italian electric power grid. Phys A 338: 92–97.
20. Ren H, Dobson I (2008) Using Transmission Line Outage Data to Estimate Cascading failure propagation in an electric power system. IEEE T Circuits-II 55: 927–931.
21. Wu JJ, Sun HJ, Gao ZY (2007) Cascading failures on weighted urban traffic equilibrium networks. Phys A 386: 407–413.
22. Freeman LC (1977) A Set of Measures of Centrality Based on Betweenness. Sociometry 40: 35–41.
23. Borgatti SP (2005) Centrality and network flow. Soc Net 27: 55–71.
24. Tolson BA, Maier HC, Simpson AR, Lence BJ (2004) Genetic Algorithms for Reliability-Based Optimization of Water Distribution Systems. J Water Resour Plann Manage 130:63–72.
25. Mays LW (2000) Water distribution systems handbook. New York: McGraw-Hill. 912p.
26. Ostfeld A, Shamir U (1993) Incorporating reliability in optimal design of water distribution networks-review and new concepts. Reliab Eng Syst Safe 42: 5–11.
27. Zhuang B, Lansey K, Kang D (2013) Resilience/availability analysis of municipal water distribution system incorporating adaptive pump operation. J Hydraul Eng 139: 527–537.
28. Bondy JA, Murty USR (1976) Graph theory with applications. New York: American Elsevier Pub. Co. 171p.
29. Su YC, Mays LW, Duan N, Lansey KE (1987) Reliability-based optimization model for water distribution systems. J Hydraul Eng 113:1539–1556.
30. Puust R, Kapelan Z, Savic DA (2000) A review of methods for leakage management in pipe networks. Urban Water J 7: 25–45.
31. Dueñas-Osorio L, Vemuru SM (2009) Cascading failures in complex infrastructure systems. Struct Saf 31: 157–167.
32. Wagner JM, Shamir U, Marks DH (1988) Water distribution reliability - Simulation methods. J Water Resour Plann Manage 114: 276–294.
33. Bao Y, Mays LW (1990) Model for water distribution system reliability. J Hydraul Eng 116: 1119–1137.
34. Rossman LA (2000) EPANET 2 user's manual, National Risk Management Research Laboratory, U.S. Environmental Protection Agency, Cincinnati, OH.
35. Islam MS, Sadiq R, Rodriguez MJ, Francisque A, Najjaran H, et al. (2011) Leakage detection and location in water distribution systems using a fuzzy-based methodology. Urban Water J 8: 351–365.
36. Islam MS, Sadiq R, Rodriguez M, Najjaran H, Hoorfar M (2013). Reliability Assessment for Water Supply Systems under Uncertainties. J Water Resour Plann Manage In press. doi: 10.1061/(ASCE)WR.1943-5452.0000349
37. Su Y, Mays LW, Duan N, Lansey KE (1987) Reliability-based optimization model for water distribution systems. J Hydraul Eng 113: 1539–1556.
38. Chai CL, Liu X, Zhang WJ (2011) Application of social network theory to prioritizing Oil & Gas industries protection in a networked critical infrastructure system. J Loss Prevent Proc 24: 688–694.

Author Contributions

Conceived and designed the experiments: QS MZ YY. Performed the experiments: QS. Analyzed the data: QS MZ YY. Contributed reagents/materials/analysis tools: MZ YY. Wrote the paper: QS.

20

Coliform Bacteria as Indicators of Diarrheal Risk in Household Drinking Water

Joshua S. Gruber*, Ayse Ercumen, John M. Colford, Jr.

Division of Epidemiology, University of California, Berkeley, California, United States of America

Abstract

Background: Current guidelines recommend the use of *Escherichia coli* (EC) or thermotolerant ("fecal") coliforms (FC) as indicators of fecal contamination in drinking water. Despite their broad use as measures of water quality, there remains limited evidence for an association between EC or FC and diarrheal illness: a previous review found no evidence for a link between diarrhea and these indicators in household drinking water.

Objectives: We conducted a systematic review and meta-analysis to update the results of the previous review with newly available evidence, to explore differences between EC and FC indicators, and to assess the quality of available evidence.

Methods: We searched major databases using broad terms for household water quality and diarrhea. We extracted study characteristics and relative risks (RR) from relevant studies. We pooled RRs using random effects models with inverse variance weighting, and used standard methods to evaluate heterogeneity and publication bias.

Results: We identified 20 relevant studies; 14 studies provided extractable results for meta-analysis. When combining all studies, we found no association between EC or FC and diarrhea (RR 1.26 [95% CI: 0.98, 1.63]). When analyzing EC and FC separately, we found evidence for an association between diarrhea and EC (RR: 1.54 [95% CI: 1.37, 1.74]) but not FC (RR: 1.07 [95% CI: 0.79, 1.45]). Across all studies, we identified several elements of study design and reporting (e.g., timing of outcome and exposure measurement, accounting for correlated outcomes) that could be improved upon in future studies that evaluate the association between drinking water contamination and health.

Conclusions: Our findings, based on a review of the published literature, suggest that these two coliform groups have different associations with diarrhea in household drinking water. Our results support the use of EC as a fecal indicator in household drinking water.

Editor: Sudha Chaturvedi, Wadsworth Center, United States of America

Funding: The authors have no funding or support to report.

Competing Interests: The authors have declared that no competing interests exist.

* Email: jsgruber@gmail.com

Introduction

Globally, drinking water has been established as a primary transmission pathway for diarrhea pathogens. [1,2] In industrialized countries, centrally treated drinking water distribution systems have largely eliminated outbreaks of waterborne diseases, such as typhoid fever and cholera. [3] In developing countries, there is a large body of evidence that improving the microbial quality of drinking water by household treatment and safe storage reduces diarrhea. [4–6] Yet, evidence directly linking diarrheal illness to measured fecal contamination in drinking water remains inconclusive. [3,7,8]

In general, it is not feasible to test water for all known waterborne pathogens to assess whether it is safe for drinking. [2,9–11] Instead, since the early 1900s there has been heavy reliance on fecal indicator organisms as measures of drinking water quality. [1,12] Current World Health Organization (WHO) guidelines recommend *Escherichia coli* (EC) and/or thermotoler-

ant ("fecal") coliforms (FC) as indicators of the effectiveness of disinfection processes, and as index organisms for the potential presence of fecal contamination and waterborne pathogens; [2,10,12,13] previous WHO guidelines use categories of EC and FC concentrations to define levels of disease risk from drinking water. [14] While EC are considered the most suitable indicator organism due to their specificity to fecal sources of contamination, FC are also recommended as an acceptable surrogate; this recommendation exists despite the recognition that the FC group includes coliform species of environmental origin, and is therefore not likely specific to fecal contamination. [2,12,13] Despite heavy reliance on EC and FC to assess the microbiological safety of drinking water, it has yet to be shown that either of these specific indicators is associated with waterborne illness. [8]

A previous systematic review and meta-analysis evaluated the evidence for the link between household drinking water quality, measured by fecal indicator organisms, and health; [8] the authors found no evidence of an association between diarrhea and indicators of drinking water contamination (EC, FC and *fecal streptococci*). This review, however, was limited by the availability, size and quality of articles published at the time the search was conducted (2001). [8] Specifically, the previous review included only three relatively small studies that reported using EC, and therefore focused mostly on FC as a measure of water quality. Given the limited number of studies available to the authors, they were also unable to evaluate the performances of EC and FC separately. Since the previous review, several studies have been published using EC as a measure of water quality (n = 6) [15–21], motivating our updated review with this newly available evidence. In addition, this larger body of evidence allows us to evaluate the performance of EC and FC separately; we hypothesized that the two coliform groups would have different associations with diarrhea given their different specificities for fecal contamination. Given the widespread use of EC and FC as measures of water quality by researchers and policy makers, and the limited evidence regarding the association of these proxy measures with actual disease outcomes, we believe that our systematic review and meta-analysis will make a significant contribution to the body of knowledge on the performance of indicator organisms in household drinking water, and have implications for their recommended use for future guidelines and research. In addition, based on the findings of our systematic review, we are able to make several recommendations for the design and reporting of future studies that assess the relationship between drinking water contamination and health.

Methods

Search Strategy

We searched MEDLINE, EMBASE, and Web of Knowledge databases for relevant articles. We used broad search terms for water, water quality and indicator organisms, diarrhea and household (point-of-use) sampling (Table 1). Titles and abstracts from the search were examined, and the full texts of relevant articles were reviewed. The bibliographies of relevant articles identified from full-text reviews were scanned to identify additional relevant studies. Searches were limited to articles published in peer-reviewed journals, in English, Spanish, German or Turkish (languages spoken by the authors). No restrictions were placed on date of publication. The final search was conducted on February 27, 2013. A protocol was not registered for this systematic review.

Inclusion and Exclusion Criteria

We reviewed articles that reported data on the association between exposure to *Escherichia coli* (EC) or thermotolerant ("fecal") coliforms (FC) in household drinking water, and the occurrence of diarrhea under non-outbreak conditions; we excluded other, less commonly used indicator organisms (e.g., total coliforms, *fecal streptococci*). Our inclusion criterion was that the study collected exposure and outcome data at the household level or at the point of use; studies were excluded if they only reported measures of source water quality (e.g., shared wells, distribution system "nodes"), or used ecologic outcome data (e.g., health ministry reports). No further restrictions were placed on: definitions of diarrhea, recall period, study design, age groups, study location (developed/developing country), study setting (urban/rural), or drinking water source (as long as exposure to indicator bacteria was measured at the point of use).

Study Reporting and Data Extraction

For each relevant study two authors (JG and AE) independently extracted basic study characteristics, including year of publication, sample size, age groups, study location, setting and design, indicator organism and enumeration methods, and diarrhea definition and enumeration methods into standardized forms, and resolved any discrepancies. Where possible, a relative risk (RR) such as odds ratio, cumulative incidence ratio, incidence density ratio or prevalence ratio was extracted from relevant studies. When no effect measure was reported by the authors, raw data were extracted to calculate an appropriate RR and confidence interval using standard methods. [22] If authors reported both raw data and unadjusted RRs we confirmed the reported estimates using the raw data; we used the RRs reported by the authors, unless otherwise noted. If authors reported results for both indicators (EC and FC) the results were treated as two separate studies in the meta-analysis. All RRs were extracted such that a value greater than unity indicated increased risk of illness among the group exposed to contaminated water. We also noted whether authors controlled for additional variables (by study design, stratified analysis or regression methods) and whether they accounted for correlation (clustering) of outcomes in their analysis.

Exposure Thresholds

There are multiple ways to define comparison groups in water quality studies. For example, authors often use a threshold to define exposed and non-exposed groups based on indicator organism counts measured in 100 ml samples of water (i.e., exposed: $\geq$ 10 EC/100 ml; non-exposed: < 10 EC/100 ml; with RRs calculated to compare these two groups); however, threshold levels can vary between studies. Current WHO guidelines recommend that drinking water that is safe for human consumption should have no detectable EC or FC in any 100 ml sample (i.e., a 1 EC/FC threshold); however, there is also evidence from the literature that much higher indicator organism concentrations are required to cause disease. For example, a study conducted in the Philippines on source water quality [23] previously reported a "threshold effect", where no association with diarrhea was found at the 1 EC, 10 EC or 100 EC cutoffs, but an association was found at the 1000 EC cutoff; it is plausible that a similar threshold effect could be present for point-of-use drinking water. Given the limited number of studies included in the previously published systematic review and meta-analysis, the authors made no attempt to differentiate indicator organism-diarrhea associations at different threshold levels. [8] In our review, we attempted to extract RRs for different thresholds at the 1 EC or FC, 10 EC or FC, 100 EC or FC and 1000 EC or FC levels, when the authors reported these data.

Categorical Exposures

As an alternative approach to exposure categorization, previous WHO guidelines defined disease risk from drinking water based on categories of indicator organism counts measured in 100 ml samples of water: 0 EC or FC, safe; 1–10 EC or FC, low risk; 11–100 EC or FC, intermediate risk; 101–1000 EC or FC, high risk; > 1000 EC or FC, very high risk. [14] Investigators commonly use these guidelines to define exposure categories in studies of drinking water quality; each category is treated as an independent exposure group, and RRs are calculated for each elevated category relative to the 0 EC or FC exposure group. In our review, we extracted RRs for these categorical exposures to explore the presence of a dose-response (increased disease risk associated with increasing exposure levels) and to evaluate previous WHO guidelines that defined health risk based on these exposure categories. Within

Table 1. Systematic Review Search Terms. [a]

Water Exposure	Household Sampling	Disease Outcome
water quality	household	Diarrhea
water microbiology	point of use	Diarrheoa
water pollution	Pou	gastrointestinal illness
water contamination	point-of-use	gastrointestinal disease
water supply	drinking water	gastrointestinal infection
drinking water	consumption	Dysentery
"indicator bacteria"	tap water	"HCGI"
"indicator organism"	well water	"highly credible gastrointestinal illness"
"thermotolerant coliforms"	domestic	"AGI"
"thermo-tolerant coliforms"		"AGII"
"fecal coliforms"		"acute gastrointestinal illness"
escherichia coli		
"e. coli"		
"fecal bacteria"		
"fecal contamination"		
"microbiological indicators"		

[a]Search terms combined using Boolean logic: within column terms combined using "or" statements; "and" statements were used to combine terms between columns.

studies that evaluated dose-responses, we extracted p-values for tests of trend. In studies that did not report p-values for tests of trend but reported results for different categorical exposure levels, we extracted the appropriate data, where possible, and conducted our own tests of trend using previously published methods that estimate a log-linear dose-response while accounting for correlations between RRs from a single study (*glst* command: STATA 12, STATA Corp, College Station Texas). [24]

Meta-analysis

For consistency with the previously published review, we conducted meta-analyses to calculate a summary measure pooled across all studies which provided an extractable RR, combining all EC and FC studies. [8] The previous review was not explicit about which results were used for meta-analyses if data could be extracted at multiple thresholds. In our review, we estimated a summary measure across all studies using the lowest extractable threshold from each study (e.g., if a study reported raw data that could be combined at the 1 EC, 10 EC and 100 EC thresholds, we included the RR calculated for the 1 EC threshold in the main meta-analysis). In addition, we conducted a sub-group analysis to explore the type of indicator organism used in each study as a source of heterogeneity in the main analysis; for this sub-group analysis we stratified studies by indicator used (EC and FC) and re-pooled summary measures separately for each indicator. In a secondary meta-analysis, we explored evidence for a threshold effect in household drinking water (see "Exposure Thresholds" above). For this analysis, we restricted meta-analyses to studies that reported extractable RRs at the 1 EC, 10 EC and 100 EC thresholds, respectively – this analysis was only feasible for studies that reported extractable data for all three threshold levels.

Summary measures were calculated by meta-analysis using random effects models with inverse variance weighting (STATA 12). [25] Heterogeneity across studies was evaluated using the Mantel-Haentzel χ^2 test; we considered a p-value on the χ^2 statistic <0.2 as evidence of heterogeneity. We used funnel plots to evaluate publication bias; plot asymmetry was interpreted as

evidence of "small study bias", for which publication bias is a likely contributor. [26,27]

Results

Systematic Review

Database searches were completed on February 27, 2013, from which we identified 5,801 titles and abstracts; 20 relevant articles were identified for review (Figure 1). Of the 20 relevant articles, 14 presented data in an extractable format (see Methods) and were included in meta-analyses. One article [28] presented results for both EC and FC; these results were treated as separate studies in the meta-analyses (resulting in 15 total studies).

Study Characteristics: All Studies

The 20 relevant studies were published between 1977 and 2012 (Table 2). Two studies were conducted in a developed country (Canada), the rest in low-resource settings; no studies were conducted among populations with access to centralized water-treatment and distribution systems. The study settings included a mix of urban/peri-urban (n = 7) and rural (n = 13), and the study populations included a range of age groups (Table 2). The included articles used cohort (n = 14) and case-control (n = 6) study designs (Tables 3 and 4). We identified one randomized control trial [16] that reported data on the direct association between water quality and diarrhea; we considered this study a cohort design for the purposes of the review since water quality was not randomized. We note that, while cohort designs traditionally measure an exposure and then follow a population over time to measure outcomes, the cohort designs we identified in this review measured exposure (EC or FC): (i) simultaneously with outcomes through cross-sectional sampling; (ii) at intervals during ongoing diarrhea surveillance, or; (iii) the timing and analysis of water samples relative to diarrhea measurement was mixed or not clearly reported (Tables 3 and 4). The included case-control studies measured exposure after disease occurrence by design; however one study [19] utilized a nested case control design with

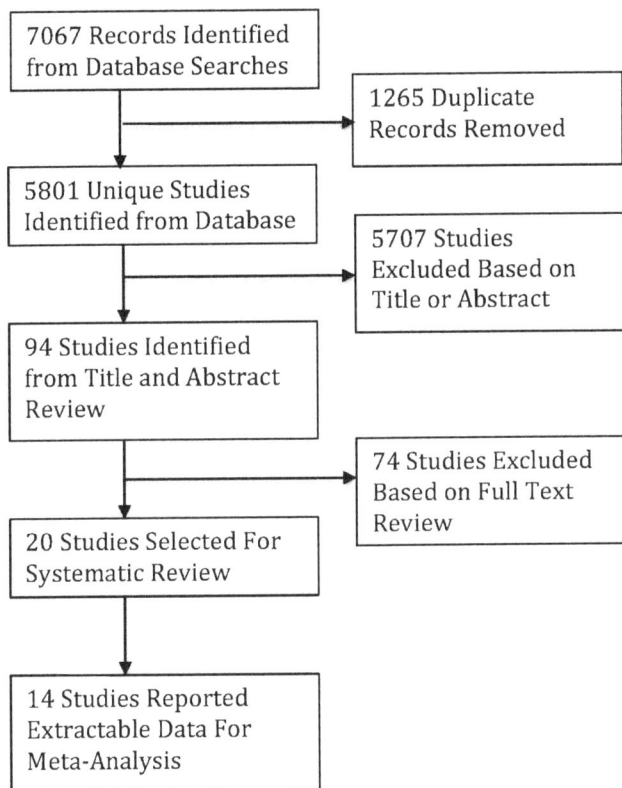

Figure 1. Systematic review flow chart.

risk set sampling in which disease and EC exposure were measured on the same day.

Included studies measured EC (n = 9), FC (n = 10), or both (n = 1) in drinking water samples collected at the point of use. Overall, studies varied in their reporting of enumeration and incubation methods, and we could not always confirm the indicator organism enumerated based on the reported information (Tables 3 and 4). In at least one study, authors reported using EC, but a description of their methods suggested they measured FC. [29] With this exception, we relied on the authors' reporting or deferred to the categorization used by the previous review. [8]

Most studies used a definition of diarrhea that is consistent with current recommendations: three or more loose or watery stools in a 24-hour period (Tables 3 and 4). [30,31] Four studies included the presence of blood or mucus in stool as part of their definition of diarrhea, [15,17,18,32] and two studies [15,17] reported dysentery as an outcome separate from "general" diarrhea; we did not include dysentery as a separate outcome in our review. Cohort studies in low-resource settings relied primarily on household surveillance and self-reported symptoms (e.g., maternal recall) for outcome assessment, using a range of recall periods (daily, up to one month; Tables 3 and 4); the two Canadian studies relied on monthly calendars maintained in the household and reported by telephone. Case-control studies mostly identified cases through hospitals and clinics, and selected controls from community samples; however, two studies identified cases and controls based on concurrent [19] or prior [33] household diarrhea surveillance conducted by the investigators.

Among the studies identified in our review, eight studies reported results that controlled for potential confounders; the other 12 studies found no evidence for confounding, or did not

report attempting to control for confounding (Tables 5 and 6) – we did not differentiate between these studies in meta-analyses. In general, reporting on the independence of observations or methods to account for clustered outcomes was poor (Tables 5 and 6). Only five of the 20 studies reported using methods to account for correlated outcomes (e.g., generalized estimating equations; GEE). In general, the RRs and confidence intervals we calculated from available raw data were consistent with those reported by the authors, whether or not they reported accounting for correlated outcomes. However, in one of these studies, that reported using GEE, we calculated confidence intervals from extracted raw data that were substantially more conservative than those reported by the authors; [15] we included the more conservative confidence intervals for this study in our meta-analyses (Table 5).

Disease-Indicator Associations

Tables 5 and 6 summarize results from EC and FC studies, respectively. Among the ten EC studies, eight reported data that quantified the relationship between indicator exposure and diarrhea (Table 5). Of the eight studies with quantitative results, six reported threshold results at the 1 EC/100 ml cutoff [15,18,20,21,28,33] and all six reported RRs greater than unity; however, only one study was able to rule random error out as an explanation for their results (Table 5). [15] The remaining two studies, among the eight that quantified a relationship between EC and diarrhea, reported significant relationships between diarrhea and linear increases in exposure to mean log10 EC concentrations; the results were very consistent across these two studies (Table 5). [16,19] The two studies that did not have extractable, quantifiable results reported that no "association" or "correlation" was found between EC and diarrhea. [17,34] However, we could not confirm whether the authors' conclusions were based on the size of the effect estimate (i.e., similar to unity), precision (i.e., "non-significant" confidence intervals or p-values), or both.

Among the 11 FC studies (including FC results from Genthe et al. 1997 as separate from their EC results), nine studies reported data that quantified the relationship between indicator exposure and diarrhea (Table 6). Results from these nine FC studies were less consistent than results from EC studies. Using the lowest extractable threshold, five studies reported estimates that suggest increases in the risk of diarrhea with exposure to FC [32,35–38], and four reported effects equivalent to unity or reduced diarrhea risk with exposure to FC. [28,39–41] Henry and Rahim (1990) reported FC effects for two distinct study areas differentiated by "adequate" versus "poor" sanitation infrastructure; while a single summary measure for this study suggested an increased risk of diarrhea with exposure to FC, the results from the two separate study populations were very different (Table 6; FC counts were associated with an increase in illness in the area with improved sanitation and a reduction in illness in the area with unimproved sanitation). Of the two FC studies that did not report extractable data, results were also mixed: one reported no difference in FC isolation from water in case and control households, [29] the other reported that households with diarrhea had higher overall geometric mean FC levels compared to control households. [42]

Summary Measures

For consistency with the previous review, we conducted meta-analyses across all studies (combining both EC and FC studies), using the lowest extractable threshold for each study (Tables 5 and 6). Across all studies, our summary estimate suggests a non-significant increase in diarrhea risk from exposure to drinking water contaminated with EC or FC (RR 1.26 [95% CI: 0.98,

Table 2. Selected Characteristics of Included Studies.

First Author	Year	Location	Setting	Study Size[a]	Ages	Indicator
deAceituno	2012	Honduras	Rural	1020	All	EC
Levy	2012	Ecuador	Rural	115	NR	EC
Gundry	2009	S. Africa/Zimbabwe	Rural	254	1–2 y	EC
Brown	2008	Cambodia	Rural	1196	All	EC
Jensen	2004	Pakistan	Rural	209	<5 y	EC
Bhargava	2003	Bangladesh	Rural	99	1–10 y	FC
Strauss	2001	Canada	Rural	647	All	EC
Raina	1999	Canada	Rural	531	All	EC
Genthe	1997	S. Africa	Urban	316	"Pre-school"	Both
Jagals	1997	S. Africa	Urban	100	NR	FC
Vanderslice	1993	Philippines	Urban	254	<2 y	FC
Knight	1992	Malaysia	Rural	196	4–59 m	FC
Han	1991	Myanmar	Urban	208	6–29 m	FC
Henry	1990	Bangladesh	Rural	92	6–18 m	FC
Henry and Rahim	1990	Bangladesh	Urban	137	1–6 y	FC
Echeverria	1987	Thailand	Urban	NR	<5 y	FC[b]
Esrey	1986	Lesotho	Rural	545	1–60 m	FC
Lloyd-Evans	1984	Gambia	Urban	20	6–36 m	EC
Black	1982	Bangladesh	Rural	40	5–18 m	EC
Rajasekaran	1977	India	Rural	1091	<5 y	FC

EC: Escherichia Coli; FC: Fecal Coliform.
[a]May not reflect the analytic sample used by the authors;
[b]Authors report EC but methods suggest FC.

1.63]; Figure 2). However, we also found evidence of significant heterogeneity ($X^2 = 36.13$, degrees of freedom (d.f.) = 12, p< 0.001), and evidence of publication bias (Figure 3).

In order to explore a potential source of heterogeneity across all studies, we stratified results by indicator organism (EC vs. FC) and re-estimated summary measures in these sub-groups. Across all EC studies, the summary measure suggests a significant association between exposure to EC in household drinking water and diarrhea (RR: 1.54 [95% CI: 1.37, 1.74]); Figure 4). While we found no evidence of heterogeneity across EC studies ($X^2 = 1.65$, d.f. = 5, p = 0.90), and minimal evidence of publication bias (Figure 5), we note that the EC summary measure was disproportionately influenced by one study [15] ("% weight", Figure 4). We therefore conducted a sensitivity analysis by repeating the EC analysis without the study by Brown et al. Removing this study did not impact the magnitude of the summary measure, but did reduce the precision of the estimate (RR 1.48 [95% CI: 1.02, 2.15]); Figure S1); there was no evidence of heterogeneity after removing Brown et al. from the analysis ($X^2 = 1.59$, d.f. = 4, p = 0.81). Across all FC studies, our summary effect measure suggests no association between FC exposure in drinking water and diarrhea (RR: 1.07 [95% CI: 0.79, 1.45]); Figure 4); we also found evidence of significant heterogeneity ($X^2 = 2.30$, d.f. = 6, p = 0.06), and publication bias (Figure 6).

Threshold Effects

Two EC studies provided extractable data that allowed us to explore threshold effects at three different levels (1 EC, 10 EC, and 100 EC per 100 ml; Table 5) [15,18]. RRs were calculated by comparing outcomes in the exposed group ($\geq$EC threshold) to

outcomes in the control group (<EC threshold). In one study, [15] increasing the threshold from 1EC to 10EC suggested a slight increase in diarrhea risk; there was no further increase in risk when the threshold was increased to 100EC (Table 5). In the other study, increasing the threshold beyond 1EC did not increase the magnitude of the effect estimates (Table 5). [18] We pooled estimates across these two studies, and found no evidence that increasing thresholds beyond WHO drinking water guidelines (1 EC/100 mL) was associated with an increased risk of diarrhea (Figure 7). [2] A third EC study reported finding no threshold effect at 1000 EC/100 ml, but did not report a RR. [19]

Categorical Exposure and Dose-Response

Four EC studies reported investigating a dose-response relationship. [15,18,19,21] Only three of these studies evaluated the presence of a dose-response using similar exposure categories (<1 EC (referent), 1–10 EC, 11–100 EC, 100+, or 101–1000 EC and 1000+; Table 5). Of these three studies, one did not report extractable data, but reported finding no evidence of a dose-response effect. [19] Results from the other two studies [15,18] suggest a qualitative dose response (increased risk of diarrhea with increasing exposure categories; Table 5); formal tests of trend, however, had mixed results (Brown et al: p<0.001 provides evidence of a dose response; Jensen et al: p = 0.26 does not provide evidence of a dose response). A fourth study [21] used different EC exposure categorizations. The results from this study qualitatively suggested increasing risk across 0.1–1.5 EC and 1.6–700 EC exposure categories; however, the authors found no evidence of a linear trend (p = 0.45; Table 5).

Table 3. Summary of EC Study Exposure and Outcome Measurements.

First Author	Year	Study Design	Outcome [a]	Recall	Measured EC Relative to Outcome	Enumeration Method	Medium	Temp (°C) [f]
deAceituno	2012	RCT	Diarrhea	7-day	After (<1 week)	MPN	IDEXX Colilert	NR
Levy	2012	CC	Diarrhea	Daily	Same Day	MF	BD Difco ml agar [e] (EPA 1604)	30
Gundry	2009	Cohort	Diarrhea	Daily	Before (<2 weeks)	MPN	IDEXX Colilert	NR
Brown	2008	Cohort	Diarrhea	7-day	After (<1 week)	MF	"selective medium" [d]	NR
Jensen	2004	Cohort	Diarrhea	7-day	Before/After (<3 days)	MF	Hach m-ColiBlue 24 Broth PourRite	35
Strauss	2001	Cohort	AGI [b]	Monthly Diary	Twice over 28-d surveillance	MF	m-FC-BCIG agar	44.5
Raina	1999	Cohort	GI Illness [c]	Monthly Diary	5 times over 1-y surveillance	NR	"standard methods" (MOE 1992)	NR
Genthe	1997	CC	Diarrhea	Hospital	After	NR	"standard methods" (APHA 1989)	NR
Lloyd-Evans	1984	CC	Diarrhea	7-day	After (1–2 months)	MF	Millipore (DHSS 1969)	NR
Black	1982	Cohort	Diarrhea	NR	After (<1 month)	Plating	MacConkey's Agar	37

EC: *Escherichia Coli*; CC: Case Control; MF: Membrane Filtration; MPN: Most Probable Number; NR: Not Reported; C: Celsius; DHSS: Department of Health & Social Security; APHA: American Public Health Association: *Standard Methods for the Examination of Water and Wastewater*; MOE: *Ministry of Environment: Ontario's Drinking Water Objectives*.

[a] Unless noted, used a definition consistent with World Health Organization recommendation of 3 or more loose/watery stools in 24 hour period;

[b] Acute Gastrointestinal Illness: 1) vomiting or liquid diarrhea or 2) nausea or soft, loose diarrhea combined with abdominal cramps;

[c] Diarrhea with or with out vomiting;

[d] selective medium: "containing chromogenic and fluorogenic b-glucuronide and b-galactoside substrates";

[e] Multiple EC enumeration methods used – we report the method with the largest sample size and for which the authors report an extractable RR;

[f] Incubation temperature.

Table 4. Summary of FC Study Exposure and Outcome Measurements.

First Author	Year	Study Design	Outcome [a]	Recall	Measured FC Relative to Outcome	Enumeration Method	Medium	Temp (°C) [e]
Bhargava	2003	Cohort	GI Morbidity [b]	1-month	After (<1 month)	MF	DIFCO m-FC media	45
Genthe	1997	CC	Diarrhea	Hospital	After	NR	"standard methods" (APHA 1989)	NR
Jagals	1997	CC	Diarrhea [c]	Clinic	After	MF	"standard methods" (APHA 1992)	44.5
Vanderslice	1993	Cohort	Diarrhea [c]	7-day	2–5 times over 1 year	MF	m-FC agar (APHA 1985)	44.5
Knight	1992	CC	Diarrhea	Clinic	After (4–14 days)	MF	Paqualab (ELE International 1986)	NR
Han	1991	Cohort	Diarrhea	7-day	Averaged over surveillance	MF	m-FC agar (WHO 1983)	NR
Henry	1990	Cohort	Diarrhea	7-day	Averaged over surveillance	NR	"standard methods" (APHA 1981)	NR
Henry and Rahim	1990	Cohort	Diarrhea	14-day	Unclear (<6 months)	dip-slides	ORION Hygicult Agar Slides [d]	37
Echeverria	1987	CC	Diarrhea	Hospital	After	MF	m-FC agar (APHA 1975)	44
Esrey	1986	Cohort	Diarrhea [c]	24-hour	Unclear	MF	Faecal Coliform Broth (AJPH 1975)	NR
Rajasekaran	1977	Cohort	Diarrhea	NR	Unclear (<15 days)	MPN	IMViC (APHA 1965)	44

FC: Fecal Coliform; CC: Case Control; GI: Gastrointestinal; MF: Membrane Filtration; MPN: Most Probable Number; NR: Not Reported; C: Celsius; APHA: American Public Health Association: *Standard Methods for the Examination of Water and Wastewater*; WHO: World Health Organization: *Guidelines for Drinking Water Quality*; IMViC: Indole, methyl red, Voges–Proskauer and citrate [a] [a] Unless otherwise noted, study used a definition consistent with current World Health Organization recommendation of 3 or more loose/watery stools in a 24 hour period; [b] Morbidity included: diarrhea, dysentery, vomiting, stomachache, acidity, typhoid, cholera; [c] Did not report a definition; [d] Report enumerating *Enterobacteriaceae* – included as FC for consistency with previous review. [e] Incubation temperature

Table 5. EC Study Results.

First Author	Year	Adjusted Analysis	RR	Threshold Results [a] RR (95% CI)	Categorical/Dose Response: [b]RR (95% CI)	Alternate Results: RR (95% CI)
deAceituno	2012	Yes	OR [c]	-	-	Log10 increase: 1.26 (1.08,1.46)
Levy	2012	NR	OR [c]	1000 EC: "no effect"	no "dose-response" (1–10 EC, 11–100 EC, 101–1,000 EC, 1,000+ EC)	Log10 increase: 1.29 (1.02,1.65)
Gundry	2009	Yes	NR	-	-	"no association"
Brown [d]	2008	No	PR [c]	1 EC: 1.55 (1.36,1.76) 10 EC: 1.70 (1.52,1.90) 100 EC: 1.67 (1.51,1.84) 1000 EC: 1.37 (1.21,1.54)	1–10 EC: 0.98 (0.81–1.20) 11–100 EC: 1.35 (1.13–1.61) 101–1000 EC: 1.83 (1.58–2.11) 1001+EC: 1.81 (1.54–2.11) [e]	-
Jensen	2004	No	IDR	1 EC: 1.32 (0.77,2.27) 10 EC: 1.25 (0.83,1.88) 100 EC: 1.23 (0.85,1.80)	1–10 EC: 1.17 (0.55–2.47) 11–100 EC: 1.26 (0.70–2.28) 100+EC: 1.45 (0.81–2.59) [f]	-
Strauss	2001	Yes	OR [c]	1 EC: 1.52 (0.33,6.92)	0.1–1.5 EC: 0.85 (0.10,7.19) 1.6–700 EC: 2.69 (0.34,21.56) [g]	-
Raina	1999	Yes	OR [c]	1 EC: 2.11 (0.90,4.94)	-	-
Genthe	1997	Yes	OR	1 EC: 1.26 (0.59,2.72)	-	-
Lloyd-Evans	1984	No	OR	1 EC: 3.86 0.33,45.57)	-	-
Black	1982	NR	NR	-	-	"no correlation found"

EC: *Escherichia Coli*; RR: Relative Risk; OR: Odds Ratio; PR: Prevalence Ratio; IDR: Incidence Density Ratio; NR: Not Reported; CI: Confidence Interval;
[a]Compares risk in groups exposed to ≥EC value, to <EC value;
[b]Reference group is <1 EC for all categories;
[c]Authors report using methods to account for correlated (clustered) outcomes;
[d]RRs and 95% CIs calculated from raw data;
[e]p-value for linear trend <0.01;
[f]p-value for linear trend = 0.26;
[g]p-value for linear trend = 0.45.

Table 6. FC Study Results.

First Author	Year	Adjusted Analysis	RR	Threshold Results [a] RR (95% CI)	Alternate Results RR (95% CI)
Bhargava	2003	Yes	NA	-	ML Coefficient: 0.204 (SE: 0.08) *p-value* <0.05
Genthe	1997	Yes	OR	1 FC: 0.84 (0.46,1.52)	
Jagals	1997	NR	NA	-	Geometric mean FC higher in homes with diarrhea (compared to control homes)
Vanderslice	1993	Yes	NA	-	Probit coefficient: −0.002 *p-value* >0.10
Knight	1992	Yes	OR	1 FC: 1.45 (0.57,4.76)	-
Han	1991	No	PR	Medium: 0.72 (0.56,0.94) [b]	Medium-High: 0.73 (0.52,1.02) [c] High: 0.72 (0.52,1.00) [c]
Henry	1990	No	IDR	Low: 1.03 (0.75,1.42) [b]	-
Henry and Rahim	1990	No	CIR	10⁴ FC: 1.26 (0.59,2.72) [d]	Improved Area: 2.58 (0.70,9.54) [d] Unimproved Area: 0.86 (0.34–2.18) [d]
Echeverria	1987	NR	NR	-	FC isolated from water as often in homes with diarrhea as in control homes
Esrey	1986	No	PR	10 FC: 1.53 (0.69,3.40) 100 FC: 1.46 (0.62,3.45)	10–100 FC: 1.41 (0.53,3.76) [e] >100 FC: 1.64 (0.64,4.18) [e]
Rajasekaran	1977	No	PR	10 FC: 2.93 (1.10,7.81)	-

FC: Fecal Coliform; RR: Relative Risk; OR: Odds Ratio; PR: Prevalence Ratio; IDR: Incidence Density Ratio; CI: Confidence Interval; ML: maximum likelihood
[a]Compares risk in groups exposed to ≥FC value, to <FC value;
[b]Based on median of samples or arbitrary cutoffs;
[c]Compared to "Low" group – Authors do not report specific FC concentrations used to determined categories;
[d]Reference group is <1000 colony forming units FC per gram;
[e]Reference group <10 FC.

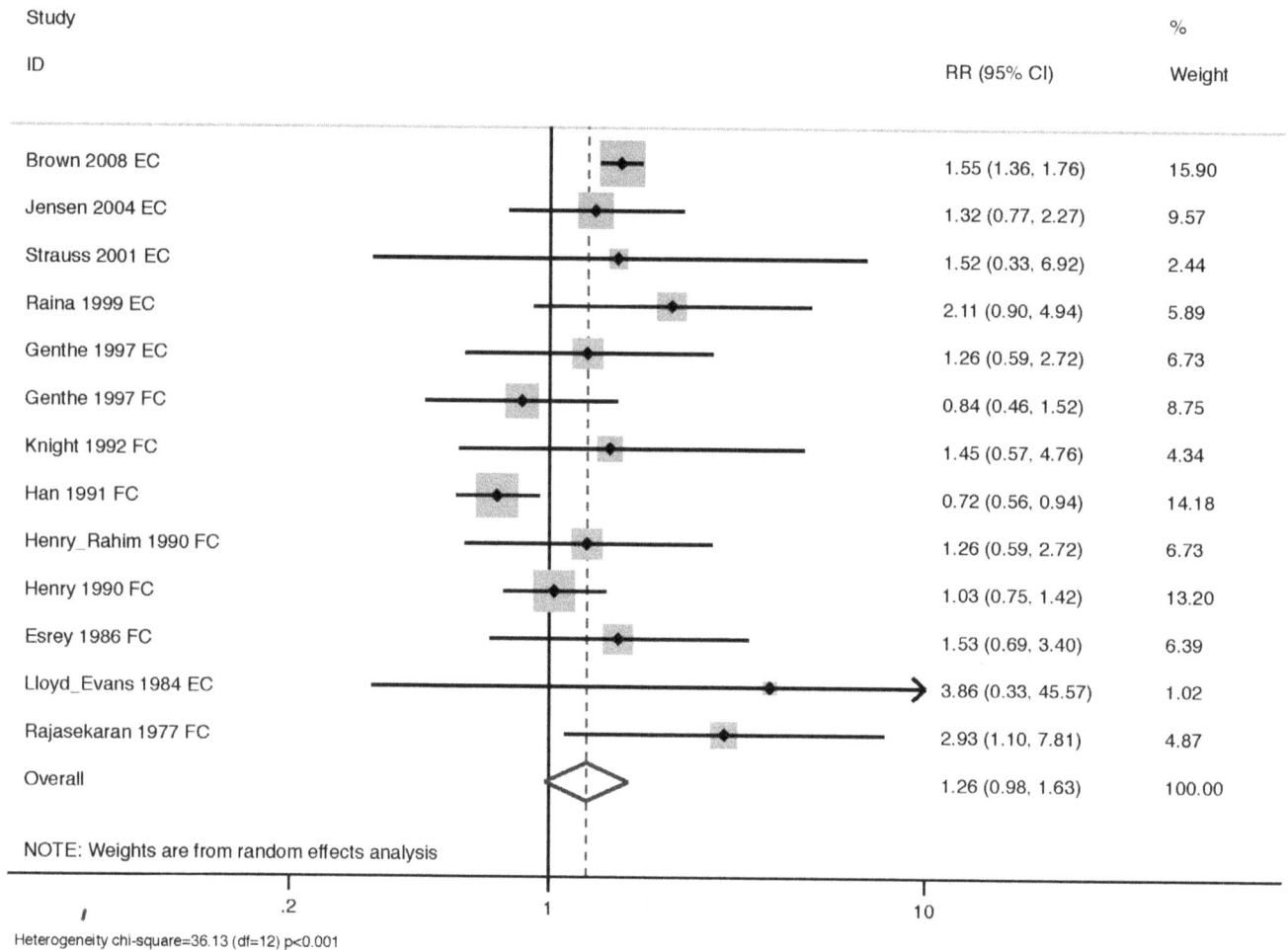

Figure 2. Forest plot for all included studies.

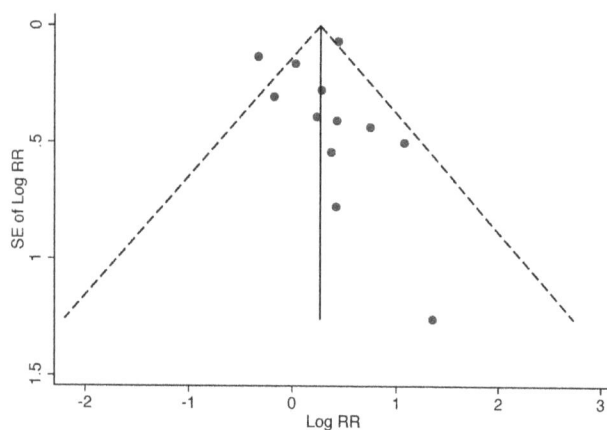

Figure 3. Funnel plot for all included studies (EC and FC studies). Studies are plotted relative to size (SE of Log RR), and reported log relative risk (Log RR); studies higher on the vertical axis are larger, and studies further to the right have larger RRs. The solid vertical line is the log of the summary RR for all studies, and the dotted lines are "pseudo 95% confidence intervals" for the summary measure. The absence of smaller studies with small or null RRs (lower left region of the plot) creates plot asymmetry, and provides evidence of possible publication bias.

Discussion

Summary of Findings

We conducted a systematic review and meta-analysis to evaluate the relationship between diarrheal illness and the presence of EC or FC indicators in household drinking water. Consistent with a prior review, [8] we did not find conclusive evidence for an association between contaminated drinking water and diarrhea when EC and FC results were combined (previous review: RR 1.12 [95% CI: 0.85, 1.48]; our review: RR 1.26, [95% CI: 0.98, 1.63]; Figure 2). We also found evidence of significant heterogeneity and publication bias across all studies (Figure 2, Figure 3). After stratifying studies by EC and FC to explore the choice of indicator organism as a source of heterogeneity, we found that EC studies reported consistent effect estimates that suggest an increased risk of diarrhea with exposure to contaminated drinking water with no evidence of heterogeneity (pooled RR 1.54 [95% CI: 1.37, 1.74]; Table 5, Figure 4). In contrast, FC studies reported mixed results with significant heterogeneity and provided minimal evidence of an association with diarrhea (pooled RR 1.07 [95% CI: 0.79, 1.45]; Table 6, Figure 4). We found no evidence that exposure thresholds greater than 1 EC/100 mL (e.g., 10 EC/100 mL, 100 EC/100 mL) were associated with increased diarrheal disease, but this evidence is based on a limited number of studies. Results regarding a dose-response relationship between EC in household drinking water and diarrhea were limited and

Study ID	RR (95% CI)	% Weight
EC		
Brown 2008	1.55 (1.36, 1.76)	89.43
Jensen 2004	1.32 (0.77, 2.27)	5.09
Strauss 2001	1.52 (0.33, 6.92)	0.64
Raina 1999	2.11 (0.90, 4.94)	2.05
Genthe 1997	1.26 (0.59, 2.72)	2.55
Lloyd_Evans 1984	3.86 (0.33, 45.57)	0.24
Subtotal	1.54 (1.37, 1.74)	100.00
FC		
Genthe 1997	0.84 (0.46, 1.52)	14.48
Knight 1992	1.45 (0.57, 4.76)	6.56
Han 1991	0.72 (0.56, 0.94)	26.61
Henry_Rahim 1990	1.26 (0.59, 2.72)	10.68
Henry 1990	1.03 (0.75, 1.42)	24.18
Esrey 1986	1.53 (0.69, 3.40)	10.07
Rajasekaran 1977	2.93 (1.10, 7.81)	7.43
Subtotal	1.07 (0.79, 1.45)	100.00

NOTE: Weights are from random effects analysis

.2   1   10

heterogeneity results: see text

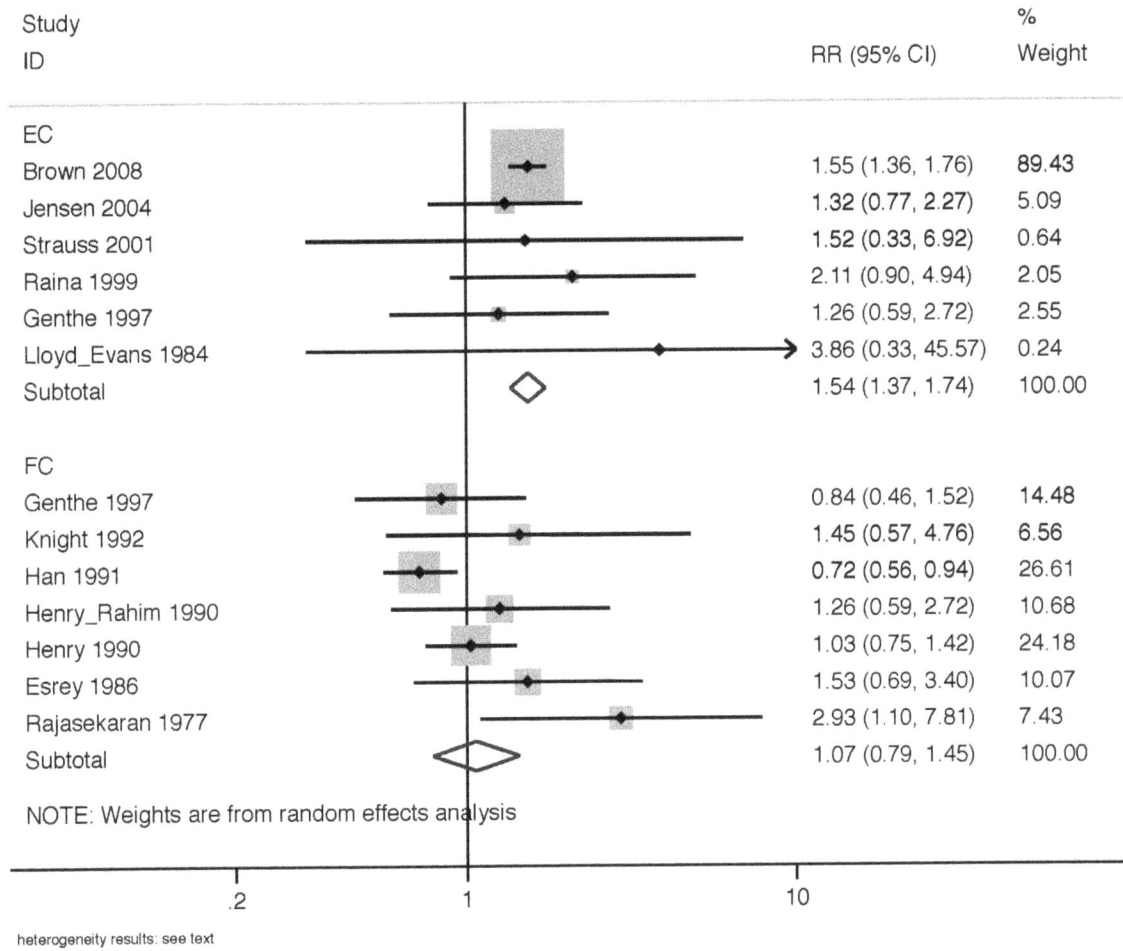

Figure 4. Forest plot for all studies stratified by indicator organism. Top panel, EC: *Escherichia coli*; Bottom panel, FC: thermotolerant ("fecal") coliforms.

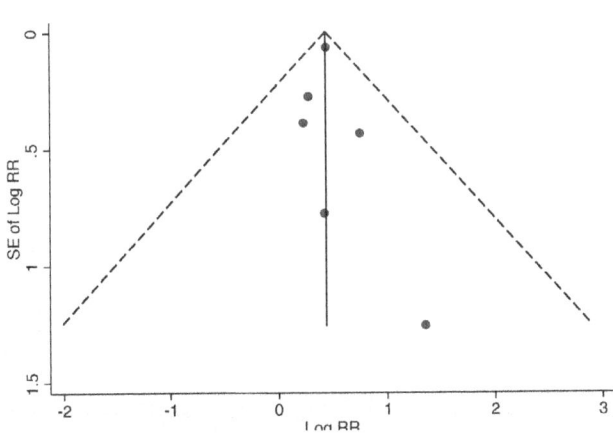

Figure 5. Funnel plot for EC studies. Studies are plotted relative to size (SE of Log RR), and reported log relative risk (Log RR); studies higher on the vertical axis are larger, and studies further to the right have larger RRs. The solid vertical line is the log of the summary RR for all studies, and the dotted lines are "pseudo 95% confidence intervals" for the summary measure. Studies on this plot are generally symmetrical, with the exception of one small study with a large RR, providing minimal evidence of publication bias.

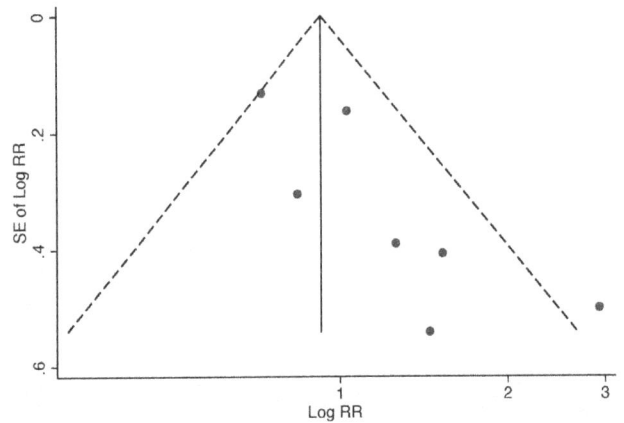

Figure 6. Funnel plot for all FC studies. Studies are plotted relative to size (SE of Log RR), and reported log relative risk (Log RR); studies higher on the vertical axis are larger, and studies further to the right have larger RRs. The solid vertical line is the log of the summary RR for all studies, and the dotted lines are "pseudo 95% confidence intervals" for the summary measure. The absence of smaller studies with small or null RRs (lower left region of the plot) creates plot asymmetry, and provides evidence of possible publication bias.

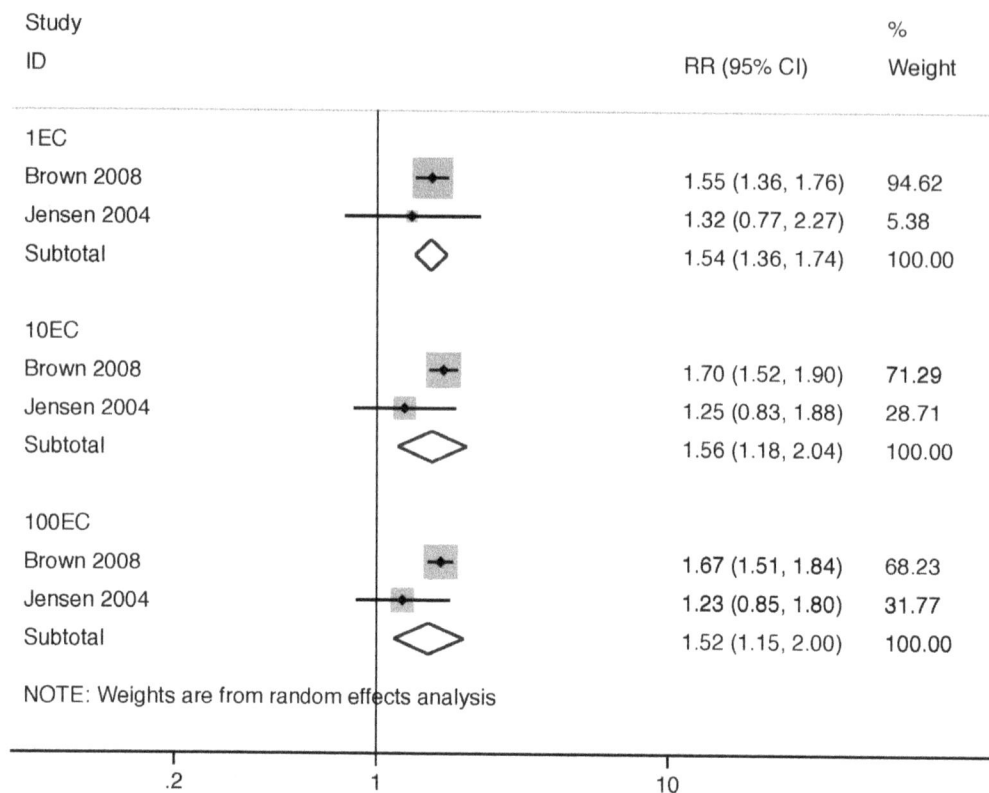

Figure 7. Forest plot for two studies with extractable data at multiple thresholds. Data were pooled across the two studies at the 1 EC/100 ml threshold (top panel), 10 EC/100 ml threshold (middle panel), and 100 EC/100 ml threshold (bottom panel). EC: *Escherichia coli*.

inconclusive. In addition to the results of the meta-analysis, we identified several elements of study design and reporting that could improve future studies evaluating the relationship between water quality and health.

Implications for Choice of Indicator Organism to Assess Microbiological Safety of Drinking Water

Current guidelines recommend two specific uses for indicator organisms: (i) fecal indicators (index organisms), used to indicate the presence of fecal contamination, potential pathogens, and possible health risk, and; (ii) process indicators used to evaluate the effectiveness of a water treatment process. [1,2,13]. While the performance of coliform bacteria as process indicators has been described previously, [2,43] and is outside the scope of this review, the results from our review have implications for the use of EC and FC as index organisms in household drinking water. Currently, EC is considered the "most suitable" fecal indicator; however, FC is generally recognized as a suitable surrogate. [1,2,12,13] Our findings support the continued use of EC as a fecal indicator (index organism) to assess water quality at the point of use, but do not support the use of FC as a surrogate as it does not demonstrate a clear association with diarrheal disease outcomes.

Our findings that EC and FC perform differently as indicators of diarrhea risk from contaminated drinking water are not surprising. While EC is generally of fecal origin, [10] and likely to be present simultaneously with diarrheagenic pathogens in recently contaminated drinking water, FC are known to include coliforms of environmental origin (e.g., *Klebsiella*), in addition to EC. [1,2,13] The consistency with which FC methods will be

specific to human pathogens will therefore vary with the environments in which water samples are being collected. Indeed, our results show that the magnitude and direction of the effect estimates for FC were highly inconsistent (Figure 4, Table 6); even within a single study [37] the authors found highly divergent results using FC in two separate study areas (Table 6). In contrast, the magnitude and direction of effect estimates across the EC studies included in our review were consistent across diverse study regions, and when pooled provided evidence of a significant association with diarrhea; these results are consistent with the higher specificity of EC for fecal contamination, compared to FC (Table 5, Figure 4). The presence of a dose-response between EC and diarrhea could have provided further evidence to support our findings that drinking water contaminated with EC is associated with diarrhea. However, dose-response results from our review were inconclusive, and represent an avenue for future research.

While our summary measures suggest that the presence of EC in household drinking water is significantly associated with diarrhea, it is important to note that only one individual EC study in our review was able to rule out chance as a possible explanation of their findings [15]. It has been demonstrated that indicator levels can vary considerably even over short periods of time in both household and source water, [44,45] and that pathogens and indicator organisms have inconsistent correlations. [46] The studies in this review relied on intermittent grab samples to classify household water quality. Even daily grab samples likely provided only a crude assessment of a household's exposure to fecal contamination and diarrheagenic pathogens, which likely contributed to the lack of precision in the individual studies included in our review. While the imprecision of results from individual studies supports a previous assertion that fecal indicator organisms are "blunt" tools to characterize water quality, [19] our findings suggest that they have continued utility in health and water research. Indeed, while recent WHO drinking water guidelines have shifted focus from recommendations using specific indicator levels to define water safety to a more integrated approach centered on water safety plans, these recommendations are largely directed towards the management of water distribution systems. [2,13] In practice, regional organizations and researchers working in areas without access to drinking water distribution systems rely heavily on indicator organisms to assess the risk of disease from drinking water, and to measure the effectiveness of household water treatment and safe storage interventions; this is particularly true in resource poor settings with limited access to advanced laboratory facilities. Our results suggest that EC have continued value in both of these applications.

Threshold Effects for Diarrhea Risk

The findings from our review do not support the existence of a "threshold effect" in contaminated household drinking water. A previous study evaluating the effect of contaminated *source* water on diarrhea [23] found no evidence of an association between EC and diarrhea at the 1 EC threshold level; however, significant associations with diarrhea were observed at the 1000 EC threshold level. The authors concluded that their findings of a "threshold effect" supported earlier assertions that relaxing drinking water guidelines (beyond the 1 EC threshold) might be acceptable in developing countries. [47] Contrary to these findings, in our review of household drinking water we found consistently elevated risks of diarrhea at the 1 EC threshold, with a significant summary measure when pooled across all studies. These results support current WHO guidelines recommending that all drinking water intended for human consumption contain zero EC/100 ml, regardless of location. [2] Beyond the 1 EC threshold, our results did not suggest increased risk of diarrhea at elevated cutoffs, but this evidence was based on a limited number of studies.

Recommendations for Future Water Quality Studies

The studies included in our review covered several decades, and varied considerably with respect to methodological quality and completeness of reporting. In this section, we highlight several findings from our systematic review that could improve the design and reporting of future studies that seek to evaluate the relationship between water quality and health.

The studies in our systematic review used very consistent definitions of diarrhea, relying primarily on self- or caregiver reporting of diarrhea symptoms (Tables 3 and 4). However, we note that diarrhea recall periods and identification strategies varied greatly (e.g., daily recall, one-month recall, cases presenting to clinics; Tables 3 and 4). Reliance on self-reported diarrhea symptoms as a health measure could have introduced bias into individual studies through two primary mechanisms, which have implications for the use of self-reported diarrhea as an outcome in future studies. First, as with any self-reported, subjective outcome, there is a potential for differential reporting relative to exposure status [48]. If study-specific reporting biases were differential with respect to drinking water contamination, it is possible that the results of individual studies (and therefore our review) could be biased. However, there is no reason to suspect that individuals would be aware of the indicator status of their drinking water when reporting diarrhea symptoms, which would decrease the potential for this bias. Second, longer follow-up periods can bias diarrhea reporting due to poor outcome recall, which could have

biased findings towards the null if non-differential with respect to exposure, and/or compounded biases associated with differential reporting in individual studies. [49] The inclusion of objective health outcome measures (e.g., pathogen-specific antibody responses) could improve future studies by reducing reliance on self-reported diarrhea symptoms and their associated biases. [50,51] In addition, while our review focused exclusively on diarrhea, expanding the range of water-related health outcomes studied and reviewed (e.g., indicators of growth and malnutrition) could also improve our understanding of the health impacts of drinking water contamination beyond gastrointestinal illnesses.

Very few studies in our review were able to establish temporality between exposures (indicator organism) and outcomes (diarrhea), and therefore mostly relied on the assumption that drinking water samples were representative of an exposure period that was relevant to measured disease outcomes. While this assumption might be reasonable, or the only option given the practicalities of field conditions, it does leave open the possibility of reverse causality (i.e., diarrhea caused by other environmental pathways led to increased fecal contamination of drinking water). The design of future water quality studies should attempt to measure exposure in advance of disease outcomes, and all studies should be explicit about the timing of exposure and disease measurements in their reporting. These recommendations would all but preclude the use of "traditional" case-control studies, where cases and controls (i.e., disease status) are identified first, followed by visits to households to collect water samples to ascertain exposure status. Instead, a cohort design that measures exposure to contaminated drinking water first, followed by surveillance visits after an appropriate incubation period to measure health outcomes, is a more suitable study design. Decisions regarding follow-up frequency, recall period, and measures of disease occurrence (i.e., incidence, prevalence) will vary depending on study context and logistics, but should be carefully considered and explicitly reported. [49,52]

In studies evaluating the relationship between drinking water quality and health, confounding bias is a concern: there are numerous factors that could cause both drinking water contamination and diarrhea in a given study setting. Overall, the studies in our review were not consistent in reporting adjusted results (Tables 5 and 6), and there was no consistent set of factors used across studies when authors did report attempting to control for confounding. Incomplete control of confounding in the included studies could have biased individual study results, and therefore the findings of our review. Future studies should be explicit about their attempts to measure and control for confounding factors. Additionally, several studies did not account for correlated outcomes in their analyses, which could have influenced the precision of summary estimates. However, for EC studies we note the consistency of the direction and magnitude of the associations reported across diverse study locations, and across studies that did and did not adjust for confounding and/or account for clustering. Regardless, reporting on these issues could be improved upon in future studies.

Our systematic review included published, peer-reviewed studies. As with any systematic review, publication bias, where smaller studies with small or null ("non-positive") results are systematically excluded from the published literature, could have impacted our findings. In our review, we found evidence of publication bias among FC studies (smaller studies with smaller effect sizes appear to be absent from the funnel plot; lower left region, Figure 6); however, inclusion of smaller FC studies with RRs at or less than unity would have strengthened our findings of no association between FC and diarrhea. We found minimal evidence to suggest publication bias among EC studies (Figure 5);

however, we cannot rule out publication bias as a possible explanation of our findings – the funnel plot suggests that most studies were balanced around the summary measure with the exception of one small study with a large positive finding. Taken together, there is evidence that smaller studies with small ("non-positive") effects sizes are systematically missing from the literature on the relationship between water quality and diarrheal illness.

In our review, we pooled RRs across relevant studies using meta-analysis to investigate the association of fecal indicator measures in drinking water with diarrhea. The limitations of meta-analysis have been discussed [53]; we caution against interpreting the summary measures from our review as "true" underlying causal effects, but rather consider them additional pieces of evidence regarding the relationship between fecal indicators in drinking water and diarrhea. The evidence from our pooled summary measures, taken together with biologic plausibility and the consistency of results across studies, provides evidence to support a relationship between diarrhea and the presence of EC in point-of-use drinking water; we do not, however, find evidence to support an equivalent relationship between diarrhea and FC.

Conclusions

We reviewed the available literature to identify studies that evaluated the relationship between microbial indicators of drinking water quality and diarrhea. We found that studies using *Escherichia coli* (EC) as an indicator of household drinking water quality reported consistent effect estimates, that when pooled suggested a significant association with increased diarrheal illness.

Results from studies using thermotolerant ("fecal") coliforms (FC), on the other hand, were inconsistent, and suggested no association with diarrhea when pooled. In this review, we also note several areas where the design and reporting of the included studies could have been improved, and make recommendations for future studies. The results from our review suggest that EC has value as a fecal indicator organism, but that use of FC should be considered carefully in contexts where an association with diarrheal disease outcomes is important.

Acknowledgments

The authors would like to acknowledge Professor Kara Nelson for generously reviewing this manuscript, and providing key feedback.

Author Contributions

Conceived and designed the experiments: JG JC. Performed the experiments: JG AE. Analyzed the data: JG. Contributed reagents/materials/analysis tools: JG AE. Wrote the paper: JG AE JC.

References

1. Dufour AP (2003) Assessing microbial safety of drinking water: Improving approaches and methods: International Water Assn.
2. WHO (2011) Guidelines for drinking-water quality, fourth edition. Geneva: World Health Organization.
3. Cutler D, Miller G (2005) The role of public health improvements in health advances: The twentieth-century United States. Demography 42: 1–22.
4. Arnold BF, Colford JM Jr (2007) Treating water with chlorine at point-of-use to improve water quality and reduce child diarrhea in developing countries: a systematic review and meta-analysis. The American journal of tropical medicine and hygiene 76: 354.
5. Clasen T, Schmidt WP, Rabie T, Roberts I, Cairncross S (2007) Interventions to improve water quality for preventing diarrhoea: systematic review and meta-analysis. bmj 334: 782.
6. Fewtrell L, Kaufmann RB, Kay D, Enanoria W, Haller L, et al. (2005) Water, sanitation, and hygiene interventions to reduce diarrhoea in less developed countries: a systematic review and meta-analysis. The Lancet infectious diseases 5: 42–52.
7. National Research Council (2004) Indicators for Waterborne Pathogens. Washington, DC: The National Academies Press.
8. Gundry S, Wright J, Conroy R (2004) A systematic review of the health outcomes related to household water quality in developing countries. J Water Health 2: 1–13.
9. Gerba C (2000) Indicator Microoogranisms. Environmental Microbiology. San Diego, CA: Academic Press.
10. Leclerc H, Mossel D, Edberg S, Struijk C (2001) Advances in the bacteriology of the coliform group: their suitability as markers of microbial water safety. Annual Reviews in Microbiology 55: 201–234.
11. Reynolds KA, K.D Mena, Gerba CP (2008) Risk of waterborne illness via drinking water in the United States. Reviews of Environmental Contamination and Toxicology 192: 117–158.
12. Tallon P, Magajna B, Lofranco C, Leung KT (2005) Microbial indicators of faecal contamination in water: a current perspective. Water, Air, and Soil Pollution 166: 139–166.
13. Ashbolt NJ, Grabow WOK, Snozzi M (2001) Indicators of microbial water quality. In: Fewtrell L, Bartram J, editors.Water Quality: Guidelines, Standards and Health. London, UK: IWA Publishing,.
14. WHO (1997) Guidelines for drinking-water quality, 2nd edition. Geneva: World Health Organization.
15. Brown JM, Proum S, Sobsey MD (2008) Escherichia coli in household drinking water and diarrheal disease risk: evidence from Cambodia. Water Science and Technology 58: 757–763.
16. de Aceituno AMF, Stauber CE, Walters AR, Sanchez REM, Sobsey MD (2012) A Randomized Controlled Trial of the Plastic-Housing BioSand Filter and Its

Impact on Diarrheal Disease in Copan, Honduras. American Journal of Tropical Medicine and Hygiene 86: 913–921.
17. Gundry SW, Wright JA, Conroy RM, Du Preez M, Genthe B, et al. (2009) Child dysentery in the Limpopo Valley: A cohort study of water, sanitation and hygiene risk factors. Journal of Water and Health 7: 259–266.
18. Jensen PK, Jayasinghe G, van der Hoek W, Cairncross S, Dalsgaard A (2004) Is there an association between bacteriological drinking water quality and childhood diarrhoea in developing countries? Trop Med Int Health 9: 1210–1215.
19. Levy K, Nelson KL, Hubbard A, Eisenberg JN (2012) Rethinking indicators of microbial drinking water quality for health studies in tropical developing countries: case study in northern coastal Ecuador. Am J Trop Med Hyg 86: 499–507.
20. Raina PS, Pollari FL, Teare GF, Goss MJ, Barry DA, et al. (1999) The relationship between E. coli indicator bacteria in well-water and gastrointestinal illnesses in rural families. Can J Public Health 90: 172–175.
21. Strauss B, King W, Ley A, Hoey JR (2001) A prospective study of rural drinking water quality and acute gastrointestinal illness. BMC Public Health 1: 8.
22. Rothman K, Greenland S, (1998) Modern Epidemiology. Philadelphia: Lippincott-Raven Publishers.
23. Moe C, Sobsey M, Samsa G, Mesolo V (1991) Bacterial indicators of risk of diarrhoeal disease from drinking-water in the Philippines. Bulletin of the World Health Organization 69: 305.
24. Greenland S, Longnecker MP (1992) Methods for trend estimation from summarized dose-response data, with applications to meta-analysis. American Journal of Epidemiology 135: 1301–1309.
25. Egger M, Smith G Davey, Altman D (2001) Systematic Reviews in Healthcare. Meta-analysis in Context. London: BMJ Books.
26. Egger M, Smith GD, Schneider M, Minder C (1997) Bias in meta-analysis detected by a simple, graphical test. bmj 315: 629–634.
27. Sterne JAC, Egger M (2001) Funnel plots for detecting bias in meta-analysis: Guidelines on choice of axis. Journal of Clinical Epidemiology 54: 1046–1055.
28. Genthe B, Strauss N, Seager J, Vundule C, Maforah F, et al. (1997) The effect of type of water supply on water quality in a developing community in South Africa. Water Science and Technology 35: 35–40.
29. Echeverria P, Taylor DN, Seriwatana J (1987) Potential sources of enterotoxigenic Escherichia coli in homes of children with diarrhoea in Thailand. Bulletin of the World Health Organization 65: 207–215.
30. Baqui AH, Black RE, Yunus M, Hoque ARA, Chowdhury H, et al. (1991) Methodological issues in diarrhoeal diseases epidemiology: definition of diarrhoeal episodes. International journal of epidemiology 20: 1057.
31. WHO (2013) Diarrhoeal disease; fact sheet.

32. Rajasekaran P, Dutt PR, Pisharoti KA (1977) Impact of Water Supply on the Incidence of Diarrhoea and Shigellosis Among Children in Rural Communities in Madurai. Indian J Med Res 66: 189–199.

33. Lloyd-Evans N, Pickering HA, Goh SGJ, Rowland MGM (1984) Food and Water Hygiene and Diarrhoea in Young Gambian Children: a Limited Case Control Study. Transactions of the Royal Society of Tropical Medicine and Hygiene 78: 209–211.

34. Black RE, Brown KH, Becker S, Alim AR, Merson MH (1982) Contamination of weaning foods and transmission of enterotoxigenic Escherichia coli diarrhoea in children in rural Bangladesh. Trans R Soc Trop Med Hyg 76: 259–264.

35. Bhargava A, Bouis HE, Hallman K, Hoque BA (2003) Coliforms in the water and hemoglobin concentration are predictors of gastrointestinal morbidity of Bangladeshi children ages 1–10 years. American journal of human biology 15: 209–219.

36. Esrey SA, Habicht J, Casella G, Miliotis MD, Kidd AH, et al. (1986) Infection, Diarrha, and Growth Rates of Young Children Following the Installation of Village Water Supplies in Lesotho. Interanation Symptosium on Water-Related Health Issues November: 11–16.

37. Henry FJ, Rahim Z (1990) Transmission of diarrhoea in two crowded areas with different sanitary facilities in Dhaka, Bangladesh. J Trop Med Hyg 93: 121–126.

38. Knight SM, Toodayan W, Caique WC, Kyi W, Barnes A, et al. (1992) Risk factors for the transmission of diarrhoea in children: a case-control study in rural Malaysia. Int J Epidemiol 21: 812–818.

39. Han AM, Oo KN, Aye T, Hlaing T (1991) Bacteriologic studies of food and water consumed by children in Myanmar: 2 Lack of association between diarrhoea and contamination of food and water. J Diarrhoeal Dis Res 9: 91–93.

40. Henry F, Huttly S, Patwary Y, Aziz K (1990) Environmental sanitation, food and water contamination and diarrhoea in rural Bangladesh. Epidemiology and infection London, New York NY 104: 253–259.

41. Vanderslice J, Briscoe J (1993) All Coliforms Are Not Created Equal - a Comparison of the Effects of Water Source and in-House Water Contamination on Infantile Diarrheal Disease. Water Resources Research 29: 1983–1995.

42. Jagals P, Grabow WOK, Williams E (1997) The effects of supplied water quality on human health in an urban development with limited basic subsistence facilities. Water Sa 23: 373–378.

43. Sobsey MD (1989) Inactivation of health-related microorganisms in water by disinfection processes. Water Science & Technology 21: 179–195.

44. Levy K, Hubbard AE, Nelson KL, Eisenberg JN (2009) Drivers of water quality variability in northern coastal Ecuador. Environmental science & technology 43: 1788–1797.

45. Levy K, Nelson KL, Hubbard A, Eisenberg JN (2008) Following the water: a controlled study of drinking water storage in northern coastal Ecuador. Environmental health perspectives 116: 1533.

46. Wu J, Long S, Das D, Dorner S (2011) Are microbial indicators and pathogens correlated? A statistical analysis of 40 years of research. J Water Health 9: 265–278.

47. Feachem RG, McGarry MG, Mara DD (1977) Water, wastes and health in hot climates: John Wiley and Sons Ltd.

48. Wood L, Egger M, Gluud LL, Schulz KF, Jüni P, et al. (2008) Empirical evidence of bias in treatment effect estimates in controlled trials with different interventions and outcomes: meta-epidemiological study. bmj 336: 601–605.

49. Arnold BF, Galiani S, Ram PK, Hubbard AE, Briceño B, et al. (2013) Optimal recall period for caregiver-reported illness in risk factor and intervention studies: A multicountry study. American journal of epidemiology 177: 361–370.

50. Crump JA, Mendoza CE, Priest JW, Glass RI, Monroe SS, et al. (2007) Comparing serologic response against enteric pathogens with reported diarrhea to assess the impact of improved household drinking water quality. American Journal of Tropical Medicine and Hygiene 77: 136.

51. Schmidt W-P, Cairncross S (2009) Household Water Treatment in Poor Populations: Is There Enough Evidence for Scaling up Now? Environmental science & technology 43: 986–992.

52. Schmidt W-P, Arnold BF, Boisson S, Genser B, Luby SP, et al. (2011) Epidemiological methods in diarrhoea studies—an update. Int J Epidemiol 40: 1678–1692.

53. Berk RA, Freedman DA (2003) Statistical assumptions as empirical commitments: Law, Punishment, and Social Control: Essays in Honor of Sheldon Messinger, 2nd edn. Aldine de Gruyter.

21

Spatio-Temporal Dynamics of Maize Yield Water Constraints under Climate Change in Spain

Rosana Ferrero[1,4]*, Mauricio Lima[2,3,4], Jose Luis Gonzalez-Andujar[1,3]

1 Departamento Protección de Cultivos, Instituto de Agricultura Sostenible, Consejo Superior de Investigaciones Científicas (CSIC), Córdoba, Spain, **2** Departamento de Ecología, Pontificia Universidad Católica de Chile, Santiago, Chile, **3** Laboratorio Internacional de Cambio Global, LINCG (CSIC-PUC), Santiago, Chile, **4** Center of Applied Ecology and Sustainability (CAPES), Santiago, Chile

Abstract

Many studies have analyzed the impact of climate change on crop productivity, but comparing the performance of water management systems has rarely been explored. Because water supply and crop demand in agro-systems may be affected by global climate change in shaping the spatial patterns of agricultural production, we should evaluate how and where irrigation practices are effective in mitigating climate change effects. Here we have constructed simple, general models, based on biological mechanisms and a theoretical framework, which could be useful in explaining and predicting crop productivity dynamics. We have studied maize in irrigated and rain-fed systems at a provincial scale, from 1996 to 2009 in Spain, one of the most prominent "hot-spots" in future climate change projections. Our new approach allowed us to: (1) evaluate new structural properties such as the stability of crop yield dynamics, (2) detect nonlinear responses to climate change (thresholds and discontinuities), challenging the usual linear way of thinking, and (3) examine spatial patterns of yield losses due to water constraints and identify clusters of provinces that have been negatively affected by warming. We have reduced the uncertainty associated with climate change impacts on maize productivity by improving the understanding of the relative contributions of individual factors and providing a better spatial comprehension of the key processes. We have identified water stress and water management systems as being key causes of the yield gap, and detected vulnerable regions where efforts in research and policy should be prioritized in order to increase maize productivity.

Editor: Pilar Hernandez, Institute for Sustainable Agriculture (IAS-CSIC), Spain

Funding: R. Ferrero gratefully acknowledges receipt of a grant from the Fundación Carolina. J. L. Gonzalez-Andujar and R. Ferrero were supported by FEDER (Fondo Europeo de Desarrollo Regional) and the Spanish Ministry of Economy and Competitiveness funds (AGL2012-33736). R. Ferrero and M. Lima acknowledge financial support from Fondo Basal-CONICYT grant FB-0002. We are grateful to LINCGlobal (Laboratorio Internacional en Cambio Global) for their support. The funders had no role in study design, data collection and analysis, decision to publish, or preparation of the manuscript.

Competing Interests: The authors have declared that no competing interests exist.

* E-mail: rferrero@ias.csic.es

Introduction

Spatio-temporal patterns of agricultural production are clearly influenced both by climate change and agricultural management practices. Recently, many studies have analyzed the impact of climate change on crop productivity [1], but comparing the performance of different crop management systems has rarely been explored (exc. [2]). To be specific, we need to evaluate how and where irrigation practices (e.g. rain-fed versus irrigated) are effective in mitigating the effects of climate change, because water constraints and crop demand in agro-systems could be increased due to climate change [3–7]. Identifying whether there are any differences in the principal bio-physical factors and mechanisms that explain both systems will enable us to improve crop productivity without expanding the cropland area and to diminish the adverse impacts of agriculture for social and ecological systems [8].

We do not know much about crop response to climate change yet, and still less about the differential response between irrigated and rain-fed systems [4]. Increases in agriculture production could potentially come from increases in irrigated crops, because higher yields could be attained with reduced production variability [9].

However, this also depends on soil and management factors that result in spatial patterns of yields [10]. Secondly, irrigation can influence local climate by inducing cooling, but this may depend on the extent of the irrigated area, the level of soil moisture alteration and cloud response to irrigation [11]. Third, average yields in rain-fed systems are commonly 50% or less of yield potential (high yield gap), suggesting ample room for improvement [12] but, again a great spatial variability has been found [13]. Yield gaps could be bigger in cropping systems that experience wider ranges of variation under climate conditions [10]. Fourth, plant population (or density) is known to affect the yield potential at a given location [12] and grain yield stability [14]. However, to our knowledge, there are no previous studies explicitly comparing endogenous processes under different water management systems. Finally, simulation at a broad scale level cannot fully explain the above process, and process-based crop models do not always relate to observed yields [15]. Finer spatial scales and historical data of irrigated versus rain-fed systems could help to compare modelled or simulated yield potentials [12].

Analyzing the sensitivity of irrigated and non-irrigated (rain-fed) crops to past climate changes is crucial to an understanding of the vulnerability of agriculture to climate change in the future, particularly in regions that already suffer from this under present conditions. This paper explores biophysical factors and water management practice constraints to maize (*Zea mays* L.) in Spain. Spatial shifts northwards have been projected for maize, due to the extremely hot, dry summers in south-central Europe [16,17], particularly in Spain [18]. The expected effects of climate change on Spain's agriculture would not be uniform. Mediterranean (arid and semiarid) regions may be particularly sensitive, where a decrease in the general availability of hydric resources and an increase in evaporative demand, especially during summer, will affect irrigation requirements [19]. Namely, it is one of the most prominent "hot-spots" in future climate change projections [20], where a mean reduction of 17% in water resources [21,22] has been predicted. For this drought-prone zone, all climate change scenarios imply the need to significantly increase the contribution of irrigation water. Therefore, identifying and quantifying the links between water management practices and food production is crucial in addressing the intensified conflicts between water scarcity and food safety.

The objective of this paper is to determine how climate variability affects maize production in Spain under irrigated and rain-fed conditions. First, we have analyzed the regulatory structure of maize production dynamics under both water management systems. Second, we have evaluated the mechanisms (in ecological parameters) underlying climate perturbations on maize yields. Third, we have assessed whether the importance of maize production structures (i.e. intrinsic regulation) and climate change perturbations (i.e. exogenous factors) could change according to the type of management (i.e. rain-fed and irrigated) and the geographical location. Fourth, we have estimated the potential yield of each region and water management using the previous models and analyzing the spatial variability of yield losses due to water stress [23]. We have combined information on spatial autocorrelation water stress patterns for maize yields to identify the importance of climate constraints at a regional scale.

Methods and Materials

Database

Provincial maize yield levels (*Zea mays*; production per hectare, *kg/ha*) for 1996–2009 were obtained from statistical yearbooks [24]. We studied selected provinces that had both rain-fed and irrigated systems (Fig. 1), and displayed trends in yield fluctuation in Fig. S1. We used Global Historical Climatology Network (GHCND) data on monthly temperature and rainfall (mean, minimum, maximum and extreme; [25]). Various summary statistics of the growing season (July to October) weather were then computed: *EMNT* extreme minimum temperature (°C), *EMXT* extreme maximum temperature (°C), *MMNT* mean minimum temperature (°C), *MMXT* mean maximum temperature (°C), *MNTM* mean temperature (°C), *EMXP* extreme maximum daily precipitation total (l/m2) and *TPCP* total precipitation (l/m2). We also examined carbon dioxide emission (CO_2), an important atmospheric gas that contributes to global warming. The annual country-level emissions of CO_2 (*kt*) were taken from the World Bank's World Development Indicators (WDI; [26]).

Diagnosis and statistical models of yields dynamics

We have analyzed and predicted maize yield responses to the impact of climate change in Spain through the use of models based on the population dynamics theory. Of course this is not a true population in the reproductive sense, but crop systems obey the same rules as all other dynamic systems, both natural and engineered.

First, where necessary, we used sequencing (i.e., splitting the series into two stationary segments) and detrending (i.e., rotating the series around the linear or quadratic trend) to generate a stationary time series. Second, we estimated the logarithmic rate of change of the yield as $R_t = Y_t - Y_{t-1}$ (the same response variable as [1,27]), where Y_t represents the provincial yield in a year *t*(the logarithm of the detrended yield) and Y_{t-1} is the same series with one year of delay (lag 1).

We were able to detect and analyze non-trivial feedback processes by examining their relationship $R_t = f(Y_{t-d})$, where the function *f* described how the crop yield change rate varied with yield level, and this has been called the *R*-function. We used the partial rate correlation function (or *PRCF*) to estimate the order of the dynamical process and determine how many time lags (*d*) should be included in the model for representing the feedback structure. This function detects the feedback order removing the confounding effect by calculating the partial correlation between R_t and Y_{t-d} with the effects of lower lags removed [28].

We then used the generalized version of the exponential form of the discrete time logistic model [29,30] in terms of the *R*-function to represent pure endogenous models in the function *f*:

$$R_t = r_{max} - \exp(a Y_{t-d} + c) \tag{1}$$

where Y_{t-d} represents the yield data at time $t-d$ (where *d* was obtained from *PRCF* function), r_{max} is a positive constant representing the maximum finite rate of change (and is estimated as the maximum rate of change from the observed data), *c* is a measure of the ratio between demand and offer of limiting resources and *a* is the nonlinearity of the curve. The nonlinearity of this model includes a biological realistic property: its net reproductive rate is bounded [29], that is, the performance of any crop must have an upper bound simply because no crop can produce an infinite number of grains that subsequently contribute to the crop yield.

Finally, we used the Royama classification of exogenous effects as a framework to deduce causal mechanisms of the climate change impact on these crop yields in a spatial-temporal study [29]. To include exogenous perturbations, we modelled r_{max} and *c* of (1) as linear functions of climate conditions, each of which has an explicit biological interpretation. In this way, we set up mechanistic hypotheses about the exogenous effects of climate on these yields data [29].

If an exogenous factor (i.e. climate or gas emissions) changes r_{max} and has an additive or independent perturbation effect on crop yield levels, it shifts the *R*-function curve along the *y*-axis ("*vertical*" perturbations):

$$r_{max}^* = r_{max} + bZ_{t-d'}$$
$$R_t = r_{max} - \exp(a Y_{t-d} + c) + bZ_{t-d'} \tag{2}$$

where $Z_{t-d'}$ is the exogenous factor (for lags or *d* 0 and 1; in logarithm scale). This model produces alterations to both r_{max} and the carrying capacity (equilibrium point of the population, $R_t = 0$), changing the level of equilibrium and its stability.

If an exogenous factor (i.e. climate or gas emissions) changes *c* and has a non-additive perturbation effect on crop yield levels, and influences the equilibrium point of the population shifting the *R*-function curve along the *x*-axis ("*lateral*" perturbations):

$$c* = c + bZ_{t-d'}$$
$$R_t = r_{max} - \exp(a Y_{t-d} + c + bZ_{t-d'}) \tag{3}$$

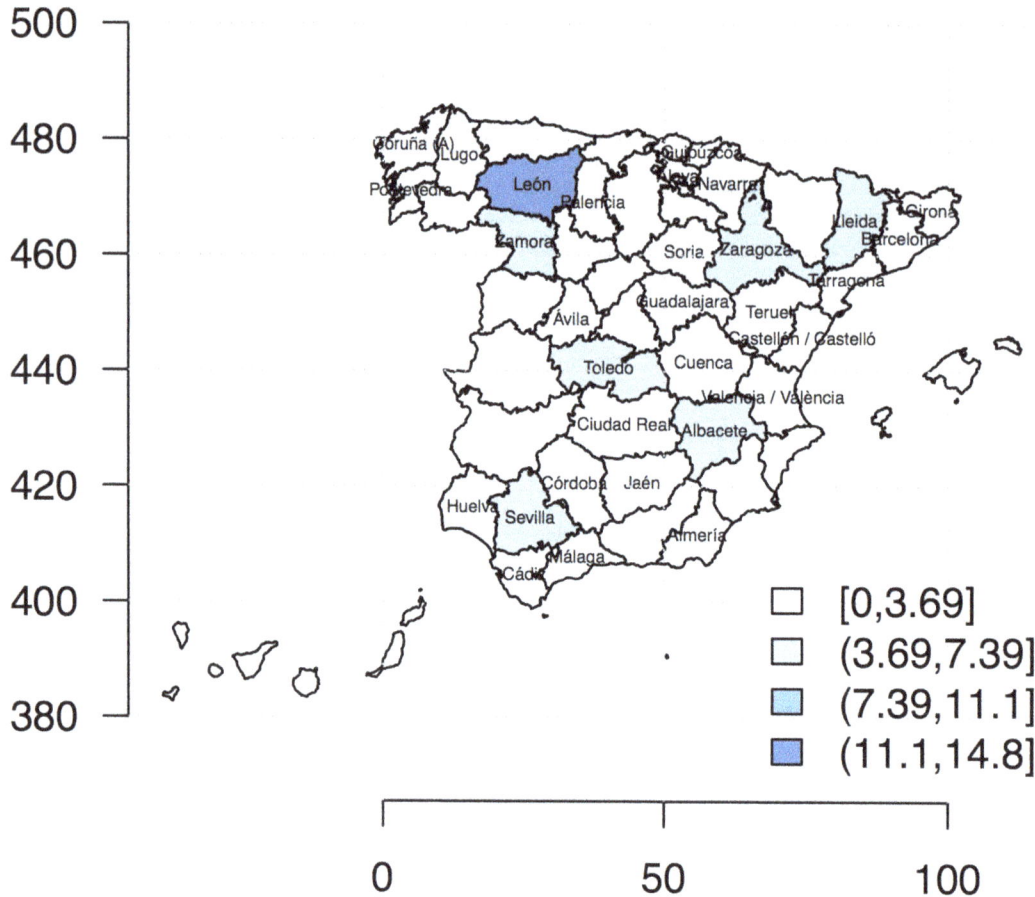

Figure 1. Definition of study regions (provinces) with percentage of total maize production for 1996–2009. Only provinces with both irrigated and rain-fed systems were analyzed.

Lateral perturbations do not change the pattern of dynamics around equilibrium because they do not change the slope at the equilibrium.

We fitted Eqs. 1–3 using nonlinear least squares regressions with the *nls* library in the software *R* [31,32]. In particular, the models were fitted by minimizing the Akaike criterion with a correction for finite sample sized (AIC_c):

$$AIC_c = (2k - 2\ln L) + \frac{2k(k+1)}{n-k-1}$$

where k is the number of parameters and L is the maximized value of the likelihood function for the model, and n denotes the sample size. Also, we maximized the pseudo R^2 measures based on the deviance residual [33]. Models were chosen on the basis of their goodness-of-fit (assessed using root mean square error RMSE and the log-likelihood values), their ability to describe the correct feedback structure, and their appropriateness.

Yield losses due to suboptimal water availability (YGRw)

We propose a new estimation of the potential yield or equilibrium productivity [34] at the provincial level as the equilibrium value of the models. By solving Eqn. (1–3) for the equilibrium dynamics $Y_t = Y_{t-1} = K$ (when $R_t = 0$), we calculated the maize yield level at equilibrium, sometimes called the carrying capacity (Mg/ha). For non-pure endogenous models we made potential yield estimations for each year as the exogenous factor

changed. Then we calculated the percentage of yield losses due to suboptimal water availability ($YGRw$; Eqn. 4; view [35]), which indicated how close the rain-fed yield potential is to the irrigated value for a given site (%).

$$YGR_w = \frac{YP_{IR} - YP_{RF}}{YP_{IR}} \tag{4}$$

We obtained some time-invariant YGR_w values when, in the same province, irrigated and rain-fed YP were estimated from pure endogenous models, so that we calculated the averaged YGR_w for each province, and studied its spatial variability without taking into account the temporal dimension of the data.

We determined whether there was any spatial autocorrelation in YGR_w with the global Moran's I (spatial correlation on average, of an entire map). At this stage, we were not yet trying to determine the causes, although the results could have motivated a hypothesis. We assumed: 1) that there was no spatial patterning due to some underlying but unmodelled factor, and 2) that the assigned spatial weights were those that generated the autocorrelation. Then we tested whether YGR_w was more spatially clustered than by chance. The matrix that represents spatial dependence (W) uses a binary indicator of neighbourhood (i.e. the spatial weights, w_{ij}, are defined as $w_{ij} = 1$ if the i and j provinces are contiguous neighbours, $w_{ij} = 0$ otherwise, based on rook contiguity; [36]). We used row-standardisation (style W) that favours observations with few neighbours. We calculated a non-parametric approach to infer-

Climate change perturbations in maize yield

Temperature effects in irrigated maize

Temperature effects in rainfed maize

Precipitation effects in irrigated maize

Precipitation effects in rainfed maize

CO2 effects in irrigated maize

CO2 effects in rainfed maize

Negative   •   Without   •   Positive

Figure 2. Effects of temperature, precipitation and CO2 emission, on maize productivity for rain-fed and irrigated crops. Provinces for both water management systems were selected for the analysis. All models are from Table S1.

ence on Moran's I using 999 simulations (Monte Carlo permutation test). Also, local indicators of spatial association (or LISA) were calculated to detect "hot spots" where there was a strong autocorrelation, and "cold spots", where there were none. The results were plotted on a Moran scatterplot: the target variable on the x-axis, and the (spatially-weighted) sum of neighbouring values on the y-axis; these are called spatially lagged values. We identified the high-influence areas.

We analyzed the environmental spatially distributed causes of averaged YGR_w through a Simultaneous Autoregressive Model (SAR; [36]) that considers spatial autocorrelation of residuals:

$$YGR_w = Z^T \beta + e \qquad (5)$$

where, for each province, YGR_w is the percentage of yield losses due to suboptimal water availability, Z is a matrix of averaged climate variables (see Database section; except country-level CO_2 emissions), $e = B(Y - Z^T \beta) + \epsilon$ is the error term, and ϵ represents residual errors (assumed to be independently distributed according to a Normal distribution with zero mean and diagonal covariance matrix σ_e^2). The error terms are modelled so that they depend on each of the other areas to account for their spatial dependence (B is a matrix that contains the dependence parameters; $B = \lambda W$, where λ is a spatial autocorrelation parameter and W is a matrix that represents the spatial dependence explained above). Global Moran's I was computed for the residuals to test if the SAR model accounts for all the spatial autocorrelations in YGR_w. For the spatial analysis we used *spdep* library in the software R [37].

Results

Regulatory structure and exogenous perturbation models

After sequencing and detrending, all the sites exhibited first-order negative feedback ($PRCF(1)$) as being the most important component of yield growth rate (Figure S2; except for irrigated maize in Vizcaya and rain-fed systems in Tarragona). Major sites showed the highly significant ($p < 0.05$) effect of endogenous processes as determinants of the structure of crop productivity regulation (Table S1).

We evaluated gas emission (CO_2) and climate factors (temperature and precipitation; see Database section), as exogenous perturbations of the production curve (R-function). Table S1 shows several models that were selected as climate change impacts on maize production for each Spanish region and management system. The stochastic versions of the step-ahead predictions of the models are shown in Fig. S3. As expected, the effects of climate on maize production were not uniform, and depended on the irrigation management system (Figure 2). Maize yields were significantly related to minimum temperatures (possibly night ones) in 11 sites and by maximum temperature in another 5 sites. Generally, there were positive effects of temperature for irrigated systems, except for Almería (for minimum temperature $-EMNT$-) and Ourense (mean temperature $-MNTM$-). However, for rain-fed systems, we detected negative effects of warming on major sites, with the exception of Málaga and Albacete (both for $EMNT$). As expected, precipitation was not important for irrigated systems (except for maximum precipitation $-EMXP$- in Navarra), but it was an important factor in some rain-fed managements. There

were positive effects of precipitation on Teruel and Soria (for a total $-TPCP$- and maximum rainfall), and negative ones in Córdoba ($TPCP$ and $EMXP$) and Zaragoza ($TPCP$) on rain-fed crops. Finally, CO_2 emissions negatively affected maize in Lugo (irrigated), Ourense and Soria (both rain-fed), and positively only in Ávila (rain-fed).

Temperature acted mainly as having non-additive (lateral) effects on maize yield dynamics, whereas CO_2 emission acted as additive (vertical) effects (Table S1; Figure 3 and S4). Finally, rainfall exerted non-additive effects when it had a negative impact on maize, but when it obtained positive responses the effects were of both types (additive or non-additive; Table S1; Figure S4). For example, Figure 3 shows positive and non-additive (lateral) effects of temperature on rain-fed maize in Albacete and irrigated maize in Sevilla. That is, the increase in temperature had a positive effect on both maize systems, and more so at high yield levels. Figure 3 also indicates a negative and additive (vertical) effect of CO_2 emission on Ourense (same strength for all yield levels), and a positive and non-additive (lateral) rainfall effect on rain-fed maize in Soria (more important for high yield levels).

Relative yield losses due to suboptimal water availability (YGR$_w$)

We first visualized the spatial relation of YGR_w (Figure 4), where several high YGR_w values were shown in central and southern provinces of Spain. The global Moran's I value ($I = 0.39$) was of an opposite sign and much larger in absolute value than the expectation ($E[I] = -0.034$); this was quite unlikely to be equal to the expectation of no spatial association. The probability of incorrectly rejecting the null hypothesis of no association (type I error) was 0.0021. The Monte Carlo approach also rejects the null hypothesis (the true value for Moran's I is zero; $I_{mc} = 0.406$, $p = 0.005$; Figure S5). The Moran scatterplot (Figure 5; the vector of values and the neighbour list with weights) showed points with a great influence, which are identified by a special symbol and their name. The highest-leverage area is marked on Almería; it has the highest YGR_w (84.56) and a zero weighted spatially-lagged proportion, because it did not have any adjacent areas in the study. Soria and Palencia had low YGR_w, and a low spatially-lagged proportion; these are the low-YGR_w neighbourhoods adjacent to low-YGR_w neighbourhoods. They have a great influence on the slope (global Moran's I). From Figure 5 it is clear that most of the global Moran's I significance comes from the local Moran's I from high YGR_w in Almería, and low YGR_w associated with low YGR_w, in the Soria and Palencia area in the north.

There was clear evidence of local clustering, 6 areas (Ciudad Real, Cuenca, Albacete, Valencia, A Coruña and Pontevedra) showed sufficiently high local Moran's I to reject the null hypothesis with less than a 5% chance of Type I error. These areas were not highlighted in the Moran scatterplot, as they did not greatly influence the global Moran's I but were locally-clustered.

There was a significant spatial correlation in the residuals, because the estimated value of lambda was 0.141 and the p-value of the likelihood ratio test 0.0354. Only averaged temperature ($MNTMt$) was significant for the SAR model, suggesting that provinces with higher temperature have larger YGR_w percentages. The model found was: $YGR_w = -64.42 + 5.69 * MNTM_t$ the SAR model, which accounted for the whole spatial autocorrelation

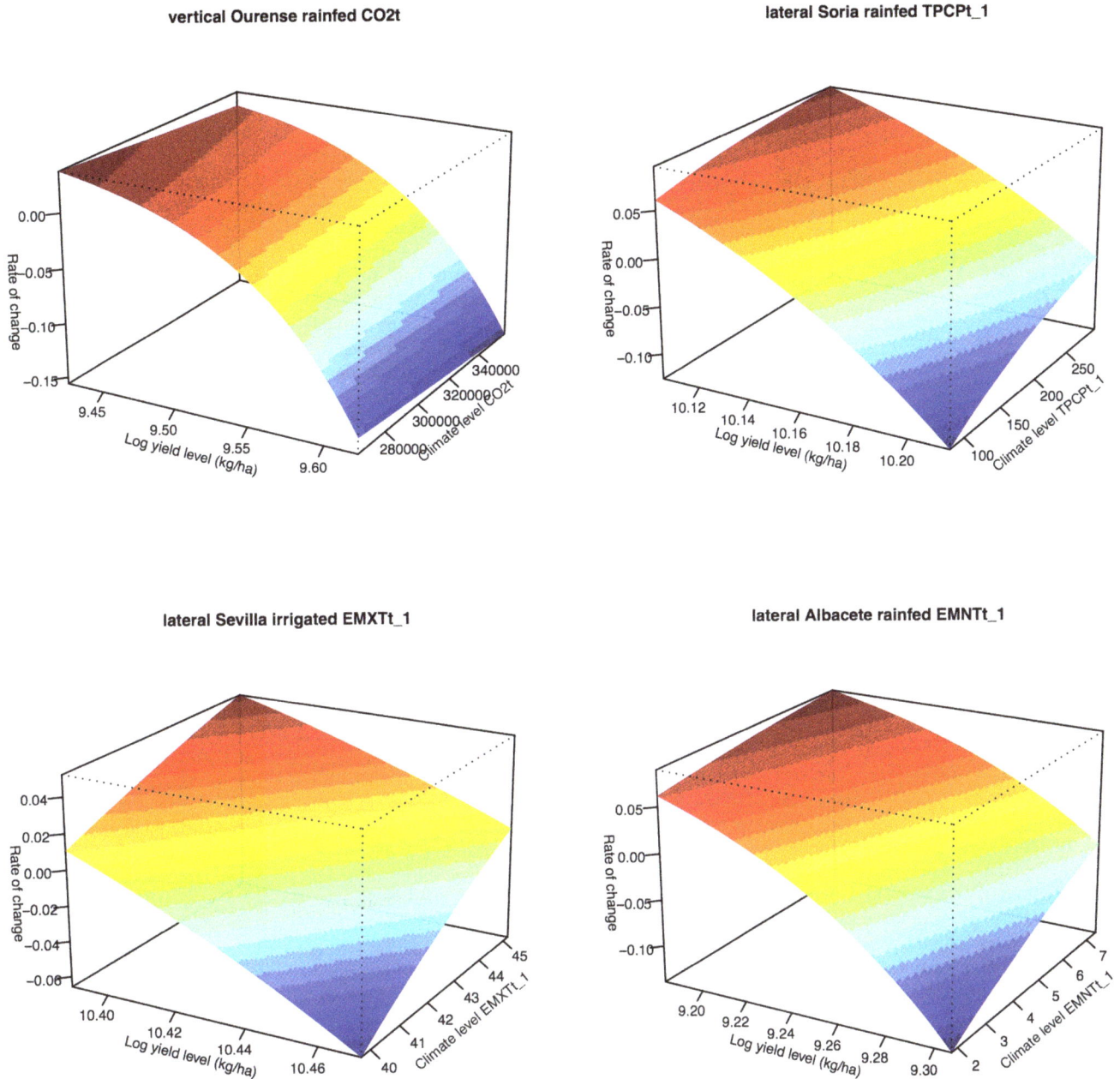

Figure 3. Yield rate of change against the log observed yield level (with one year of delay) and the exogenous factor that perturbs the productivity function (*R*-function). Exogenous factors include carbon emissions ($CO2_t$), precipitation ($TPCP_{t\_1}$), and maximum and minimum temperature ($EMXT_{t\_1}$ and $EMNT_{t\_1}$). Additive (vertical) and non-additive (lateral) perturbation effects were detected. Colours indicate the *R*-function value. See Table S1 for description of models and Figure S4 for their graphs.

in YGR_w (global Moran's test for residuals was $I = -0.00811$, $p = 0.422$). Thus, the autocorrelation in the linear model residuals was explained.

Discussion

In the present study, the impact of climate variability on maize yields in Spanish rain-fed and irrigated systems was investigated for the period 1996–2009. We explored the endogenous structure (regulation) and the exogenous perturbations effects on maize production at a regional scale.

Regulatory structure: endogenous feedback

We found that maize productivity had a persistently negative effect on crop yields for a one year time delay (first order negative feedback, *PRCF(1)*). Maize productivity was characterized by negative first-order feedback structure in major sites and in both irrigation systems. Namely, there were biomass or density-induced feedback loops in the growth, survival rates, seed germination or grain production rates of individual plants, tending to stabilize their dynamics [38]. In Spain, the seeds produced are used for the next year and, therefore, a year's crop performance could change seed viability and vigour, which also affects the performance of the following crops (changing the demand for resources). Also, a crop

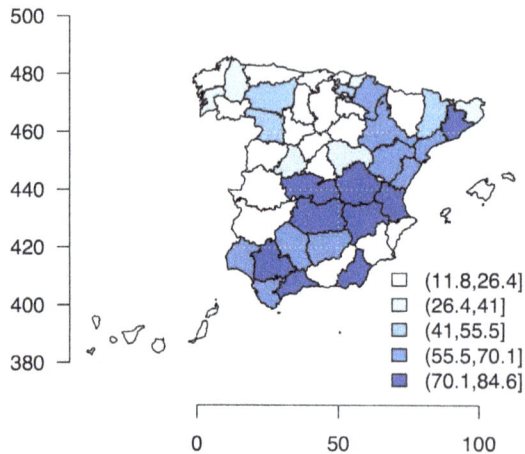

Figure 4. Relative yield losses due to suboptimal water availability (YGRw; %). The percentage of yield losses due to suboptimal water availability indicates how close rain-fed yield potential are to the irrigated value for a given site.

system could alter habitat conditions; in fact, the frequent practice of crop rotation is a testimony to the importance of negative feedbacks in agricultural systems (i.e. it modifies resource supplies). This produces high-frequency dynamics due to year-to-year endogenous variability in maize yields. Our logistic models appear to capture the essential features of the fluctuations observed, and suggest a mechanistic explanation for the latter. This implies that, to understand the response of maize productivity to climate, we must also know the endogenous feedback structure of the system.

Our models are important to conceptualizing the problem of regulated versus unregulated systems. If a system were to be controlled entirely by an exogenous process (unregulated systems), then the series would perform a random walk and we saw no sign of the generated series becoming stabilized, but it drifted increasingly away from the origin with the passing of time [29]. However, persistence implies regulation (but not necessarily vice versa) and, therefore, the rate of change in a persistent crop productivity system is not statistically independent of the yield level and should be bounded (i.e. regulated systems).

Climate change effects: exogenous perturbations

In line with previous studies, temperature during the growing season was the most important weather variable influencing maize yields [39]. However, we deciphered the effects of climate on maize productivity providing new interpretations. First, diagnostic analysis suggested that temperature acts mainly as a non-additive (lateral) perturbation in maize productivity. Therefore, the relationship between temperature and maize yields was nonlinear and could not be captured adequately by a linear or quadratic functional relation as in previous studies [40]. Our analysis suggests a biological reason for the nonlinear interaction between climate and maize yield level. Temperature had no direct impacts on yield rate of change (affecting r_{max}; additive or vertical effects), but influenced the availability or requirements of some limiting factor or resource (changing c; non-additive or lateral effects). There is probably a relationship between extreme heat and plant water stress, increasing water demand and/or soil water content in rain-fed systems, in agreement with the recent results of Lobell et al. [41]. This is because, the effects of high temperature are experienced only when the maize yield level is close to equilibrium [29]. This kind of perturbation exerts strong effects on the average

level of yield but few on the intrinsic periodicity induced by endogenous feedback.

Secondly, rain-fed maize yields are negatively affected by temperature increases, but irrigated systems may gain from warming in some regions. As expected, rain-fed crop damage may result from greater water and heat stress during hot growing seasons. However, unexpected positive effects of temperature in irrigated systems are possibly a consequence of heat tolerance, which is consistent with other studies on local adaptation to hot temperatures being able to minimize stress effects [40] or the cooling effect of irrigation [42]. Therefore, we detected some adaptation to heat stress that could mitigate the projected heat-related losses, at least in a few regions with irrigated systems.

Thirdly, climate variability and extreme events are more important than averages. Thus, we detected that minimum temperature was the dominant factor in maize production, in agreement with other recent studies for maize [43–45] and rice [46]. Currently, a new paradigm has been originated: crop yields have declined with a higher minimum or night temperature [46,47] or when there was a marked asymmetry between maxima and minima [48]. One possible explanation includes the facts that the grain-growth rate has increased and that the duration of grain-filling has been shortened as the temperature increased, producing lower crop production (yield levels) [49]. Mohammed & Tarpley [47] proposes a list of the effects of high night temperatures on crop production. Also, our findings are in line with the results of recent research which argue that global minimum temperatures are increasing faster than maximum temperatures, and the need to explore the ecological consequences of this phenomenon [41,50,51]. Therefore, we wish to highlight the importance of considering extreme climate variables in crop production studies, and limiting the use of averages or accumulative climate data which ignore inter-annual variability of climate and extreme events. Our results differ from those of most studies which do not take into account food production structure regulation, and those which use degree-days [40] concepts which assume a cumulative or additive effect of temperature on crop yield and do not adequately account for the effects of extreme temperatures (high or low) either.

In the study period, precipitation was not a major abiotic factor limiting maize yield of cultivated rain-fed crops in Spain. We only detected positive effects of precipitation for irrigated maize in Navarra, Teruel and Soria. Also, growing season rainfall negatively affected rain-fed maize yield in Córdoba and Zaragoza, possibly due to flood and waterlogging problems causing production losses. Again, we agree with Lobell et al. [41], who argue that the apparent paradox of the scant effect of precipitation on rain-fed maize yield whereas, on the contrary, there is a water stress effect of temperature, can be solved with the following reflection "*large precipitation changes are required to rival the effect of temperature on water stress, because high temperature affects both water demand and supply*".

As in the study of Long et al. 2006 [52], ours study indicates that there was a smaller CO_2 effect on maize yield than previously presumed. Impacts of higher CO_2 on maize yield were reduced probably because it is a C4 plant, and also because of the national scale of the variable in our study.

Spatial variability of yield losses due to water stress

We found that the global spatial pattern of yield losses due to water stress is not a random one (Figure 4); there was a high influence in Palencia, Soria (lowest) and Almería (highest). We detected clusters of "cold spots" in northern Spain (A Coruña and Pontevedra) and "hot spots" in central provinces (Ciudad Real,

Cuenca, Albacete, Valencia; Figure 5). Neither cluster greatly influenced the global Moran's *I* but they were locally-clustered. Moreover, we modelled spatial YGR_w values with climate variables and found that the mean temperature was the highest constraint of maize productivity due to water stress. In conclusion, policy action to decrease the relative yield gap due to water stress on maize productivity has the potential to geographically target high YGR_w areas. Future work will help determine other non-climatic causal relationships between YGR_w and an array of factors that could influence water management practices in maize (e.g. access to water, management technology, soil conditions, etc.).

A recent comparison of simulated and observed yield patterns highlights the value of data in the spatial distribution of yields for understanding the causes of landscape yield variability [10].

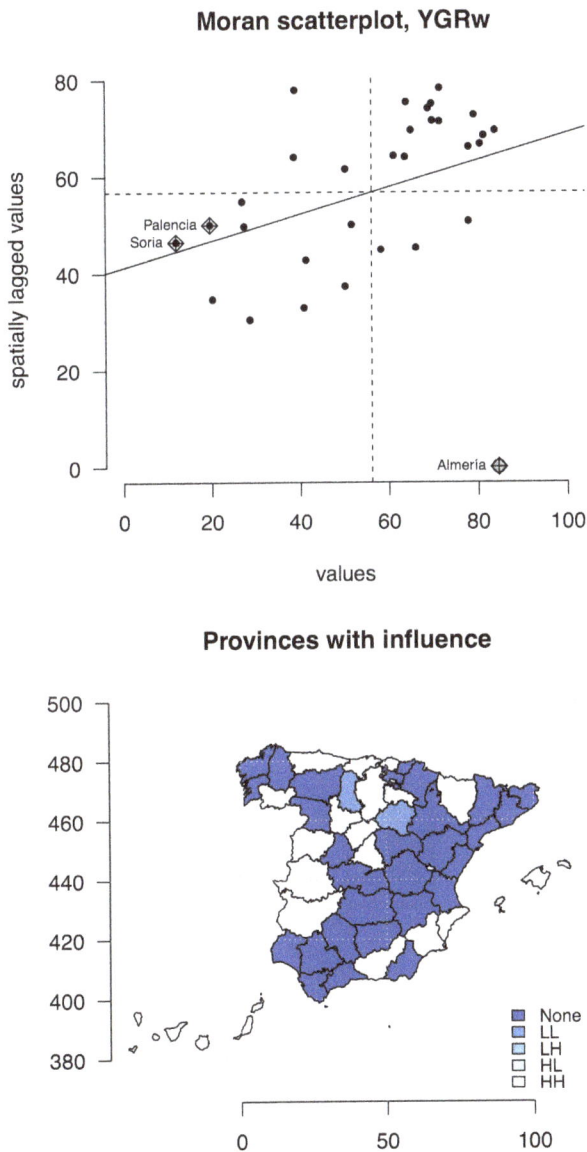

Figure 5. Spatial autocorrelation analysis of the relative yield losses due to suboptimal water availability (*YGRw*). Top: Moran scatterplot; bottom: high-influence areas neighbours: no influence (*None*), high proportion with low proportion neighbours (*HL*), the reverse (*LH*), and both high (*HH*). We define the break between "low" and "high" as the third quartile.

However, to our knowledge, this is the first study explicitly evaluating the spatial pattern of real relative yield gaps due to a water management system and its sensitivity to underlying climate factors. The results demonstrate that spatial patterns of yield loss due to water stress possess substantial information on the relative importance of water management factors for maize productivity.

The need for an analysis to identify and implement adaptation options in agriculture emphasizes the importance of regional scales (federal, provincial, and territorial governments). Global and non-spatial studies can provide only a very partial and potentially misleading insight into the true impact of climate change, where aggregation can indeed conceal vulnerability and climate change costs [53]. However, individual regions (provinces) allow a better analysis of uncertainty and risks, thus providing practical recommendations to farmers.

Conclusions

We identified the same regulation structure for both management systems, i.e. a negative first-order feedback process that tends to stabilize the crop's dynamics. We analyzed the underlying mechanisms of the interaction between climate variation and regulatory structure on maize production. Different climate variables appear to operate differently on maize productivity. We found that the effect of temperature (mainly extreme values) cannot be evaluated independently of crop productivity as in previous studies, because its consequences are experienced only when maize yield level is close to equilibrium (lateral perturbation). We suggest that high maize yield crops are especially vulnerable to weather-related yield variations. These data support the belief that lower yields are more suitable for low-input conditions, because climate might be more severe in crops that interact strongly with productivity [14].

Our results also indicate that it may be important to consider explicitly the irrigation system and spatial variability. Rain-fed agriculture may be at risk as heat waves will be more intense, more frequent and longer (particularly in Seville, Cádiz, Almería, Navarra and Ávila; see Fig. 2). Irrigation seems to allow some tolerance to warming but future levels of water availability would be compromised if water restrictions and irrigation costs increased, as climate change projections indicate. We propose a new framework to estimate yield potential as the equilibrium yield or yield carrying capacity. Climate change is not uniform over Spain and the effectiveness of irrigated and rain-fed management varies with the location, producing different regional vulnerabilities and potential yields. Accordingly, the general strategies for adapting maize productivity to climate change will vary between different zones in Spain.

Supporting Information

Figure S1 Time series of maize yield level for rain-fed (red) and irrigated (blue) systems. Each provinces of Spain were analyzed for 1996–2009.

Figure S2 Partial rate correlation function (*PRCF*).

Figure S3 Comparison of observed crop yield levels (points, *obs*) for the period 1997–2009 with stochastic predictions from models fitted to the data until the year 1996 (broken line, *sim*) and 95% confidence intervals for forecasts (shaded area, *95PPU*). P-factor is the percent of observations that are within the given uncertainty bounds and R-factor represents the average width of the given uncertainty bounds divided by the standard deviation of the observations. See Table S1 for description of models and variables.

Figure S4   _R_-functions: yield rate of change against the log observed yield level (with one year of delay). Climate factors had vertical (additive) and lateral (non-additive) perturbations on the _R_-function. Colors indicate the value of the _R_-function. See Table S1 for description of models and variables.

Figure S5   A non-parametric approach to inference on Moran's _I_ using 999 simulations (Monte Carlo permutation test).

Table S1   Summary statistics of nonlinear logistic models, 1996–2009. We evaluated pure Endogenous models (E), and additive (or Lateral, L) and non-additive (or Vertical, V) models that also represent the effect of exogenous perturbations. Different crop management systems were analyzed ($IR=$ irrigated and $RF=$ rainfed). _%Total_ percentage of total crop production in Spain, K carrying capacity or potential yield, $rmax$ maximum finite reproductive rate, a non-linearity coefficient, c the ratio between demand and offer of limiting resources, b coefficients for different

exogenous effects, R^2 pseudo-coefficient of determination, _logLIK_ log-likelihood, _RMSE_ root-mean-square error and _AICc_ corrected Akaike information criterion. NOTE: *$p<0.05$, **$p<0.01$, Number of not avaiable data (NA) were indicated by I. _CO2_ carbon dioxide emission (kt, country-level emissions), and summary statistics of the growing season weather: _EMNT_ extreme minimum temperature (°C), _EMXT_ extreme maximum temperature (°C), _MMNT_ mean minimum temperature (°C), _MMXT_ mean maximum temperature (°C), _MNTM_ mean temperature (°C), _EMXP_ extreme maximum daily precipitation total (l/m2), _TPCP_ total precipitation (l/m2).

Acknowledgments

We thank the editor and two anonymous reviewers for their constructive comments, which helped us to improve the manuscript.

Author Contributions

Conceived and designed the experiments: RF ML JLG-A. Performed the experiments: RF. Analyzed the data: RF. Contributed reagents/materials/analysis tools: RF. Wrote the paper: RF ML JLG-A.

References

1. Lobell DB, Schlenker W, Costa-Roberts J (2011) Climate trends and global crop production since 1980. Science 333: 616–620.
2. Licker R, Johnston M, Foley JA, Barford C, Kucharik CJ, et al. (2010) Mind the gap: how do climate and agricultural management explain the "yield gap" of croplands around the world? Glob Ecol Biogeogr 19: 769–782.
3. Döll P (2002) Impact of climate change and variability on irrigation requirements: a global perspective. Clim Change 54: 269–293.
4. Turral H, Burke J, Faurès JM, Faurés JM (2011) Climate change, water and food security. Rome Food Agric Organ United Nations: 204.
5. Easterling WE, Aggarwal PK, Batima P, Brander KM, Erda L, et al. (2007) Food, fibre and forest products. In: Parry ML, Canziani OF, Palutikof JP, Linden PJ van der, Hanson CE, editors. Climate Change 2007: Impacts, Adaptation and Vulnerability. Contribution of Working Group II to the Fourth Assessment Report of the Intergovernmental Panel on Climate Change. Cambridge University Press, Cambridge, UK. pp. 273–313.
6. Smith P, Martino D, Cai Z, Gwary D, Janzen H, et al. (2007) Agriculture. In: Metz B, Davidson OR, Bosch PR, Dave R, Meyer LA, editors. Climate Change 2007: Mitigation. Contribution of Working Group III to the Fourth Assessment Report of the Intergovernmental Panel on Climate Change. Cambridge University Press, Cambridge, United Kingdom and New York, NY, USA. pp. 497–540.
7. Medici LO, Reinert F, Carvalho DF, Kozak M, Azevedo RA (2014) What about keeping plants well watered? Environ Exp Bot 99: 38–42.
8. Zou X, Li Y, Gao Q, Wan Y (2011) How water saving irrigation contributes to climate change resilience—a case study of practices in China. Mitig Adapt Strateg Glob Chang 17: 111–132.
9. Mueller ND, Gerber JS, Johnston M, Ray DK, Ramankutty N, et al. (2012) Closing yield gaps through nutrient and water management. Nature 490: 254–257.
10. Lobell DB, Ortiz-Monasterio JI (2006) Regional importance of crop yield constraints: Linking simulation models and geostatistics to interpret spatial patterns. Ecol Modell 196: 173–182.
11. Lobell D, Bala G, Mirin A, Phillips T, Maxwell R, et al. (2009) Regional Differences in the Influence of Irrigation on Climate. J Clim 22: 2248–2255.
12. Lobell DB, Cassman KG, Field CB (2009) Crop Yield Gaps: Their Importance, Magnitudes, and Causes. Annu Rev Environ Resour 34: 179–204.
13. Meng Q, Hou P, Wu L, Chen X, Cui Z, et al. (2013) Understanding production potentials and yield gaps in intensive maize production in China. F Crop Res 143: 91–97.
14. Tokatlidis IS (2013) Adapting maize crop to climate change. Agron Sustain Dev 33: 63–79.
15. Reidsma P, Ewert F, Boogaard H, van Diepen K (2009) Regional crop modelling in Europe: The impact of climatic conditions and farm characteristics on maize yields. Agric Syst 100: 51–60.
16. Olesen JE, Carter TR, Díaz-Ambrona CH, Fronzek S, Heidmann T, et al. (2007) Uncertainties in projected impacts of climate change on European agriculture and terrestrial ecosystems based on scenarios from regional climate models. Clim Change 81: 123–143.
17. Wolf J, Diepen CA (1995) Effects of climate change on grain maize yield potential in the european community. Clim Change 29: 299–331.
18. Iglesias A, Mínguez MI (1995) Prospects for maize production in Spain under climate change. In: Rosenzweig C, editor. Climate Change and Agriculture: Analysis of Potential International Impacts. American Society of Agronomy, Vol. asaspecial. pp. 259–273.
19. Moreno JM, Aguiló E, Alonso S, Cobelas MÁ, Anadón R, et al. (2005) A Preliminary General Assessment of the Impacts in Spain Due to the Effects of Climate Change A Preliminary Assessment of the Impacts in Spain due to the Effects of Climate Change. Available: http://www.magrama.gob.es/en/cambio-climatico/temas/impactos-vulnerabilidad-y-adaptacion/plan-nacional-adaptacion-cambio-climatico/eval_impactos_ing.aspx. Accessed 2014 May 8.
20. Rodríguez-Puebla C, Ayuso S, Frías M, García-Casado L (2007) Effects of climate variation on winter cereal production in Spain. Clim Res 34: 223–232.
21. Iglesias A, Garrote L, Quiroga S, Moneo M (2011) A regional comparison of the effects of climate change on agricultural crops in Europe. Clim Change 112: 29–46.
22. Iglesias A, Mougou R, Moneo M, Quiroga S (2010) Towards adaptation of agriculture to climate change in the Mediterranean. Reg Environ Chang 11: 159–166.
23. Neumann K, Verburg PH, Stehfest E, Müller C (2010) The yield gap of global grain production: A spatial analysis. Agric Syst 103: 316–326.
24. MAGRAMA (2012) Ministerio de Agricultura, Alimentación y Medio Ambiente (MAGRAMA). MAGRAMA Stat Databases. Available: http://www.magrama.gob.es/es/estadistica/temas/publicaciones/anuario-de-estadistica/. Accessed 1 April 2012.
25. Lawrimore JH, Menne MJ, Gleason BE, Williams CN, Wuertz DB, et al. (2011) An overview of the Global Historical Climatology Network monthly mean temperature data set, version 3. J Geophys Res 116: D19121. Available: http://doi.wiley.com/10.1029/2011JD016187. Accessed 1 April 2012.
26. World Bank, World Development Indicators (WDI) (2012). ESDS Int Univ Manchester. Available: http://data.worldbank.org/data-catalog/world-development-indicators/wdi-2012. Accessed 1 April 2012.
27. Lobell DB, Bänziger M, Magorokosho C, Vivek B (2011) Nonlinear heat effects on African maize as evidenced by historical yield trials. Nat Clim Chang 1: 42–45.
28. Berryman A, Turchin P (2001) Identifying the density-dependent structure underlying ecological time series. Oikos 92: 265–270.
29. Royama T (1992) Analytical population dynamics. London, New York: Chapman & Hall. Springer.
30. Ricker WE (1954) Stock and recruitment. J Fish Res Board Canada 11: 559–623.
31. R Development Core Team (2011) R: A language and environment for statistical computing. R Found Stat Comput Vienna. Available: http://www.r-project.org/. Accessed 2014 May 8.
32. Bates D, Chambers JM (1991) Nonlinear Models. In: Chambers JM, Hastie TJ, editors. Statistical Models in S. Wadsworth & Brooks/Cole, Pacific Grove, California.
33. Cameron AC, Windmeijer FAG (1996) R-Squared Measures for Count Data Regression Models With Applications to Health-Care Utilization. J Bus Econ Stat 14: 209–220.

34. De la Maza M, Lima M, Meserve PL, Gutierrez JR, Jaksic FM (2009) Primary production dynamics and climate variability: ecological consequences in semiarid Chile. Glob Chang Biol: 1116–1126.

35. Liu Z, Yang X, Hubbard K, Lin X (2012) Maize potential yields and yield gaps in the changing climate of Northeast China. Glob Chang Biol 86.

36. Bivand RS, Pebesma E, Gómez-Rubio V (2008) Applied Spatial Data Analysis with R. New York, NY: Springer New York.

37. Bivand R, Altman M, Anselin L, Assunçao R, Berke O, et al. (2013) spdep: Spatial dependence: weighting schemes, statistics and models. R package version 0.5-56.

38. Berryman AA (1999) Principles of population dynamics and their application. Stanley Thrones (Publishers) Limited.

39. Sun BJ, Van Kooten GC (2013) Weather effects on maize yields in northern China. J Agric Sci: 1–11.

40. Butler EE, Huybers P (2012) Adaptation of US maize to temperature variations. Nat Clim Chang 3: 68–72.

41. Lobell DB, Hammer GL, McLean G, Messina C, Roberts MJ, et al. (2013) The critical role of extreme heat for maize production in the United States. Nat Clim Chang 3: 497–501.

42. Lobell DB, Bonfils CJ, Kueppers LM, Snyder MA (2008) Irrigation cooling effect on temperature and heat index extremes. Geophys Res Lett 35: L09705.

43. Muchow RC, Sinclair TR, Bennett JM (1990) Temperature and Solar Radiation Effects on Potential Maize Yield across Locations. 82: 338–343.

44. Chen C, Lei C, Deng A, Qian C, Hoogmoed W, et al. (2011) Will higher minimum temperatures increase corn production in Northeast China? An analysis of historical data over 1965–2008. Agric For Meteorol 151: 1580–1588.

45. Lobell DB (2007) Changes in diurnal temperature range and national cereal yields. Agric For Meteorol 145: 229–238.

46. Peng S, Huang J, Sheehy JE, Laza RC, Visperas RM, et al. (2004) Rice yields decline with higher night temperature from global warming. Proc Natl Acad Sci U S A 101: 9971–9975.

47. Mohammed A, Tarpley L (2007) Effects of High Night Temperature on Crop Physiology and Productivity: Plant Growth Regulators Provide a Management Option. Global Warming Impacts – Case Studies on the Economy, Human Health, and on Urban and Natural Environments. doi: 10.5772/24537.

48. Rosenzweig C, Tubiello FN (1996) Effects of changes in minimum and maximum temperature on wheat yields in the central US A simulation study. Agric For Meteorol 80: 215–230.

49. Muchow RC (1990) Effect of high temperature on grain-growth in field-grown maize. F Crop Res 23: 145–158.

50. Alward RD, Detling J, Milchunas D (1999) Grassland Vegetation Changes and Nocturnal Global Warming. Science (80-) 283: 229–231.

51. Katz RW, Brown BG (1992) Extreme events in a changing climate: Variability is more important than averages. Clim Change 21(3):289–302.

52. Long SP, Ainsworth EA, Leakey ADB, Nösberger J, Ort DR (2006) Food for thought: lower-than-expected crop yield stimulation with rising CO2 concentrations. Science 312: 1918–1921.

53. Bosello F, Carraro C, De Cian E (2009) An Analysis of Adaptation as a Response to Climate Change. SSRN Electron J.

Permissions

List of Contributors

Colin A. Simpfendorfer, Beau G. Yeiser, Tonya R. Wiley and Michelle R. Heupel
Center for Shark Research, Mote Marine Laboratory, Sarasota, Florida, United States of America

Gregg R. Poulakis and Philip W. Stevens
Florida Fish and Wildlife Conservation Commission, Fish and Wildlife Research Institute, Charlotte Harbor Field Laboratory, Port Charlotte, Florida, United States of America

Matthew R. Kasper, Carmen Lucas and Erik J. Reaves
U.S. Naval Medical Research Unit 6, Lima, Peru

Andres G. Lescano
U.S. Naval Medical Research Unit 6, Lima, Peru
Universidad Peruana Cayetano Heredia, Lima, Peru

Duncan Gilles
Madigan Healthcare System, Tacoma, Washington, United States of America

Brian J. Biese and Gary Stolovitz
452nd Combat Support Hospital, U.S. Army Reserve, Milwaukee, Wisconsin, United States of America

Michael Kiparsky
Wheeler Institute for Water Law & Policy, University of California, Berkeley, California, United States of America

Brian Joyce, David Purkey and Charles Young
Stockholm Environment Institute, Davis, California, United States of America

Jason A. L. Jeffery, Peter A. Ryan and Brian H. Kay
Queensland Institute of Medical Research, PO Royal Brisbane Hospital, Brisbane, Queensland, Australia

Archie C. A. Clements
Queensland Institute of Medical Research, PO Royal Brisbane Hospital, Brisbane, Queensland, Australia
Infectious Disease Epidemiology Unit, School of Population Health, University of Queensland, Herston, Queensland, Australia

Yen Thi Nguyen, Le Hoang Nguyen and Son Hai Tran
National Institute of Hygiene and Epidemiology, Hanoi, Vietnam

Nghia Trung Le
Institute Pasteur, Nha Trang, Vietnam

Nam Sinh Vu
General Department of Preventive Medicine and Environmental Health, Ministry of Health, Hanoi, Vietnam

Alina Nescerecka
Department of Water Engineering and Technology, Riga Technical University, Riga, Latvia
Department of Environmental Microbiology, Eawag, Swiss Federal Institute for Aquatic Science and Technology, Dübendorf, Switzerland

Janis Rubulis and Talis Juhna
Department of Water Engineering and Technology, Riga Technical University, Riga, Latvia

Marius Vital and Frederik Hammes
Department of Environmental Microbiology, Eawag, Swiss Federal Institute for Aquatic Science and Technology, Dübendorf, Switzerland

Martina G. Vijver
Institute of Environmental Sciences (CML), Leiden University, Leiden, The Netherlands

Paul J. van den Brink
Alterra, Wageningen University and Research centre, Wageningen, The Netherlands
Wageningen University, Wageningen University and Research centre, Wageningen, The Netherlands

Elizabeth Ailes and Sarah Collier
International Health Resources Consulting, Atlanta, Georgia, United States
National Center for Emerging and Zoonotic Infectious Diseases, Centers for Disease Control and Prevention, Atlanta, Georgia, United States

Philip Budge, Manjunath Shankar, Melissa Chen, Andrew Thornton, Michael J. Beach and Joan M. Brunkard
National Center for Emerging and Zoonotic Infectious Diseases, Centers for Disease Control and Prevention, Atlanta, Georgia, United States

William Brinton
Regional Epidemiologist, San Luis Valley Public Health Emergency Preparedness and Response Program, Alamosa, Colorado, United States

Alicia Cronquist
Disease Control and Environmental Epidemiology Division, Colorado Department of Public Health and Environment, Denver, Colorado, United States

Meha Jain, Yili Lim and María Uriarte
Department of Ecology, Evolution and Environmental Biology, Columbia University, New York, New York, United States of America

Javier A. Arce-Nazario
Department of Biology, University of Puerto Rico in Cayey, Cayey, Puerto Rico, United States of America
Anders S. Huseth
Department of Entomology, Cornell University, New York State Agricultural Experiment Station, Geneva, New York, United States of America

Russell L. Groves
Department of Entomology, University of Wisconsin-Madison, Madison, Wisconsin, United States of America

Paul R. Hunter and Helen Risebro
Norwich School of Medicine, University of East Anglia, Norwich, United Kingdom

Marie Yen, Hélène Lefebvre and François Jaquenoud
1001 fontaines pour demain, Caluire et Cuire, France

Chay Lo
Teuk Saat 1001, Phnom Penh, Cambodia

Philippe Hartemann
Département Environnement et Santé Publique, Faculté de médecine de Nancy – Université de Lorraine, Nancy, France

Christophe Longuet
Fondation Me´rieux, Lyon, France

Mia Catharine Mattioli, Alexandria B. Boehm and Angela R. Harris
Environmental and Water Studies, Department of Civil and Environmental Engineering, Stanford University, Stanford, California, United States of America

Jennifer Davis and Amy J. Pickering
Environmental and Water Studies, Department of Civil and Environmental Engineering, Stanford University, Stanford, California, United States of America
Woods Institute for the Environment, Stanford University, Stanford, California, United States of America

Mwifadhi Mrisho
Ifakara Health Institute, Bagamoyo Research and Training Unit, Bagamoyo, Tanzania

Katri Jalava, Jukka Ollgren, Marja Palander, Pia Räsänen, Sallamaari Siponen, Outi Nyholm, Aino Kyyhkynen, Markku Kuusi and Anja Siitonen
Department of Infectious Disease Surveillance and Control, National Institute for Health and Welfare, Helsinki, Finland

Hanna Rintala, Sirpa Hakkarainen, Juhani Merentie and Martti Pärnänen
Siilinjärvi municipality, Finland

Leena Maunula, Marja-Liisa Hänninen and Joana Revez
Department of Food Hygiene and Environmental Health, University of Helsinki, Helskinki, Finland

Vicente Gomez-Alvarez, Hodon Ryu and Jorge W. Santo Domingo
Office of Research and Development, United States Environmental Protection Agency, Cincinnati, Ohio, United States of America

Jenni Antikainen and Raisa Loginov
Laboratory HUSLAB, Helsinki University Hospital, Helsinki, Finland

Ari Kauppinen, Ilkka Miettinen and Tarja Pitkänen
Department of Environmental Health, National Institute for Health and Welfare, Kuopio, Finland

Julianne L. Baron
Department of Infectious Diseases and Microbiology, University of Pittsburgh, Graduate School of Public Health, Pittsburgh, Pennsylvania, United States of America
Special Pathogens Laboratory, Pittsburgh, Pennsylvania, United States of America

Amit Vikram
Department of Civil and Environmental Engineering, University of Pittsburgh, Swanson School of Engineering, Pittsburgh, Pennsylvania, United States of America

Scott Duda
Special Pathogens Laboratory, Pittsburgh, Pennsylvania, United States of America

Janet E. Stout
Special Pathogens Laboratory, Pittsburgh, Pennsylvania, United States of America

Department of Civil and Environmental Engineering, University of Pittsburgh, Swanson School of Engineering, Pittsburgh, Pennsylvania, United States of America

Kyle Bibby
Department of Civil and Environmental Engineering, University of Pittsburgh, Swanson School of Engineering, Pittsburgh, Pennsylvania, United States of America
Department of Computational and Systems Biology, University of Pittsburgh Medical School, Pittsburgh, Pennsylvania, United States of America

John J. Kelly and Nicole Minalt
Department of Biology, Loyola University Chicago, Chicago, Illinois, United States of America

Alessandro Culotti and Aaron Packman
Department of Civil and Environmental Engineering, Northwestern University, Evanston, Illinois, United States of America

Marsha Pryor
Pinellas County Utilities Laboratory, Largo, Florida, United States of America

Yushi Ye, Xinqiang Liang, Liang Li and Yuanjing Ji
Institute of Environmental Science and Technology, College of Environmental and Resource Sciences, Zhejiang University, Hangzhou, China

Yingxu Chen and Chunyan Zhu
Zhejiang Province Key Laboratory for Water Pollution Control and Environmental Safety, Hangzhou, China

Jaana I. Halonen, Tuula Oksanen and Jaana Pentti
Finnish Institute of Occupational Health, Helsinki, Finland

Mika Kivimäki
Finnish Institute of Occupational Health, Helsinki, Finland
Department of Epidemiology and Public Health, University College of London, London, United Kingdom

Pekka Virtanen
Tampere School of Health Sciences, University of Tampere, Tampere, Finland

Mikko J. Virtanen
Epidemiologic Surveillance and Response Unit, National Institute for Health and Welfare, Helsinki, Finland

Jussi Vahtera
Finnish Institute of Occupational Health, Helsinki, Finland
Department of Public Health, University of Turku, and Turku University Hospital, Turku, Finland

Dehua Zhao, Hao Jiang, Ying Cai and Shuqing An
Department of Biological Science and Technology, Nanjing University, Nanjing, China

Justin Stoler
Department of Geography and Regional Studies, University of Miami, Coral Gables, Florida, United States of America
Department of Public Health Sciences, Miller School of Medicine, University of Miami, Miami, Florida, United States of America

John R. Weeks
Department of Geography, San Diego State University, San Diego, California, United States ofAmerica

Richard Appiah Otoo
Ghana Urban Water Limited, Head Office Operations, Accra, Ghana

Qing Shuang, Mingyuan Zhang and Yongbo Yuan
Department of Construction Management, Dalian University of Technology, Dalian, Liaoning, China

Joshua S. Gruber, Ayse Ercumen and John M. Colford, Jr.
Division of Epidemiology, University of California, Berkeley, California, United States of America

Rosana Ferrero
Departamento Protección de Cultivos, Instituto de Agricultura Sostenible, Consejo Superior de Investigaciones Científicos (CSIC), Córdoba, Spain
Center of Applied Ecology and Sustainability (CAPES), Santiago, Chile

Mauricio Lima
Departamento de Ecología, Pontificia Universidad Católica de Chile, Santiago, Chile
Laboratorio Internacional de Cambio Global, LINCG (CSIC-PUC), Santiago, Chile
Center of Applied Ecology and Sustainability (CAPES), Santiago, Chile

Jose Luis Gonzalez-Andujar
Departamento Protección de Cultivos, Instituto de Agricultura Sostenible, Consejo Superior de Investigaciones Científicos (CSIC), Córdoba, Spain
Laboratorio Internacional de Cambio Global, LINCG (CSIC-PUC), Santiago, Chile

Index